Advances in Science, Technology & Innovation

IEREK Interdisciplinary Series for Sustainable Development

Advances in Science, Technology & Innovation (ASTI) is a series of peer-reviewed books based on important emerging research that redefines the current disciplinary boundaries in science, technology and innovation (STI) in order to develop integrated concepts for sustainable development. It not only discusses the progress made towards securing more resources, allocating smarter solutions, and rebalancing the relationship between nature and people, but also provides in-depth insights from comprehensive research that addresses the **17 sustainable development goals (SDGs)** as set out by the UN for 2030.

The series draws on the best research papers from various IEREK and other international conferences to promote the creation and development of viable solutions for a **sustainable future and a positive societal** transformation with the help of integrated and innovative science-based approaches. Including interdisciplinary contributions, it presents innovative approaches and highlights how they can best support both economic and sustainable development, through better use of data, more effective institutions, and global, local and individual action, for the welfare of all societies.

The series particularly features conceptual and empirical contributions from various interrelated fields of science, technology and innovation, with an emphasis on digital transformation, that focus on providing practical solutions to **ensure food, water and energy security to achieve the SDGs.** It also presents new case studies offering concrete examples of how to resolve sustainable urbanization and environmental issues in different regions of the world.

The series is intended for professionals in research and teaching, consultancies and industry, and government and international organizations. Published in collaboration with IEREK, the Springer ASTI series will acquaint readers with essential new studies in STI for sustainable development.

ASTI series has now been accepted for Scopus (September 2020). All content published in this series will start appearing on the Scopus site in early 2021.

More information about this series at https://link.springer.com/bookseries/15883

Hesham El-Askary · Zeynal Abiddin Erguler · Murat Karakus · Helder I. Chaminé
Editors

Research Developments in Geotechnics, Geo-Informatics and Remote Sensing

Proceedings of the 2nd Springer Conference of the Arabian Journal of Geosciences (CAJG-2), Tunisia 2019

Springer

Editors
Hesham El-Askary
Schmid College of Science and Technology
Chapman University
Orange, CA, USA

Murat Karakus
School of Civil, Environmental and Mining Engineering
University of Adelaide
Adelaide, SA, Australia

Zeynal Abiddin Erguler
Kütahya Dumlupınar Üniversitesi
Kütahya, Turkey

Helder I. Chaminé
Lab of Cartography and Applied Geology
School of Engineering
Instituto Superior de Engenharia do Port (ISEP)
Polytechnic of Porto
Porto, Portugal

ISSN 2522-8714 ISSN 2522-8722 (electronic)
Advances in Science, Technology & Innovation
IEREK Interdisciplinary Series for Sustainable Development
ISBN 978-3-030-72895-3 ISBN 978-3-030-72896-0 (eBook)
https://doi.org/10.1007/978-3-030-72896-0

This Springer imprint is published by the registered company Springer Nature Switzerland AG
The registered company address is: Gewerbestrasse 11, 6330 Cham, Switzerland

About the 2nd Springer Conference of the Arabian Journal of Geosciences (CAJG-2), Tunisia, 2019

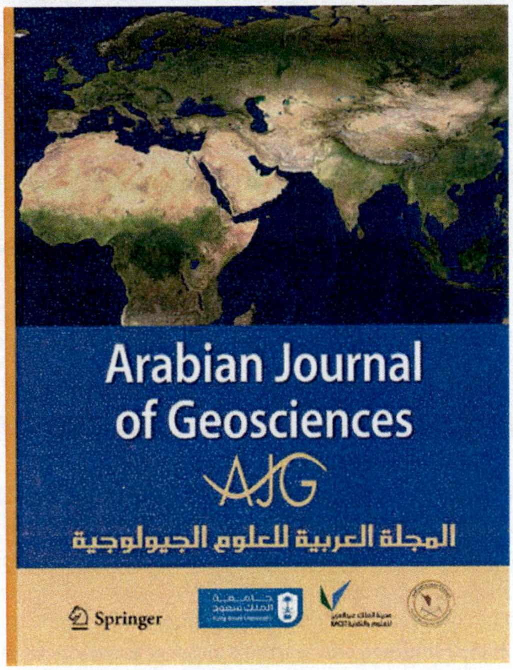

The Arabian Journal of Geosciences (AJG) is a Springer journal publishing original articles on the full range of earth sciences in partnership with the Saudi Society for Geosciences. The journal focuses on, but is not limited to, research themes which have regional significance for the Middle East, the Euro-Mediterranean, Africa, Asia and some other regions of the world. The journal receives on average 4000 submissions a year and accepts around 1000 papers for publication in its 24 annual issues (acceptance rate around 25%). It benefits from the participation of an editorial team of 100 international Associate Editors who generously help in evaluating and selecting the best papers.

In 2008, Prof. Abdullah Al-Amri, in close partnership with Springer, founded the Arabian Journal of Geosciences (AJGS). In 2018, the journal celebrated its 10th anniversary. To mark the event, the founder and Editor-in-Chief of the AJGS organized the 1st Conference of the Arabian Journal of Geosciences (CAJG) in close collaboration with Springer on November 12–15, 2018. The conference was an occasion to endorse the journal's long-held reputation and brought together 450 authors from 70 countries, who work in the wide-ranging fields of earth sciences. The dynamic four-day conference in a stimulating environment in Hammamet, Tunisia, provided attendees with opportunities to share their latest unpublished findings and learn about the latest geosciences studies. The event also allowed attendees to meet and talk to the journal's editors and reviewers. Three field trips were organized alongside the conference, and many participants enjoyed the wonders of the geology of Tunisia.

In a continuation of the successful 1st CAJG, the 2019's conference aimed to bring geoscientists from all over the world to present and discuss their most recent findings. The 2nd CAJG was an occasion to publish the newest findings in its proceedings by Springer and a special issue in the AJGS, with a clear mission to drive greater North-South (Europe-Africa) scientific cooperation and to open doors to new and enriching collaborations with geoscientists based in Asia and the Americas. The 2nd CAJG devoted a special session (workshop) to studies focusing on unraveling the undiscovered oil and gas resources in the Mediterranean and North Africa. Many international experts took part in the discussion.

The conference covered all cross-cutting themes of geosciences, and focused principally on the following 15 tracks:

- Track 1. Atmospheric Sciences, Meteorology, Climatology, Oceanography
- Track 2. Biogeochemistry, Geobiology, Geoecology, Geoagronomy
- Track 3. Earthquake Seismology and Geodesy
- Track 4. Environmental Earth Sciences
- Track 5. Exploration & Theoretical Geophysics, Seismic & Well Logging Methods, Mathematical Geosciences
- Track 6. Geo-Informatics and Remote Sensing
- Track 7. Geochemistry, Mineralogy, Petrology, Volcanology
- Track 8. Geological Engineering, Geotechnical Engineering
- Track 9. Geomorphology, Geography, Soil Science, Glaciology, Geoarchaeology, Geoheritage
- Track 10. Hydrology, Hydrogeology, Hydrochemistry
- Track 11. Marine Geosciences, Historical Geology, Paleoceanography, Paleoclimatology
- Track 12. Numerical and Analytical Methods in Mining Sciences and Geomechanics
- Track 13. Petroleum and Energy Engineering, Petroleum Geochemistry
- Track 14. Sedimentology, Stratigraphy, Paleontology, Geochronology
- Track 15. Structural Geology, Tectonics and Geodynamics, Petroleum Geology

The dynamic four-day conference provided more than 400 attendees with opportunities to share their latest unpublished findings and learn the newest geosciences studies. The event also allowed attendees to meet and discuss with the journal's editors and reviewers.

More than 710 short contributing papers to the conference were submitted by authors from more than 74 countries. After a pre-conference peer review process by more than 500 reviewers, 462 papers were accepted. These papers are published as chapters in the conference proceedings which consist of four edited volumes, each edited by the following group of Arabian Journal of Geosciences (AJGS) editors and other guest editors:

Proceedings Volume 1: New Prospects in Environmental Geosciences and Hydrogeosciences

Haroun Chenchouni: University of Tebessa, Tebessa, Algeria
Helder I. Chaminé: School of Engineering—ISEP, Polytechnic of Porto, Porto, Portugal
Md Firoz Khan: Department of Chemistry, Faculty of Science, University of Malaya, Kuala Lumpur, Malaysia
Broder J. Merkel: TUBAF, Freiberg, Germany
Zhihua Zhang: Shandong University, Jinan, China
Peiyue Li: School of Water and Environment, Chang'an University, Xi'an, China

Amjad Kallel: Laboratory of Water, Energy and Environment (Lab 3E), Sfax National School of Engineers, University of Sfax, Tunisia
Nabil Khélifi: Springer, a part of Springer Nature, Heidelberg, Germany

Proceedings Volume 2: Advances in Geophysics, Tectonics and Petroleum Geosciences

Mustapha Meghraoui: Institut de Physique du Globe, Université de Strasbourg, Strasbourg, France
Narasimman Sundararajan: Sultan Qaboos University, Muscat, Oman
Santanu Banerjee: Indian Institute of Technology Bombay, Mumbai, India
Klaus-g. Hinzen: University of Cologne, Germany
Mehdi Eshagh: University West, Trollhättan, Sweden
François Roure: IFP—Energies Nouvelles, France, France
Helder I. Chaminé: School of Engineering—ISEP, Polytechnic of Porto, Porto, Portugal
Said Maouche: Center for Research in Astronomy and Astrophysics Geophysics, Algeria
André Michard: Paris-Sud University, France
Abdullah Al-amri: King Saud University, Saudi Arabia

Proceedings Volume 3: Recent Research on Geomorphology, Sedimentology and Geochemistry

Attila Ciner: Istanbul Technical University, Turkey
Stefan Grab: School of Geography, Archaeology and Environmental Studies, University of the Witwatersrand, South Africa
Etienne Jaillard: Université Grenoble Alpes, France
Domenico M. Doronzo: National Institute of Geophysics and Volcanology, Rome, Italy, Spain
André Michard: Paris-Sud University, France
Marina Rabineau: CNRS, Univ Brest, Laboratoire Géosciences Océan, Institut Universitaire Européen de la Mer, France
Helder I. Chaminé: School of Engineering—ISEP, Polytechnic of Porto, Porto, Portugal

Proceedings Volume 4: Research developments in Geotechnics, Geo-Informatics and Remote Sensing

Hesham El-askary: Schmid College of Science and Technology at Chapman University, Orange, CA, USA
Zeynal Abiddin Erguler: Kütahya Dumlupınar Üniversitesi, Kütahya, Turkey
Murat Karakus: School of Civil, Environmental and Mining Engineering, the University of Adelaide, SA, Australia
Helder I. Chaminé: School of Engineering—ISEP, Polytechnic of Porto, Porto, Portugal

About the Conference Steering Committee

General Chair

Abdullah Al-Amri
Founder and Editor-in-Chief
Arabian Journal of Geosciences
King Saud University, Saudi Arabia

Conference Supervisor

Nabil Khélifi
Senior Publishing Editor, MENA program
Journal Publishing Manager
Arabian Journal of Geosciences
Springer, a part of Springer Nature, Germany

Advisory Co-chair

Walter D. Mooney
Guest of Editorial Board
Arabian Journal of Geosciences
United States Geological Survey Western Region, USA

Advisory Co-chair

Dorrik Stow
Guest of Editorial Board
Arabian Journal of Geosciences
Heriot-Watt University, Edinburgh, Scotland

Scientific Committee Co-chair

François Roure
Chief Editor—Track 15
Arabian Journal of Geosciences
IFP—Energies Nouvelles, France

Scientific Committee Co-chair

Biswajeet Pradhan
Chief Editor—Track 6
Arabian Journal of Geosciences
University of Technology Sydney, Australia

Local Organizing Co-chair

Mohamed Soussi
Former Associate Editor
Arabian Journal of Geosciences
Tunis El Manar University, Tunis, Tunisia

Local Organizing Co-chair

Samir Bouaziz
Former Associate Editor
Arabian Journal of Geosciences
University of Sfax, Sfax, Tunisia

Publications Co-chair

Beatriz Bádenas
Chief Editor—Track 14
Arabian Journal of Geosciences
University of Zaragoza, Zaragoza, Spain

Publications Co-chair

Marina Rabineau
Chief Editor—Track 11
Arabian Journal of Geosciences
University of Brest, Brest, France

Program Co-chair

Amjad Kallel
Chief Editor—Track 4
Arabian Journal of Geosciences
ENIS, University of Sfax, Tunisia

Program Co-chair

Sami Khomsi
Former Associate Editor
Arabian Journal of Geosciences
King Abdulaziz University, Jeddah, Saudi Arabia

Proceedings Editorial Manager

Mourad Amer
Editor of Springer/IEREK ASTI Series
Guest of Editorial Board of AJGS
IEREK, Alexandria, Egypt

Communication Chair

Zakaria Hamimi
Associate Editor
Arabian Journal of Geosciences
Benha University, Benha, Egypt

Public Relations Chair

Faïez Gargouri
Director of the Higher Institute of Computer Science and Multimedia
President of the AIG
University of Sfax, Tunisia

Conference Manager

Mohamed Sahbi Moalla
Journal Coordinator
Arabian Journal of Geosciences
ISET, University of Sfax, Tunisia

Preface

In comparison with many disciplines in which the algorithm can be created in a few steps to reach the intended purposes, the applicability and success in the earth sciences based scientific studies, particularly engineering projects, require region-specific case studies in accurately achieving the petrophysical and hydro-mechanical properties of the complex and challenging geological units. It may be specified that the engineering problems in geosciences cannot be solved in a few algorithm-based steps. Therefore, the discussions on new ideas, causes and consequences of natural disasters occurred in different regions, knowledge and experiences at international conferences are becoming more significant approaches for further understanding on mechanical and deformation characteristics of anisotropic and heterogeneous geomaterials. By considering these requirements, the 2nd Springer Conference of the Arabian Journal of Geosciences (CAJG-2), was successfully organized in 2019 in Tunisia. The latest various trending and significant earth sciences related issues expected to be clarified were discussed during this conference. This proceedings book contains the best peer-reviewed papers accepted for presentation at CAJG-2 in the field of the below given sections:

1. Geological and geotechnical engineering,
2. Geomechanical studies based on numerical and analytical methods,
3. Geo-informatics and remote sensing.

The section of "geological engineering and geotechnical engineering" starts with a keynote on the physical and mechanical characteristics of the most sensitive sedimentary rocks known as mudrocks. It should be noted that mudrocks are currently significant for shale gas production and are thought to be of strategic importance in hosting nuclear waste disposal in future. After comprehensive revision processes, totally 36 papers are presented in this section with four different subsections entitled as "slope instabilities and landslides," "ground stabilization with functional geomaterials," "geophysical approaches in engineering geology studies" and "predictive models in geological engineering." It is believed that the content of these papers would provide new scientific knowledge for further understanding on landslides, new stabilization techniques, importance of geophysics for engineering geology investigations as well as new empirical approaches for easily predicting some physical and hydrogeomechanical properties of geomaterials.

The book is of interest to all researchers, practitioners and students in the fields of geological and mining engineering, geotechnical engineering, hydrogeomechanics, engineering geology, geotechnologies and natural hazards.

Los Angeles, Orange, USA — Hesham El-Askary
Kütahya, Turkey — Zeynal Abiddin Erguler
Adelaide, Australia — Murat Karakus
Porto, Portugal — Helder I. Chaminé
July 2020

Contents

About the Editors

Prof. Hesham El-Askary received his Ph.D. in Computational Sciences and Informatics from George Mason University in 2004. He is the 2015 recipient of the Chapman University's elite Senior Wang-Fradkin Professorship Award. In 2016, he was named as the regional coordinator on a $3 million Euro grant from the European Union's (EU) Horizon 2020. The three year project, known as GEO-CRADLE, deals with Coordinating and integrating state-of-the-art Earth Observation Activities in the regions of North Africa, Middle East and Balkans and Developing Links with GEO-related initiatives toward GEOSS. Through this work, he with the research team were able to deliver the first analytical solar Atlas of Egypt that is now considered to be the official document of the government for solar investment. This work was recently presented at Planet Earth Institute at the Royal Society in London during a seminar that discussed the future of solar energy in Africa. His research interests include dust storms monitoring and detection using different remote sensing technologies as well as studying other extreme events. He is also involved in studying air pollution problems over mega cities due to natural and man-made effects as well as climate change and its impacts on sea level rise and coral reefs for coastal areas. His research also included using earth observations in studying impact of sever dust storms anomalous chlorophyll outbreaks in the marine environment, hurricanes intensification as well as transport of microbes' causing Kawasaki disease outbreaks. Recently, he has been focusing on using earth observations for water resources management, precision agriculture along the sustainable development goals. Today, he views himself as an Earth System Scientist with a major interest in natural hazards, atmospheric events and using renewable energy as the only way to address global climate change issues. He has published over a 100 refereed research publications, conferences full paper and book chapters in these research areas. His research has been supported by National Science Foundation, NASA, United States Department of Agriculture and European Union. He has received the Saudi Arabia Award hosted by the Arab Administrative Development Organization (ARADO) affiliated with the League of Arab states for the best published article in

Environmental Management among 150 articles in 2006. He is also Member of the Institute of Electrical and Electronics Engineers (IEEE), AGU, EGU, COSPAR, and Phi Beta Delta Honor Society.

Dr. Zeynal Abiddin Erguler is a full professor at the Geological Engineering Department at Kutahya Dumlupınar University (Turkey). He obtained all of his degrees from Hacettepe University located in Ankara, Turkey. His research interests are primarily in the areas of rock mechanics, engineering geology, environmental geology and soil mechanics. His current investigation is to experimentally and numerically model thermo-, hydro-, and mechanical behavior of shale rocks in the area of shale gas production. In addition to doing many researches and performing industry-funded projects, he has also been teaching and supervising undergraduate and graduate students. In 2017, he joined the *Arabian Journal of Geosciences* (AJGS) as Associate Editor responsible for evaluating submissions in the fields of rock mechanics, engineering geology, environmental geology and soil mechanics. In 2019, he was appointed Chief Editor of AJGS for submissions on "Geological Engineering, Geotechnical Engineering."

Prof. Murat Karakus (University of Adelaide, Australia) Dr. Karakus holds a B.Sc. in Mining Engineering (1991) from Hacettepe University (Turkey) and a Ph.D. degree in Geotechnical Engineering (2000) from the University of Leeds, UK. He is currently a chair professor at the China University of Geosciences, Wuhan, and Professor of the School of Civil, Environmental and Mining Engineering. His research area is the field of experimental rock mechanics; fracture propagation and damage assessment by acoustic emissions, damage mechanics, plasticity and constitutive modeling in rock mechanics, numerical analysis in tunneling and mining, flow modeling in cave mining. He has been also conducting research on strain burst/rock burst mechanisms for deep tunnels and mining excavations. He has directed and involved in many research projects including ARC—Australian Research Council grants, CRC-Deep Exploration Technologies, ACARP—Australian Coal Association Research, and Industry supported projects. He has published several chapters and over a hundred papers in international Journals and Conferences. He is a member of the editorial board of several journals including the *International Journal of Rock Mechanics and Mining Sciences, Rock Mechanics and Rock Engineering, Tunnelling and Underground Space Technology, International Journal of Mining, Reclamation, and Environment, Bulletin of Engineering Geology and the Environment.* He is currently serving as a chief editor of Track-12; *Numerical and Analytical Methods in Mining Sciences and Geomechanics* of *Arabian Journal of Geosciences.*

Helder I. Chaminé is skilled Geologist and Professor of Engineering Geosciences at School of Engineering (ISEP) of the Polytechnic of Porto, with over 29 years' experience in multidisciplinary geosciences research, consultancy and practice. He studied geological engineering and geology (B.Sc., 1990) at the Universities of Aveiro and Porto (Portugal), respectively. He received his Ph.D. in geology at the University of Porto in 2000 and spent his postdoctoral research in applied geosciences at the University of Aveiro (2001–2003). In 2011, he received his Habilitation (D.Sc.) in Geosciences from Aveiro University. Before joining academy, he worked over a decade in international projects for mining, geotechnics and groundwater industry and/or academia related to geodynamics and regional geology, hard-rock hydrogeology and water resources, engineering geosciences and applied geomorphology, rock engineering and georesources. His research interests span over fundamental to applied fields: GIS mapping techniques for applied geology, structural geology and regional geology, engineering geosciences and rock engineering, slope geotechnics, mining geology and hydrogeomechanics, hard-rock hydrogeology, exploration hydrogeology, urban groundwater and hydromineral resources. He has interests on mining geoheritage, history of cartography, military geosciences and higher-education dissemination, skills and core values. Presently, he is Head of the Laboratory of Cartography and Applied Geology (LABCARGA | I SEP), Senior Researcher at Centre GeoBioTec | U. Aveiro and Centre IDL | U. Lisbon, as well as belongs to the executive board of the M. Sc.+B.Sc. Geotechnical and Geoenvironmental Engineering (OE+EUR-ACE Label) and the Department of Geotechnical Engineering (ISEP). Currently, he belongs to the board of the Portuguese Geotechnical Society (SPG) and IAH—Portuguese Chapter. He was a board member of the APGeom—Portuguese Association of Geomorphologists (2009–2013). He was consultant and or responsible over 70 projects of rock engineering, applied geology, hydrogeomechanics, slope geotechnics, mining geology, exploration hydrogeology, hard-rock hydrogeology, water resources, urban groundwater and applied mapping (Mozambique, Portugal, and Spain). He has been co-authored over 200 publications in indexed journals, conference proceedings/full papers, chapters, technical and professional papers. He co-edited over 13 special volumes, as well as is presently evolved in editing themed issues for three international journals (Geotechnical Research ICE, Springer Nature Applied Sciences, Water MDPI). He has a wide activity as a referee for several international journals. He served as invited Expert Evaluator of Bologna Geoscience program for DGES (Portugal) and Scientific Projects Evaluation for NCST, 2017–2019 (Kazakhstan) and NRF | RISA, 2019 (South Africa), as well as Coordinator of "Geology on Summer/Ciência Viva" program at ISEP, since 2005, for geosciences dissemination. He has been also active with teaching and supervising of many Ph.D., M.Sc. and undergraduate students. He has been on the editorial board, among others, of *Arabian Journal of Geosciences* (SSG+Springer), *Hydrogeology Journal* (IAH +Springer), *Euro-Mediterranean Journal for Environmental Integration* (Springer), *Springer Nature Applied Sciences* (Springer), *Mediterranean Geoscience Reviews* (Springer), *Geotechnical Research* (ICE), *Geosciences* (MDPI), *Revista Geotecnia* (Portugal), and *Geología Aplicada a la Ingeniería y al Ambiente* (Argentina). He integrates as moderator or session chair in several conferences, workshops and meetings. Currently, he is Co-chair of the International Conference "Geoethics and Groundwater Management" (IAH+IAPG, Porto, May 2020) and in organizing/scientific committee of the 3rd International Workshop on Natural Hazards—NATHAZ'21 (Terceira Island, Azores, May 2021).

Geological Engineering, Geotechnical Engineering (T8): Slope Instabilities and Landslides

Slope Instability Mechanisms in Karst Rocks and Their Impact on Sustainable Development, Egypt

Bosy Abdelaziz El-Haddad, Ahmed Mohamed Youssef, Abdel-Hamid El-Shater, and Mohamed H. El-Khashab

Abstract

Recently, Egypt has faced unprecedented development with rapid urban and infrastructure expansion. Vast areas of Egypt are underlain by karst rocks (carbonate and evaporite). Highways and roads constructed along these karst areas are susceptible to slope stability problems. In this work, remote sensing imagery, field investigation, and laboratory studies are applied to analyze and identify unstable areas and their characteristics. Five slope instability mechanisms are categorized including collapses of weak materials from filled cavities, breakdown of cavities, instability due to differential erosions, dissolution along discontinuities, and sliding along weak surfaces. This study could help to understand the causes of slope instability problems, to maintain the sustainability of the infrastructure, and to design appropriate prevention and remediation measures to avoid future problems.

Keywords

Karst • Instability mechanisms • Egypt • Sustainable development

1 Introduction

Karst rocks (carbonate and evaporite) cover vast areas of Egypt (Halliday 2003). Problems associated with karst terrains have substantially increased in the last decade due to rapid development in these areas underlain by soluble rocks. Many authors documented karst landforms in Egypt (Wanas et al. 2009; Abdeltawab 2013). Slope instability is the downslope movement of different materials by the action of gravity leading to the development of landslides (Blasio 2011; Duan et al. 2019). Worldwide, transportation networks are exposed to various types of landslides, especially wherever they cut across or skirt along mountains, posing a risk to travelers, damaging properties, and interrupting the service in the infrastructure. These problems may severely affect local economies (Ferlisi et al. 2012). Different factors such as the geological structure, lithology, water, rock mass characteristics, geomechanical properties of rocks, and karst development are believed to be the most influencing factors for engineering purposes and triggering factors of slope instability (Cheng et al. 2017; Santo et al. 2017; Andriani and Parise 2017; Palma et al. 2017). Several studies have dealt with different slope stability problems associated with various karst features (e.g., cavities, solutionally widened fractures, differential weathering, filled caves, sinkholes at the foot of the slopes) (Gutiérrez et al. 2014; Feng et al. 2014; Parise 2015; Martinotti et al. 2017; Guarino et al. 2018). The current study aims to categorize different mechanisms of landslides associated with karst rocks using field and laboratory investigations, as well as remote-sensed imagery.

2 Study Area and Methodology

Egypt is characterized by four geological zones (Fig. 1a). Zones 1–3 include the Nubian Shield (Precambrian rocks), exposed Paleozoic rocks (sandstone and shale), and outcrops of Mesozoic rocks (sandstone and some Cretaceous carbonates) characterized by rare karst features. Cenozoic rocks zones (zone 4) are characterized by the presence of karst rocks (carbonate and evaporite), covering large areas of Egypt: the eastern and western limestone plateaus overlooking the Nile valley, from Luxor to Cairo, and evaporites

B. A. El-Haddad · A. M. Youssef (✉) · A.-H. El-Shater · M. H. El-Khashab
Faculty of Science, Geology Department, Sohag University, Sohag, 82524, Egypt

A. M. Youssef
Geological Hazards Center, Saudi Geological Survey, Jeddah, 21514, Kingdom of Saudi Arabia

H. El-Askary et al. (eds.), *Research Developments in Geotechnics, Geo-Informatics and Remote Sensing*, Advances in Science, Technology & Innovation,
https://doi.org/10.1007/978-3-030-72896-0_1

deposits and sabkha environments located along the Red Sea coast (Fig. 1a). Recently, Egypt has faced unprecedented infrastructure and urbanization expansion. Roads network is established to facilitate transportation for people and commodities. A significant number of highways are built on or cutting through karst rocks and are affected by slope stability problems. Several field investigations were carried out by the authors to understand and identify different slope instability mechanisms. Field samples were collected and tested to identify their uniaxial compressive strength, density, absorption, durability, swelling characteristics, and Atterberg properties. Besides, visual interpretation of high-resolution satellite images (QuickBird images of June 2015 and Professional Google Earth images) was used to identify the instability features.

3 Results

3.1 Laboratory Analysis Results

Various tests including compressive strength, density, and water absorption of massive limestone, bedded limestone, marl limestone, shale, and block-in-matrix deposit were conducted in this study. The results are shown in Table 1.

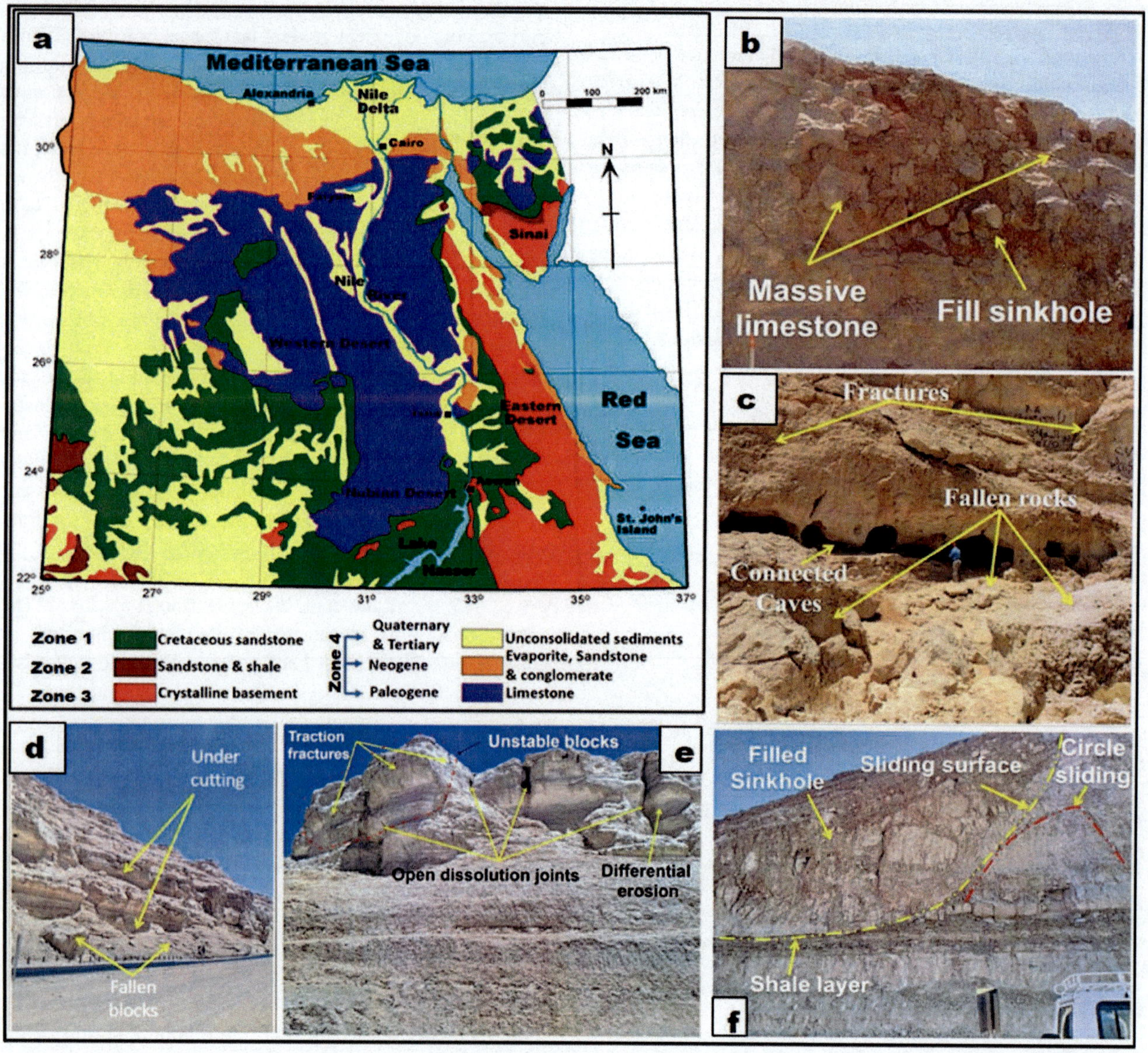

Fig. 1 **a** Modified generalized geological map of Egypt (http://www.eeesci-ence.utoledo.edu/Faculty/Harrell/Egypt); **b–f** are photographs showing examples of different slope instability mechanisms in karst rocks

Table 1 Laboratory tests results

Rock type	Compressive strength (MPa)	Density (kg/m^3)	Water absorption (%)
Massive limestone	51–54	2.4–2.5	2.5–2.9
Bedded limestone	26.1–29.7	2.35–2.45	6.1–7.9
Marl limestone	10.9–12.3	2.3–2.4	10.3–12.6
Gray shale Yellow shale	1.3–2.1	1.7–1.9	25.1–30.2 60.9–73.9
Block-in-matrix deposit	<1.25	–	–

The tests indicated that water absorption increased with decreasing strength. Jar slake and free swell tests were carried out on ten shale samples, and slake durability test was carried out on five samples of marl limestone. Jar slake test result indicated that yellow shale samples degraded to form a pile of flakes and that gray shale samples broke to form chips. Free swell test indicated that the yellow shale had a free swelling value of 17.5% with a high liquid limit ranging from 32 to 36%, and that the gray shale had a free swelling value of 11.4% with a liquid limit varying from 26 to 28.5%. For the slake durability test, the marl limestone samples were subjected to four cycles of slake durability test. The findings demonstrated that the slake durability index ranges from 88% after the fourth cycle to 99% after the first cycle.

3.2 Slope Instability Mechanisms

According to field investigations, laboratory results, and analysis of high-resolution satellite images, five different slope instability mechanisms were identified: (1) rockfalls related to the collapse of cave fill deposits (Fig. 1b). Caves-filling deposits were characterized by large blocks up to several tens of meters long embedded in a fine-grained matrix with a compressive strength lower than 1.25 MPa. This weak matrix could be easily eroded by seepage water, runoff, and wind action. (2) Landslides associated with the collapse of caves (Fig. 1c)—mass wasting was the main process involved in the evolution of limestone caves at the timescale of an engineering project, given the low dissolution rate of these rocks. Breakdown processes were controlled by various types of discontinuities (e.g., bedding, joints) that might be accompanied by secondary deposits (e.g., clay fills, breccias). The strength of these discontinuities and that of the rock mass might decrease by dissolution and internal erosion processes. (3) Landslides induced by differential erosion and slope undermining (Fig. 1d)—weak layers of shale and marl limestone might recede rapidly by weathering and erosion processes, resulting in the lack of basal support for the overlying overhanging rocks. (4) Rock falls controlled by fractures widened by dissolution (Fig. 1e). (5) Rocks slides controlled by discontinuities (e.g., bedding, joints, faults, shear zones) including subcircular, planar, and wedge failures (Fig. 1f).

4 Conclusions

Five different landslide mechanisms in karst rocks in Egypt were identified and addressed. Field and laboratory investigations and remote sensing analysis indicated that the main factors caused slope and cliff failures in karst rocks were related to the weathering process, differential erosion, presence of discontinuities affected by dissolution, and karst features. Understanding of these landslide mechanisms will help decision-makers, planners, the department of transportation, and engineers to choose the proper prevention and remediation measures.

References

Abdeltawab, S.: Karst limestone geohazards in Egypt and Saudi Arabia. Int. J. Geoeng. Case Hist. **2**, 258–269 (2013). https://doi.org/10.4417/IJGCH-02-04-02

Andriani, G.F., Parise, M.: Applying rock mass classifications to carbonate rocks for engineering purposes with a new approach using the rock engineering system. J. Rock Mech. Geotech. Eng. **9**, 364–369 (2017)

De Blasio, F.V.: Introduction to the Physics of Landslides: Lecture Notes on the Dynamics of Mass Wasting. Springer, Netherlands, 408 p (2011). ISBN 978-94-007-1122-8

Cheng, W.C., Cui, Q.L., Shen, J.S.L., Arulrajah, A., Yuan, D.J.: Fractal prediction of grouting volume for treating karst caverns along shield tunneling alignment. Appl. Sci. Basel **7**(7), 652 (2017). https://doi.org/10.3390/app7070652

Duan, Z., Cheng, W.C., Peng, J.B., Wang, Q.Y., Chen, W.: Investigation into triggering mechanism of loess landslides in south Jingyang platform, Shaanxi province. Bull. Eng. Geol. Env. **7**, 4919–4930 (2019). https://doi.org/10.1007/s10064-018-01432-8

Feng, Z., Li, B., Yin, Y.P., He, K.: Rockslides on limestone cliffs with subhorizontal bedding in the southwestern calcareous area of China. Nat. Hazards Earth Syst. Sci. **14**, 2627–2635 (2014). https://doi.org/10.5194/nhess-14-2627-2014

Ferlisi, S., Cascini, L., Corominas, J., Matano, F.: Rockfall risk assessment to persons travelling in vehicles along a road: the case study of the Amalfi coastal road (southern Italy). Nat. Hazards **62** (2), 691–721 (2012). https://doi.org/10.1007/s11069-012-0102-z

Guarino, P.M., Santo, A., Forte, G., De Falco, M., Niceforo, D.M.A.: Analysis of a database for anthropogenic sinkhole triggering and zonation in the Naples hinterland (Southern Italy). Nat. Hazards **91**, 173–192 (2018)

Gutiérrez, F., Parise, M., DeWaele, J., Jourde, H.: A review on natural and human-induced geohazards and impacts in karst. Earth Sci. Rev. **138**, 61–88 (2014)

Halliday, W.: Caves and karsts of northeast Africa. Int. J. Speleol. **32** (1/4), 19–32 (2003)

Martinotti, M.E., Pisano, L., Marchesini, I., Rossi, M., Peruccacci, S., Brunetti, M.T., Melillo, M., Amoruso, G., Loiacono, P., Vennari, C., Vessia, G., Trabace, M., Parise, M., Guzzetti, F.: Landslides, floods and sinkholes in a karst environment: the 1–6 September 2014 Gargano event, southern Italy. Nat. Hazard. **17**, 467–480 (2017)

Palma, B., Parise, M., Ruocco, A.: Geomechanical characterization of carbonate rock masses by means of laser scanner technique. IOP Conf. Ser. Earth Environ. Sci. **95**, 062007 (2017). https://doi.org/10.1088/1755-1315/95/6/062007

Parise, M.: Karst geo-hazards: causal factors and management issues. Acta Carsologica **44**(3), 401–414 (2015)

Santo, A., Budetta, P., Forte, G., Marino, E., Pignalosa, A.: Karst collapse susceptibility assessment: a case study on the Amalfi Coast (Southern Italy). Geomorphology **285**, 247–259 (2017)

Wanas, H.A., Soliman, H., Pickford, M., Ségalen, L., Mein, P.: Late Miocene karst system at Sheikh Abdallah, between Bahariya and Farafra, Western Desert, Egypt: implications for palaeoclimate and geomorphology. Geol. Acta **7**, 475–487 (2009)

Application of a New Generalized Method to Assess Alert Thresholds for Landslides

Andrea Segalini, Alessandro Valletta, Roberto Savi, Andrea Carri, and Corrado Carini

Abstract

Within the framework of natural risks mitigation, early warning systems (EWSs) are among the most valuable tools. The application of effective procedures aimed to identify a landslide time-of-failure and to assess alert thresholds is a key component in every EWS design process. For this reason, a wide number of methodologies have been proposed over the years, each focusing on a specific case study. This paper presents the application of a new generalized method, which is intended as a tool to assess alert thresholds for different landslide typologies. In particular, the proposed approach is developed to overcome the site-specific nature that characterizes the other methods, by identifying a common behavior in several different events. The generalized criterion features also the possibility to evaluate a theoretical time-of-failure by applying the inverse velocity method (IVM) to monitoring data. The proposed method is applied to a displacement monitoring dataset related to a slope collapse, in order to present an example of its implementation in a real case scenario.

Keywords

Landslides • Monitoring • Early warning system • Alert thresholds

1 Introduction

An early warning system (EWS) can be defined as an integrated system of monitoring procedures, forecasting models, disaster risks assessment, and awareness activities, aimed to enable threatened communities to act timely and to appropriately reduce the possibility of damages and losses induced by a natural event (United Nations General Assembly 2016). The great variability of parameters such as hazard types, event frequencies, and involved communities contributes to the high complexity of an EWS design (Garcia and Fearnley 2012).

Among the different components of an alert system, models related to failure forecasting and threshold assessment represent the most complicated challenge from a scientific point of view. For this reason, several pieces of research and studies have been carried out, aiming to test different approaches to slope collapse prediction (Intrieri et al. 2019; Federico et al. 2012) and define meaningful alarm levels related to landslides (Crosta and Agliardi 2002; Manconi and Giordan 2016; Carlà et al. April 2017; Segalini et al. 2018). The criterion applied in this paper was developed in order to provide a generalized procedure for alert threshold assessment, regardless of the landslide features, e.g., failure mechanism and materials (Segalini et al. 2018).

2 Materials and Methods

The method presented here relies on the empirical parameter α, typical of the inverse velocity method (Fukuzono 1985; Voight 1988), to describe the landslide displacement trend approaching to failure. After a calibration procedure, IVM equations were implemented to derive a series of theoretical velocity curves corresponding to the 30 days preceding the slope collapse. At last, a normalization process allowed to obtain a dimensionless velocity trend (Segalini et al. 2018). The final result, reported in Fig. 1, is a single graph that can

A. Segalini · A. Valletta · R. Savi (✉)
University of Parma, 43121 Parma, PR, Italy
e-mail: roberto.savi@aseltd.eu

A. Carri
ASE—Advanced Slope Engineering S.r.l., 43124 Parma, PR, Italy

C. Carini
Earth System S.R.L., 43123 Parma, PR, Italy

H. El-Askary et al. (eds.), *Research Developments in Geotechnics, Geo-Informatics and Remote Sensing*, Advances in Science, Technology & Innovation,
https://doi.org/10.1007/978-3-030-72896-0_2

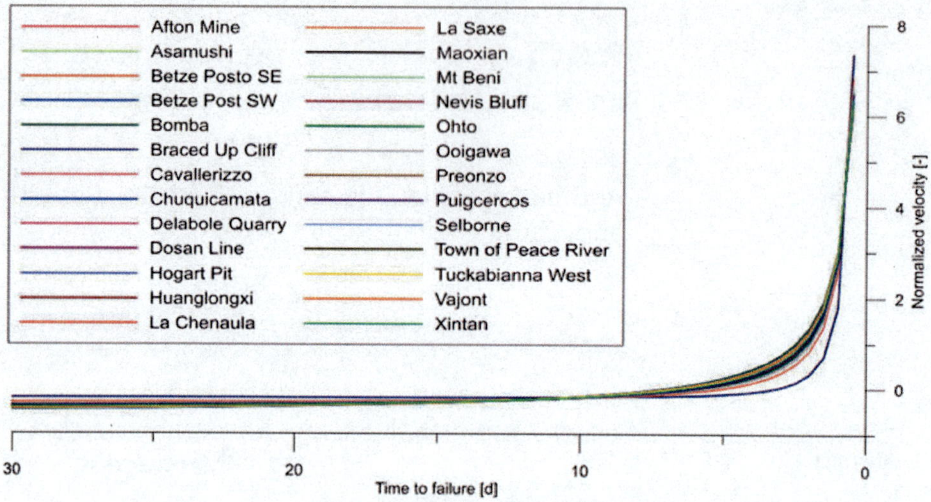

Fig. 1 Normalized velocity versus time plot (Segalini et al. 2018)

be used as a theoretical reference to define alert levels depending on the value of α parameter.

This section presents a simulated real-time application of the generalized method aiming to provide a realistic test of the proposed criterion. Following the identification of the onset-of-acceleration point, it is possible to define a starting 4-value dataset. The α parameter is then calibrated to represent the studied event better, and to choose the corresponding theoretical velocity curve, allowing to define a set of generalized velocity thresholds. After a normalization procedure, the last monitoring data available is compared to the alert levels previously defined. In the following passage, the dataset is expanded with the addition of new data, and the whole procedure is repeated to follow the phenomenon evolution. In this example, the chosen dataset was acquired with total station surveys. It refers to a rock slope collapse in an open-pit mine involving approximately $1 \times 106\ m^3$ of material (Rose and Hungr 2007).

3 Results

The model calibration provided values of α parameter close to 2.00; thus, the corresponding curve was selected to assess meaningful alert thresholds. Figure 2 displays the good correlation between normalized monitoring data and theoretical trend.

In a real case scenario, alert thresholds should be defined on the basis of the specific case study knowledge and experts' judgment. In this paper, the authors have chosen two warning levels for demonstration purposes. In particular, thresholds refer to velocity values $v_{thr1} = -0.0426$ and

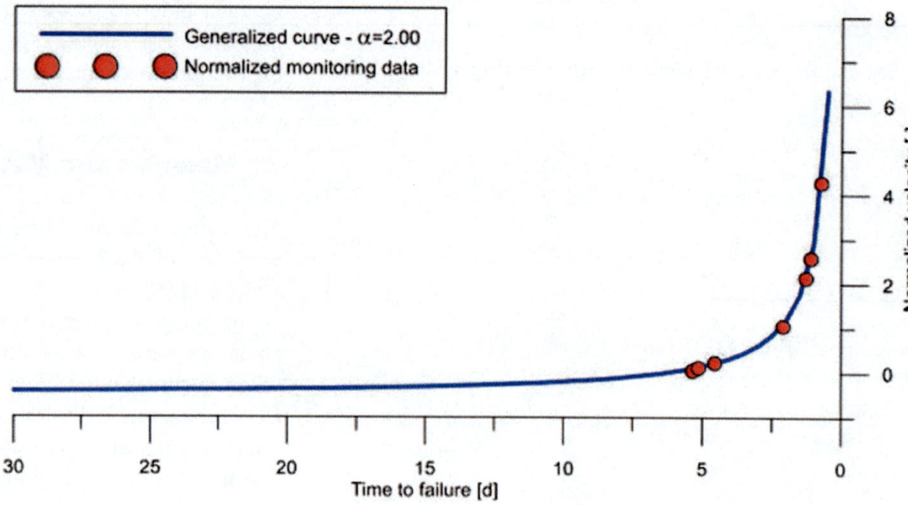

Fig. 2 Comparison between normalized velocity data and theoretical curve featuring $\alpha = 2.00$

Table 1 Normalized monitoring data and thresholds overcoming

Time [d]	Normalized monitoring data [−]	Threshold #1	Threshold #2
9.5	0.0501	X	–
10.0	0.1139	X	–
10.5	0.2269	X	–
13.0	1.0555	X	X
13.8	2.1381	X	X
14.0	2.5824	X	X
14.5	4.2756	X	X

v_{thr2} = 0.6187, corresponding to a 7-day and 3-day temporal distances from the theoretical collapse, respectively. Table 1 reports the comparison with monitored data, evidencing the threshold overcoming as the event develops over time.

4 Discussion

As it is possible to appreciate in Fig. 2, the theoretical velocity curve computed with the generalized approach provides an accurate representation of the phenomenon evolution. Specifically, the calibration of the IVM parameter for this landslide returned a value close to 2.00 for each monitoring data. This value represents a linear relationship between inverse of velocity and time, and it is generally considered a good approximation in the evaluation of a landslide time-of-failure (Fukuzono 1985; Voight 1988; Rose and Hungr 2007).

In this case, the relatively rapid evolution of the phenomenon caused the exceedance of the 7-day threshold with the first monitoring data. Following displacement measures allowed to study the event evolution until the overcoming of the second warning level, which occurred at the fourth monitoring data. Additionally, the continuation of the monitoring activity made it possible to carry on the event observation, providing new measures that consistently exceeded the chosen thresholds and confirmed the collapsing trend predicted by the model.

5 Conclusions

This paper describes the application of a new generalized criterion, developed to define alert thresholds for early warning purposes. A case study is presented, simulating a real-time sampling of landslide displacements data, in order to test the effectiveness of the proposed method.

The theoretical velocity curve selected based on the value of empirical parameters after a calibration procedure allowed to represent the landslide evolution correctly. The comparison with elaborated monitoring data permitted to identify the overcoming of previously assessed thresholds, effectively identifying the gradual evolution of the landslide toward its definitive collapse.

References

Carlà, T., Intrieri, E., Di Traglia, F., Nolesini, T., Gigli, G., Casagli, N.: Guidelines on the use of inverse velocity method as a tool for setting alarm thresholds and forecasting landslides and structure collapses. Landslides **14**(2), 517–534 (2017)

Crosta, G.B., Agliardi, F.: How to obtain alert velocity thresholds for large rockslides. Phys. Chem. Earth **27**(36), 1557–1565 (2002)

Federico, A., Popescu, M., Elia, G., Fidelibus, C., Internò, G., Murianni, A.: Prediction of time to slope failure: a general framework. Environ. Earth Sci. **66**, 245–256 (2012)

Fukuzono, T.: A new method for predicting the failure time of a slope. In: Proceedings of the Fourth International Conference and Field Workshop on Landslides (Tokyo; 1985).Tokyo University Press, pp 145–15 (1985)

Garcia, C., Fearnley, C.J.: Evaluating critical links in early warning systems for natural hazards. Environ. Hazards **11**, 123–137 (2012)

Intrieri, E., Carlà, T., Gigli, G.: Forecasting the time of failure at a slope-scale: a literature review. Earth Sci. Rev. **193**, 333–349 (2019)

Manconi, A., Giordan, D.: Landslide failure forecast in near-real time. Geomat. Nat. Hazards Risk **7**(2), 639–648 (2016)

Rose, N.D., Hungr, O.: Forecasting potential rock slope failure in open pit mines using the inverse-velocity method. Int. J. Rock Mech. Min. Sci. **44**(2), 208–320 (2007)

Segalini, A., Valletta, A., Carri, A.: Landslide time-of-failure forecast and alert threshold assessment: a generalized criterion. Eng. Geol. **245**, 72–80 (2018)

United Nations General Assembly: Report of the open-ended intergovernmental expert working group on indicators and terminology relating to disaster risk reduction. A/71/644 Seventy-first session, Agenda item 19 (c) (2016)

Voight, B.: A method for prediction of volcanic eruptions. Nature **332**, 125–130 (1988)

Geological, Hydrological and Geotechnical Impact for the Landslides Susceptibility in Mila Basin (Northeast Algeria)

Nadira Bounemeur, Riad Benzaid, and Souad Atoub

Abstract

Landslides are ranked third after earthquakes and floods in terms of material loss of life and damage to loose formations. This sliding phenomenon is prevalent in the Mila basin, particularly the northeastern part of Oued El-Kebir and Rhumel. It is established that landslides are caused by the spatiotemporal conjunction of several factors. These factors may be quasi-static variables such as geology, geomorphology and geotechnical characteristics of the site, or dynamic variables such as precipitation, earthquakes and anthropogenic activities. The present work is a hydrological and geotechnical geological study aiming to propose effective solutions and to determine areas with high potential for instability.

Keywords

Mila Basin • Landslides • Proposed solutions

1 Introduction

The basin of Mila covers an area of 3500 km^2. It is also known under the name of the Mila-Constantine basin. It forms a vast depression of about 40 km in width and 120 km in length (Marmi et al. 2008; Bounemeur et al. 2019).

Among the geological formations of the study site (Fig. 1a), only the geological characteristics of the Mio-Pliocene formation is defined here, since the study site belongs geologically to this area. These formations cover most of the study site and form the cover of a substratum composed of a flysch layer stack and a generally non-native Tellian series. These marly-clay formations are susceptible to provoke the phenomenon of landslide (Fig. 1b) manifested in disorders affecting the road network and buildings (Atmania 2010).

2 Materials and Methods

This work is also based on various cartographic documents such as topographic maps of the study area at the scale of 1/50.000 and geological maps at 1/50.000 scale. The data is supplemented by GPS measurements and observations on the field. The software Arc Gis10.2.2, Map Info and Global Mapper 17.0 are used as part of image processing and georeferencing of the data (Bennia et al. 2007).

3 Results and Discussions

3.1 Slope Map and Hydrographic Network

Slopes represent an essential element in characterizing aspects of relief. They are also related to the dynamics of the environment going into action as factors playing a fundamental role in morphodynamic analysis and the explanation of certain forms of erosion and mass movements (Bennia et al. 2007; Kherrouba et al. 2019). Thus, the hydrological study passes mainly by the treatment of hydroclimatic measures and consequently to understand the mechanisms of the circulation and the infiltration of waters inducing the instability of the grounds.

The ombrothermal diagram presented in Fig. 2 and the precipitation histogram displayed in Fig. 3 indicate that the rainiest season is the winter where the potential for the risk of instability increases considerably, thus the feeding of rivers and streams. Groundwater by precipitation is high in winter season, while summer is the driest season.

N. Bounemeur (✉) · R. Benzaid · S. Atoub
Laboratoire de Génie Géologique, Université de Jijel,
Cité Ouled Aissa, B.P. 98 18000 Jijel, Algeria

H. El-Askary et al. (eds.), *Research Developments in Geotechnics, Geo-Informatics and Remote Sensing*, Advances in Science, Technology & Innovation,
https://doi.org/10.1007/978-3-030-72896-0_3

Fig. 1 Examples of landslides in the Mila Basin; **a** slip zone, **b** break in the road

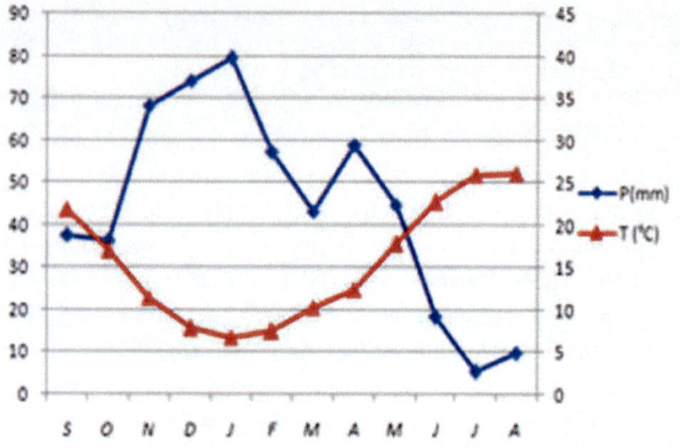

Fig. 2 Ombrothermic diagram (Constantine station, period 1990/2003)

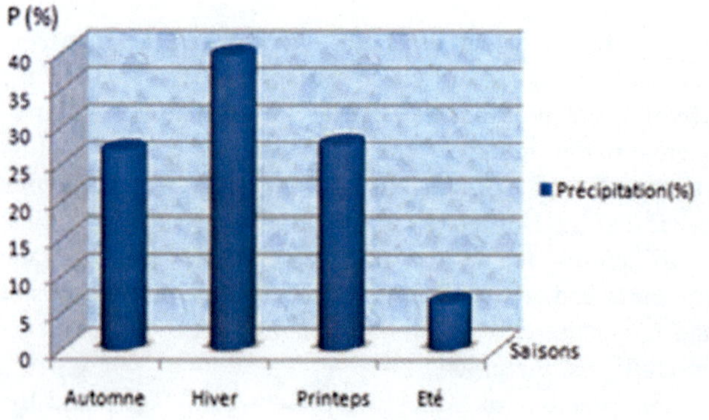

Fig. 3 Histogram of seasonal mean precipitation (ANRH station, Constantine)

3.2 Geotechnical Identification Tests

The measurement of the physical parameters (Table 1) allows an approach to the identification of the nature of the soil in place. The values of the main physical parameters measured for each survey at different depths are represented as follows according to the Technical Study Company of Sétif (2014) (Belabbas 2012). The results of the physical parameters of the study site and the Casagrande diagram summarize the nomenclature of our soil considered as clay soil (Figs. 4 and 5).

4 Discussion

The study of the landslide in the Mila Basin aimed to contribute to better diagnosis of this hazard and to identify the factors behind it. This study was based on the combination of geological, hydrological and geotechnical data. After identifying the causes, it is recommended to adopt the following strategies for site stabilization and road comfort.

- The drainage of the surface waters by the setting up of concrete ditches along the road.
- Prohibiting any discharge of material on the slopes.
- Rehabilitation of the pavement body and its coating on a geotextile-reinforced earthen embankment to ensure good road holding.

Table 1 Physical parameters of the study site

SC. No.		Identification				Atterberg limits		Sedimentometry	Classification
	Depth (m)	γ_d	γ_h	W	Sr	LL	I_o	% < 80 μm	
01	1.20–1.70	1.45	1.85	28.08	90.22	67.54	33.56	94	At
	3.50–3.80	1.54	1.92	25.17	92.73	69.99	34.76	100	At
	5.60–6.00	1.67	2.03	22.15	99.20	77.38	39.34	100	At
	7.60–8.00	1.63	2.01	23.52	99.22	73.40	36.45	100	At
	9.45–9.75	1.70	2.05	21.12	99.49	63.41	34.22	100	At
02	1.25–1.60	1.53	1.92	26.13	94.55	–	–	–	–
	2.45–2.80	1.64	2.02	23.36	99.81	69.42	34.64	100	At
	3.60–4.00	1.53	1.95	27.56	99.33	–	–	–	–
	4.45–4.75	1.54	1.95	26.98	99.57	71.71	34.97	100	At
	5.70–6.00	1.57	1.97	25.79	99.92	–	–	–	–
	6 75–6.95	1.63	2.01	23.48	99.93	66.7	33.74	100	At
	7.70–7.95	1.54	1.95	27.01	99.83	64.80	32.38	100	At
	8.75–9.00	1.62	2.01	24.16	99.97	–	–	–	–
	9.70–10.0	2.18	2.35	08.10	97.96	46.89	23.20	100	Ap

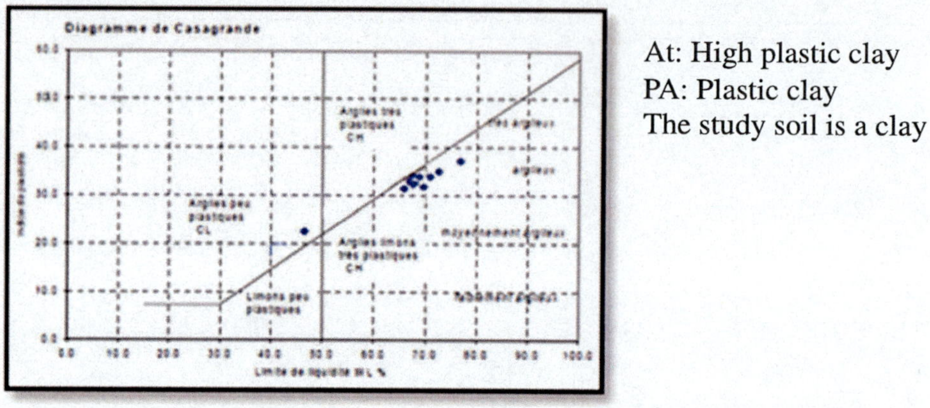

Fig. 4 Classification of soils with the Casagrande diagram

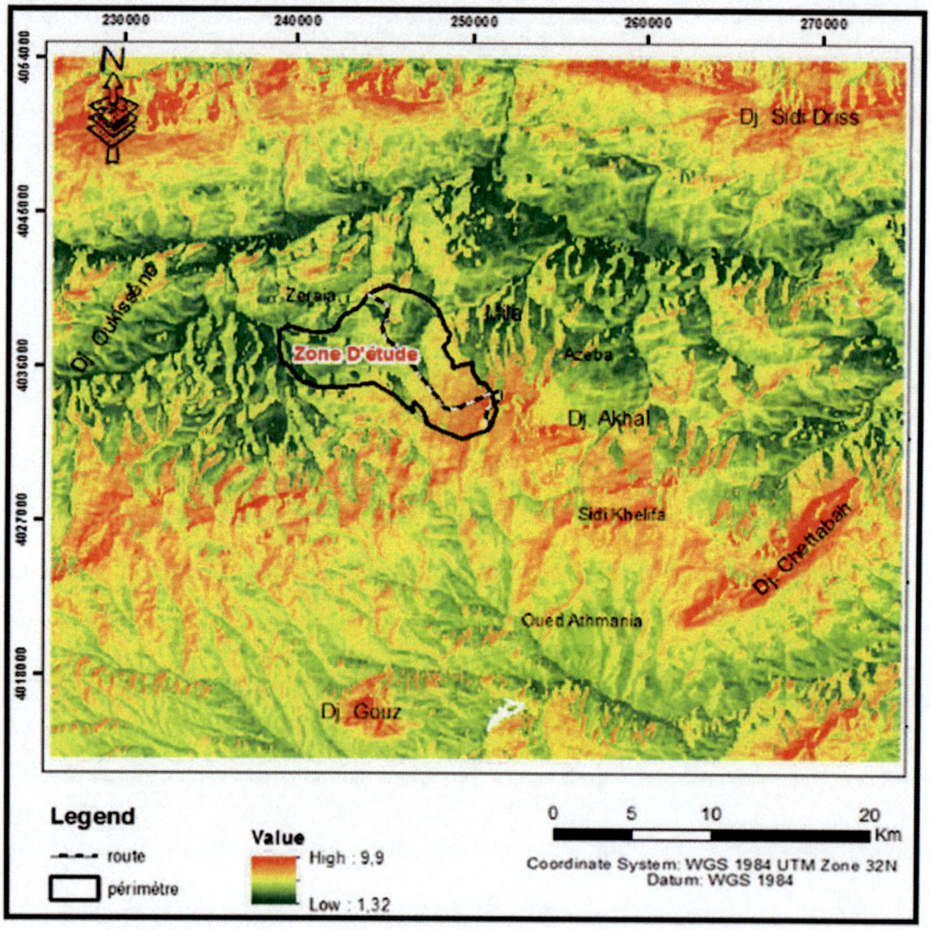

Fig. 5 Susceptibility map of landslide in the Mila Basin

References

Atmania, D.: Clay mineralogy and shrinkage-swelling phenomenonin the Mila basin (North Constantinois). Thesis in Science, Constantine 1 University (Algeria), 172 p (2010)

Belabbas, S.: Inventory and characteristics of active accidents (impact on seismicity of the Constantine Region). Magister thesis, Ferhat Abbas-Sétif University (Algeria), 120 p (2012)

Bennia, A., Saad, A.Z., Mesbah, C.: 1/50.000 geological hazard mapping from high resolution satellite imagery (ALSAT-2A) and geographic information systems (SIG): the case of the pilot area of Bejaia. Geogr. Sci. Bull. **30**, 7P (53–59) (2007)

Bounemeur, N., Benzaid, R., Atoub, S.: Landslides in the Mila Basin-AGIS Approach. In: Kallel, A. et al. (eds.) Recent Advances in Geo-Environmental Engineering, Geomechanics and Geotechnics, and Geohazards. CAJG 2018. Advances in Science, Technology & Innovation (IEREK Interdisciplinary Series for Sustainable Development). Springer, Cham (2019). https://doi.org/10.1007/978-3-030-01665-4_100

Kherrouba, H., Lamara, M., Benzaid, R.: Contribution of electrical tomography to the study of landslides in Texenna Region (North-east Algeria). In: Sundararajan N., Eshagh, M., Saibi, H., Meghraoui, M., Al-Garni, M., Giroux, B. (eds.) On Significant Applications of Geophysical Methods. CAJG 2018. Advances in Science, Technology & Innovation (IEREK Interdisciplinary Series for Sustainable Development). Springer, Cham (2019). https://doi.org/10.1007/978-3-030-01656-2_11

Marmi, R., Kacimi, M., Boularak, M.: Land movements in the Mila (Nord-EastAlgeria)—impact on infrastructure. Revista de geomorfologie **10**, 51–56 (2008) S.E.T.S: Société d'Etude Technique de Sétif. Rapport geotechnique dédoublement de pénétrante sud de la ville de Mila, wilaya de Mila. Lot N°1, 92 p (2014)

Analysis of the Causes of Landslide in the Village Podi, Montenegro

Slobodan Zivaljevic, Zvonko Tomanovic, and Milan Radulovic

Abstract

The disposal of construction waste materials above the location of the village Podi was followed by the occurrence of a landslide which affected 3.5 ha and damaged 27 structures. In the course of legal proceedings initiated against the contractor, a question arose of whether the deposited material could have triggered the landslide. Response to this complex question required a multidisciplinary approach, including also a numerical parametric analysis. The results of such analysis indicated that the improper depositing of material over the upper part of a conditionally stable slope induced the landslide activation for two reasons: an increase in shear stress and a change in the natural groundwater regime, which in turn affected elevation in the pore water pressure.

Keywords

Landslide • Numerical • FEM • Hydrogeology

1 Introduction

The uncontrolled disposal of construction waste materials above the location of the village Podi (coordinates: 42.466061° N, 18.539521° E) in the Municipality of Herceg-Novi, Montenegro (Fig. 1), contributed to the occurrence of the landslide which affected 3.5 ha and damaged 27 structures. The material was mainly disposed in 2007 and January 2008, while the landslide to a larger extent, followed by displacement of decimeters and damages to the structures, occurred in mid-February 2008, immediately after the first major rainfall. In the course of legal proceedings initiated against the contractor, a question arose of whether the disposed material could have triggered the landslide. To provide a response to this complex question, an expert team composed of geologists, hydrogeologists and geotechnical engineers performed an analysis (Tomanovic et al. 2013) based on the available data. A summary of the analysis is presented in this paper. Landslide incidents have resulted in significant economic losses and human casualties (Meng et al. 2018). This study aims to contribute to a better understanding of landslide causes and triggering mechanisms (Duan et al. 2019).

2 Methodology

The adopted approach involved a thorough examination of the available documentation (geodetic, geological and hydrogeological maps and input data), site visits, producing additional reports and conducting a numerical parametric analysis. In the separate reports, geology specialists within the team provided their critical analysis of the two independent geological reports (Ivanovic 2008; Lokin 2010) which have different assumptions in terms of depth, shape, strike and dip of sliding surface and hydrogeological parameters. A numerical parametric analysis of stability was performed on numerous terrain models, using software package Phase2. Rocscience (2019) aimed to determine the representative values of parameters of the soil shear strength (back analysis), possible failure mechanisms and to quantify the effects of water table on the decrease of factor of safety (FoS). The shear strength reduction (SSR) method is applied to determine the value of factor of safety of the slope.

3 Results

3.1 Results of the Analysis of Geological and Hydrogeological Reports

In the past, the terrain of the village Podi was naturally unstable. This conclusion can be drawn based on the sliding

S. Zivaljevic (✉) · Z. Tomanovic · M. Radulovic
Faculty of Civil Engineering, University of Montenegro, Podgorica, 81000, Montenegro

H. El-Askary et al. (eds.), *Research Developments in Geotechnics, Geo-Informatics and Remote Sensing*, Advances in Science, Technology & Innovation,
https://doi.org/10.1007/978-3-030-72896-0_4

zone depths obtained through the conducted soil investigation in the form of boreholes and also based on the morphology of the slope concerned. Two gullies enclose the steep slope terrain with seasonal high water (Fig. 1). It can be concluded that at the time of the material disposal, the terrain was in a conditionally stable state of limit equilibrium. One geological report (Ivanovic 2008) accounted for one sliding body of complex shape (Fig. 2a), while the other report (Lokin 2010) introduced the assumption of the sliding surface being discontinued by the bedrock outcrops—flysch (Fig. 2b), i.e. suggesting the existence of two separate landslides. According to Ivanovic (2008) after the uncontrolled waste disposal, the water level raised from the initial level illustrated in Fig. 2a to the terrain surface, followed by the occurrence of diffuse springs in numerous locations, while the western stream went dry.

3.2 Results of the Numerical Analysis of Slope Stability

For this analysis, over ten numerical models were developed to include and confirm different assumptions in terms of the shape, depth and orientation of the sliding surface. All models were subjected to the final analysis which provided for the shear strength parameters of individual soil layers and the sliding surface itself, counting with the value of factor of safety of near 1.00 (F_s, Table 1). Consequently, the verification of the shear strength parameters provided in the geological report was accomplished. It was quite challenging to quantify the effects of the applied weight of the disposed material on the slope stability due to the lack of accurate data about the slope geometry from the period preceding the disposal of material. Therefore, the final analysis took into

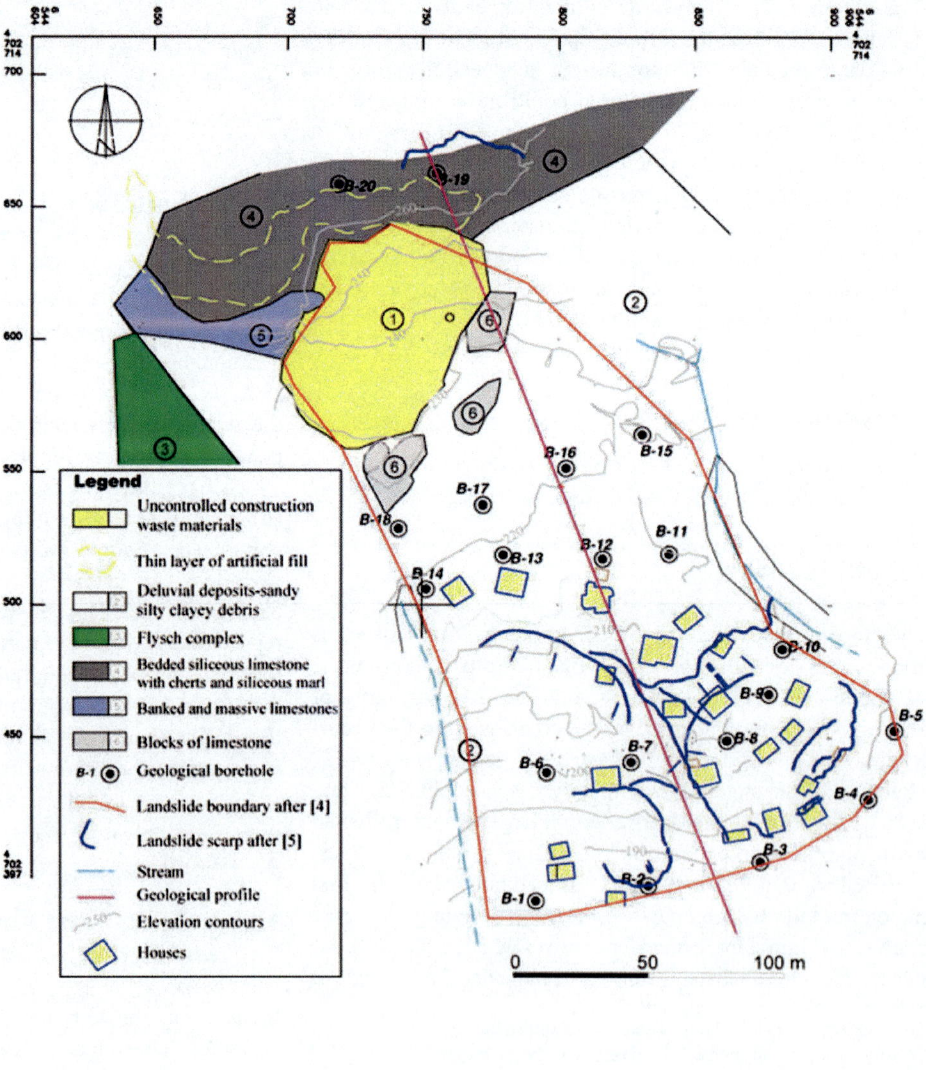

Fig. 1 Geological map of landslide at Podi after (Ivanovic 2008)

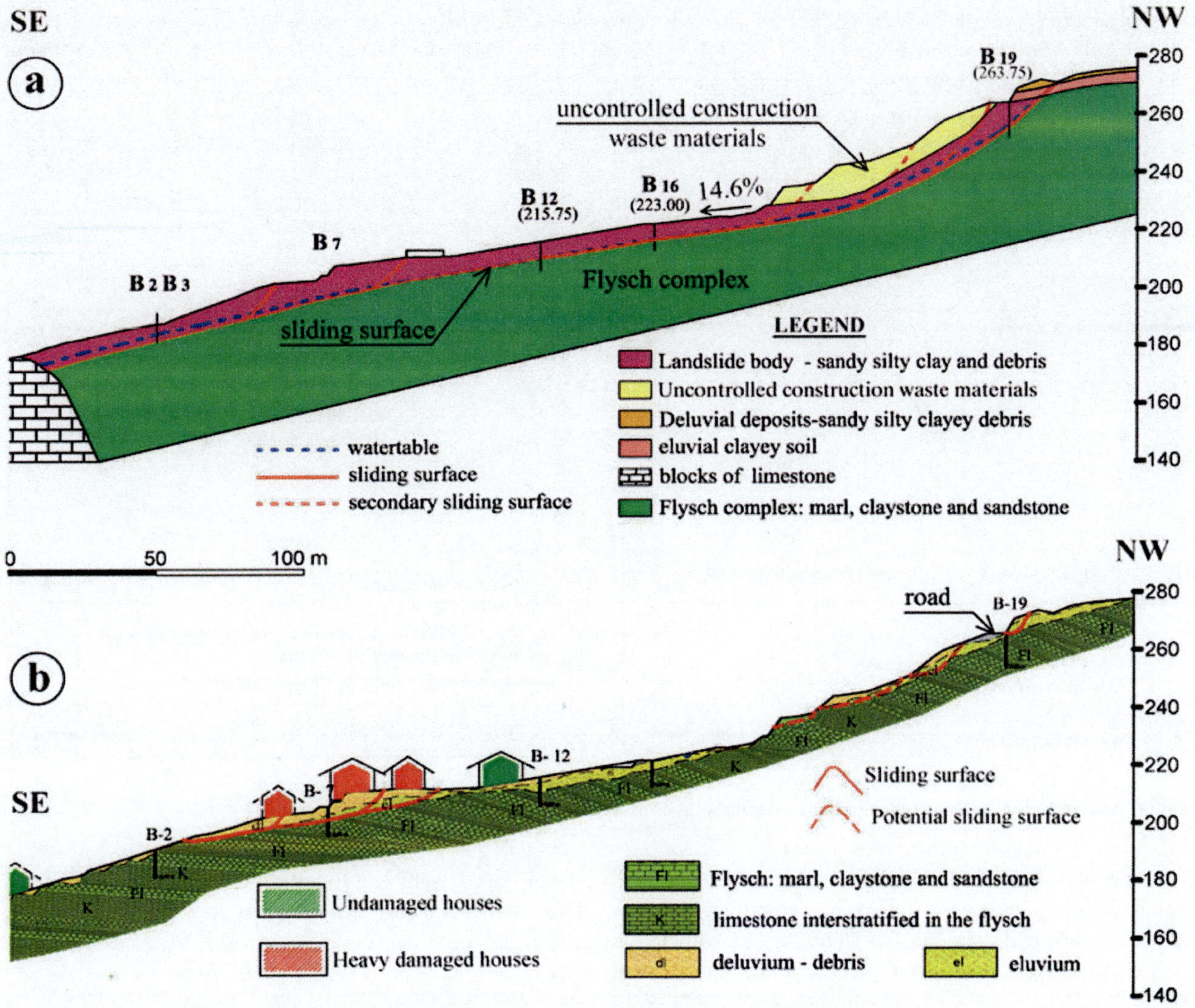

Fig. 2 Engineering–geological profile: **a** one sliding body after Ivanovic (2008), **b** after Lokin (2010)

account the geometry and state of the slope together with the body of construction waste material, and the quantification of the effects of water table rises on the global factor of safety (F_{sw}, Table 1). The geometry has been reconstructed based on the analysis of all available data.

4 Discussion

General assessment implies that the unplanned development activities on a conditionally stable terrain of the village Podi could induce some small landslides. However, this is not the case with the landslide concerned, where the change in the stress–strain state occurred along the slope in its full extent. The geological report (Ivanovic 2008) offers proper conclusions on the landslide cause, with the presented sliding surface being somewhat more profound than the actual one. The main deficiency of the geological report (Lokin 2010) lies in a misassumption on a sliding surface being discontinued by the flysch outcrops. These are, in fact, blocks transported by the colluvial–eluvial process. In the Podi landslide area, limestones are weathered and fractured. The limestone blocks represent their remains which were detached from the underlying stratum and moved together with the deluvial sediments. Therefore, it can be concluded that the Podi landslide is a complex landslide—a combination of rotational slide in the upper part of the slope and translational slide in the central and lower part of the slope. Both reports reveal that the landslide concerned is dormant-old, which indicates relatively small global and

Table 1 Selected results of the numerical analysis

Model	Model description	Back-analysis $\phi' c'$ (kPa) F_s			F_{sw}	ΔF_s (%)
3A	Lower part of sliding body after Ivanovic (2008)	9°	0	0.99	0.80	−19.2
4A	Upper part of sliding body after Lokin (2010)	13°	0	1.02	0.74	−27.5
5A	Total slope after Ivanovic (2008)	17°	0	0.97	0.79	−18.6

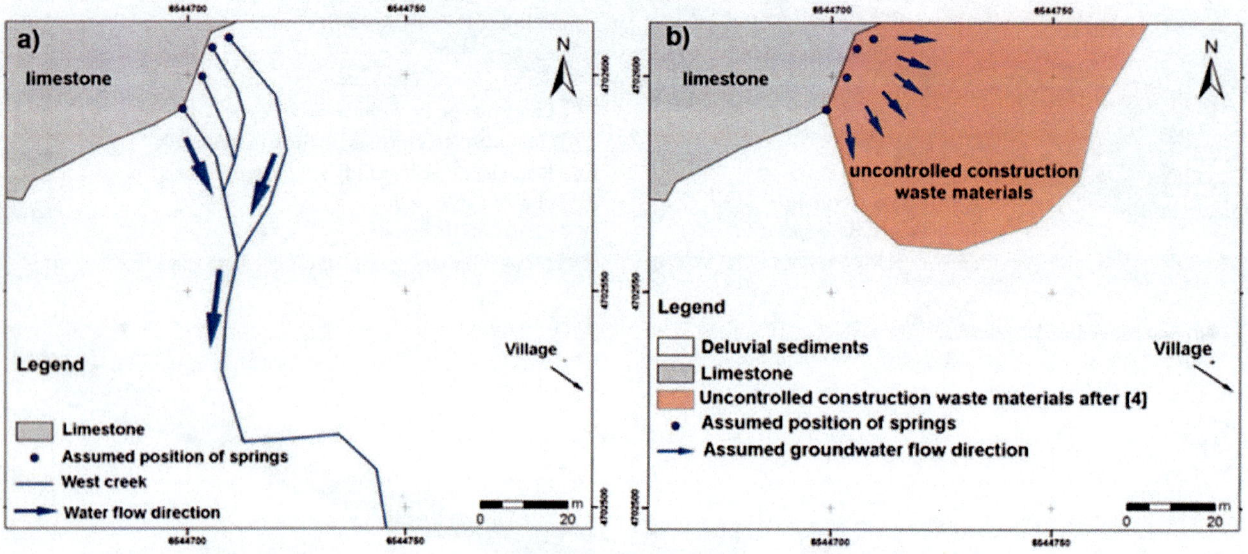

Fig. 3 Assumed water outflow: **a** before and **b** after disposal of material

local factors of safety of the slope in its natural state, i.e. before construction waste disposal. In the northwestern part of the investigated area, the limestone contact with flysch sediments is identified. Such contact is to some extent masked by natural deluvial material that creates a western stream in times of the intensive rainfall karstic springs. However, after the backfilling of the aquifer discharge zone with construction waste material, the naturally concentrated water flow towards the western stream (Fig. 3a) was prevented and replaced with an artificially established diffuse discharge of the karst aquifer through the body of the disposed material (Fig. 3b). This led to disposed material saturation by water, which probably caused a part of the former aquifer water to start flowing towards the landslide body, in other words to the inhabited village. The fact that the depositing of the material was consequently followed by drying up of the western stream (Ivanovic 2008), the over-saturation of the landslide body by water and the occurrence of diffuse springs in many locations further confirm this explanation. Numerical analyses indicated that with a moderate water table rising, the factor of safety decreased by some 20% (ΔF_s, Table 1), which in case of a conditionally stable slope is sufficient to initiate the terrain movement and formation of a landslide on a wider area.

5 Conclusions

This paper summarizes the results of the analysis performed to determine the causes of the landslide in village Podi. The available hydrogeological and geological maps and input data and series of numerical simulations have been investigated. The findings suggest that the improper depositing of material in the upper part of the conditionally stable slope initiated the landslide due to two reasons: an increase in shear stress (additional load in the upper part of the sliding body) and a change in the groundwater regime which gave rise to the pore pressure. Possible limitations of this study are related to the insufficient level of research in the broader area of the landslide location, notwithstanding that the location has already been classified as a conditionally stable slope. Also, the lack of more precise geodetic, geological and hydrogeological data from the period preceding the landslide activation affected the research.

References

Duan, et al.: Investigation into triggering mechanism of loess landslides in south Jingyang platform, Shaanxi province. Bull Eng Geol Environ (2019). https://doi.org/10.1007/s10064-018-01432-8(2019)

Ivanovic G (2008) Geological report—geotechnical characteristics of the landslide "Podi" (Book 2). "MI" d.o.o Podgorica

Lokin P (2010) Report on additional geotechnical investigations of landslide in the village Podi near Herceg-Novi. "Geotehnika" d.o.o, Bijelo Polje

Meng et al: Landslide events on June 24 in Sichuan Province, China: preliminary investigation and analysis. Geosciences **8**(2), 39. https://doi.org/10.3390/geosciences8020039(2018)

Roscience Homepage. https://www.rocscience.com. Last accessed 2019/06/14

Tomanovic, Z., Anagnosti, P., Cadjenovic, N., Radulovic, M., Zivaljevic, S.: Expert findings on causes of the landslide occurrence in "Podi"—Herceg-Novi. University of Montenegro Faculty of Civil Engineering, Podgorica (2013)

Seismic Internal Stability of Reinforced Earth Retaining Walls in Saturated Soils

Hicham Alhajj Chehade, Daniel Dias, Marwan Sadek, Orianne Jenck, and Fadi Hage Chehade

Abstract

This study concerns the seismic internal stability analysis of geosynthetic reinforced soil retaining walls using the discretization technique with the upper-bound theorem of limit analysis. A pseudo-dynamic approach is adopted instead of the commonly used pseudo-static one. Knowing that the majority of the failure cases encountered in the literature were caused by the water presence, the pore-water effect within the backfill soil is considered together with a possible crack opening in case of cohesive soils. The reinforcement strength required to ensure the reinforced saturated soil wall stability is searched through an optimization process. The effect of pore-water pressure induces an increase of the required reinforcement strength.

Keywords

Pore-water effect • Limit analysis • Discretization technique • Earth retaining wall • Pseudo-dynamic approach

H. A. Chehade · O. Jenck
University Grenoble Alpes, CNRS, Grenoble INP, 3SR, 38000 Grenoble, France

H. A. Chehade · M. Sadek · F. H. Chehade
Doctoral School of Sciences and Technologies, Lebanese University, Beirut, Lebanon

D. Dias (✉)
School of Automotive and Transportation Engineering, Hefei University of Technology, Hefei, China
e-mail: daniel.dias@anteagroup.com

D. Dias
Antea Group, Antony, France

1 Introduction

Geosynthetic reinforced earth retaining walls are now a mature technology. They exhibit a good performance during strong earthquakes. However, failure of this type of structures has been examined in the literature. Among these studies, Koerner and Koerner (2018) who investigated 320 failure cases of geosynthetic reinforced earth retaining walls, reported that 73% of these failures occurred when a cohesive soil (clay or silt) was used as backfill in the reinforced zone and 63% were caused by the water presence.

Some of the most important design codes (e.g., FHWA-NHI-11-024, NF P 94-270) suggest a cohesionless granular soil as a backfill in the reinforced zone with an appropriate drain system. However, the use of a geosynthetic reinforcement which does not have a corrosion risk gives the opportunity to use poorly draining cohesive soils when granular soils are not available or are expensive. Nevertheless, the use of cohesive soils can lead to a crack formation in the backfill and therefore, reduce the system stability (Abd and Utili 2017).

Limit analysis is an efficient method to analyze the seismic internal stability of reinforced earth retaining walls. It permits to obtain more rigorous solutions considering the stress–strain relation of soils than the limit equilibrium ones.

In this study, the kinematic theorem of limit analysis is used for the seismic stability assessment of geosynthetic reinforced earth retaining walls in saturated soils. The discretization technique is coupled with the limit analysis method in order to consider the seismic loading with the pseudo-dynamic approach. Considering a saturated cohesive soil as backfill, the influence of pore-water pressures and the possible formation of cracks on the seismic internal stability of a geosynthetic reinforced soil retaining walls are investigated.

H. El-Askary et al. (eds.), *Research Developments in Geotechnics, Geo-Informatics and Remote Sensing*, Advances in Science, Technology & Innovation,
https://doi.org/10.1007/978-3-030-72896-0_5

2 Methodology

The kinematic theorem of limit analysis based on the work rate balance of a kinematically admissible velocity field is applied. A rigid block assumption is considered, and the soil is described by an ideally rigid perfectly plastic behavior governed by the normality flow rule. A toe log-spiral failure mechanism is assumed (Abd and Utili 2017; Alhajj Chehade et al. 2019a,b). The discretization technique is used to generate this failure mechanism as shown in Fig. 1a. It allows to overcome the main limitation of the traditional kinematic approach (Alhajj Chehade et al. 2019b).

The whole failure mechanism ABCD is limited by the log-spiral part BC and the vertical crack CD. This mechanism is assumed to rotate rigidly around the center O with an angular velocity Ω. The detailed steps and equations used in the generation process are presented in Alhajj Chehade et al. (2019a,b).

The pseudo-dynamic approach is adopted to represent the seismic loading. It is more realistic than the pseudo-static one since it accounts for the amplification, duration and frequency of the earthquake. A sinusoidal acceleration is considered here for simplicity. The expressions of horizontal and vertical acceleration at any depth y and time t with soil amplification factor f are expressed as follows:

$$\begin{cases} a_{\mathrm{h}} = \left[f + \frac{y}{H}(f-1)\right] \cdot k_{\mathrm{h}} g \sin\frac{2\pi}{T}\left(t - \frac{H+y}{V_s}\right) \\ a_{\mathrm{v}} = \left[f + \frac{y}{H}(f-1)\right] \cdot k_{\mathrm{v}} g \sin\frac{2\pi}{T}\left(t - \frac{H+y}{V_p}\right) \end{cases} \tag{1}$$

where H is the wall height, g the acceleration due to gravity, T the waves period, k_{h} and k_{v} the horizontal and vertical seismic coefficients, respectively, V_{s} and V_{p} the shear and primary wave velocities, respectively.

The internal energy dissipation takes place along the reinforcement by tensile or pullout failure and along the log-spiral part of the failure surface. Taking into consideration the crack formation as a part of the failure mechanism, the internal energy dissipation along the vertical crack CD is also accounted. The external work rate comes from the work rate of the soil block weight, the inertia forces and the work rate produced by the pore pressure. The computations of the work rates are performed by summation of the elementary work rate by considering the elementary trapezoidal surface shown in Fig. 1b. The detailed equations to calculate the work rates are presented in Alhajj Chehade et al. (2019b), except the calculation of the pore-water pressure work rate which is not considered and is presented in the next paragraph.

For saturated soils, the effect of water must be included as an external loading applied to the soil skeleton and the boundary. As presented in Viratjandr and Michalowski (Viratjandr and Michalowski 2006), the work of both seepage and buoyancy forces must be included in the analysis. In order to include both forces works in the kinematic approach of limit analysis, the pore-water pressure is employed by using the so-called pore-water pressure coefficient r_u introduced by Bishop et Morgenstern (Bishop and Morgenstern 1960). The pore-water pressure u at a depth h below the surface is then assumed to be equal to $r_u \gamma h$ where γ is the soil unit weight. The work rate of the pore-water pressure $\dot{W}_u$ can be expressed as a summation of both forces' works as follows:

$$\dot{W}_u = -\gamma_w \int_V \frac{\partial h}{\partial x_i} \cdot v_i \mathrm{d}V + \gamma_w \int_V \frac{\partial Z}{\partial x_i} \cdot v_i \mathrm{d}V \tag{2}$$

where the first term represents the seepage force work rate over the entire collapse volume and the second one is the buoyancy force work rate, γ_{w} is the water unit weight, h is the hydraulic head, Z is the elevation head, v_i the velocity vector in the collapse block and V is the volume of the failure block. Alternatively, Viratjandr and Michalowski (Viratjandr and Michalowski 2006) by developing the derivative $\frac{\partial}{\partial x_i}$ of the product uv_i showed that the expression of the water pressure work rate can be written as a sum of the pore pressure work rates on skeleton expansion and the boundary as follows:

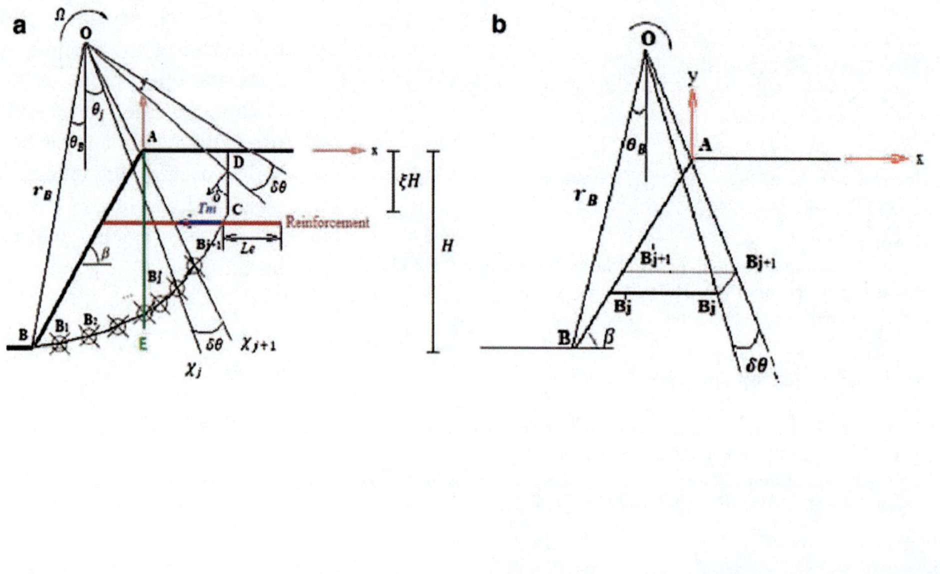

Fig. 1 a Discretization technique for the failure mechanism, b elementary trapezoidal surface

$$\dot{W}_u = -\int_V u \cdot \dot{\varepsilon}_{ii} dV - \int_S u \cdot n_i \cdot v_i dS \quad (3)$$

where $\dot{\varepsilon}_{ii}$ is the volumetric strain rate, S the boundary surface of the failure block, n_i the outward unit vector normal to the surface S.

The first term in Eq. (3) is assumed to be zero since a rigid block is considered. In the framework of the discretization technique to generate the failure surface, the kinematical admissibility condition must be satisfied, and therefore, the angle between the velocity vector and the failure surface must be equal to the friction angle φ. Hence, the angle between the vector n_i and the velocity vector is equal to $\frac{\pi}{2} + \varphi$. Knowing that the velocity at a point B_j on the failure surface is equal to $\Omega \cdot L_j$ where L_j is the length of $[OB_j]$, the work rate of the pore pressure effect along the discontinuity surface is then given as the summation of the elementary work rates as follows:

$$\begin{cases} \dot{W}_u = \sum_j r_u \cdot \gamma \cdot (x_j \cdot \tan\beta - y_j) \cdot l_j \cdot \Omega \cdot L_j \cdot \sin\varphi & \text{along the boundary BE} \\ \dot{W}_u = \sum_j r_u \cdot \gamma \cdot (-y_j) \cdot l_j \cdot \Omega \cdot L_j \cdot \sin\varphi & \text{along the boundaryED} \end{cases} \quad (4)$$

where x_j and y_j are the coordinates of point B_j and l_j is the length of $[B_{j-1}B_j]$.

The pullout force should account for the presence of water. It can then be written as:

$$T_p = 2\gamma z^*(1 - r_u)L_e f^* \quad (5)$$

where z^* is the overburden depth; L_e is the effective length; f^* corresponds to the apparent friction coefficient at the soil/reinforcement interface.

According to the upper bound theorem, the energy balance equation provides the objective function to be optimized in order to determine the required reinforcement strength. Four variables are considered in the optimization process, namely $[r_B, \theta_B, \xi, t]$ where r_B is the length of OB, θ_B is the angle between the y axis direction and the line OB, ξ is a parameter to define the crack depth (Fig. 1a) and t the time involved in the pseudo-dynamic approach.

3 Results and Discussion

To validate the robustness of the proposed discretization method, a comparison between the existing results of the conventional limit analysis method obtained by Abd and Utili (2017) and the results obtained by the discretization method is presented. Abd and Utili (2017) used the conventional kinematic theorem of limit analysis to investigate the internal stability of geosynthetic reinforced slopes with cracks under static loading and considering only the reinforcement tensile failure to calculate the required reinforcement strength. They assumed both linear (uniform) and triangular distribution of the reinforcement. The effect of pore-water pressure on the required level of reinforcement was analyzed by employing the same approach used here through the so-called pore-water pressure coefficient r_u.

In order to validate the application of the discretization method in this study, the results of the discrete method are obtained under the same conditions used in the work of Abd and Utili (Abd and Utili 2017). The case of crack formation as a part of the failure mechanism with tension cut-off is considered. The comparisons of results are depicted Fig. 2 for different slopes inclinations β and coefficient r_u. A total agreement can be observed between the two methods.

To investigate the pore-water pressure effect, tension crack and seismic loading, the results in terms of required reinforcement strength in a normalized form are presented for two wall inclinations $\beta = 75°$ and $\beta = 90°$. The two reinforcement failures modes are considered simultaneously. The following parameters are considered: H = 7 m, γ = 18 kN/m^3, φ = 25°, c = 12.6 kPa, $\delta\theta$ = 0.01°, λ = 0.5, $L_r = H$, n = 10, f_0^* = 1.2, f_1^* = 0.6,f = 1.2, V_S = 150 m/s, V_P = 280.5 m/s, and T = 0.3 s, where $\lambda = k_v/k_h$, n the reinforcement number, L_r the reinforcement length and c the soil cohesion.

According to Fig. 3, it is observed that the required reinforcement strength increases with the horizontal seismic coefficient, the pore-water ratio and the wall inclination (β). In addition, the normalized reinforcement strength values for the reinforced backfill with cracks are greater than the values obtained when considering an intact soil.

It is evident that the required reinforcement strength increases with the increase of k_h. Moreover, as expected, the presence of the pore-water pressure effect is unfavorable to

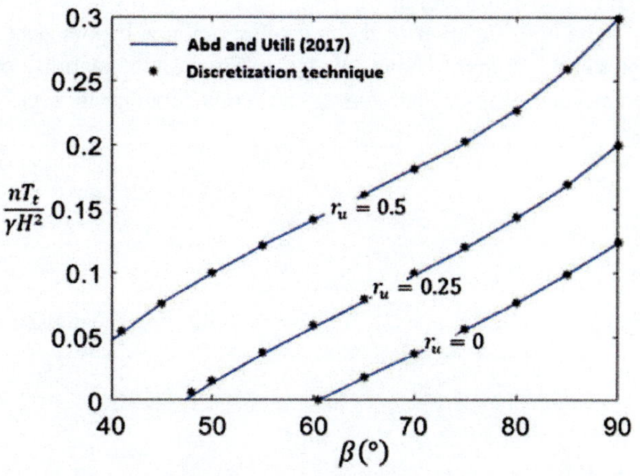

Fig. 2 Comparison of results of the conventional limit analysis and the discretization technique ($\varphi = 20°$ and $c/\gamma H = 0.1$)

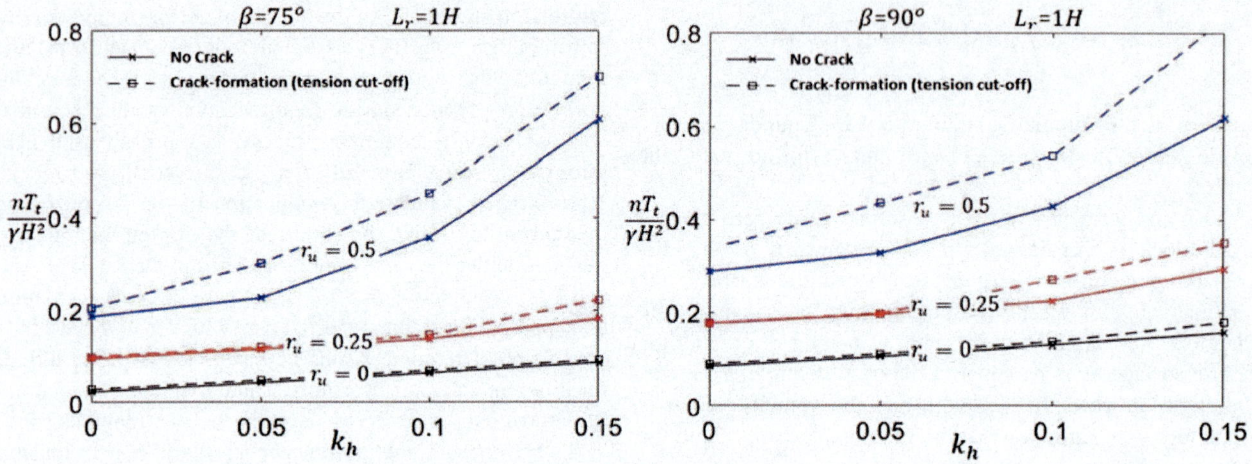

Fig. 3 Required reinforcement strength against seismic coefficient for intact and cracked backfill

the reinforced wall stability. The presence of cracks leads to an increase in the required normalized reinforcement strength. In particular, when the value of the pore-water pressure or the seismic loading is important, the increase in the required reinforcement strength is more obvious. Hence, it is critical to consider these effects in regions where poorly draining cohesive soils are used in seismic zones as backfill materials for economic reasons.

4 Conclusion

In this study, the kinematic approach of limit analysis combined with the discretization method is adopted. This method represents a suitable alternative to assess the seismic internal stability of geosynthetic reinforced soil retaining walls considering the possibility of a crack formation. This method also allows the use of the pseudo-dynamic approach for the seismic loading.

The consideration of the crack formation and pore-water pressure is found to have a negative effect on the stability of reinforced walls. Therefore, their consideration is highly recommended particularly in seismic regions where the lack or the price of granular materials leads to the use of poorly draining soils.

References

Abd, A.H., Utili, S.: Design of geosynthetic-reinforced slopes in cohesive backfills. Geotext. Geomembr. **45**, 627–641 (2017)

Alhajj Chehade, H., Dias, D., Sadek, M., et al: Seismic analysis of geosynthetic-reinforced retaining wall in cohesive soils. Geotext. Geomembr. **47**, 315–326 (2019a)

Alhajj Chehade, H., Dias, D., Sadek, M., et al: Pseudo-static analysis of reinforced earth retaining walls. Acta Geotech (2019b) (under review)

Bishop, A.W., Morgenstern, N.R.: Stability coefficients for earth slopes. Géotechnique **10**, 129–153 (1960)

Koerner, R.M., Koerner, G.R.: An extended data base and recommendations regarding 320 failed geosynthetic reinforced mechanically stabilized earth (MSE) walls. Geotext Geomembr. **46**, 904–912 (2018)

Viratjandr, C., Michalowski, R.L.: Limit analysis of submerged slopes subjected to water drawdown. Can. Geotech. J. **43**, 802–814 (2006)

Contribution Parametric Optimization Study of Landslides Movements Using Statistical Tools in the Region of Souk Ahras (Algeria)

Nouar Charef, Yacine Berrah, and Abderrahmane Boumezbeur

Abstract

The effects of key factors in the initiation of landslide movements in recent events in Algeria are evaluated using stochastic-based parametric optimization. This paper describes the different parameters of causality effect on landslides movements based on geological, hydrogeological, mechanical and geomorphological factors affecting the stability of different sectors in the region of Souk Ahras in Algeria causing significant damage to build infrastructure and natural/social environment. These mass movements are varied in nature, generated by multiple factors that reflect the variability of the behavior of geological structures composing the different areas in movement. Practical observations of the lands in movement can establish a correlation between mentioned parameters and the landslide phenomenon. This study combines parametric and statistical tools to optimize the effect of the different factors behind the behavior of landslides in major sectors around the wilayah of Souk Ahras territory.

Keywords

Landslides • Souk Ahras • Parametric optimization • Statistical tools • Pore water pressure

1 Introduction

Landslides mass movements increasingly threaten the growing development in the region of Souk Ahras (Algeria). Sustainability of structures and in particular roads is threatened by this phenomenon widespread in the region. These mass movements are varied in nature, generated by multiple factors that reflect the variability of the behavior of the geological materials in movement. Practical observations of the lands in movement can establish a correlation between geological, hydrogeological conditions and the landslide phenomenon. However, it is challenging to estimate the exact influence of a certain factor on the instability process separately.

This study combines geology with soil mechanics in order to investigate the causes of landslides in most important sectors (Mechroha, Zaarouria, Hammam Tassa, Sedrata …) all over the wilayah of Souk Ahras territory. The present paper describes the different parameters of causality effect on landslides movements through a statistic study with a focus of the most threatening factors to landslides.

2 General Setting

2.1 Geography and Topography

Being part of the Medjerda's Mounts, the region of Souk Aharas is situated in northern Algeria (Fig. 1). It is drained by the main Medjerda wadi and its two tributaries, Laghdir and Echaref. Souk Ahras region is characterized by a continental moderate climate. During the three last decades, the annual mean precipitation was about 763.27 mm (Hadji et al. 2013).

The topography is asymmetric with a dominant SW slope. Two major structural directions characterize the region: NE–SW, marked by Triassic extrusions; and NW–SE to WNW–ESE, subsequent to post Miocene distensive collapse (Rouvier 1990).

2.2 Types and Causes of Landslides

The dominant type of landslides in the region of Souk Ahras is rotational rock slump. It is very similar to the rotational

N. Charef (✉)
University of Annaba, Annaba, Algeria

Y. Berrah · A. Boumezbeur
University of Tebessa, Tébessa, Algeria

H. El-Askary et al. (eds.), *Research Developments in Geotechnics, Geo-Informatics and Remote Sensing*, Advances in Science, Technology & Innovation,
https://doi.org/10.1007/978-3-030-72896-0_6

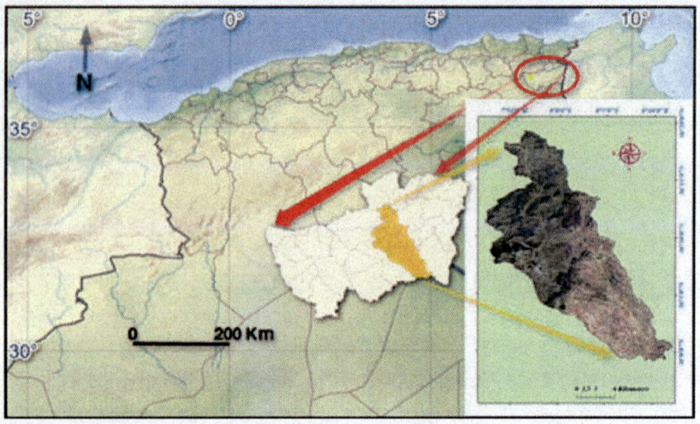

Fig. 1 Geographic position of the studied zone

slide in its circular failure surface, but the vertical section of the sliding surface is irregular (Colas and Pilot 1976).

Several factors may be responsible for these landslides:

- The moving or creating a road embankment (anthropic action).
- The slope (gravity action).
- Geological nature of soil (particularly swelling-shrinkage of the clay).
- Rain and exceptional winter rainstorms (increased interstitial pressure).
- Torrential flood and rapid river flood (surface erosion).
- Infiltration and drainage in the soil (decrease of soil strength).
- Intense temperature change (large daily variations in temperature, exfoliation).

3 Materials and Methods

The lab tests were performed in the public works laboratory of Annaba (LTPE of Annaba), our research collaborator. According to the French Standards, 151 soil samples from the studied sectors were analyzed in terms of: particle size analysis, dry density, natural water content, Atterberg limits, direct shear-box tests and oedometer tests. In-situ tests were presented by Standard Penetration Test (SPT) along a main profile located in the axis of sliding (Costet and Sanglerat 1975).

4 Results and Interpretations

Depending on the granulometry and Atterberg limits, the results of soil analysis demonstrate that the fine grains of unstable terrain are composed of inorganic clays (of illite dominance). The plasticity of these clays is generally

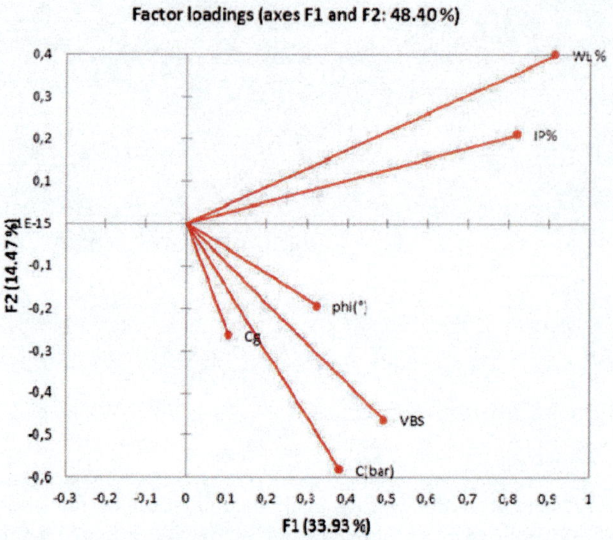

Fig. 2 The factors of loading of variables in the first factorial space (F1F2)

medium to high for some soils; Wl varies between 21 and 77%, while Ip varies between 05 and 54%.

Using principal component analysis PCA tools the following results were obtained (Fig. 2; Table 1).

The results of the principal component analysis PCA use six of the soil parameters. They reveal a good correlation on the first major axes that absorb more than 59% of the total variance. The PCA allows the grouping of the best correlated parameters as the first group composed of (methylene blue value of soil (VBS), C (bar), phi (°) and swelling coefficient (Cg); the second group included Wl and Ip (Fig. 3).

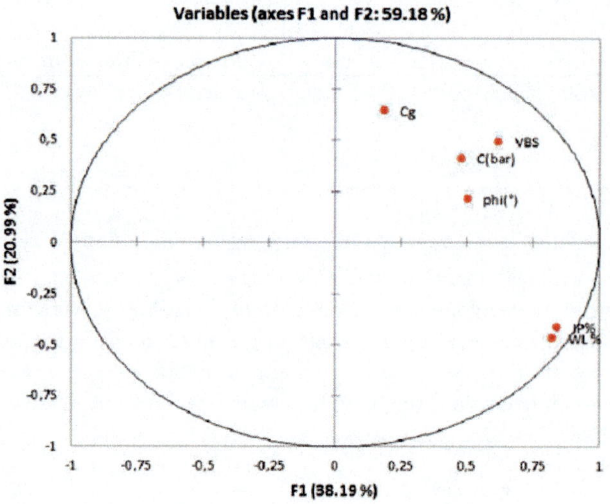

Fig. 3 Circles of correlation obtained for the soil studied

Table 1 Summary statistics

Variable	Observations	Minimum	Maximum	Mean	Std. deviation
WL%	151	21.0000	77.0000	47.7152	12.8682
IP%	151	5.0000	54.0000	25.3907	9.9170
VBS	151	0.5800	65.0000	4.9591	7.1744
C (bar)	151	0.0290	5.0000	0.8916	0.9339
Phi (°)	151	5.0000	48.0000	20.1681	8.4403
Cg	151	0.0160	0.9600	0.0743	0.1433

The Eigenvectors and the factors of variable loading in the first factorial space (F1F2) indicate that four parameters (VBS, C, phi and Cg) have a positive contribution, whereas two parameters (Wl and Ip) contribute negatively to the weakening of the lands in risk of sliding. The water content contributes to the decrease of the soils mechanical parameters; thus, it weakens its resistance.

5 Conclusion

This study investigates the factors that can influence landslides in important sectors of the wilayah of Souk Ahras (northern Algeria) by combining of geology with soil mechanics. Principal component analysis PCA is applied on 151 samples in order to highlight the contribution of physico-mechanical parameters to the weakening of the lands in risk of sliding. The results suggest that four parameters (VBS, C, phi and Cg) have a positive contribution, whereas two parameters (Wl and Ip) contribute negatively to the land sliding risk. The water is the factor favoring the movement of ground; hence, it must be controlled. In other words, an increase in water content results in a precarious ground set to move at the presence of other unfavorable conditions.

References

Colas, G., Pilot, G.: Description and classification of landslides. Bull. Liaison Lab. P. et Ch. special 21–30 (1976) (in French)

Costet, J., Sanglerat, G.: Practice Course of Soil Mechanics, Paris. Edition Dunod, pp. 265, 269, 273 (1975) (in French)

Hadji, R., Boumazbeur, A., Limani, Y., Baghem, M., Chouabi, A., Demdoum, A.: Geologic, topographic and climatic controls in landslide hazard assessment using GIS modeling: a case study of Souk Ahras region, NE Algeria. Quat. Int. **302**, 224–237 (2013)

Rouvier, H.: Polymetallic concentrations linked to evaporitic diapirs of the Algerian-Tunisian borders. EREM internal report. Boumerdès-Algeria, 71p (1990) (in French)

GIS-Based Mapping of Geotechnical Properties of Residual Soil in Kibawe, Bukidnon, Philippines: Implications to Slope Stability

Vera Karla Caingles and Glen Lorenzo

Abstract

The soil properties must be extensively analyzed and understood in order to categorize their significant effects on slope stability problems. Providing interpreted soil information is one of the on-going challenges for geotechnical engineers and soil investigators. This study attempts to develop thematic maps on the important geotechnical properties of soil by using GIS software based on the in situ investigation conducted and existing soil exploration reports in the study sites located at Kibawe, Bukidnon, Philippines. The methodology included reconnaissance survey; field sampling (20 boreholes) and gathering of existing soil exploration reports (30 boreholes); laboratory testing such density, moisture content, grain size analyses, and fall cone tests; and GIS mapping. Test results revealed that mostly inorganic but highly expansive fine-grained soils underlie the study sites. The developed thematic maps provide visualized insights on the spatial variation of soil properties in the study area, which are essential for slope stability analysis.

Keywords

GIS • Thematic maps • Plasticity • Expansiveness • Collapsibility

V. K. Caingles (✉)
Mindanao State University—Iligan Institute of Technology, 9200 Iligan City, Philippines
e-mail: vera.karla@ustp.edu.ph

G. Lorenzo
Mindanao State University, 9700 Marawi Campus, Marawi City, Philippines

1 Introduction

The characteristics of underlying soil formation determine the stability of natural slopes. Thus, soil properties must be analyzed thoroughly and understood in order to categorize their significant effect on slope stability problems. Due to time-consuming and expensive soil sampling and laboratory testing, it is an on-going challenge to the geotechnical engineers on how to provide interpreted soil information, of a particular area, for various end-users who need the data. Geographic Information System (GIS), a computer-based technique, can capture, store, analyze and display geographically referenced information such as the data identified according to a particular location/region (Mhaske and Choudhury 2009). GIS allows the users to produce maps of soil classifications, obtain engineering information regarding soil type as well as an initial indication of soil properties and create soil profile for any location (Jamassi 2013). Thus, this work attempts to develop thematic maps for Kibawe, Bukidnon particularly for Barangay New Kidapawan and Barangay Kiorao on important geotechnical properties of soil by means of GIS software based on the in situ investigation and existing soil exploration reports.

2 Study Sites and Methods

The study sites are the Barangay New Kidapawan and Barangay Kiorao located in the Municipality of Kibawe, Bukidnon, Philippines. The center of Barangay New Kidapawan is geographically situated at 7° 33′ 24.12″ N latitude and 124° 55′ 18.12″ E longitude with an elevation of 347.8 m above mean sea level. The center of Barangay Kiorao is geographically situated at 7° 31′ 42.24″ N latitude and 124° 55′ 8.76″ E longitude with an elevation of 303.0 m above mean sea level. The areas were chosen for having experienced tremendous landslides which caused vast damage to properties and natural environment in 2014. Further, the study areas had existing soil exploration reports with 30 boreholes.

H. El-Askary et al. (eds.), *Research Developments in Geotechnics, Geo-Informatics and Remote Sensing*, Advances in Science, Technology & Innovation,
https://doi.org/10.1007/978-3-030-72896-0_7

The study comprises the following stages: reconnaissance survey; field sampling (in 20 boreholes) and gathering of existing soil exploration reports (30 boreholes); laboratory testing such density, moisture content, grain size analyses and fall cone tests and data analysis. The soil thickness was measured from the soil surface down to the interface of soil and the bedrock. The prepared data sets were imported into a GIS framework to produce interactive maps that show the spatial distribution of the variables and identify their characteristics. The output thematic maps were based on the predetermined 20 slope ranges obtained from the slope map through a processed digital elevation model (DEM) at approximately 5-m resolution.

3 Results

3.1 Thickness of Residual Soil

During field investigation, the thickness of residual soil was found to decrease as the slope angle increases. It ranges from 0.24 to 3.25 m in which the average thickness for gentler slopes (<28°) is 2.45 m. It decreases to about 1.0 m for the modal class slopes (28°–32°) and to more or less 0.5 m for the steep slopes (>32°).

3.2 Physical and Index Properties

The data on grain size analysis and fall cone tests of fifty (50) soil samples are summarized in Table 1. The results of the particles size distribution test show that soil samples from the respective sites were generally fine-grained soil, with more than 50% of the material passing the 0.075 mm sieve. The engineering properties of fine-grained soils are typically assessed through the Atterberg limits: liquid limit, plastic limit and plasticity index. The typical thematic maps for the liquid limit, plastic limit and plasticity index are presented in Figs. 1, 2 and 3, respectively. The color-coding system was used to indicate levels of indices or ranges of properties of soils.

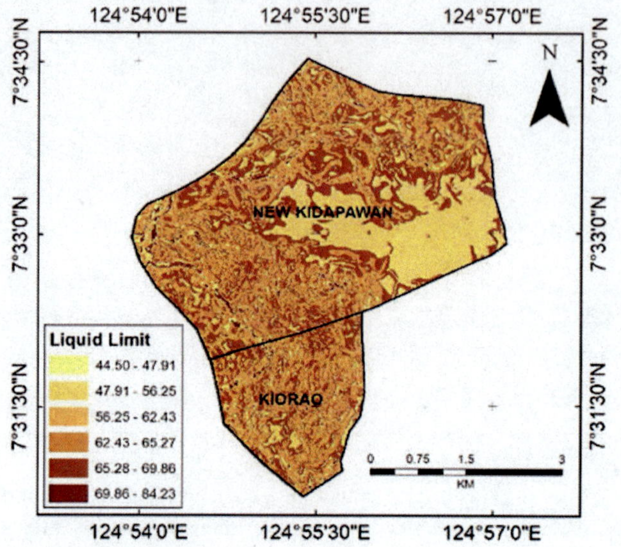

Fig. 1 Liquid limit map

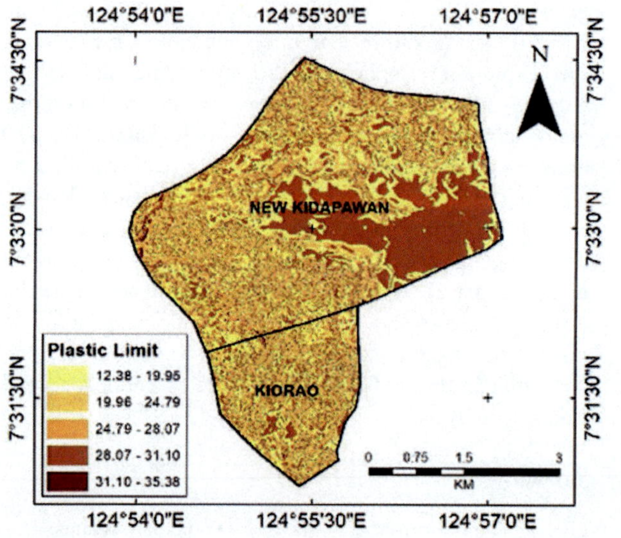

Fig. 2 Plastic limit map

Table 1 Physical and index properties of residual soil

Physical properties	Value ranges		Index properties	Value ranges	
	From	To		From	To
Gravel (%)	0.00	49.17	Liquid limit, LL (%)	40.54	94.17
Sand (%)	0.42	58.52	Plastic limit, PL (%)	11.33	46.82
Clay (%)	18.28	47.57	Plasticity index, PI (%)	13.38	71.28
Silt (%)	1.15	73.79			

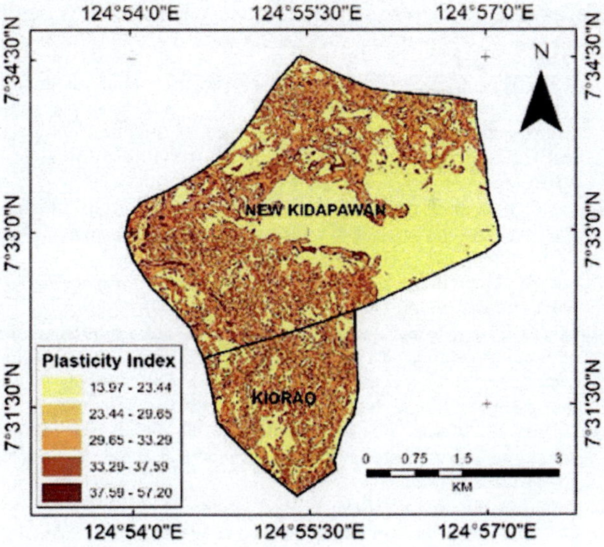

Fig. 3 Plasticity index map

3.3 Plasticity, Expansiveness and Collapsibility of Soils

In assessing the plasticity (Fig. 4), expansiveness (Fig. 5) and collapsibility (Fig. 6) of soils, the results can vary depending on the input parameter used. In this study, the range of soil's plasticity is defined based on the Unified Soils Classification System (USCS) plasticity chart. The degree of expansion is ascertained from the liquid limit and plasticity index results based on IS 1498. The common criterion for determining collapsibility potential of soil is through an empirical relationship between dry unit weight and liquid limit.

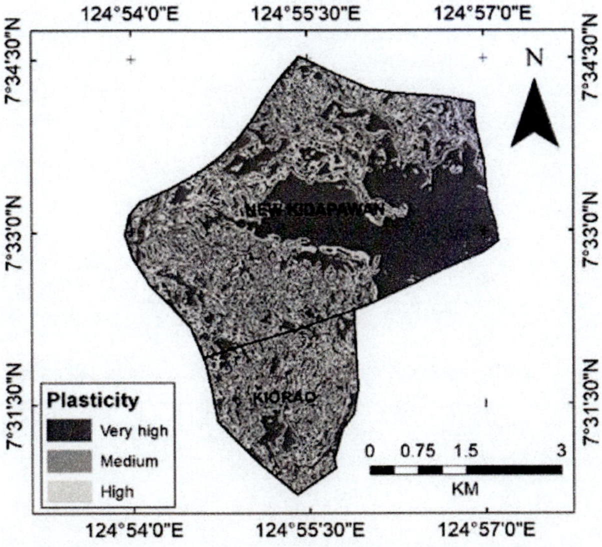

Fig. 4 Soil plasticity map

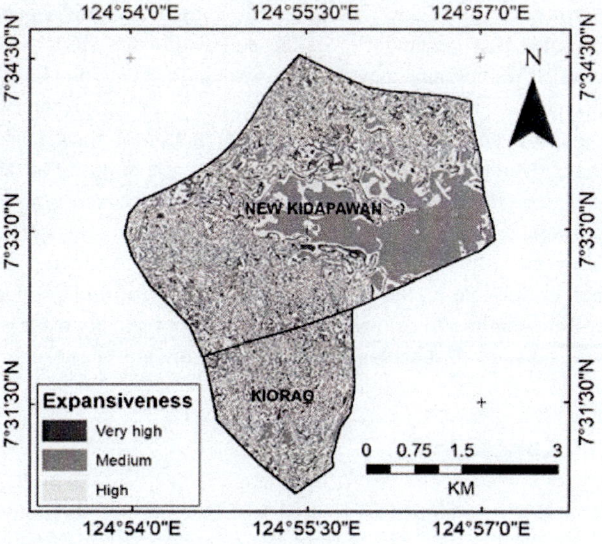

Fig. 5 Soil expansiveness map

4 Discussion

Findings on the relationship between residual soil thickness and slope angle agreed to the observation of Salciarini et al. (2008). According to Mukhlisin and Taha (2009), during the major rainfall event, a deeper soil depth increased the weight of solids, soil moisture conditions and pore water pressure, which tend to increase the likelihood of slope failure. Grain size analyses of the soils indicate that the study area are mostly composed of fine-grained soils with high clay content ranging from 18.28 to 47.57%, thus, with high void ratio, low

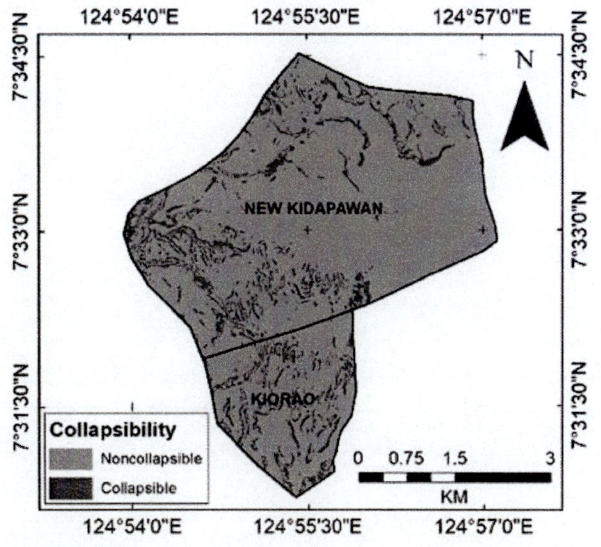

Fig. 6 Soil collapsibility map

permeability and low strength. The Atterberg limit values revealed that the majority of the soil within the study area consists of inorganic clays and silts with high plasticity. Highly plastic inorganic soils are prone to sliding during prolonged rainfall event due to the subsequent reduction of shear resistance (Muggaga et al. 2010). Furthermore, the majority of the soil samples at the study sites are highly expansive soils, unstable upon saturation during rainfall event (Hou et al. 2015). However, the soils at the study sites are considered as non-collapsible as they are mostly comprised of fine-grained soils, in contrast to collapsible soils which are typically made of silt and sand sizes with sometimes small amount of clay.

5 Conclusion

This study developed GIS-based thematic maps on the liquid limit, plastic limit, plasticity index, plasticity, expansiveness and collapsibility of soil for Barangay New Kidapawan and Barangay Kiorao in Kibawe, Bukidnon, Philippines. These thematic maps, which are essential for slope stability analysis, provide potential visual display and the spatial variation of soil properties in the study area.

References

Hou, D., Tao, Z., Hao, Z., Wang, J.: Numerical analysis of rainfall saturated-unsaturated seepage and stability of expansive soil slope with fissures. In: International Conference on Structural, Mechanical and Materials Engineering 2015, pp. 103–106. Atlantis Press (2015)

Jamassi, A.: Using geographic information systems (GIS) in soil classification and analysis in Gaza City, Palestine. Environ. Nat. Resour. Res. **3**(2), 146–159 (2013)

Mhaske, S., Choudhury, D.: Application of GIS-GPS for mapping soil index properties. Int. Geotech. Soc. 36–39 (2009)

Muggaga, F., Kakembo, V., Buyinza, M.: A characterisation of the physical properties of soil and the implications for landslide occurrence on the slopes of Mount Elgon, Eastern Uganda. J. Int. Soc. Prevent. Mitigat. Nat. Hazards **52**(1) (2010)

Mukhlisin, M., Taha, M.: Slope stability analysis of a weathered granitic hillslope as effects of soil thickness. Eur. J. Sci. Res. **30**(1), 36–44 (2009)

Salciarini, D., Godt, J.W., Savage, W.Z., Baum, R.L., Conversini, P.: Modeling landslide recurrence in Seattle, Washington, USA. Eng. Geol. **102**, 227–237 (2008)

Reliability Analysis to Locate the Propagation Failure Direction in Case of Slopes Progressive Failure

Houcine Djeffal and Smain Belkacemi

Abstract

Slope stability analysis is traditionally carried out by using the limit equilibrium methods. In these approaches, the global assessment of slope stability is obtained through the evaluation of the safety factor. Slope stability analysis considers strength parameters that allow the evaluation of resistant forces along the hypothetical slip surface unique and constant. This work discusses (i) the influence of uncertainties on the parameters of peak and post-peak strength and (ii) the impact of a progressive failure on the failure probability of a hypothetical slip surface. The conventional Morgenstern–Price method of slope stability analysis was used to locate the critical slip surface by evaluating the minimum safety factor. The failure probability was estimated using three different techniques—first-order reliability method (FORM), first-order second moment (FOSM) and the technique of Monte Carlo simulation. This study allowed better assessing the impact of progressive failure on the risk of slope failure. The reliability analysis of slices allowed locating the failure area and observing the failure propagation direction.

Keywords

Limit equilibrium • Slope stability • Progressive failure • Uncertainties • Reliability index • Numerical simulation

1 Introduction

The limit equilibrium method is prevalent in slope stability analysis for its simplicity and high efficiency. The assumptions and simplifications of conventional limit equilibrium slope stability methods may not be sufficient to represent the behavior of complex slope problems. The classical limit equilibrium method only considers the ultimate limit state of the slope and provides no information on the development of progressive failure (Cheng and Lau 2008). Some researchers (Khan et al. 2002; Miao et al. 1999; Zhang et al. 2013) extended the limit equilibrium method to analyze the stability of strain-softening slopes, assuming that the soil strength decreases directly to the residual value from the peak value. With the recent advancements in computational approaches, obtaining more realistic models of the slope stability problems has become possible by adopting numerical simulation methods (Cundall 1971; Eberhardt 2003; Eberhardt et al. 2004; Griffiths and Fenton 2004).

The progressive failure mechanism is also dominant for the slope stability because laboratory shear test results indicate that many interfaces between components in slopes exhibit strain-softening behavior (Gilbert and Byrne 1996; Griffiths and Fenton 2004; Jiang et al. 2015).

This paper examines an issue of slope stability in a probabilistic framework and aims to quantify the influence of uncertainties of strength parameters and incidence of progressive failure on the stability. The deterministic method of Morgenstern–Price was adopted in the analysis. Probabilistic methods FORM—first-order reliability method, FOSM—first-order second moment and the technique of Monte Carlo simulation MCS—were used to perform the reliability analysis of slopes at different uncertainty levels of the basic parameters.

2 Methodology

The evaluation of the reliability index of slopes, taking into account the progressive failure, contains two stages. In the first step, the limit equilibrium method of Morgenstern–Price is applied to locate the critical slip surface. This method decomposes the soil mass above the slip surface into a finite number of slices. With the Morgenstern–Price method, both force and moment equilibrium conditions are satisfied. The

H. Djeffal (✉) · S. Belkacemi
Ecole Nationale Polytechnique, 16200 El-Harrache, Algeria

H. El-Askary et al. (eds.), *Research Developments in Geotechnics, Geo-Informatics and Remote Sensing*, Advances in Science, Technology & Innovation,
https://doi.org/10.1007/978-3-030-72896-0_8

method also assumes that the inclination of the inter-slice force is known.

In the second step, a probabilistic analysis of the critical slip surfaces is carried out by means of physical models of the object function for each of the developed considered deterministic methods. The variation of the strength parameters is taken into consideration. The impact of a progressive failure through an evolution of strength parameters of their initial values (C_p, ϕ_p) at their peak to residual values (C_r, ϕ_r) is included as well. Probabilistic methods of first order FORM, FOSM and MCS were applied.

To illustrate the method developed, we consider the example of a slope. The geometrical parameters are assumed deterministic with their average values are shown in Table 1.

Reliability index is obtained by (i) keeping the average random variables constant, (ii) setting the variation coefficient in other variables random an arbitrary fixed value to 30% and (iii) changing the variation coefficient value of the variable into account. Random variables distributions are considered normal. Strength parameters to peak and post-peak are either independent.

The performance function G_j* for progressive failure given in Eq. (1) is defined as the conditional safety margin related to failure propagation until the *j*th slice of the slip surface

$$G_j^* = (R_p)_j + (R_r)_{1,j-1} - S_{1,j} \quad (1)$$

where $(R_P)_j$ is the shear strength of the slice j in peak, $(R_r)_{1,j-1}$ is the sum of the residual shear resistance of units 1 to $(j-1)$, S_j is the shear stress of the slice j and $S_{1,j}$ is the sum of shear stresses along the slices 1 to j.

3 Results and Discussion

The results of the reliability analysis of each slice are shown in Fig. 1. These figures show, for each value of the variation coefficient of strength parameters, variations in the reliability index of each slice of the system depending on the physical number of slices and the number of slices ordered in ascending order of reliability index. The main results are as follows:

- The change in the shape of the reliability index as a function of the physical number of slices can locate the

Table 1 Parameters and mean values considered in the slope stability analysis

Parameter	Symbol	Variable	Mean value
Cohesion in peak	C_p	Random	20 kN/m^2
Friction angle in peak	$Ø_p$	Random	25°
Cohesion in post-peak	C_r	Random	10 kN/m^2
Friction angle in post-peak	$Ø_r$	Random	10°
Unit weight	γ	Deterministic	18 kN/m^3
Slope angle	β	Deterministic	26.56°
Slope height	H	Deterministic	20 m

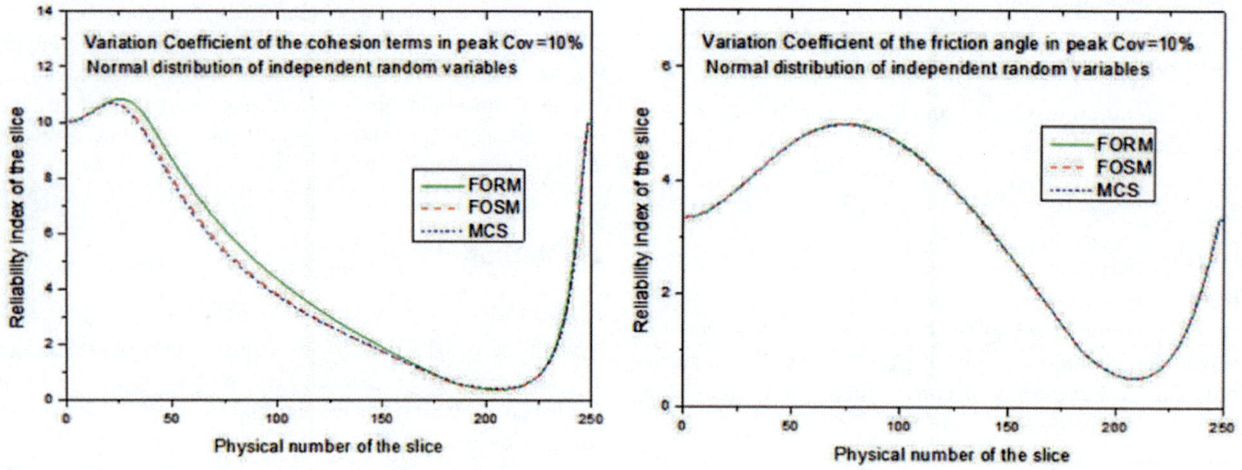

Fig. 1 Reliability index of the slice as a function of the physical number of the slice

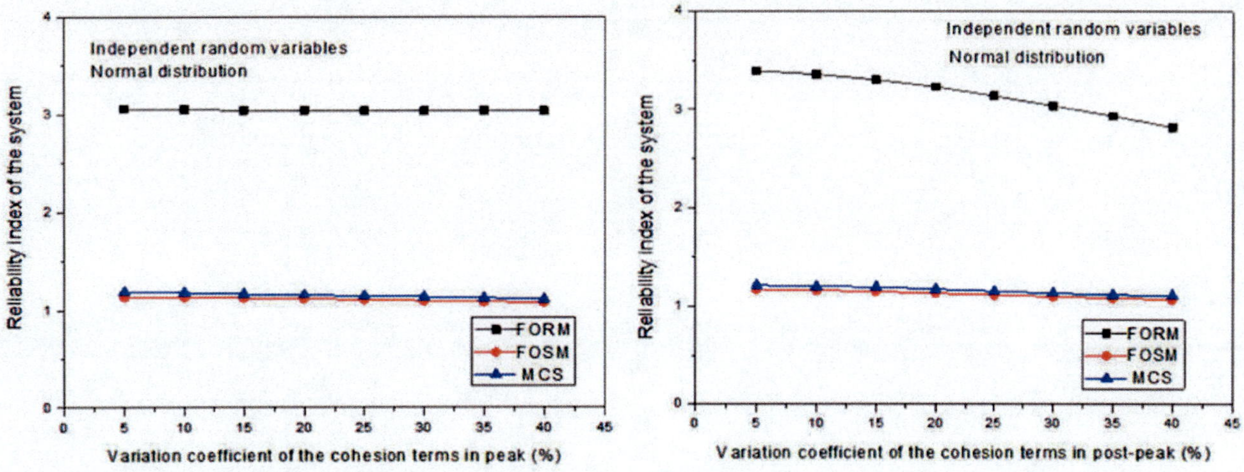

Fig. 2 Reliability index of the system as a function of a variation coefficient of the cohesion

first slice passed in the failure state, it corresponds to the slice of the smallest reliability index.

- The change of the reliability index takes place simultaneously in both directions, relative to the first phase triggering failure.

The results of the system reliability analysis are presented in Fig. 2. These figures show, for each strength parameter of the limit state function, the variations in the reliability index of the system as a function of a variation coefficient. The main results are as follows:

- The reliability index is strongly dependent on the strength parameters in post-peak relative to peak strength parameters.
- In post peak, the cohesion term has a strong influence on the reliability index of system with respect to the friction angle.

4 Conclusions

In the case of a progressive failure, the reliability analysis of slope stability allows the following conclusions:

- The integration of the residual strength parameters in the proposed model improved the stimulation of the progressive failure process of slopes.
- The reliability analysis of slices allowed locating the area triggering a failure and observing the direction of failure propagation.
- The existence of the subsystem which limits the evaluation of system behavior to that of the subsystem.
- The residual strength parameters of soil are the most dominant variables, and the order of influence of other variables depends on the choice of the variation coefficient of strength parameters.

References

Cheng, Y.M., Lau, C.K.: Slope Stability Analysis and Stabilization, New Methods and Insight, pp. 18–19. Taylor & Francis Group, London (2008)

Cundall, P.: A computer model for simulating progressive, large scale movements in blocky rock systems. In: Proceedings of the International Symposium Rock Fracture, ISRM. Paper No. II–8 (1971)

Eberhardt, E.: Rock slope stability analysis—utilization of advanced numerical techniques (2003)

Eberhardt, E., Stead, D., Coggan, J.: Numerical analysis of initiation and progressive failure in natural rock slopes—the 1991 Randa rockslide. Int. J. Rock Mech. Min. Sci. **41**(1), 69–87 (2004)

Gilbert, R.B., Byrne, R.J.: Strain-softening behavior of waste containment system interfaces. Geosynth. Int. IFAI **3**(2), 181–203 (1996)

Griffiths, D.V., Fenton, G.A.: Probabilistic slope stability analysis by finite elements. J. Geotech. Geoenviron. Eng. **130**(5), 507–518 (2004)

Jiang, S.H., Li, D.Q., Cao, Z.J., Zhou, C.B., Phoon, K.K.: Efficient system reliability analysis of slope stability in spatially variable soils using Monte Carlo simulation. J. Geotech. Geoenviron. Eng. **141**(2), 04014096 (2015)

Khan, Y.A., Jiang, J.C., Yamagami, T.: Progressive failure analysis of slopes using non-vertical slices. J. Jpn. Landslide Soc. **39**(2), 203–211 (2002)

Miao, T., Ma, C., Wu, S.: Evolution model of progressive failure of landslides. J. Geotech. Geoenviron. Eng. **125**(10), 827–831 (1999)

Zhang, K., Cao, P., Bao, R.: Progressive failure analysis of slope with strain-softening behaviour based on strength reduction method. J. Zhejiang Univ. SCIENCE A **14**(2), 101–109 (2013)

Assessment and Stabilization of Natural Landslide Hazards by a Reinforcement System

Dounia Amrani, Chafia Merah, and Fethi Baali

Abstract

The phenomenon of landslides is a constant global natural hazard that occurs in most countries of the world. Algeria, particularly the northern region, is a case in point. In this respect, the present study examines the phenomenon of landslides in the Setif region, in order to protect and preserve the multilayer soil that supports the reinforced structures. Numerical modeling predicts the risk of sliding pillars against the wall. This study establishes an analytical calculation with a reference model. The investigations are carried out as a comparative study by three programs. The solution proposed is to reinforce the soil with an adequate system. The present study suggests that the use of different software within reasonable ranges has considerable effects on the total ground motion. Finally, it is concluded that the use of numerical analysis procedures could be more time and cost-effective in the design of geotechnical projects.

Keywords

Natural hazards • Landslide • Reinforcement • Multilayered soil • Numerical modeling

1 Introduction

The growing population has increased the demand for land development for house construction, infrastructure, and other facilities. The availability of land to be urbanized has been limited by the great need to protect natural areas and farmlands. Many constructions in marginal areas present signs of instability (cracks, collapse); they are threatened by natural phenomena such as slope movements. The risk of sliding can be estimated by analyzing the variation of the safety factor as well as the geotechnical characteristics of the terrain (Coppola et al. 2006; Rotaru 2007; Bilgin 2010). Numerical modeling methods in geotechnical engineering are common to solve practical engineering problems (Kumar and Dey, GeoInnovations. 2014; Krajewski and Reul 2004; Saeterbo et al. 2004; Bilgin and Erten 2009) with emphasis on numerical analysis of the computation of displacements and stresses in soils, structure, and interfaces. The finite element and the limit equilibrium (Saeterbo et al. 2004; Bilgin and Erten 2009; Song 2012) are among the most useful numerical methods for solving stability problems.

In this study, a detailed evaluation of the classical methods has been developed, in two cases: natural hazards and slope stabilization with concrete wall reinforced by a group of piles. This late analytical step has been compared by numerical computation with the use of a finite element in Plaxis program and limit equilibrium with Geoslope and Slide. The example used in this case study is the stability of natural slopes close to an area reserved for the construction of a new city in Setif region (Algeria). In addition, this work includes a contribution to the calculation of the safety factor of the combined system of reinforcement and stabilization in this slope compared to many works in the literature (Rotaru 2007; Saeterbo et al. 2004; Song 2012). A comparative study of finite elements and limit equilibrium in terms of displacement has also been carried out in the present case. The proposed solution to stabilize the slope is the reinforcement with a pile-reinforced retaining wall. This proposition has been detailed, and the obtained results seem to give good predictions of the soil behavior.

D. Amrani (✉) · C. Merah
Mining Institute, University of Larbi Tebessi, Tebessa, Algeria

F. Baali
Geology Institute, University of Larbi Tebessi, Tebessa, Algeria

H. El-Askary et al. (eds.), *Research Developments in Geotechnics, Geo-Informatics and Remote Sensing*, Advances in Science, Technology & Innovation,
https://doi.org/10.1007/978-3-030-72896-0_9

2 Materials and Methods

The soil of the area studied is a mixture of clayey soil and silty clay containing large blocks. It has a thickness varying from 5 to 8.40 m and extending over a length of 177 m, a weathered marl gray of 2 m thickness, substrate indurate and sometimes fractured (Fig. 1) This study aims to explore the different cases of slope stability with and without a reinforcement system, using numerical modeling with the Plaxis code, Geoslope and Slide software. The physico-mechanical characteristics of the different soil layers used in the modeling are presented in Table 1:

3 Modeling

See Fig. 2.

4 Results and Discussion

The problem was solved by three software (Fig. 2):

- Plaxis in several stages (using the model Mohr–Coulomb).
- Geoslope (using different methods).
- and Slide (in two stages).

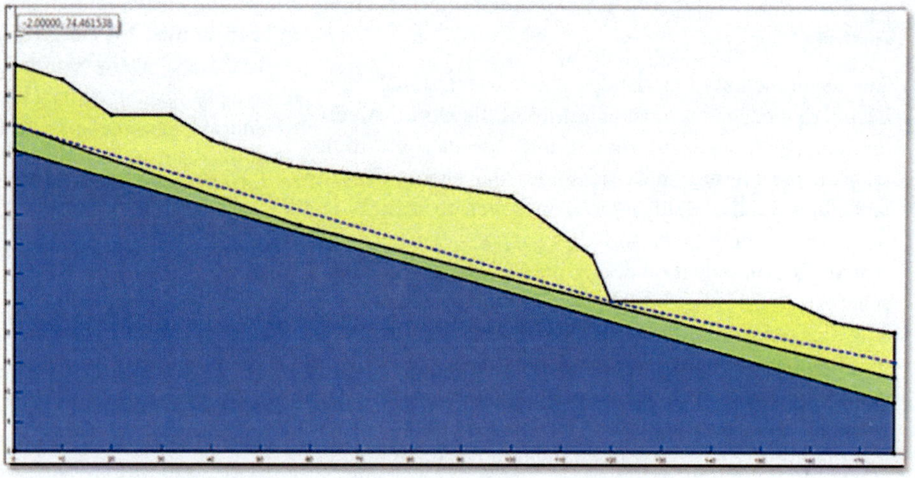

Fig. 1 Geometrical model

Table 1 Geotechnical parameters of model

Settings	Boulder slope	Altered Marl	Indurated Marl	Pile
Saturated unit weight kN/m^3	20	20	21	2500
Cohesion C kPa	0	4	12	200
Friction angle ϕ (°)	40	25	30	
Young's modulus E kPa	150,000	50,000	1,000,000	40
Poisson's ratio ν	0.4	0.25	0.35	0.2

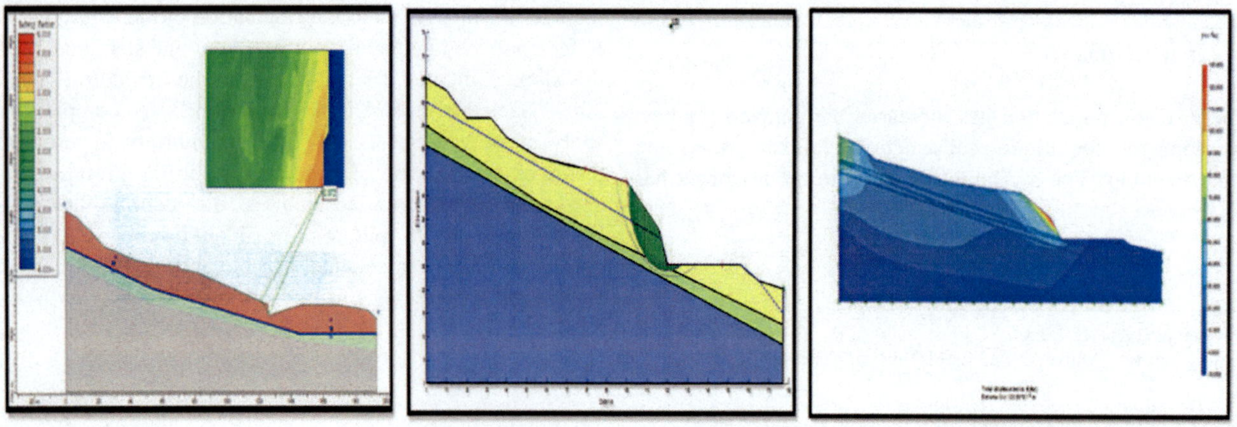

Fig. 2 Modeling after earthworks by Slide, Geoslope, and Plaxis

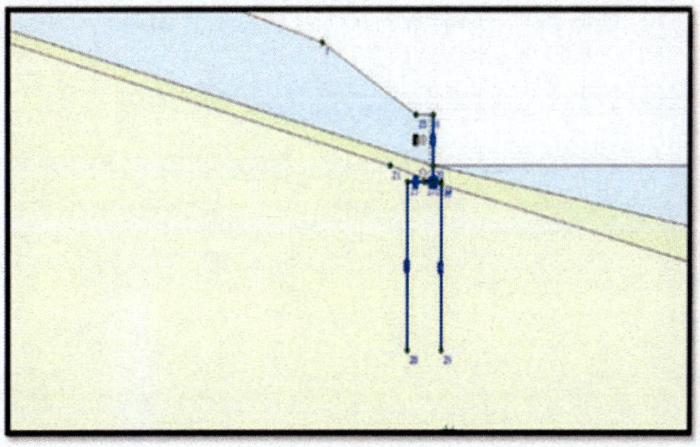

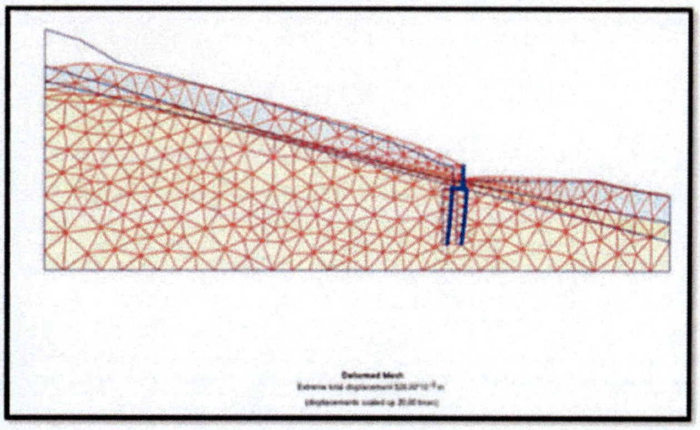

Fig. 3 Geometry and deformation mesh of the reinforced model

Table 2 Results of safety factors Sf calculated by three methods

Software	Methods	Sf
Slide	Janbu	1.348090
Geoslope	Janbu	0.6819
Plaxis	Fi-Redaction	0.52

Or the first concerning the natural stability of the slope, and the second seeking stability with the type of reinforcement chosen as a wall with group of piles reinforcing the whole system (Fig. 3).

The findings at all modeling stages prove the accuracy of the calculations at the construction stages. It can be concluded that numerical procedures provide an acceptable prediction of ground landslides motions in all phases of construction. The results of Sf safety factors are presented in Table 2.

5 Conclusions

The geological and geotechnical study of the landslide allowed the identification of the causes of the phenomenon and the determination of the measures to be taken to stabilize the phenomenon in question. The results obtained in this comparative study by numerical modeling (Plaxis, Geoslope, and Slide) depended on the parameters used, namely the displacement of a multilayer soil. The system, compared to the results obtained with the reinforcement, presented a total displacement (U_{tot} = 138, 15 × 10^{-3} m) and remarkable and important bending moments; the proposal of the reinforcement stabilization system as a solution has a very remarkable influence on the stability of the system and the total stability of the entire structure (Sf = 3) in order to reduce natural and environmental risks.

References

Bilgin, O.: Numerical studies of anchored sheet pile wall behavior constructed in cut and fill conditions. Comput. Geotech. **37**, 399–407 (2010)

Bilgin, Ö., Erten, M.B.: Analysis of anchored sheet pile wall deformations. Contemporary topics in ground modification, problem soils, and geo-support (GSP 187). In: Proceedings, International Foundation Congress & Equipment Expo (2009)

Coppola, L., Nardone, R., Rescio, P., Bromhead, E.: Reconstruction of the conditions that initiate landslide movement in weathered silty clay terrain: effects on the historic and architectural heritage of Pietrapertosa, Basilicata, Italy. Landslides **3**, 349–359 (2006)

Krajewski, W., Reul, O.: Deep excavation in stiff clay: comparison between numerical analyses and in-situ-measurements. Ottawa, Canada, 25–27 Aug 2004

Kumar, N., Dey, A.: GeoInnovations.: Finite Element Analysis of Flexible Anchored Sheet Pile Walls: Effect of Mode of Construction and Dewatering (2014)

Rotaru, A., Daniel, O., Paulică, R.:. Analysis of the landslide movements. Int. J. Geol. **1**, 70–79 (2007)

Saeterbo, M.G., Nordal, S., Emdal, A.: Slope stability evaluations using the finite element method. NGM, XIV Nordic Geotechnical Meeting, vol. 1, pp. A49–A61 (2004)

Song, Y.-S., Won-Pyo, H., Kyu-Seok, W.: Behavior and analysis of stabilizing piles installed in a cut slope during heavy rainfall. Eng. Geol. 129, 56–67 (2012)

Ground Stabilization with Functional Geomaterials

Effect of Sewage Sludge Ash on Collapsible Soil

Amir Mosallaei, Hasan Eteraf, Balázs Kovács, and Viktória Mikita

Abstract

The bio-environmental problems caused by sewage sludge and the destructive effects created by disposal of this dangerous poisoning substance are attracting attention. However, there is a remarkable growth of the sewage sludge volume. Therefore, establishing a proper cycle for application and use of this material is a very convenient way to reduce its bio-environmental risks. Inspection of the stability and improvement of collapsible soil are now crucial due to the vast territory of collapsible soil, the growth of urbanizations and the expansion of big cities over that kind of soil. This project aims to assess the effect of sewage sludge ash on collapsible soil shear strength parameter and to analyze SEM micrographs. To this end, the use of sewage sludge ash as an additive is set on the agenda. After the primary assessments, shear strength parameter of under investigation soil is determined without additive and with additive in different weight percentages, including 2, 4, 6 and 8%. The final results show that increasing the percentage of sewage sludge ash as an additive will also increase soil cohesion and decrease the internal friction angle ϕ.

Keywords

Collapsible soil • Sewage sludge ash • Shear strength • Cohesion • Internal friction angle

1 Introduction

Soil stabilization is a common engineering technique used to improve the physical properties of weak soil and enable it to achieve the desired engineering requirements. Many mechanical and chemical methods were used in soil stabilization, such as the addition of lime, fly ash, cement, rice husk fly ash and coal fly ash. In recent years, researchers and engineers have started testing some waste materials that could be employed in order to solve many geo-environmental problems and disposable issues.

Important factors should be considered in choosing the stabilizing agents, such as abundance, cost and easy availability. Sludge is available in vast amounts, and it is likely to be more available in the near future due to the wide growth of wastewater treatment plants in the urban areas. (Ai-sharif and Attom 2014).

There are different types of weak soils all over the world; the collapsible soil chosen for this project is one of them. The collapsible soil, also known as problematic soil, usually causes considerable problems to engineering projects. Soil structure collapse caused by increase in moisture or rising level of underground water triggers large settlements. (Kavandi et al. 2016).

Collapsible soils such as loess are usually found in arid and semi-arid regions (Haeri et al. 2015).

Loess soil is a uniformly sized, wind-deposited soil. The worldwide distribution area covers 10%, approximately 13 × 106 km^2 (Xing and Liu 2018).

2 Materials and Methods

Stabilization progress is defined for weak soils in general and collapsible soils in particular. According to the collapsibility index, the soil chosen for this study is collapsible. Sewage sludge samples were first collected from Tehran, Ikbatan, waste water treatment plant. Then, they were

A. Mosallaei (✉)
Budapest University of Technology and Economics, Budapest, Hungary
e-mail: mosallaei@edu.bme.hu

H. Eteraf · B. Kovács · V. Mikita
University of Miskolc, 3515 Miskolc, Hungary

H. El-Askary et al. (eds.), *Research Developments in Geotechnics, Geo-Informatics and Remote Sensing*, Advances in Science, Technology & Innovation,
https://doi.org/10.1007/978-3-030-72896-0_10

incinerated in an electric furnace at 800 °C and ground into fine particles to pass through a #200 sieve. Tests of pH values, in accordance regulated by ASTM, was performed to analyze the properties of incinerated sewage sludge ash (ISSA). Sewage sludge ash was also applied as an additive in this study. Hence, the admixture was manufactured with ISSA and collapsible soil. Five samples with portions of 0, 2, 4, 6 and 8% (by weight) were proposed for mixture. Experiments on the pH values, Atterberg's limits, compaction test, direct shear test and SEM test were carried out for the ISSA soil specimens in order to understand the effects of admixtures on the strength of soil (Tables 1 and 2).

3 Results

Direct shear tests were carried out to obtain the cohesion (c) and the internal friction angle (Φ) of samples, based on ASTM D3080. Speed of loading was 0.048 mm per minute (Table 3).

4 Discussion

According to the Coulomb failure envelope plotted in Fig. 1, C parameter is higher in the mixtures than the sample without an additive. Figure 1 shows that although the amount of cohesion is increasing by adding ISSA, the ratio of this growth is decreasing. The percentage of variation $[(tf - t0)/t0] \times 100$ describes the difference of shear strength values between the mixtures and specimen without additive.

The growth in soil cohesion can be explained by the oxidation of the main elements of sewage sludge ash (CaO, Al_2O_3 and SiO_2). Pozzolanic reactions between oxide elements and water, based on Eqs. 1, 2 and 3, can produce cohesive particles, including calcium hydrosilicate (CSH) and calcium hydroaluminate (CAH).

$$CaO + H_2O \rightarrow Ca(OH)_2 \quad (1)$$

$$Ca(OH)_2 + SiO_2 \rightarrow C - S - H \quad (2)$$

$$Ca(OH)_2 + Al_2O_3 \rightarrow C - A - H \quad (3)$$

For an accurate result, scanning electron microscopy (SEM) tests were performed.

Figure 2b reveals that pores of soil were permeate by sewage sludge ash. This implies that the amount of soil void ratio is decreasing. Particles of ISSA are fine-grained and round-cornered. This explains the decrease of the angle of internal friction.

5 Conclusions

At last, it can be concluded that adding incinerated sewage sludge ash as an additive has a significant influence on shear strength parameters. Increasing of ISSA causes an increase in the value of cohesion and a decrease in the value of internal friction angle.

Table 1 Result of XRF test

Soil		Sewage sludge ash	
Oxide elements	Weight percentage	Oxide elements	Weight percentage
SiO_2	45.902	P_2O_5	22.370
CaO	17.382	SiO_2	20.363
Al_2O_3	11.970	CaO	15.105
Fe_2O_3	4.798	Fe_2O_3	7.438
K_2O	2.227	Al_2O_3	6.415
Na_2O	0.801	MgO	3.709
TiO_2	0.640	K_2O	3.040
P_2O_5	0.125	SO_3	1.482
MnO	0.083		
SrO	0.045		

Table 2 Common soil parameters

PI	PL	LL	G_S	pH	USCS	Soil
5	20	25	2.67	7.9	CL	Collapsible soil

Table 3 Shear strength parameters of soil and mixture

Ø (°)	*C* (kPa)	Additive (%)
18.89	13.48	0
15.64	25.5	2
13.13	36.2	4
16.81	33.16	6

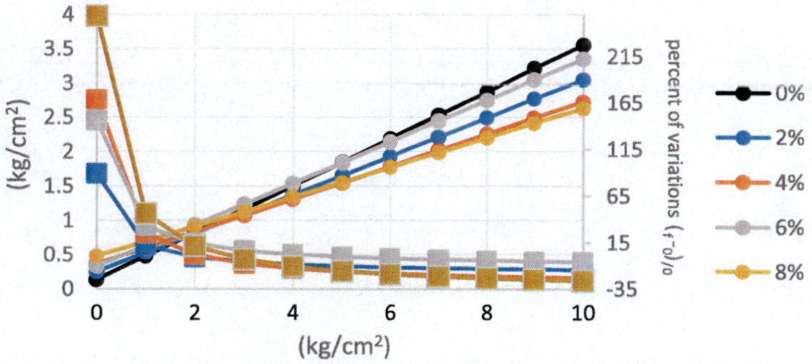

Fig. 1 Coulomb failure envelope and percentage of changes in shear strength

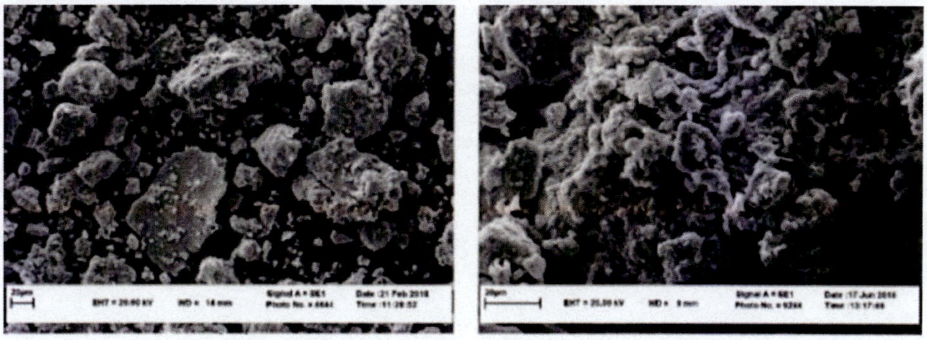

Fig. 2 SEM test. **a** Sample of soil without additive **b** Soil with ISSA

Considering the Mohr–Coulomb line, shear strength increases for the small quantities of stress and decreases for the larger ones. ISSA can increase the shear strength of collapsible soil, but it seems that 4% additive is the optimum portion. In other words, adding more than 4% additive does not have considerable influence on shear strength.

Acknowledgements The described article/presentation/study was carried out as part of the EFOP-3.6.1-16-2016-00011 "Younger and Renewing University—Innovative Knowledge City—institutional development of the University of Miskolc aiming at intelligent specialisation" project and the GINOP-2.3.2-15-2016- 00031 "Innovative solutions for sustainable ground-water resource management" project implemented in the framework of the Szechenyi 2020 program. The realization of this project is supported by the European Union, co-financed by the European Social Fund and the European Structural and Investment Funds.

References

Ai-sharif MM, Attom MF: A geoenvironmental application of burned wastewater sludge ash in soil stabilization. Environ. Earth Sci. **71** (5), 2453–63 (2014)

Haeri, S.M., Saberia, S., Beigia, M., Garakania, A.A.: Study on the behavior of a collapsible soil under constant matric suction loading implementing an unsaturated framework. In: 15th Pan-American Conference on Soil Mechanics and Geotechnical Engineering, At Buenos Aires, Argentina (2015)

Kavandi, P., Firoozfar, A., Hemmati, M.A.: Bearing capacity assessment of collapsible soils improved by deep soil mixing using finite element method. Open J. Geol. 6(09), 1055 (2016)

Xing H, Liu L (2018) Field tests on influencing factors of negative skin friction for pile foundations in collapsible loess regions. Int. J. Civil Eng. 16(10), 1413–22 (2018)

Soil Stabilization Using Local Waste

Inas Berdi, Salah Messast, Riad Benzaid, and Imane Idoui

Abstract

The growth of human activities requires expanding the land used for building and construction. The nature of the soil, however, is not guaranteed to be suitable for that purpose. In this context, the improvement of the soil characteristics is recommended as a remedy. To this end, certain techniques of stabilization and or reinforcement are applied. Each year, the phenomenon of swelling affects various types of structures (light buildings, roadways), causing considerable disorders that require costly rework. Besides, the implementation of the reconstruction works is often not well-mastered. This article aims to characterize certain types of problem soils of an Algerian site. It also proposes several solutions for the improvement of the characteristics of these soils through the use of simple techniques and the valorization of the addition of local materials (waste). The interest of our contribution is in the valorization of the local materials, by an experimental approach of stabilization vis-à-vis the swelling soils.

Keywords

Local waste • Swelling • Stabilization • Clay

1 Introduction

Soil stabilization involves modifying one or more properties of the soil, by mechanical or chemical means, to create an improved soil material with the desired technical properties.

I. Berdi (✉) · R. Benzaid · I. Idoui
Geological Engineering Laboratory, Mohammed Seddik Benyahia University, Jijel, Algeria
e-mail: ines.berdi@univ-jijel.dz

S. Messast
LMGHU Laboratory, Skikda University, 20 Aout 1955, Skikda, Algeria

2 Literature Review

Production is still a key driver of economic growth, which generates waste. In order to limit negative impacts of these substances on the environment and save natural resources, many researchers have tried to develop them and use them as a material to stabilize the swelling soils. Some works are mentioned in the following parts.

2.1 Soil Stabilization Using Industry Waste

Plastic: This study aimed to stabilize black cotton soil in the Andhra Pradesh capital region of India. The soil had a low lift capacity and a problem of shrinkage swelling, not to mention a low resistance to shear. Stabilization was provided by plastic strips with the present percentages 2, 4, 6, 8%. Based on tests conducted by Proctor and CBR, Mallikarjuna and Bindumani (Mallikarjuna and Bindu Mani 2016) showed that the values started to increase up to 4% and then decreased, so that the best result was obtained with the percentage of 4% plastic.

Granulated slag: Granulated slag is used as an adjuvant to treat soil problems. Zemouli and Chelghoum (Zemouli and Chelghoum 2018) created a laboratory floor (85% kaolin + 15% bentonite) and treated them with the different percentages 5, 10, 15, 20, and 25% slag granules only and the mixture of slag with lime. In order to evaluate the influence of the adjuvant, a series of physico-mechanical tests was carried out (Atterberg limits, Proctor). The compressive strength results showed that the% milk percentage increase reduced the plasticity index from 38 to 25% and increased the compressive strength from 600 to 1000 KN/m. The results also revealed that the percentage of 20% was the optimal content to stabilize the fine soil.

Ceramic dust: The study investigated the behavior of Ceramic Dust on a marly clay soil that was purely swelling

H. El-Askary et al. (eds.), *Research Developments in Geotechnics, Geo-Informatics and Remote Sensing*, Advances in Science, Technology & Innovation,
https://doi.org/10.1007/978-3-030-72896-0_11

from the Bhubaneswar region of India. The various tests carried out on the soil with different percentages of 0% to 30% of the addition improved the limits of Atterberg (decrease in the plasticity index), optimum water content (decrease), dry density (increase), compressive strength (Increase), and the swelling pressure (decrease) (Sabat 2012).

Glass powder: Black cotton soil has high swelling potential and low resistance. The researchers performed an experimental program to improve soil properties by combining fly ash, coconut fiber (CCF) and crushed glass (CG) with a percentage variation of fly ash (FA) at 10, 15, 20, 25%, coconut coir (CCF) at 0.25, 0.5, 0.75, 1% and crushed glass (CG) at 3, 5, 7% (crushed glass). The best results obtained by optimal combination were 20% FA + 5% CG + 1% CCF. This additive decreased the swelling pressure (Tiwari and Mahiyar 2014). Moreover, other researchers (Olufowobi et al. 2014) stabilized the clay soil by a mixture of percent between 1, 2, 5, 10%, and 15% glass powder and 15% cement by weight of the soil sample.

2.2 Soil Stabilization Using Agricultural Waste

Rice husk ash: The peat soil is characterized by low bearing capacity and very high compressibility. For that reason, Yulianto and Mochtar (2010) improved the geotechnical characteristics of a 10% stabilizer of the addition (30% lime + 70% rice husks) with the cure time 1, 10, 20, 30 days. The findings indicated that the shear strength increased almost 25% of its initial state, while the compressive strength decreased. Other researchers, namely Soni et al. (2011) managed to treat the black cotton floor of India by combining three stabilizers (fly ash waste rice and lime). The combination was more effective than ash and rice waste and lime alone. As a result, the plasticity ratio decreased as the resistance increased. The best result of the combination was registered at 10%.

Bagasse ash: Bagasse fiber is an agricultural waste produced by grinding sugar cane for extraction. Dang et al. (2017) succeeded in strengthening the swelling soil (Queensland, Australia) with 0.5% Bagasse fibers at 2% combined with lime. The outcome made the soil more resistant to compression and reduced the swelling potential efficiently as it approaches zero with the percentage increase of the addition.

2.3 Soil Stabilization Using Commercial Waste

Sawdust: Improving the geotechnical characteristics of a lateritic soil extracted from a burrow of the Federal University of Technology Akure Nigeria was achieved by the combination of lime and sawdust with a variation of 2, 4, 6, 8. 10%. The results of the tests (Atterberg limits, Proctor, and compressive strength) showed that the percentage of 6% was sufficient to ensure this stabilization; the resistance increased from 38 to 119%. Meanwhile, the plasticity index decreased from 13.7 to 12% (Nochiri et al. 2017).

Woodash: The use of additives improves the stabilization of plastic floors. Samples were collected from a pit dug at Awgu, in south-eastern Nigeria. Okagbue (2007) investigated the behavior of wood-clay ash with the different percentages of 5, 10, 15, and 20%. The study revealed that the plasticity was reduced by 35%, while the resistance increased by almost 50% with the increasing of the addition percentage. It also proved that wood ash played the same role as lime (chemical stabilization).

Egg Shell Powder: The technique of the stabilization of a lateritic soil by the shell of ground eggs is performed by Oluwatuyi et al. (2018) who treated the stabilization by three additions eggs shell, cement, and a mixture between the two with a variation of 2, 4, 6, and 8%. The 8% mixing percentage greatly reduced the plasticity index and the dry density as it increased the water content and compressive strength; therefore, it was considered the highest percentage.

Polymers: Rehab Bekkouche (2018) and Rehab Bekkouche and Boukhatem (2016) studied the soil-polymer behavior on two terrains of the region of Skikda (*Elhadeik, Djamel Ramadan*) North East of Algeria by the different stabilizers (PVC, PHDE, Plastic) with different percentages of treatment. The mechanical tests revealed that the swelling sensitivity was reduced from 6 to 2%.

Salts: Hachichi et al. (2007) launched a study on two sites near the city of Oran (Sidi Shahmi, Mers-el Kebir) north-west of Algeria known for a high potential for swelling. The treatment phase was based on the salts (KCl, NaCl, $CaCl_2$) with different concentrations (0.05, 0.5, 1, and 2 mol/l) and on polymers, as well. The results indicated that the mixture of salts and polymers significantly reduced swelling. It was more effective than using salts alone. Also, Azzouz (2015) applied the chemical salt stabilization technique to two Tlemcen swelling clays. They concluded that salts had a positive effect on swelling reduction that exceeded 50%.

3 Conclusion

The literature review on soil stabilization by local waste allows gathering a maximum of information:

- The waste used above significantly improves the geotechnical properties of soils, thus offers better results.
- The choice of waste is based on the economic, ecological, and technical aspects.
- Previous research set the ground to focus more on this topic that is only recently beginning to spike interest in the field.

References

Azzouz, F.Z.: Clay soil stabilization of the région Tlemcen by salts. ElWahat pour les Recherches et les Etudes **8**(n°1), 108–117 (2015). ISSN:1112–7163

Dang, L.C., Khabbaz, H., Fatahi, B.: An experimental study on engineering behaviour of lime and bagasse fibre reinforced expansive soils. In: ICSMGE 2017–19th International Conference on Soil Mechanics and Geotechnical Engineering (2017)

Hachichi, A., Bourokba, S., Fleureau, J.M.: Chemical stabilization of two expansive soils from the Oran region, French Geotech. Rev. **118**, 3–11 (2007)

Mallikarjuna, V., Bindu Mani, T.: Soil stabilization using plastic waste. Int. J. Res. Eng. Technol. eISSN, pp. 2319–1163 (2016)

Nochiri, E.S., Emeka, H.O., Tanimola, M.: Geotechnical characteristics of lateritic soil stabilized with sawdust ash-lime mixtures. StavebníObz.-Civ. Eng. J. **26**(1), 66–76 (2017)

Okagbue, C.O.: Stabilization of clay using woodash. J. Mater. Civ. Eng. **19**(1), 14–18 (2007)

Olufowobi, J., Ogundoju, A., Michael, B., et al.: Clay soil stabilisation using powdered glass. J. Eng. Sci. Technol. **9**(5), 541–558 (2014)

Oluwatuyi, O.E., Adeola, B.O., Alhassan, E.A.: Ameliorating effect of milled eggshell on cement stabilized lateritic soil for highway construction. Case Stud. Construct. Mater. **9**, e00191 (2018)

Rehab Bekkouche, S.: Mechanical behavior of clay reinforced by layers of polymer. Int. Inven. Sci. J. **2**(4) (2018)

Rehab Bekkouche, S., Boukhatem, G.: Experimental characterization of clay soils behavior stabilized by polymers. J. Fondam. Appl. Sci. **8**(3), 1193–1205 (2016)

Sabat, A.K.: Stabilization of expansive soil using waste ceramic dust. Electron. J. Geotech. Eng. **17** (2012)

Soni, S.R., Dahale, P.P., Dobale, R.M.: Disposal of solid waste for black cotton soil stabilization. Int. J. Adv. Eng. Sci. Technol. **8**(1), 113–120 (2011)

Tiwari, A., Mahiyar, H.K.: Experimental study on stabilization of black cotton soil by fly ash, coconut coir fiber & crushed glass. Int. J. Emerg. Technol. Adv. Eng. **4**(11), 330–333 (2014)

Yulianto, F.E., Mochtar, N.E.: Mixing of rice husk ash (RHA) and lime for peat soil stabilization. In: Proceedings of the First Makassar International Conference on Civil Engineering (2010)

Zemouli, S., Chelghoum, N.: Use of ground granulated blast furnace slag in soils stabilization. Synthèse: J. Sci. Technol. **36**(1), 103–114 (2018)

Strength Properties of Expansive Clayey Soil Stabilized with Nano-scoria and Nano-lime

Aref al-Swaidani and Ayman Meziab

Abstract

This paper investigates the effect of adding nano-volcanic scoria (NVS) and nano-lime (NL) to the strength properties of expansive clayey soils. To this end, three expansive clayey soil samples quarried from three different sites in the south of Syria are analysed. Three replacement levels of NVS are used, i.e. 0, 1 and 2%. Atterberg limits, compaction, unconfined compressive strength (UCS) and California bearing capacity (CBR) are investigated. The test results reveal the positive effect of NVS when added to the original soil. Plasticity index (PI) is reduced by more than 50% when 2% NVS is added to the original soil. In addition, 0.6% NL is added to further investigate the combined effect of NL and NVS on the strength properties of the clayey soil. All investigated properties are significantly improved when NVS and NL are added together, i.e. UCS and CBR values increase to more than twice and 20 times when compared with those of the original soil, respectively.

Keywords

Nano-volcanic scoria • Nano-lime • UCS • CBR • Expansive clay soil • Supplementary cementitious materials

1 Introduction

Expansive clayey soils cover a wide area of Syria. They occupy a surface area of about 12% of the country (Al-Swaidani et al. 2016). Their undesirable properties such as high compressibility, high moisture susceptibility and low bearing capacity have caused serious damage to infrastructure. Many attempts were made to stabilize the expansive clayey soils using a wide variety of additives (Al-Swaidani et al. 2016, 2018; Harichane et al. 2011). One of the most common attempts was using lime, which is considered the oldest stabilizer for clayey soil (Mallela et al. 2004). Volcanic scoria was widely used as cement replacement (Massazza 1993; Al-Swaidani 2018). However, the authors think that very little work or even no detailed work on using nano-VS as a soil stabilizing agent is found in the literature.

Nano-materials have recently been used in many various engineering applications including the geotechnical engineering (Ghasabkolaei et al. 2017; Changizi and Haddad 2015; Majeed and Raihan 2016; Lou et al. 2012; Pham and Nguyen 2014; Mohammadi and Niazian 2013). For instance, nano-additives, being extremely small in size, react very actively with other particles in the soil matrix (Ghasabkolaei et al. 2017). Therefore, adding nano-stabilizers to the original soil not only increases strength and compressibility, but also decreases swelling and plasticity index (Ghasabkolaei et al. 2017). Furthermore, nano-additives reduce the porosity by filling the space between soil particles and, thus, bond them together (Majeed and Raihan 2016).

The objective of the current study is to investigate the effect of adding NVS and NL on the strength properties of expansive clayey soil. The examined properties of the modified clayey soil are: Atterberg limits, compaction, UCS, and CBR. The authors claim that the current study can be beneficial for other areas of similar geology and may have a great importance in the rebuilding stage in Syria.

2 Materials and Methods

The investigated clayey soils were quarried from three different sites in the south of Syria. Table 1 shows some of the important properties of the investigated clayey samples. In addition, all of the soil samples were analysed using the wet

A. al-Swaidani (✉) · A. Meziab
Arab International University, Damascus, Syria

A. Meziab
Damascus University, Damascus, Syria

H. El-Askary et al. (eds.), *Research Developments in Geotechnics, Geo-Informatics and Remote Sensing*, Advances in Science, Technology & Innovation,
https://doi.org/10.1007/978-3-030-72896-0_12

Table 1 Some important properties of the studied clayey soil

Soil	Depth (m)	Specific gravity	Passing (%) sieve nr. 200	Plasticity Index (%)	Soil classification (USCS)	Optimum moisture content (OMC) (%)	Maximum dry density (MDD) (kN/m^3)
S1	1	2.69	93	29	CH (very plastic clay)	25.5	14.2
S2	0.75	2.67	91.5	35	CH	23	13.8
S3	1.25	2.72	95	42	CH	26	13.0

method. Figure 1 presents the XRD analysis of the investigated soils conducted by SATOE STADI X-ray diffractometer (XRD).

Quick lime obtained from Hama city was used in the experimental part. It was calcined up to 950 °C. Its CaO content was about 94%. The obtained quick lime was ground in a laboratory ball mill (Retsch, S100, Germany) for six hours, three hours in a "wet form" followed by drying the sample in an oven for 18 h, and then ground for three hours in a "dry form".

A local volcanic scoria quarried from the south of Syria (Tal Shihan site) was used in the experimental part. Scanning electron micrograph (SEM) and atomic force micrograph (AFM) of NVS obtained after 6 h-grinding are shown in Fig. 2. Six clay mixtures for each expansive clay soil were prepared. The mixture ingredients percentages are shown in Table 2. The mixtures were designated according to the replacement level. For instance, NVS2NL0.6 refers to the clay mixture, which contains 2% NVS and 0.6% NL.

The laboratory tests conducted for this work can be summarized as follows:

(i) Atterberg limits test (ASTM D4318).
(ii) Standard Proctor compaction test (ASTM D698). The soil mixtures, with or without nano-additives, were thoroughly mixed for 1 h prior to compaction.
(iii) Unconfined compressive strength (UCS) (ASTM D2166).
(iv) California bearing ratio (CBR) test (ASTM D1883). The specimens were tested after being soaked in water for 96 h.

3 Results and Discussion

3.1 Atterberg Limits

Figure 3 shows the results of plasticity index of the clay soil mixtures. The decrease in PI values indicates the improvement of soil workability. As illustrated in Fig. 3, adding 0.6% NL led to a significant decrease in PI values for all studied soils. Similar behaviour was reported in the literature for soils with a similar classification (Harichane et al. 2011; Al-Swaidani et al. 2018; Mallela et al. 2004). This improvement can be ascribed to the flocculation and aggregation of colloidal clay particles when NL is added. The addition of NVS enhanced the workability because of a reduction in the plasticity of the soils. Reductions of about 45 and 65% in PI values were observed when NVS was added at replacement levels of 1 and 2%, respectively. However, the lowest PI values were observed when 2% NVS and 0.6% NL were added together.

3.2 Compaction Test Results

Figures 4 and 5 present the effect of NVS, NL and their combinations on the compaction properties of the investigated soils. It can be clearly seen that adding 0.6 NL alone to the three soils decreased MDD and increased OMC. This can be attributed to the lower specific gravity of lime and the higher water retention property of lime, when compared with those of soils (Al-Swaidani et al. 2016, 2018). However, adding NVS caused a decrease in OMC and an increase in MDD with the increase of NVS content. This can be attributed to the lower affinity to water and the higher specific gravity of NVS, when compared to the original soil. The combination of NL and NVS kept the values of the maximum dry density (MDD) and the optimum moisture content (OMC) around their values in the original soils as shown in Figs. 4 and 5, respectively.

3.3 Strength Properties

The results of UCS test are shown in Fig. 6. The results revealed that UCS increased with an increase in NVS content. However, the highest values were observed when 2% NVS and 0.6NL were added together. This increase in UCS values can be attributed to the cementing materials formed because of the pozzolanic reactions occurring between the soil and the nano-additives, as clearly seen in Fig. 8; para. 3.4.

The results of CBR test are presented in Fig. 7. According to Fig. 7, the CBR increased from about 3 to

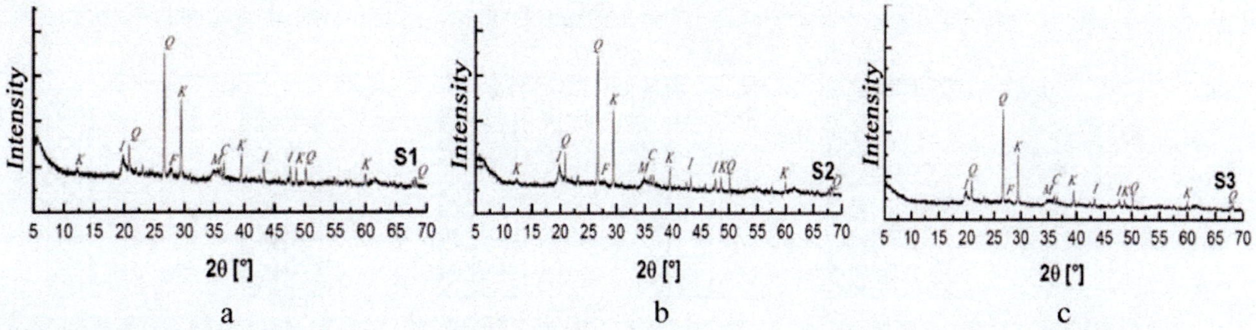

Fig. 1 XRD analysis of the investigated clayey soils; Soil 1 (**a**), Soil 2 (**b**), and Soil 3 (**c**). (I: illite; Q: quartz; M: montmorillonite; K: kaolin; F: feldspar; C: calcite)

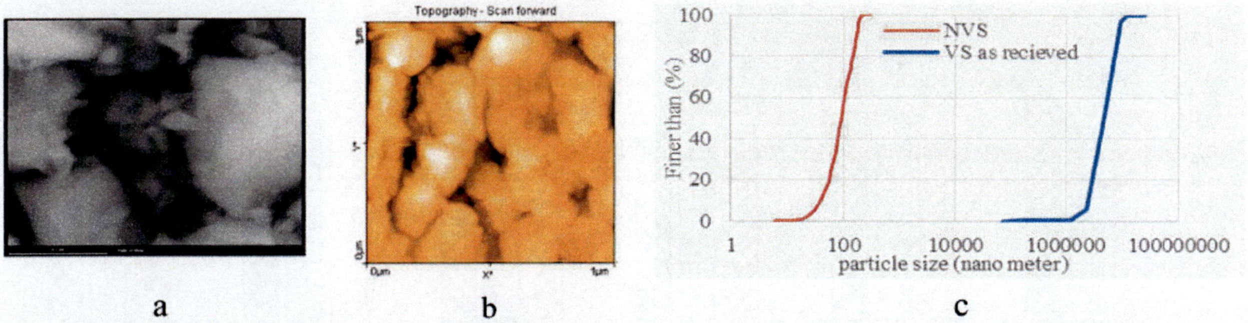

Fig. 2 SEM (**a**) and AFM (**b**) micrographs of the obtained NVS with its particle size distribution (**c**). (Its main oxides (%): SiO_2: 45; CaO: 9; Al_2O_3: 17; Fe_2O_3: 9; MgO: 8; alkalis (Na_2O and K_2O): 4.5)

Table 2 Clay mixtures used in the study

Clay mixture	Percentage (%)		
	Original clayey soil	Nano-VS	Nano-lime
NVS0NL0 (control)	100	0	0
NVS0NL0.6	99.4	0	0.6
NVS1NL0	99	1	0
NVS2NL0	98	2	0
NVS1NL0.6	98.4	1	0.6
NVS2NL0.6	97.4	2	0.6

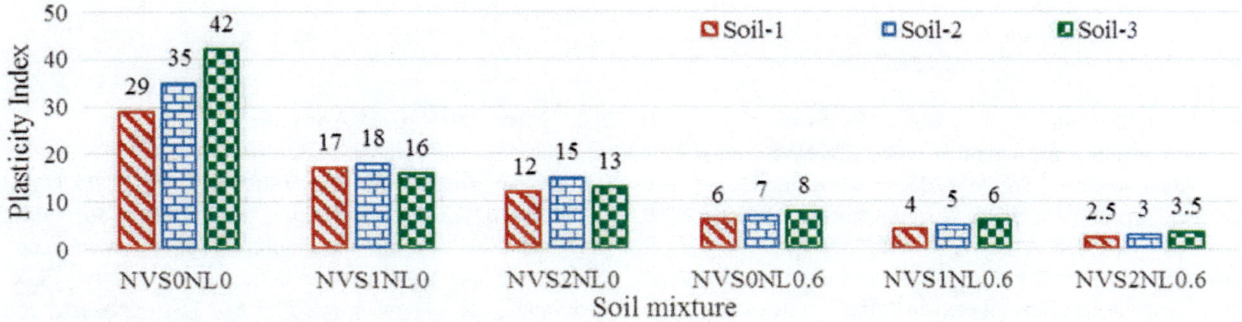

Fig. 3 Plasticity index values for the investigated clay mixtures

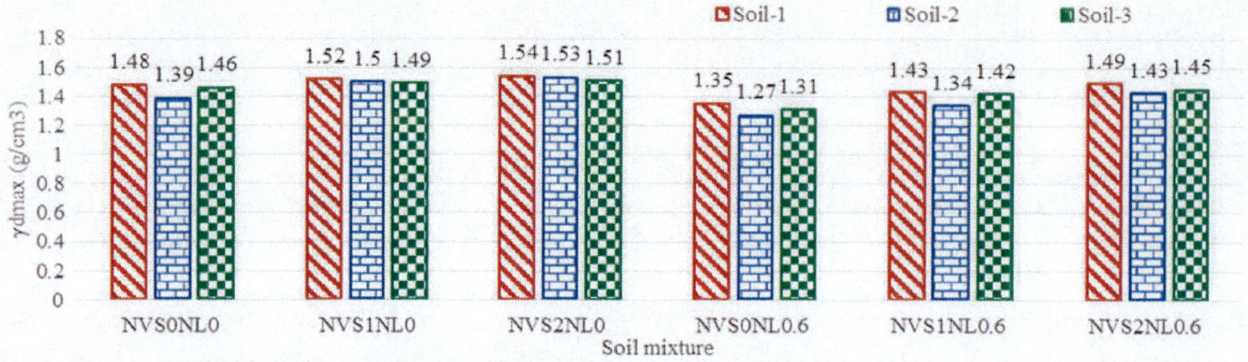

Fig. 4 Maximum dry density of the investigated clay mixtures

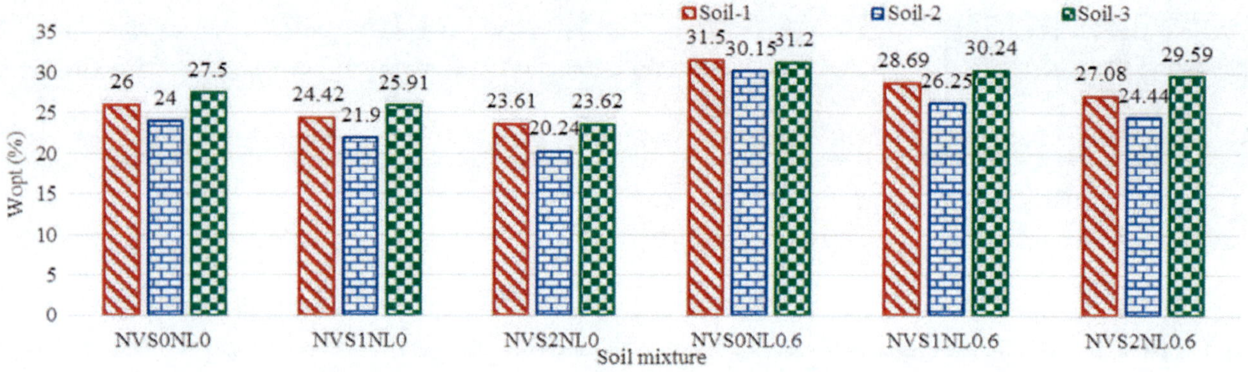

Fig. 5 Optimum moisture content of the investigated clay mixtures

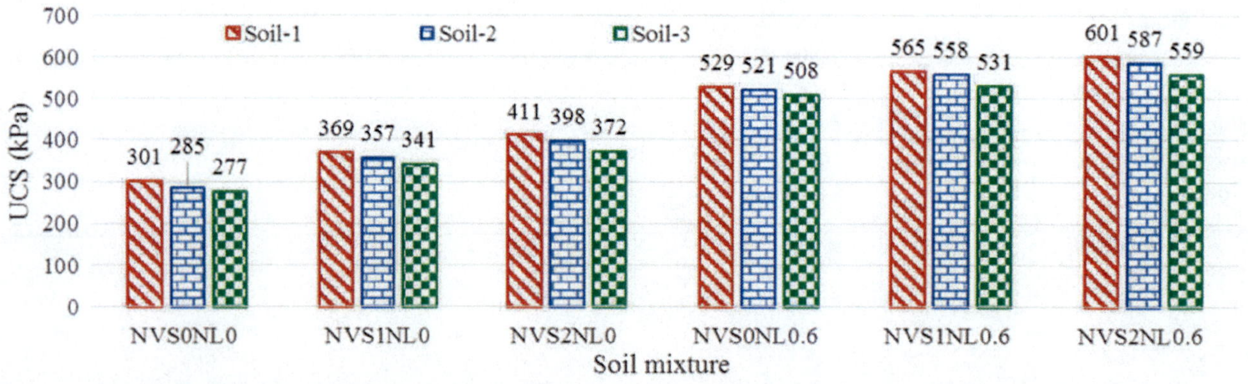

Fig. 6 UCS test results of the investigated clayey soils

more than 30 when 2% NVS was added to the clayey soil. The best results were obtained when 2% NVS and 0.6NL were added together. Soils of CBR values of more than 70% can be considered suitable for pavement subgrade. The increment in CBR values can be ascribed to the gradual formation of cementing materials in the soil as a result of the pozzolanic reactions between the soil and the nano-additives.

3.4 Microstructural Analysis

In the current study, NVS0NL0, NVS2NL0 and NVS2NL0.6 “soil 2” samples prepared and cured for 7 days were analysed using scanning electron microscopy (SEM) and energy dispersive X-ray spectroscopy (EDX) techniques, as shown in Fig. 8. The original soil (NVS0NL0), as clearly seen in Fig. 8a, has a discontinuous

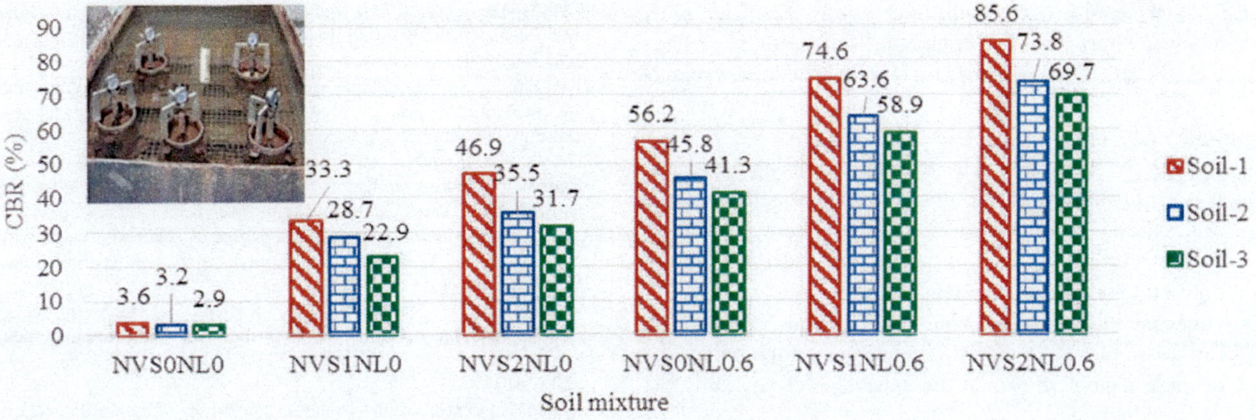

Fig. 7 Results of CBR test for the three expansive clayey soil

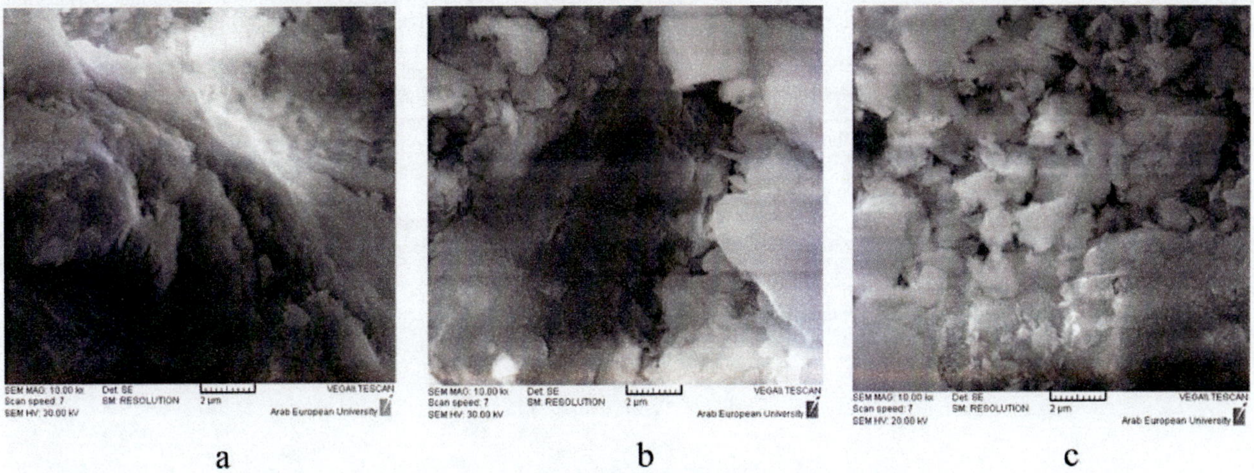

Fig. 8 SEM micrographs of NVS0NL0 (**a**), NVS2NL0 (**b**) and NVS2NL0.6 soil mixtures cured for 7 days

structure with no detected hydration compounds. However, Fig. 8b, c shows that the microstructure of the original soil has been significantly modified when either NVS or NVS and NL were added. It can be easily observed in Fig. 8c that cementing compounds such as calcium aluminosilicates hydrate (C-A-S-H) and calcium silicates hydrate (C-S-H) were formed. These cementing compounds have been formed as a result of pozzolanic reactions occurred between the calcium hydroxide (CaO) present in NL particles and $SiO_{2active}$ and $Al_2O_{3active}$ occurring in each of soil and NVS.

4 Conclusions

- Plasticity index decreased with an increase in NVS content.
- Use of NVS alone led to a reduction in compressibility of the original clay soil. Further reduction was also observed when NL was added.
- The strength characteristics (UCS and CBR) were significantly improved by adding NVS, NL or both.

- The test results were confirmed using SEM and EDX analysis. Formation of cementing compounds such as C-S-H and C-A-S-H led to significant improvements in the geotechnical properties of the examined clayey soils.
- Adding NVS could be considered a promising approach. It not only improved the geotechnical properties of the original clay soil but also saved natural sources. Nevertheless, further research is needed in order to get the optimum contents of the nano-additives.
- Other techniques such as TGA, FTIR and XRD can be further employed to thoroughly understand the behaviour of these nano-additives in the expansive clayey soils.

References

Al-Swaidani, A., Hammoud, I., Meziab, A.: Effect of adding natural pozzolana on geotechnical properties of lime-stabilized clayey soil. J. Rock Mechan. Geotech. Eng. **8**(5), 714–725 (2016)

Al-Swaidani, A.M., Hamood, I., Meziab, A.: Thermally Treated Clay as a Stabilizing Agent for Expansive Clayey Soil: Some Engineering Properties. Studia Geotechnica Et Mechanica **40**(3), 220–232 (2018)

Al-Swaidani, A.: Volcanic scoria as cement replacement. In: Aeillo, G. (ed.) Volcanos. Intechopen (2018)

ASTM D2166: Standard Test Method for Unconfined Compressive Strength of Cohesive Soil. American Society for Testing and Materials, West Conshohocken

ASTM D4318: Standard Test Methods for Liquid Limit, Plastic Limit, and Plasticity Index of Soils. American Society for Testing and Materials, West Conshohocken

ASTM D698: Standard Test Method for Laboratory Compaction Characteristics of Soil Using Standard Effort. American Society for Testing and Materials, West Conshohocken

ASTM D1883: Standard test method for CBR (California Bearing Ratio) of laboratory-compacted soils. American Society for Testing and Materials, West Conshohocken

Changizi, F., Haddad, A.: Strength properties of soft clay treated with mixture of nano-SiO_2 and recycled polyester fiber. J. Rock Mechan. Geotech. Eng. **7**(4), 367–378 (2015)

Ghasabkolaei, N., Choobbasti, A.J., Roshan, N., Ghasemi, S.E.: Geotechnical properties of the soils modified with nanomaterials: a comprehensive review. Archiv. Civil Mechan. Eng. **17**(3), 639–650 (2017)

Harichane, K., Ghrici, M., Missoum, H.: Influence of natural Pozzolana and lime additives on the temporal variation of soil compaction and shear strength. Front Earth Sci. **5**(2), 162–169 (2011)

Lou, H.L., Hasiao, D.H., Lin, C.K.: Cohesive soil stabilized using Sewage Sludge Ash/cement and nano aluminium oxide. Int. J. Transport. Sci. Technol. **1**(1), 83–100 (2012)

Majeed, Z.H., Raihan, M.: The effects of using nanomaterials to improvement soft soils. Saudi J. Eng. Technol. **1**(3), 58–63 (2016)

Mallela, J., Harold Von Quintus, P., Smith, K.L.: Consideration of lime-stabilized layers in mechanistic-empirical pavement design. Arlington, Virginia, USA. The National Lime Association (2004)

Massazza, F.: Pozzolanic Cements. Cement Concr. Compos. **15**(4), 185–214 (1993)

Mohammadi, M., Niazian, M.: Investigation of nano-clay effect on geotechnical properties of Rasht clay. J. Adv. Sci. Technol. **3**(3), 37–46 (2013)

Pham, H., Nguyen, Q.P.: Effect of silica nanoparticles on clay swelling and aqueous stability of nanoparticle dispersions. J. Nanopart. Res. **16**(1), 2137 (2014)

Soil Stabilization Using Biopolymer: A Study on Its Gelling and Hydration Mechanism to Improve Strength

Evangelin Ramani Sujatha and Saisree Sivaraman

Abstract

Soil improvement and treatment represent a necessity off-late in response to the growing demand for construction spaces. Biopolymers present an eco-friendly alternative material for soil treatment. This study investigates the viability of using guar gum, a polysaccharide for improving the geotechnical properties of clay soil. Guar gum is solvable, non-ionic and hydrocolloidal. Its tendency to form viscous gels, hydrate and form hydrogen bonds helps in improving the stiffness of the soil matrix and thereby enhances its strength. The treated soil behaves like clay in stiff consistency, failing at higher loads in lower strains. Also, the stiffness modulus of the soil-guar mixtures increases significantly. Formation of hydrogen bonds is a function of time, leading to an increase in strength of the treated soil-guar mixtures with age. Though biopolymers are susceptible to decay and degradation, the rate of degradation is minimal and strength gain is observed for the investigated period of 90 days.

Keywords

Biopolymer • Viscosity • Hydrogen bonds • Gel plug • Bio-film

1 Introduction

Soil stabilization is a mandatory pre-construction activity off-late. Many different traditional techniques are used in line with the site conditions, economy and time constraints. But the environmental impacts of methods using admixtures like lime, cement, fly ash and other chemicals are a cause for concern due to the alarmingly high energy consumption related to issues like carbon sequestration, hazardous waste disposal, infrastructure rehabilitation, brown fields clean-up, global warming and water resources protection (De Jong 2010). Biological soil stabilization proves to be a promising technique as it is not only environmentally friendly but also economical and durable (Chang et al. 2016). The biological alternatives are sustainable, carbon-neutral and always renewable, given they are extracted from available crops (Stupp and Braun 1997). Biopolymers are naturally available polymeric molecules extracted from endosperm of seeds, cell walls of plants, exoskeletons of crustaceans, cell walls of fungi. Little research has been carried out on the effect of various biopolymers like xanthum gum, betaglucan, gellan gum, guar gum, cellulose, modified starch and chitosan on the soil modification, mainly to stabilize mine tailings, to control dust, provide erosion resistance on slopes and stabilize sidewalls of trenches against collapse (De Jong 2010; Chang et al. 2016, 2015; Stupp and Braun 1997; Khatami and O'Kelly 2013; Aminpour and O'Kelly 2015; Ayeldeen et al. 2016; Latifi and Majid 2016; Chang and Cho 2014; Puppala and Pedarla 2017). The literature argues that biopolymer treatment has a profound effect on the geotechnical properties of the soil (De Jong 2010; Chang et al. 2016, 2015; Stupp and Braun 1997; Khatami and O'Kelly 2013; Aminpour and O'Kelly 2015; Ayeldeen et al. 2016; Latifi and Majid 2016; Chang and Cho 2014; Puppala and Pedarla 2017). Guar gum shows a better potential for use as a soil stabilizer in different types of soil for various applications. This study, therefore, investigates the possibility of using guar gum to (i) modify the compaction characteristics of the soil, (ii) improve the strength of the soil and (iii) understand the mechanism of soil modification based on its gelling and hydration properties.

E. R. Sujatha (✉)
Centre for Advanced Research in Environment, School of Civil Engineering, SASTRA Deemed University, Thanjavur, 613401, India

S. Sivaraman
Larson & Tubro, Mumbai, India

H. El-Askary et al. (eds.), *Research Developments in Geotechnics, Geo-Informatics and Remote Sensing*, Advances in Science, Technology & Innovation,
https://doi.org/10.1007/978-3-030-72896-0_13

2 Materials and Methods

The clay soil sample used in this study had a grey colour and a pungent odour. The soil was highly compressible with a liquid limit of 58% and a plasticity index of 28%. Therefore, it was classified as a highly compressible silt—clay mixture based on the Unified Soil Classification System. It did not exhibit a tendency to swell and the organic content in the soil was less than 1%. Guar gum is a naturally occurring biopolymer extracted from the guar bean, a polysaccharide that consists of galactose of sugar and mannose. It can hydrate in cold water to produce a highly viscous solution. It exhibits a non-ionic and hydrocolloidal nature (Ayeldeen et al. 2016). Guar gums are stable in solutions over a wide range of pH. The non-ionic, uncharged nature of the macromolecules makes it stable in a wide range of solutions too (El-Daw 1994). The specific gravity of the guar gum used is 1.3.

The soil samples were dried in a thermostatically controlled oven at a temperature between 100 and 108 °C for 24 h. Guar gum was replaced in quantities of 0.5, 1, 1.5 and 2% by weight of soil and was added to the soil in gel form, mixed at the required water content (Chang et al. 2015). Guar gel was left to hydrate for 24 h to avoid the absorption of water from the soil matrix during the initial stages of hydration. Ample care was taken to mix the soil and guar gel to ensure that the gel was coated uniformly over the soil and also the monomeric threads were not broken (El-Daw 1994). The prepared soil-guar mixtures were allowed to sit for at least 6 h in air-tight bags to attain equilibrium moisture content. Geotechnical properties of the soil were determined following procedures outline in IS: 2720. The samples for unconfined compressive strength test were prepared at the optimum moisture content and maximum dry unit weight.

3 Results and Discussion

3.1 Compaction Characteristics

The compaction characteristics of the soil depend on the viscosity of the guar gum gel, which in turn is a function of the guar gum content. The effect of guar gum addition on the compaction characteristics of the soil indicates a minimal change. Compaction curves on the addition of guar gum show a shift towards the left (Fig. 1).

There is a slight change in the maximum dry unit weight, it increases from 15.17 kN/m^3 to a maximum of 15.95 kN/m^3 at 1% guar gum addition and then decreases to 15.47 kN/m^3 at 2% guar gum content. The viscous nature of the guar gel and its nature to hydrate and form hydrogen bonds resist the packing of the soil particles under compaction. The dry unit weight initially increases with the increase in the viscosity of the solution, reaches its peak at a viscosity of approximately 0.47 Ns/m^2 and then decreases for high viscosity solution marginally (i.e.) from 15.95 kN/m^3 at a viscosity of 0.47 Ns/m^2 (1% guar gum concentration) to 15.47 kN/m^3 at 1.52 Ns/m^2 viscosity (2% guar gum concentration) as shown in Fig. 2a. High viscosity solutions tend to push the soil particles apart leading to reduced unit weight.

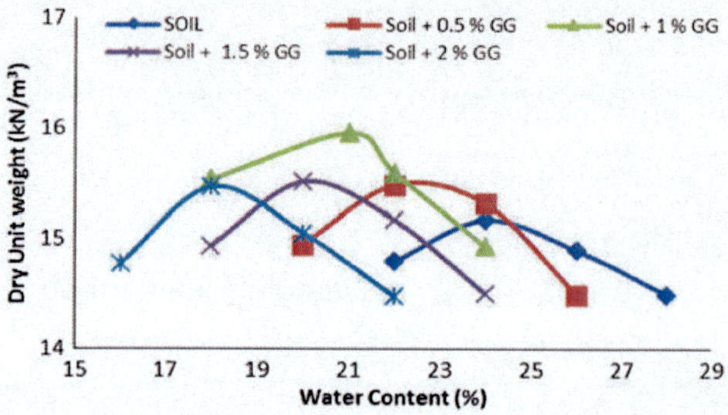

Fig. 1 Compaction characteristics of the guar gum treated soil

The decrease in the optimum moisture content (OMC) can be attributed to the absorption and utilization of water for the hydration and bonding process (De Jong 2010). The water requirement increases with the increase in the guar gum content as the more viscous colloidal dispersions with numerous galactose and mannose branches that strengthen the soil-guar matrix entail more water. Addition of guar gels clogs the void spaces in the soil matrix and reduces the pore sizes (Fig. 2b) as the guar accumulates in the voids spaces, the cross-linking bonds occur. This results in a stiffened soil-guar matrix which resists the change in packing on the application of compaction energy.

3.2 Strength Behaviour

The guar gum treated soil show a considerable increase in the undrained shear strength. It tends to increase with the increase in guar gum content for all the investigated percentages of guar gum (Fig. 3). However, the increase rate is observed to decline when the addition exceeds 1%. The viscosity of the guar gum solutions controls the strength gain mechanism. The ability of guar gum to form high viscous colloidal solutions can also be attributed to the increase in strength. The viscous solution of guar gum makes the particles move away, thereby increasing the adhesion between the soil matrix. This reorientation of the particles is reflected in the increase in the undrained shear strength. Also, as the viscosity increases, the hydrogen bonding becomes stronger

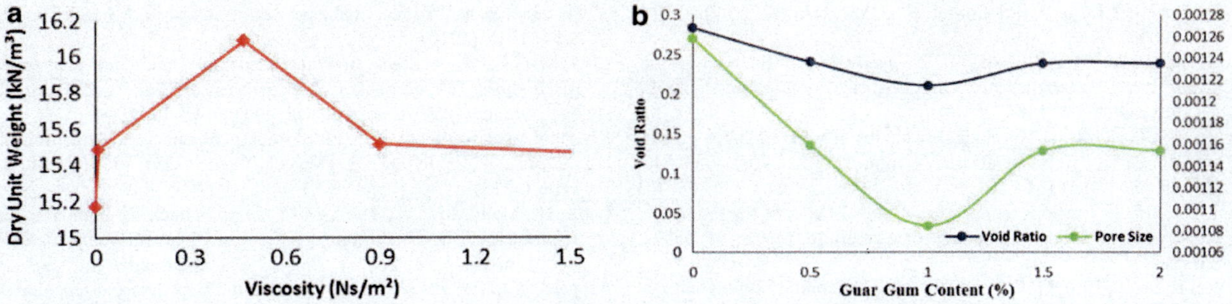

Fig. 2 **a** Effect of guar gum on dry unit weight. **b** Effect of guar gum on void ratio

Fig. 3 Effect of guar gum content on the days of curing and strength

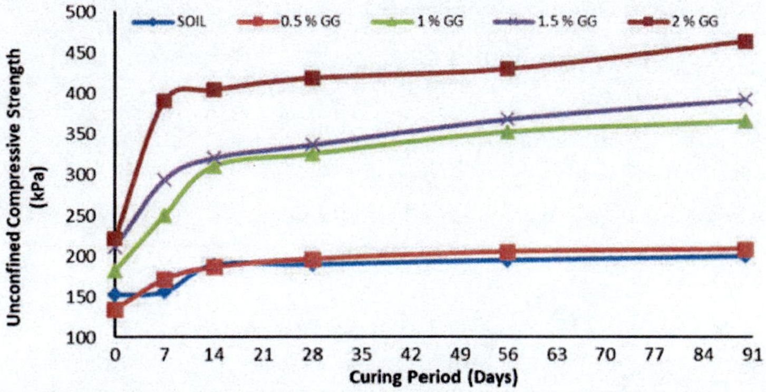

due to the formation of links between the soil particles which resist load better. At 0.5% addition of guar gum, the viscosity is nearly equal to that of water and there is a slight decrease in the undrained strength. But, the viscosity sharply increases beyond 0.5% guar gum and significant strength gain is observed. The slight decrease in strength at 0.5% guar gum addition can be attributed to random orientation of the low viscosity guar gum that is susceptible to undergo constant deformation with the increase in stress. But, with further addition of guar gum, the high viscosity of guar gum results in high shearing resistance. Guar gum is non-ionic and charge neutral polysaccharide with numerous hydroxyl groups. This enables the formation of hydrogen bonds within the soil-guar matrix between the guar gel and clay particles. The bonding inhibits the formation of aggregates and thereby controls the percentage of voids in the treated matrix. As a result, the unconfined compressive strength of the treated soil increases. The strength also depends on the curing period. The strength of the treated matrix depends on the number of links that exist in the pore space. With time, these links become thicker and wider, filling the void space in the soil matrix. Consequently, they become stronger and denser, which enables them to resist greater loads. It is observed that the links, with time, turn to bundle like formations that offer more friction to resist the loads. Maximum strength gain is attained at early days of curing (i.e.) in the seven days of curing. The increase in strength over a period of 90 days indicates that within a soil—guar matrix, the rate of deterioration of the biopolymer is controlled. Also, the water present in the soil—gaur matrix enables the hydration and formation of hydrogen bonds completely.

4 Conclusions

The study shows that guar gum treatment successfully modifies soil properties. This modification primarily lies in the ability of the guar gum to dissolve in water to form highly viscous gels that hydrate and form strong hydrogen bonds able to hold the soil matrix together. This process continues with time rendering the soil-guar matrix stronger and stiffer. Guar gum stiffens the soil matrix improving its toughness, thereby improving its strength and reducing its compressibility. Nevertheless, long term behaviour of guar gum treatment and the stability of guar-soil mixes in aggressive environments must be studied.

References

Aminpour, M., O'Kelly, B.C.: Applications of biopolymers in dam construction and operation activities. In: Proceedings of 2nd

International Dam World Conference, Portugal, vol. 1, pp. 937–946, (2015)

Ayeldeen, M.K., Negm, A.M., El Sawwaf, M.A.: Evaluating the physical characteristics of biopolymer/soil mixtures. Arab. J. Geosci. **9**(4), 371–384 (2016)

Chang, I., Cho, G.-C.: Geotechnical behavior of a beta-1,3/1,6-glucan biopolymer-treated residual soil. Geomech. Eng. **7**(6), 633–647 (2014)

Chang, I., Im, J., Prasidhi, A.K., Cho, G.-C.: Effects of xanthan gum biopolymer on soil strengthening. Constr. Build. Mater. **74**, 65–72 (2015)

Chang, I., Im, J., Cho, G.-C.: Introduction of microbial biopolymers in soil treatment for future environmentally-friendly and sustainable geotechnical engineering. Sustainability **8**(3), 251–273 (2016)

De Jong, J.T., Soga, K., Banwart, S.A., Whalley, W.R., Ginn, T.R., Nelson, D.C., Mortensen, B.M., Martinez, B.C., Barkouki, T.: Soil engineering in vivo: harnessing natural biogeochemical systems for sustainable, Multi-functional engineering solutions. J. R. Soc. Interf. **8**(54),1–15 (2010)

El-Daw, G.E.-A.: A Study of Guar Seed and Guar Gum Properties (Cyamopsis Tetragonolabous), Department of Food and Technology, University of Khartoum, vol. 17, pp. 21–13 (1994)

Khatami, H.R., O'Kelly, B.C.: Improving mechanical properties of sand using biopolymers. J. Geotech. Geoenv. Eng. **139**(8), 1402–1406 (2013)

Latifi, N., Majid, M.Z.A., Tahir, M.M., Horpibulsuk, S.: Improvement of problematic soils with biopolymer—an environmentally friendly soil stabilizer. J. Mater. Civ. Eng. **29**(2), 16–24 (2016)

Puppala, A.J., Pedarla, A.: Innovative ground improvement techniques for expansive soils. Innov. Infrastruct. Solut. **2**(24), 21–37 (2017)

Stupp, S.I., Braun, P.V.: Molecular manipulation of microstructures: biomaterials, ceramics, and semiconductors. Sci. N. Y. **277**(5330), 1242–1248 (1997)

Use of Soil Densification Process to Resolve Soil Instabilities in Chlef City (Algeria)

Zohra Boutaraa, Ahmed Arab, Mohammed Chemmam, and Abed Elkadder Brahimi

Abstract

In this work, soil densification using a dynamic compaction process is proposed to solve soil instabilities in Chlef city (Algeria), which is a prone zone to seismic risk. In this respect, a laboratory investigation is conducted using the triaxial apparatus to test the efficiency of the use of densification process on the improvement of Chlef silt-sandy soil's mechanical properties. Thus, soil samples with various relative densities D_r are tested in drained and undrained conditions and under both static and cyclic loading. Results show a reduction of the soil deformations and an increase of the shear strength versus of D_r. Therefore, this process represents an efficient tool that can be used by the local authorities to solve soil deformations due to settlement and liquefaction, especially in road pavements adjacent to Chlef River.

Keywords

Soil densification • Relative density • Deformations • Shear strength • Chlef city

1 Introduction

Soil densification processes allow the resolution of soils deformations, especially in seismic zones. This technique consists of increasing the soil relative density D_r using for instance dynamic compaction, and gives positive results in thin saturated sandy soil layers. It has the main advantages of increasing the bearing capacity and of reducing the soil settlement and liquefaction potential (Feng et al. 2013) and of increasing the friction angle and the elastic modulus of granular soils (Bo et al. 2014).

In Chlef city (formerly known by Al Asnam), which is an area prone to seismic risk, several soil instability problems caused by settlement, especially on national express roads (RN4 and RN19) pavement (> 35.000 Vehicles/day), and by liquefaction phenomenon that can occur in the case of seismic event need to be treated. To this end, this study investigates the effect of the relative density that can be increased using densification process on the behavior of Chlef sandy soil under drained and undrained conditions and using both static and cyclic loading. The soil in Chlef city is an alluvial Quaternary deposit with low compactness, mainly formed of sand, gravel, and clay. The superficial layers have low shear wave velocity (230–500 m/s) and are classified as soft soils by Algerian seismic code (RPA 2003). Note that several experimental and numerical studies were carried out to study Chlef sandy soil behaviour under monotonic and cyclic loadings without reinforcement or using different reinforcement processes, such as geosynthetic and densification (Chemmam et al. 2015; Arab et al. 2014; Boutaraa et al. 2019).

2 Study Methodology

In order to highlight the soil properties improvement after densification, a laboratory study is carried out using a triaxial apparatus with a control strain device. Soil samples, constituted by sand and silt which densities are respectively equal to 268 and 2.7 g/cm^3 are reconstituted and saturated before testing. The saturation degree significantly affecting the soil liquefaction resistance (Arab et al. 2014) is evaluated by measuring the Skempton's coefficient B after consolidation (Skempton 1954). Three soil states (loose,

Z. Boutaraa (✉) · A. Arab
Hassiba Benbouali University, 02000 Chlef, Algeria
e-mail: z.boutaraa@univ-chlef.dz

M. Chemmam
Ahmed Zabana University, 48000 Relizane, Algeria

A. E. Brahimi
Abdelhafidh Boussouf University, 43000 Mila, Algeria

H. El-Askary et al. (eds.), *Research Developments in Geotechnics, Geo-Informatics and Remote Sensing*, Advances in Science, Technology & Innovation,
https://doi.org/10.1007/978-3-030-72896-0_14

medium, and dense) and several relative densities (D_r = 10, 18, 50, 65, and 80%) are experimented. The under-compaction method is used to prepare samples (Ladd 1978). Cylindrical specimens having an H/D rapport of 1 (70 mm in height and 70 mm in diameter) are used. The sample density is defined by the relative density parameter D_r given by Eq. (1):

$$D_r = (e_{\max} - e)/(e_{\max} - e_{\min}) \quad (1)$$

where $e_{\min}$ and $e_{\max}$ are respectively the minimum and the maximum void ratio, and e is the target void ratio in place at the state of natural compactness.

3 Results

3.1 Relative Density Effect on Shear Strength in Drained Monotonic Conditions

At first, the D_r effect on the soil shear strength in drained monotonic conditions is studied. Figure 1 shows the soil stress deviator q (Fig. 1a) and the volumetric strain E_v (Fig. 1b) evolutions versus of ε for three D_r values (10, 65 and 80%).

3.2 Relative Density Effect on Shear Strength in Undrained Monotonic Tests

Secondly, the effect of D_r on the soil shear strength in undrained conditions is investigated.

Figure 2 shows the stress deviator q evolution that characterizes the soil shear strength (Fig. 2a), and the pore pressure u evolution (Fig. 2b) of the studied soil, versus to the samples axial deformation ε for three D_r values (18, 50 and 80%).

3.3 Relative Density Effect on Cyclic Shear Strength

The last investigation stage concerns the D_r effect on the soil cyclic behavior. Figure 3 shows the results of undrained cyclic tests conducted for an amplitude value of 50 kPa and for a D_r used values equal to 10, 18, 50, 65, and 80%.

4 Discussion

The laboratory investigation results allow us to note the following remarks:

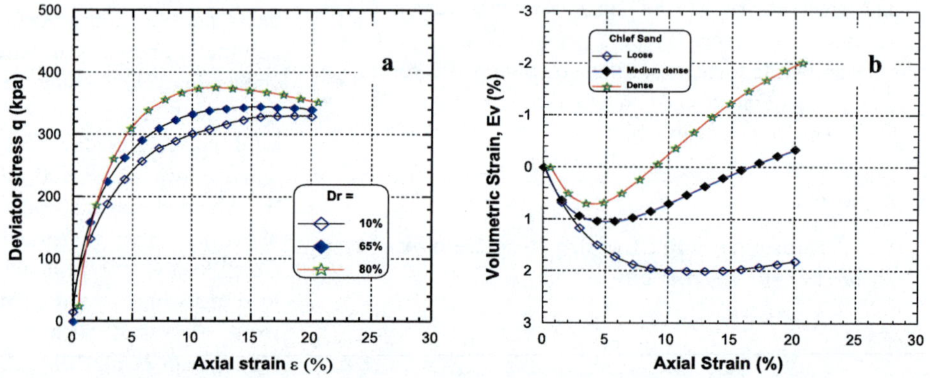

Fig. 1 Soil drained behavior: stress deviator (**a**) and volumetric strain (**b**) evolutions

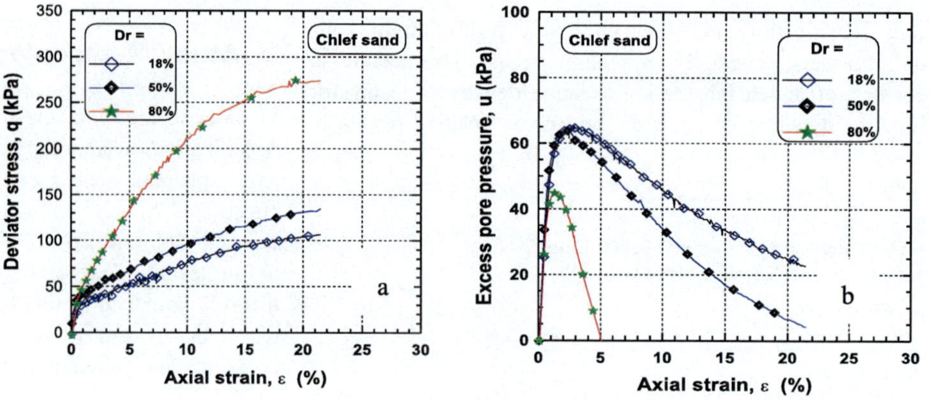

Fig. 2 Soil undrained behavior: stress deviator (**a**) and pore pressure (**b**) evolutions

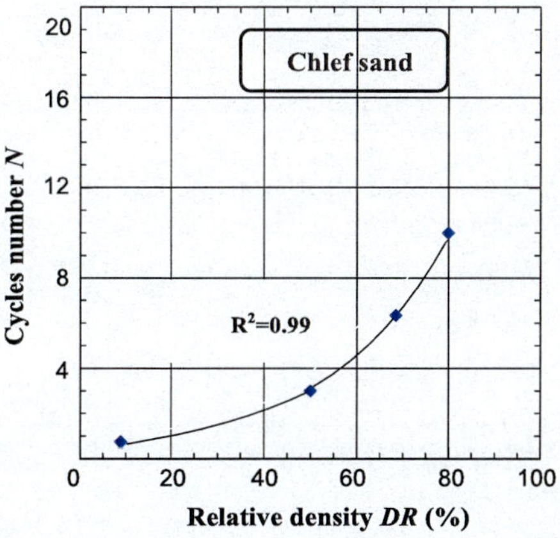

Fig. 3 Number of cycles necessary to produce soil liquefaction for the different D_r values

First, from Fig. 1a, it can be clearly noticed that the stress deviator q characterizing the soil shear strength increases versus of D_r. The soil resistance reaches the maximum values of 320, 340, and 390 kPa for D_r = 10, 65, and 80%, respectively. It is found that the dense sample (D_r = 80%) shows a peak in the stress deviator at 10% of axial deformation ε_1, reflecting a maximum soil dilatancy rate. Figure 1b reveals a contracting behavior of the loose simple that reaches the maximum volumetric deformation E_v, while medium dense and dense samples behave differently. Indeed, after a contracting behavior at the beginning and up to a value of axial strain ε = 5% and 2.5% for the medium dense and the dense sand samples respectively, both samples show a dilatant character, which is more pronounced for the latter. These findings highlight the positive effect of D_r increasing the shear strength and on the reduction of the volumetric strain in drained conditions.

In undrained conditions, the deviator stress q is higher for the dense sample compared to those of the two others (Fig. 2a). For ε = 21%, the shear strength of the loose sample reaches a maximum value of only 110 kPa, while the medium dense sample reaches a resistance of 135 kPa for the same value of ε. Finally, the dense sample grasps a maximum q value equal to 270 kPa. These findings reveal the positive effect of D_r increase on the shear strength in undrained conditions. In Fig. 2b, the three pore pressure curves are experiencing an increase to maximum values of: 45, 65, and 67 kPa for D_r values equal to 80, 50, and 18%, respectively. Then the curves decrease gradually with a much more pronounced decrease of the dense sample, for which ε = 5% and is at least, four times less than ones of the other two samples (ε = 21%). The sudden decrease of u in the dense sand sample is due to the very dilating nature of this soil. It is clear that u is inversely proportional to D_r, indicating the positive effect of increasing D_r on the undrained behavior of the soil. The excessive increase of u can induce the sample liquefaction. However, with a high value of D_r the sample rupture can be avoided.

Finally, under cyclic loading, the soil liquefaction is reached quickly for the loose sample (D_r = 15%), whereas the dense sample (D_r = 65%) took a relatively higher cycle's number. Liquefaction is reached at 1, 5, and 12 cycles for loose, medium dense, and dense samples, respectively (Fig. 3). The increase of D_r delays significantly the increase of u and consequently, the cancelation of the effective stress.

5 Conclusions

This paper presented the main results of a laboratory study concerning the influence of the relative density on the behavior of Chlef sandy soil. The tests results showed that the soil shear strength increased significantly versus the relative density D_r. Besides, the decrease of soil deformations was proportional to the increase of D_r.

These laboratory investigation findings indicated the positive effect of the increase of the relative density on the soil mechanical proprieties improvement in drained and undrained conditions and both under static and cyclic loadings. Therefore, it is safe to conclude that soil densification using a dynamic compaction process could be used as a valuable tool to resolve soil deformation problems in Chlef city due to settlement and to liquefaction in the case of a seismic event, mainly on the RN4 and the RN19 express roads pavements adjacent to Chlef River.

References

Arab, A., Sadek, M., Belkhatir, M., Shahrour, I.: Monotonic preloading effect on the liquefaction resistance of Chlef Silty Sand: a laboratory study. Arab. J. Sci. Eng. **2014**(39), 685–694 (2014)

Bo, M.W., Arulrajah, A., Horpibulsuk, S., Leong, M., Disfani, M.M.: Densification of land reclamation sands by deep vibratory compaction techniques. J Mater Civ Eng **26**(8) (2014). https://doi.org/10.1061/(ASCE)MT.1943-5533.0001010

Boutaraa, Z., Arab, A., Chemmam, M.: Effect of the geotextiles reinforcement area on the stabilization of soil deformations in Chlef city (Algeria). In: Proceedings of the 3rd International Conference Smart Underground Space & Infrastructure, October 8–10, Lille, France (2019)

Chemmam, M., Arab, A., Belkhatir, M., Bouferra, R.: Behavior of loose silty sand of Chlef River: effect of low plastic fine contents and other parameters. Mar. Georesour. Geotechnol., 1–11 (2015). https://doi.org/10.1080/1064119X.2015.1014983

Feng, S.J., Tan, K., Shui, W.H., Zhang, Y.: Densification of desert sands by high energy dynamic compaction. Eng. Geol. **157**, 48–54 (2013)

Ladd, R.S.: Preparing test specimens using under compaction. Geotech. Test. J. ASTM **1**(1), 16–23 (1978)

RPA: Règles Parasismiques Algériennes. Centre National de Recherche Appliquée en Génie Parasismique, Alger (2003)

Skempton, A.W.: The pore pressure coefficients A and B. Géotechnique **4**(4), 143–147 (1954)

Experimental Determination of Pullout Rate Reinforcements of Reinforced Earth Structures

Med Salah Laouar, Adel Djellali, Rafik Boufarh, Salah Messasset, and Abdelkader Houam

Abstract

Soil reinforcement techniques by reinforced earth or geosynthetics allow the building of flexible and aesthetically more attractive structures, faster and at a lower cost than conventional retaining structures made of stone or reinforced concrete. In this respect, this research aims to present an experimental device designed to simulate the behavior of reinforcements under static and cyclic loadings. Galvanized reinforcement smooth and ribbed metal strips are tested and different load scenarios are used in order to make a distinction between the behaviors of each type of reinforcement and to study the reinforcement behavior under the effect of static, cyclic, and cyclic loading succeeding static. As for the fill material used, it respects the recommendations of the company of reinforced earth. The findings show that the reliability of the test device used allows to understand the effect of the loadings applied on each type of reinforcement and the corresponding pulling out rate and thus demonstrate the complexity of the reinforcing elements' response to repeated loads.

Keywords

Reinforcement • Friction • Reinforced earth • Pullout • Apparatus

M. S. Laouar (✉) · A. Djellali · R. Boufarh
Civil Engineering Department, Larbi Tebessi University, 12002 Tebessa, Algeria
e-mail: medsalah.laouar@univ-tebessa.dz

S. Messasset
Civil Engineering Department, University 20 Aout 1955, 21000 Skikda, Algeria

A. Houam
Civil Engineering Department, College of Engineering Abha, KSA, King Khalid University, Abha, Saudi Arabia

1 Introduction

The construction of structures on reinforced earth does not require scaffolding or heavy construction equipment. In addition to the conventional earthwork equipment used to set up and compact the embankment, the steel siding can be set up manually. As for prefabricated concrete facings, a light crane is sufficient for their handling. The combination of backfill and metal or geosynthetic reinforcements builds up a composite material whose ductility, stability, and flexibility have made it the preferred solution for supporting structures. The interaction between the embankment and the reinforcement allows efficient absorption of vibrations, such as those generated by train convoys or heavy vehicles. Hatami (2005) tested walls reinforced with geosynthetic strips. He observed that the distribution of the tensile forces along the reinforcements as a function of the depth of the wall was trapezoidal and not linear as assumed in the simplified methods. The strength and stability of the composite structure, therefore, provide an important bearing capacity, provided a sufficient mobilization of the ground-reinforcement friction.

2 Description of the Experimental Device

The experimental model consists of a tank made of sheet steel of 6 mm thick mechanically welded, parallelepiped shape and of dimensions: Length = 1.00 m, Width = 0.20 m, Depth = 0.24 m. The lid of the tank also made of sheet metal of 6 mm, must receive in its lower part an inflatable air chamber. The upper side should contain an inflation valve, a 4 bar pressure gauge, a safety valve, and a water supply valve. The orifices are made by appropriate drilling of the metal sheet, Fig. 1. The tank is intended to be filled with sand coating an armature and must be firmly bolted to its cover before testing. The apparatus is equipped with a system for exerting pulling out efforts and as many resuming

H. El-Askary et al. (eds.), *Research Developments in Geotechnics, Geo-Informatics and Remote Sensing*, Advances in Science, Technology & Innovation,
https://doi.org/10.1007/978-3-030-72896-0_15

moments per minute. The displacement of the armature, or possibly its rupture under the effect of pulling out will be observed for pressures and unevenly variable pulling out forces. The measurements of the displacements of the reinforcement are carried out with a comparator of sensitivity of 1/1000 mm.

3 Materials, Reinforcements, and Methods

The material used is the same as that used for the construction of real reinforced earth structures; it is sand consisting of silica grains. The specific gravity of the solid grains $\gamma_s = 2.65$ kN/m^3. The dry specific weight ranges from $\gamma_d$min = 14.34 kN/m^3 to $\gamma_d$max = 17.84 kN/m^3. Minimum and maximum void ratio range from = 0.48 to 0.88, the internal friction angle φ is 36°.

The armatures used as reinforcement are strips cut from sheets of high-strength galvanized steel, of dimensions 1000 × 30 × 0.8 mm, they have an elastic limit of 2400 kg/cm^2, and permissible stress of 1600 kg/cm^2. The justification method used is that of the current French standard for structures reinforced with soft and inextensible reinforcement (NF P 94 220-0-1998). The ribbed reinforcements have ribs perpendicular to their axis on both sides, made by fixing ribs by weld points on a smooth armature. The dimension of each rib is 3 × 0.8 mm; their distribution on each armature face follows the distribution of Schlosser (1993). The armature is fully embedded in the tank between two perfectly horizontal grained soil beds and passes through the lateral face of the tank through an opening of dimensions greater than the cross section of the armature. In order to understand the effect of static and cyclic loading on the behavior of smooth and ribbed reinforcements and the rate pullout of each type of reinforcement; the experimental program includes three series of tests: application of static loading, application of the cyclic loading, and application of the static load after cyclic loading. In all tests, the confining pressure applied on the top of the reinforcement takes the same value equal to: 0, 25 Bars.

4 Results and Discussion

4.1 Behavior of Reinforcements Under Static Loading

Concerning the behavior of smooth and ribbed reinforcements under static loading, Fig. 2, the load-displacement curves, show that for a pulling out force of less than or equal to 0.4 kN, the reinforcements are characterized by stability followed by sudden pullout. The ribbed armatures are characterized by slow gradual displacements until rupture.

4.2 Behavior of Reinforcements Under Cyclic Loading

The study of the behavior of reinforcements under cyclic loading suggests that the displacements are proportional to the increase in the number of cycles, Fig. 3. For a loading amplitude of 30%, the smooth reinforcement remains stable, a small displacement of 0.035 mm recorded after 10,000 cycles. As for large reinforcements, the smooth reinforcement is characterized by the first phase of stability, followed by a sudden rupture. The relation number of cycles-displacements for the ribbed reinforcement shows two forms of curves: The first corresponds to the amplitude of loading of 30% with weak displacements; despite a large number of cycles applied the reinforcement does not pull out. The second, relative to the amplitudes 50%, 60%, and 80%, is characterized by gradual displacements. The number of cycles corresponding to the pullout decreases with the increase of the load amplitude.

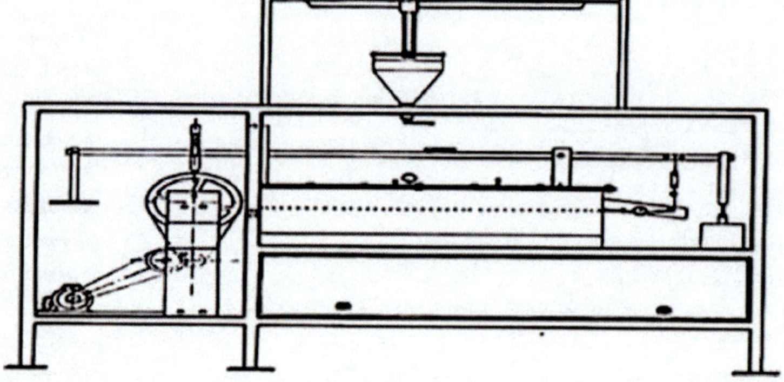

Fig. 1 Experimental device «pull-out apparatus»

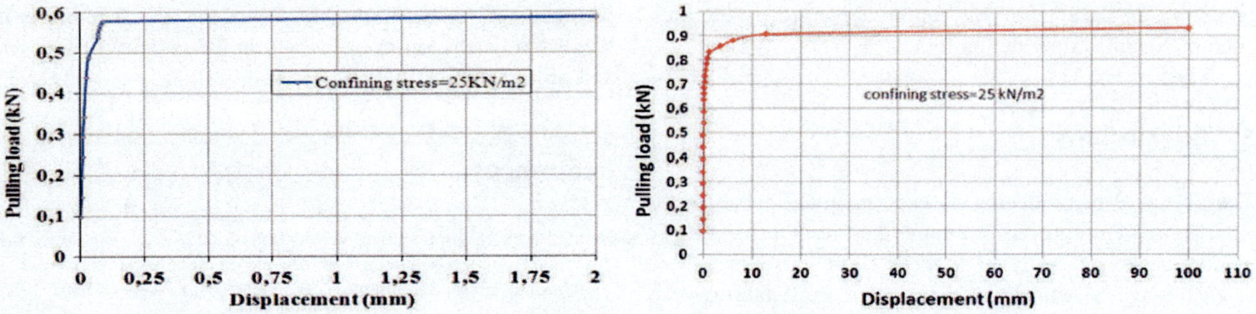

Fig. 2 Static loading test: Smooth reinforcement. Static loading test: Ribbed reinforcement

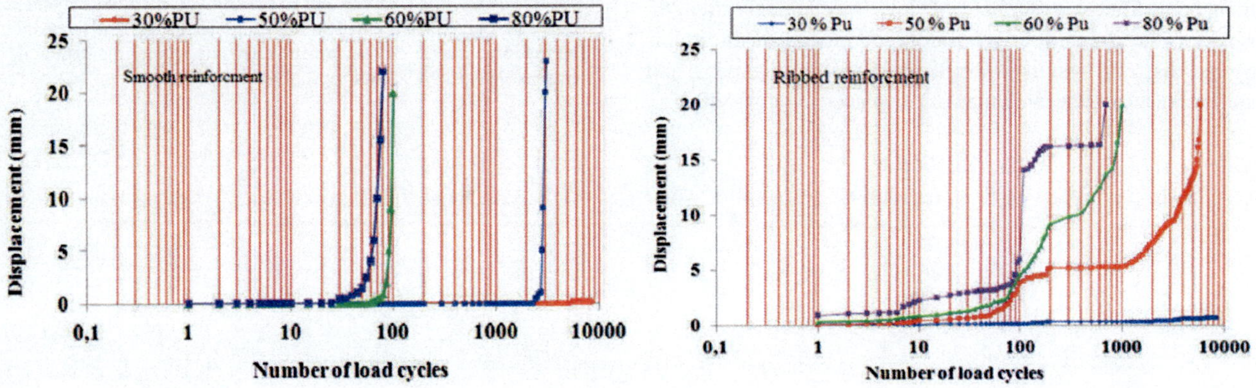

Fig. 3 Displacement versus number of load cycles

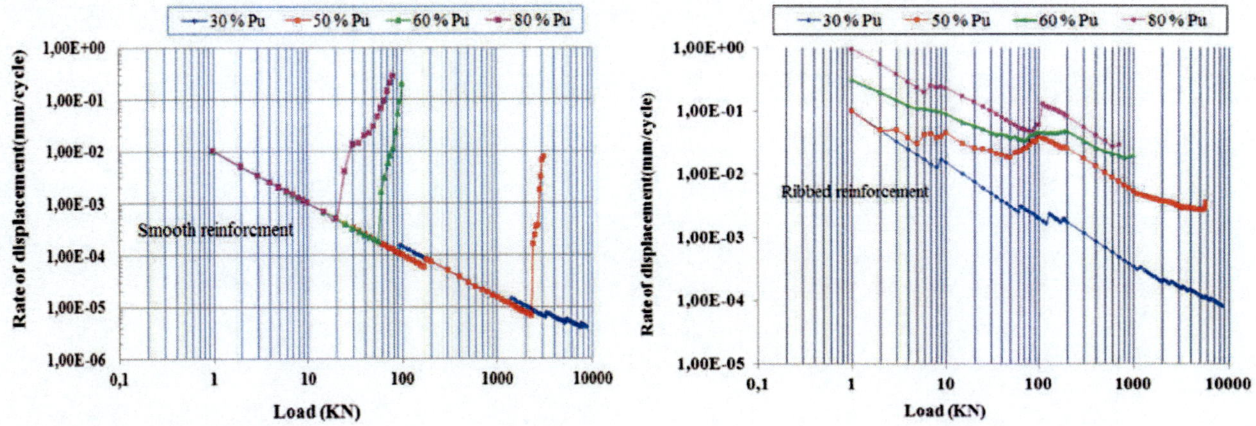

Fig. 4 Rate of displacement versus load cycles

4.3 The Rate of Tearing of the Reinforcement

To better interpret the behavior of reinforcements under repeated loadings, a presentation giving the rate of displacement according to the number of cycles is proposed. According to Fig. 4, the smooth reinforcements remain stable for loading amplitude of 30%, which is justified by the inverse proportionality between the pulling out rate and the number of loading cycles. For the other loading amplitudes, the smooth reinforcements are characterized by a decrease then increase of the rates of pullout with the increase of cycle of loading; so stability followed by abrupt pullout.

The curves of the ribbed reinforcements have almost the same pace, they fit into each other and go through gradual phases of displacement. This can be explained by the presence of ribs that slow the pullout.

5 Conclusions

The objective of producing an experimental device for the simulation of reinforced structures has been achieved. To avoid mistakes made by the models, efforts have been made at each stage of equipment realization. Repeatability tests confirm the reliability of the tests. The reason for different behaviors lies in the design state of each type of reinforcement. Indeed, the perfect flatness of the smooth reinforcement constitutes a surface of low mechanical strength. Besides, the ribbed reinforcement tends to retain the grains by its ribs. This results in displacement, rearrangement, and an increase in the volume of displaced grains, dilation phenomenon caused in the vicinity of the ribs. Therefore, it leads to a densification of the soil in the vicinity of the ribs and an increase in friction.

References

Hatami, K.: Development and verification of a numerical model for the analysis of geosynthetic reinforced soil segmental walls under working stress conditions. Can. Geotech. J. **42**(4), 1066–1085 (2005)

https://www.boutique.afnor.org/norme/a05-252/corrosion-par-les-sols-aciers-galvanises-ou-non-mis-au-contact-de-materiaux-naturels-de-remblai-sols/article/755257/fa021801

Schlosser, F., Hoteit, N., Pierce, D.: Expérimentation en vraie grandeur d'un mur Freyssisol-Websol en sol renforcé. In: C. R. Symposium International sur le Renforcement des Sols. Expérimentation des années 1980, ENPC, Paris, pp. 299–320 (1993)

Effect of Dune Sand on the Swelling Soils Stabilization; Case Study: The Region of Adrar (Algeria)

Khedidja Belaidi, Belkacem Mekerta, Mohamed Benabdelfattah, and Mohamed Amine Allal

Abstract

Clayey soil is the foundation for many buildings. Construction on this kind of soil requires a good companion for the recognition of identification and characterization of their swelling potential. The present research focuses on the addition of dune sand to stabilize this phenomenon. The choice of this material is justified by its availability at a local level, in the region of Adrar, thus, the low cost. Several tests are carried out for the identification of expansive soil in the region of Adrar in order to stabilize it and asses the influence of dune sand in different percentages. Finally, it can be concluded that free swelling decreases with increasing percentage of sand.

Keywords

Clay • Dune sand • Swelling • Proctoroptimum • Stabilization

1 Introduction

The phenomenon of swelling is responsible for severe damage and disturbances around the world. Several solutions are proposed as remedies, among them: chemical, mechanical, or physical stabilization. In order to limit disturbances in construction, several studies have been carried out on this phenomenon, and solutions based on various stabilization techniques have led to reduced swelling potential.

The study presented is a contribution to the technique of improvement and stabilization by adding dune sand. Research on the matter suggest that this stabilizer has a key role in reducing the parameters of swelling. Accordingly, several studies confirm that sand addition has a positive impact on the stiffness in compressible soils. However, several authors studied the influence of the sand addition on swelling (Mououx, Didier, El Sayed El Sohby; Kaoua, DerricheetLaradi; Bengraa et al.; Lamara et al.), and deduced that the addition of sand reduced the plasticity of the clay-sand mixture, and consequently its potential for swelling. Among the various kinds of soil stabilization, the present study focuses on one kind of expansive soil stabilization. The purpose of this work is to assess the effects of adding dune sand to this soil, aiming to offer a general understanding of the physicochemical and mechanical characteristics of a particular clay, abundant in the Adrar region.

2 Materials and Methods

2.1 Site Location

Two local materials were investigated in this work: clayey soil and the dune sand. The soil under study is an expansive soil from southwestern Algeria known as Adrar clay.

2.2 Clayey Soil Identification

The physical properties of the soil are given in Table 1. According to the classification of minerals, our clay can be put in the class of the Illite. According to classification

K. Belaidi (✉) · B. Mekerta · M. Benabdelfattah
Ahmed Draia University, 01000 Adrar, Algeria
e-mail: khedidja.belaidi@univ-adrar.edu.dz

B. Mekerta
e-mail: bel.mekerta@univ-adrar.edu.dz

M. Benabdelfattah
e-mail: Benabdelfattah_med@univ-adrar.dz

M. Benabdelfattah
LAE Laboratory of EPAU, LDDI Laboratory of Adrar, Adrar, Algeria

M. A. Allal
Aboubekr Belkaid University, BP 230, Tlemcen, Algeria

H. El-Askary et al. (eds.), *Research Developments in Geotechnics, Geo-Informatics and Remote Sensing*, Advances in Science, Technology & Innovation,
https://doi.org/10.1007/978-3-030-72896-0_16

LCPC and USCS, the clay of Adrar is a very argillaceous CH (At) and very plastic. Its activity (Ac) is normal. Total specific surface is deduced from the test to the methylene blue by the formula of Tran Ngoc Lan. The values obtained for clay (199.5 m^2/g) are high but correspond to the usual values of Illite L.C.P.C. According to the classification of Williams and Donalds, the clay of Adrar has a potential of very high swelling.

2.3 Dune Sand

The dune sand is used as an addition stabilizer with various percentages. From these results (Table 1), and according to the classification of soils LCPC (central laboratory of the High ways Departments) based on the coefficients of uniformity and curve, the dune sand is classified among clean sands and graduated good symbolized by the letters Sb, its curve is uniform and is spread out. The value with blue confirms that our sand is a soil muddy sand and sensitive to water (Table 2).

3 Results and Discussion

3.1 Effect of Sand on Proctor Parameters

The modified Proctor tests determine two principal parameters: the maximum dry voluminal weight (g_{dmax}) and water content optimal (w_{opt}). The results of these tests are indicated in Table 3. The clayey soil was dried in an oven at 105 °C during 24 h before being grinded. To prepare the mixtures of clayey soil and sand, the required amounts of clayey soil and sand were first measured and mixed together. Secondly, the clayey soil and the sand were mixed with the required amount of water to reach the modified Proctor optimum (MPO) water content of the clay-sand mixture, and the samples were compacted to the corresponding maximum dry density (Table 3). Mixing was done manually, and proper care was taken to prepare homogeneous mixtures.

From Table 3, we see a small decrease in dry density en function of sand percentage (except 10%).

3.2 Stabilization Swelling by Sand

During the free swelling test, the sample should be subjected to a low pressure corresponding to the weight of the piston and porous stone and left in contact with water at atmospheric pressure. The potential or (rate) of swelling corresponds to the relative variation of the volume or the variation of the height G% = (H_f/H_0 in %).

Same procedure is used to study the swelling of the samples in the presence of sand. The reduction of final swelling $\Delta G/G_0$ expressed in % is the difference between swellings with water G_0 final swelling in the presence of water G_0. The results of the tests of the potential of swelling are summarized in Table 4.

Figure 1 shows the graphic representation of the change in the rate of swelling as a function of time for clay-sand

Table 1 Physical parameters of the studied clayey soil

Samples	γs KN/m³	LL %	PL %	PI %	SL %	SI %	≤ 2 µm	Ac	MO %	V_B %	TSS M²/g	γ_{dmax} KN/m³	W_{opt} %
Clay	26.5	70	25.49	44.51	10.09	59.91	40	1.113	0.086	9.5	199.5	17.40	14.33

Table 2 Physical parameters of the studied dune sand

Samples	≤ 80 µm	≤ 2 µm	Cu	Cc	Esv	Esp	V_B	SST m²/g	MO
Sand	2	0	4	0.56	39.99	35.22	0.2	4.2	0.043

Table 3 Effect of dune sand on Proctor parameters

% Sand	0%	2%	4%	6%	8%	10%	12%
γ_{dmax} (g/cm³)	1.74	1.73	1.73	1.66	1.68	1.74	1.73
W_{opt} (%)	14.33	15.26	13.83	20.52	17.73	17.67	14.71

Table 4 Swelling results according to the clay + sand mixture oedometric test

% Sand	0%	2%	4%	6%	8%	10%	12%
G%	29.7	25.2	21.55	20.5	23.5	24	20.1
(ΔG/G) %	/	15.15	27.44	19.19	21.88	31.64	32.32

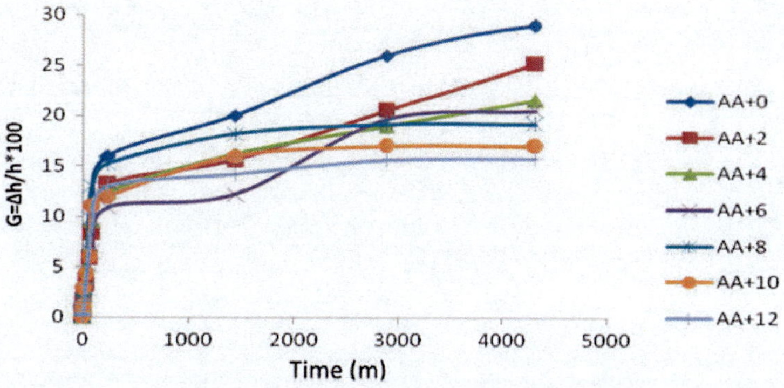

Fig. 1 Evolution of the free potential of swelling according to time for the mixture of Adrar clay + sand of dune

mixtures. The development of swelling over time is the same for all mixtures. However, as the percentage of sand increases, so does the potential for swelling. With a small percentage, 2% of sand, the reduction of swelling is almost remarkable after one will have a weak reduction. While, with a strong percentage of 12% sand, the reduction in swelling is about 32.32%.

4 Conclusions

The stabilization of the clay soils was largely investigated. Many methods and equipment were developed in an attempt to minimize the phenomenon of swelling. In this respect, the resent study came to show the possibility for improvement and valorization of the mechanical characteristics of local materials by techniques of compaction. The findings of soil analyses allowed to note that the improvement by the sand addition to percentages of 2.4, 6, 8, 10, and 12%, modified the physical characteristics considerably, and mechanical as well as classification. The results also showed an increase in the water content optimal, and a remarkable reduction in the optimum dry voluminal weight according to the percentage of sand with addition. The sand of dune decreased the potential of swelling. Furthermore, this reduction was proportional to the added percentages of sand. For future work, it is recommended to determine the possibility for improvement and valorization of the mechanical characteristics of other local materials of the area of Adrar like the Clay of Timimoun, Regane, etc.

Bibliography

Bengraa, L., Hachichi, A., Bourokba, S.A., Benaissa, A.: Etude de la stabilisation des argiles gonflantes par ajout de sable de carrière. In: 2éme Journée d'études sur les sols gonflants, pp. 101–112. Universitéde Tlemcen, Algérie (2005)

Didier, G.: Gonflement Cristallin et Macroscopique des Montmorillonites sa Prévision. PhD these. Publisher, Université Claude Bernard, Lyon (1972)

El-Sobhy, M.A., El-Sayed, A.: Some Factors Affecting Swelling of Clayey Soils. PhD these. Faculty of Engineering Department, University of Al-Azhar, Cairo, Egypt (1981)

Kaoua, F., Derriche, Z., Laradi, N.: Contribution à l'étude de la stabilisation des sols gonflants par ajout de sable. Algérie Equipement, Revue Technique de l'Ecole NationaldesTravauxPubliques **15**, 12–15 (1994)

Lan, T.N.: Nouvel essai d'identification des sols : l'essai au bleu de méthylène. Bulletin de Liaison des Ponts et Chaussées **88**, 136–137 (1977)

Lamara, M., Gueddouda, M.K., Benaissa, A., Hachichi, A.: Stabilisation des sols gonflants par ajout de sable de dune. 2éme Journée d'études sur les sols gonflants. In: 2éme Journée d'études sur les sols gonflants, pp. 129–139. Universitéde Tlemcen, Algérie (2005)

Mouroux, P.: Contribution à l'étude des problèmes de retrait et de gonflement des sols. PhD these. Faculté des Sciences de l'Université de Grenoble (1969)

Williams, A.B., Donaldson, C.W.: Developments related to building on expansive soils in South Africa. In: Proceedings of 4th international conference on expansive soils, pp. 834–844. Denver (1980)

Study of the Fracturing of the Kissir Dam Site and Risks of Water Leakage—Jijel, North-East of Algeria

Mustapha Tekkouk, Riad Benzaid, and Chahra Yellas

Abstract

Jijel region, known for its Mediterranean climate and huge reserves of water resources, has been chosen to be the location of the—currently in operation—Kissir Dam which will serve to mobilize a volume of 68 Hm^3 of the water of the Oued eponym. This study aims to contribute to the understanding of the hydrogeological conditions of the site that no previous work has realized yet. To this end, the fracking of the site is investigated and the data of various tests that the companies performed is treated. Therefore, the present work is expected to identify areas that could cause water leaks and this, by lack of perfect sealing of the studied site.

Keywords

Kissir Dam • Dam plate • Fracking • Water testing • Sealing • Water leaks

1 Introduction

The reality of the phenomenon, several times reported, about the sealing of hydraulic reservoirs is well established. The studies carried out so far have shown that this is a domain where geological and geotechnical studies—not ignoring any possibility—are required, seeing as some simplistic rules applied do not guarantee sufficient safety. It is in this spirit that we have undertaken our work. The different points treated highlight the importance of knowing the fracking state of the dam site and the risks it may cause once the impoundment of the dam is launched.

2 Geographic Location

In the coastal zone, about ten kilometers west of the city of Jijel, the earth dam Kissir was built (Fig. 1). The 68 Hm^3 of water from the eponymous Oued that is mobilized is used for drinking water supply and irrigation.

3 Geological Formations

Of all the civil works, the dams request in the most complex and complete ways the natural environment (soils, rocks, water) on which they are established. Geology is therefore a key factor in the assessment of the conditions of realization of these works, and the geologist must be a full partner of the studies.

From south to north, all along its course, the Kissir Oued crosses the molasse and various geological formations (Fig. 2) that compose it. Those are as follows (Tekkouk 2019a, b).

- The molasse: Aquitano–Serravalian age, also called OMK (Fig. 3a); the molasse contains blocks of various origins and sizes, conglomerates, and sub-Numidian clays (Tekkouk 2019a, b; Boulkhiout and Bouchemella 2006; Delga 1955).
- Elements of the molasse. The Mauritian flysch (Fig. 3a) or flysch of Guerrouch, the flysch gréso-micacé (Fig. 3a), the flysch Numidian (Fig. 3b) (Tekkouk 2019a, b; Delga 1955).

4 Fracturing

4.1 Fracturing of the Dam Site

Three families of NE–SW, SE–NW, and E–W faults affect the Kissir region. The plate of the dam forming part of the

M. Tekkouk (✉) · R. Benzaid · C. Yellas
Laboratoire de Génie Géologique (LGG), University of Jijel, BP. 98, Ouled Aissa, 18000 Jijel, Algeria
e-mail: mustapha.tekkouk@univ-jijel.dz

H. El-Askary et al. (eds.), *Research Developments in Geotechnics, Geo-Informatics and Remote Sensing*, Advances in Science, Technology & Innovation,
https://doi.org/10.1007/978-3-030-72896-0_17

Fig. 1 Geographic location of the Kissir Dam

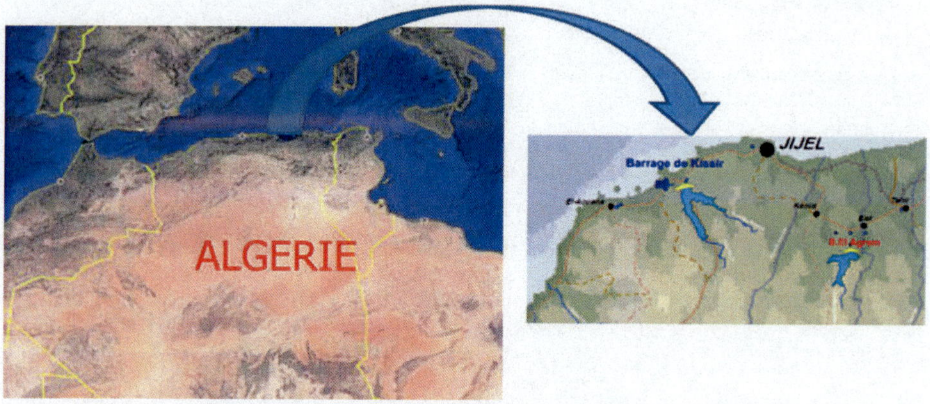

Fig. 2 Geological sketch of the Kissir dam site

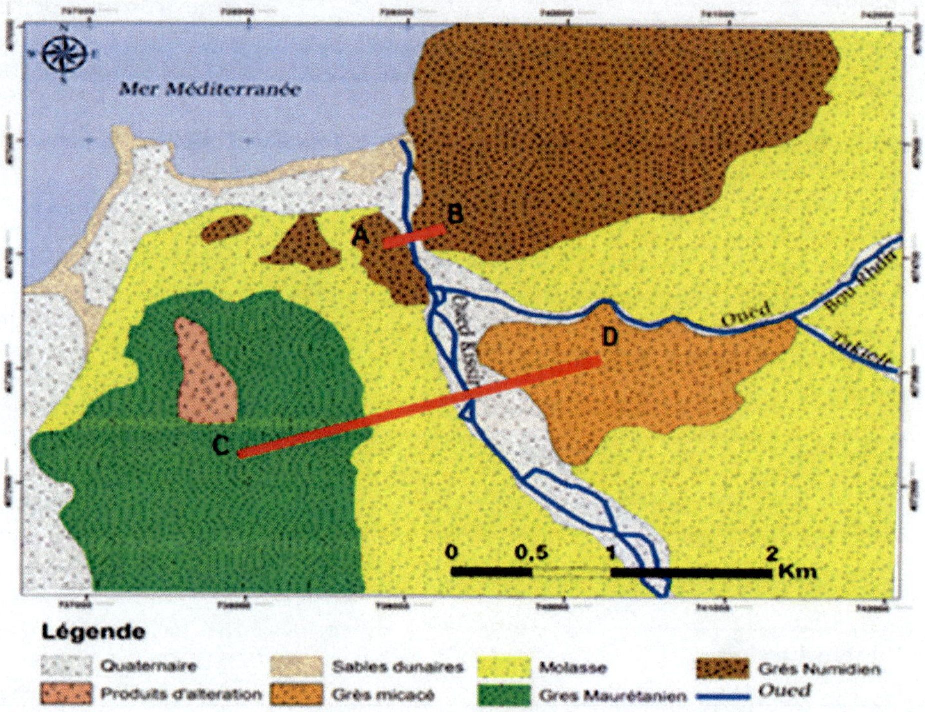

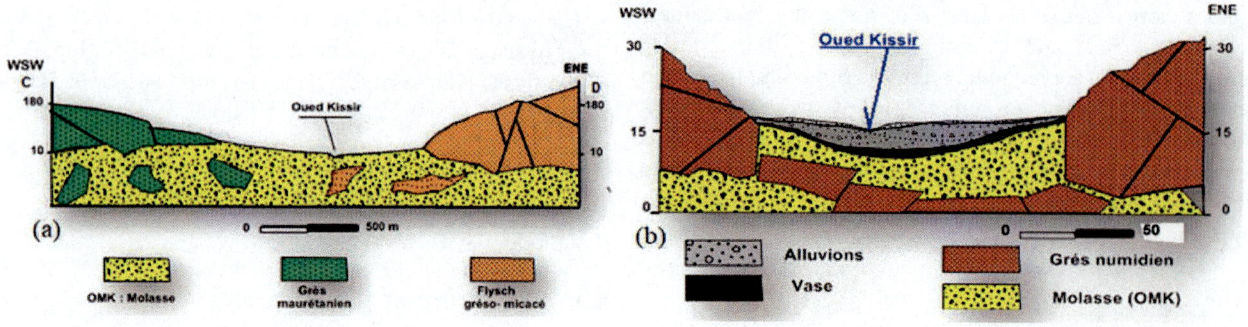

Fig. 3 Geological sections of the Kissir site

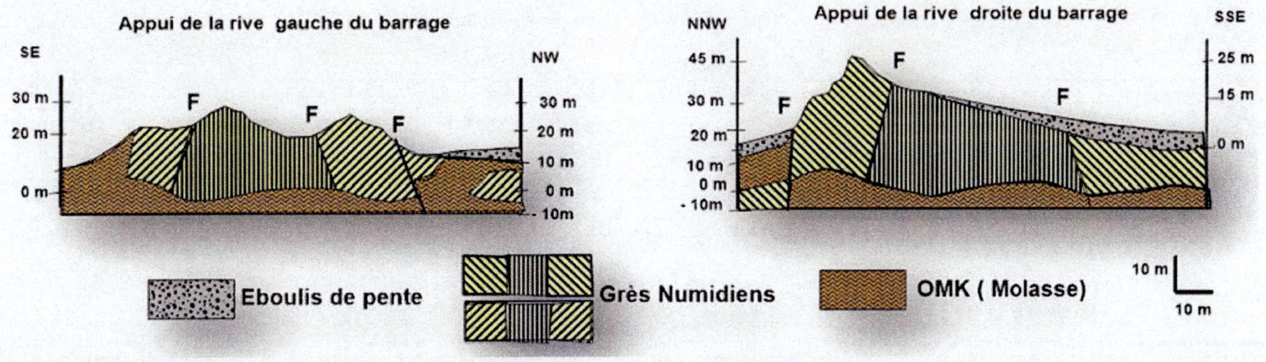

Fig. 4 Fracturing at the supports of the Kissir Dam

region, in turn, is not spared by this fracturing (Tekkouk 2019a, b).

4.2 Fracturing at the Supports of the Dam

The faults of the three families previously mentioned affecting the Numidian sandstones (dam supports) are in their majority, sub-vertical (Fig. 4). In addition to the layering joints, the identified faults affecting the supports can cause leaks, thus affecting the sealing of the reservoir (Tekkouk 2019a, b).

5 Core Surveys

The cored soundings of the right (SR3) bank, left (SR2) bank, and those at the level of the wadi (SR6) yield RQD ranging from a few percent to reaching and exceeding 80% (Fig. 5). Hundred percentage RQD is only rarely achieved.

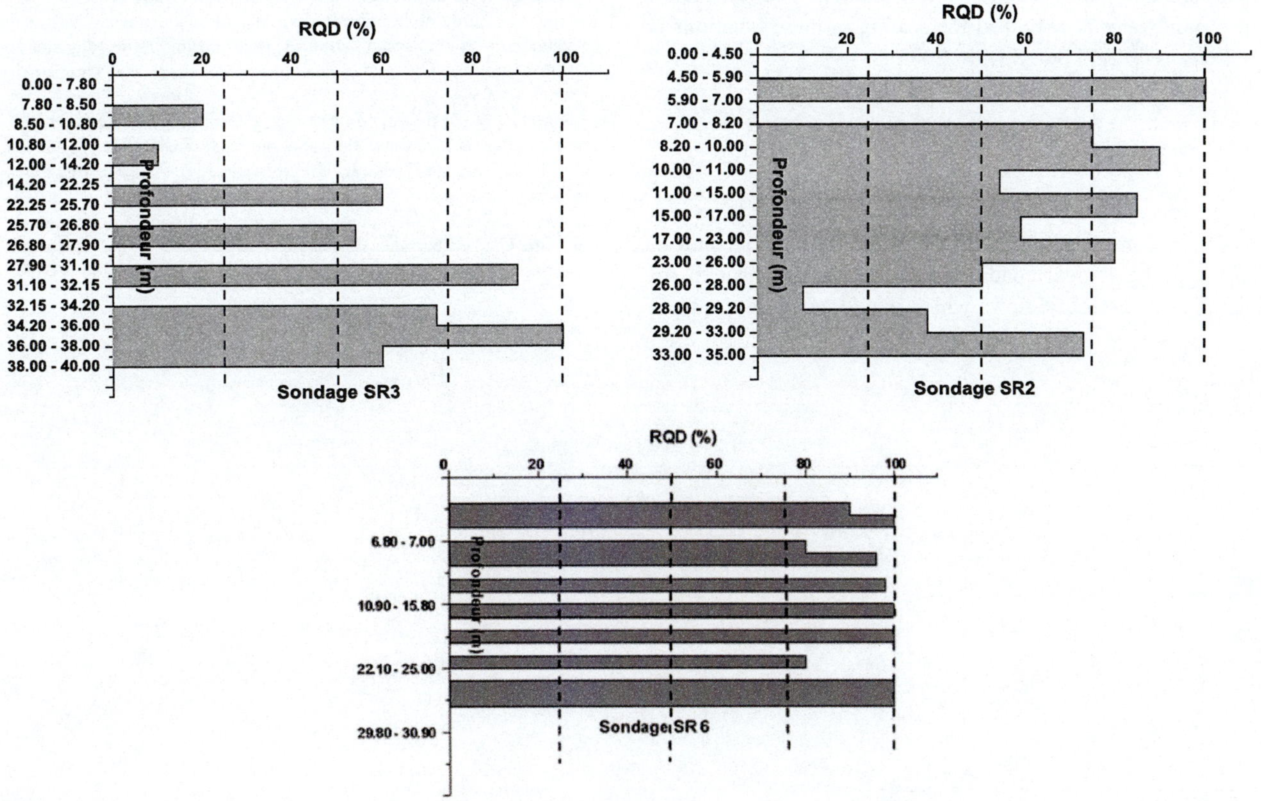

Fig. 5 RQD of coring surveys (Tekkouk 2019a)

Table 1 Permeability values at the supports

Zone	Permeability	Unit	Depth (m)			
			0–20	20–40	40–60	0–60
Right shore	Medium	UL	1.8	7.6	4.6	4.7
		cm/s	$2.30.10^{-5}$	$9.94.10^{-5}$	$6.05.10^{-5}$	$6.10.10^{-5}$
Valley	Medium	UL	/		0.01	/
		cm/s	$6.53.10^{-3}$	$2.75.10^{-3}$	$1.6.10^{-6}$	$3.09.10^{-3}$
Left shore	Medium	UL	14.1	22.2	1.5	12.6
		cm/s	$1.84.10^{-4}$	$2.88.10^{-4}$	$1.92.10^{-5}$	$1.64.10^{-4}$

The RQD value variations as a function of depth for core holes are indicative of fracturing indices (hence probable water leaks) at depth (Tekkouk 2019a, b; Boulkhiout and Bouchemella 2006).

6 Water Testing

The results of the Lugeon's tests carried on the supports and the base are summarized in the following (Table 1). The results obtained of the water tests perfectly confirm the mechanical characteristics of the geological formations and the fracturing state of the latter. The average value of the permeability (10^{-5} and 10^{-3} cm/s) reveals that the sites (supports and plate) have good permeability. The circulation of groundwater is easy, and the sealing in these locations is not ensured (Tekkouk 2019b, 2005).

7 Conclusion

It can be concluded that the sealing of the Kissir Dam was really compromised. Indeed, the geological, geotechnical, and hydrogeological conditions of the site all pleaded for possible underground flows which would be at the origin of serious leaks of water as well at the level of the supports as within the basin. Aware of this risk and the importance of the losses that were going to take place, the engineers and managers in charge of the project reacted by installing a sealing membrane under the body of the dam.

References

Boulkhiout, F.Z., Bouchemella, W.: Contribution à l'étude hydrogéologique et géotechnique du Barrage de Kissir-Wilaya de Jijel, Mémoire d'ingéniorat en Géologie de l'ingénieur. Université de Jijel, Algérie, in French (2006)

Delga, M.D.: Etude géologique de l'ouest de la chaîne numidique. Thèse de Doctorat, Bulletin du service de la carte géologique de l'Algérie, Série 2, Stratigraphie, Paris, France, 533p (1955)

Tekkouk, M.: Etude de la pollution par les sulfates des eaux du bassin versant de l'oued Djendjen (Jijel, Algérie). Mémoire de Magister en Hydrogéologie. Université Mentouri Constantine, Algérie, in French (2005)

Tekkouk, M.: Etude hydrogéologique et géotechnique de deux sites de barrage. Cas des barrages de Tabellout et de Kissir, Jijel, Nord-Est algérien. Thèse de Doctorat, Univesité de Jijel, Algérie, In French (2019)

Tekkouk, M.: Etude hydrogéologique et géotechnique de deux sites de barrage. Cas des barrages de Tabellout et de Kissir—Jijel - Nord–est algérien. Thèse de Doctorat d'Etat, Univ. de Jijel, Algérie (2019)

A Literature Review of Experimental Studies on Effect of Biopolymers on Soil Stabilization

Imane Idoui, Souhila Rehab Bekkouche, Riad Benzaid, and Inas Berdi

Abstract

Soil improvement methods enable engineers to solve the stability problems they encounter when carrying out a project. Soil stabilization focuses on improving the natural properties of the soil, instead of traditional soil treatment techniques. In this respect, polysaccharide biopolymers such as xanthan gum, guar gum, welan gum, beta glucan and chitosan are currently used in the field of soil behavior improvement in order to minimize environmental problems. Biopolymers are considered very auspicious in soil improvement in terms of shear strength and subsidence potential at water contents. The treatment efficiency depends on several factors, such as the nature of the soil and the type of biopolymer.

Keywords

Improvement • Stabilization • Soil • Biopolymers

1 Introduction

Soil stabilization aims primarily at improving the natural properties of the soil and increasing its density and resistance. Besides, it has a dual effect on its permeability. It can reduce its compressibility, minimize its sensitivity to water and modify the fraction of fines by providing practical and economical solutions. Biopolymers are replacing the traditional soil treatment techniques to improve the behavior of problematic soils and to minimize environmental problems. This work presents the main biopolymers used in the field of soil stabilization.

2 Measuring Methods

Biopolymers are natural polymers obtained from biomass, i.e., produced by living beings (plants, algae, animals). Polysaccharides are considered the most common family of biopolymers. They are long chains of simple sugars. Each of these links is represented by a hexagon and is made up of a loop of six carbon atoms, often carrying an oxygen atom and a hydrogen atom. These include cellulose and starch. Other sources such as xanthan lignin, guar gum, gellan gum, welan gum, beta glucan and chitosan can serve to produce a biopolymer (Guo 2014; Chang and Cho 2012; Chen et al. 2013; Khatami and O'Kelly 2013; Chang et al. 2015; Ayeldeen et al. 2016). Different types of biopolymers (such as xanthan gum, guar gum, modified starches, agar and glucan) were employed in Ayeldeen et al. (2017). It was concluded that the mechanical properties and behavior of typical soils (sand, silt and clay) could be improved, in particular, the shear strength.

2.1 Xanthan Gum

Xanthan gum is a microbial exopolysaccharide produced from the action of a bacterium, Xanthomonas campestris, by fermenting glucose, sucrose or other carbohydrate sources. This biopolymer is used in several fields as a thickening, stabilizing or emulsifying agent and can, in combination with other gums, act as a gelling agent. Chen et al. (2013), Chang et al. (2015), Swain (2015) and Cabalar et al. (2017) reported that the effect of strengthening xanthan gum with biopolymer was complex. It depended on the type of soil, the content of the xanthan gum, the water content and the mixing technique. Moreover, the presence of xanthan acted as a rheological modifier increasing the resistance to

I. Idoui (✉) · R. Benzaid · I. Berdi
Geological Engineering Laboratory (LGG), Mohammed Seddik Benyahia University, Jijel, Algeria
e-mail: imane.idoui@univ-jijel.dz

S. R. Bekkouche
Civil Engineering Department, 20 Aout 1955-Skikda University, Skikda, Algeria

H. El-Askary et al. (eds.), *Research Developments in Geotechnics, Geo-Informatics and Remote Sensing*, Advances in Science, Technology & Innovation,
https://doi.org/10.1007/978-3-030-72896-0_18

compression through the interaction between particles. Chang et al. (2015) performed a study on the reinforcement effect of xanthan gum on different types of sandy soil, natural soil, red soil, yellow soil and clay as well as different water and xanthan gum contents and two blending methods (dry and wet) to conclude that (i) the treatment depended on these last four factors and (ii) gum was more effective with fine soils than with grained soils.

2.2 Guar Gum

Ayeldeen et al. (2017) studied a natural collapsible soil sampled from New *Borg El Arab* City, Egypt, in the presence of two types of biopolymers (xanthan gum and guar gum) at different rates from 0 to 4%. The findings showed that a high concentration had the effect of increasing the water content from 12.4 to 15.3% for guar gum and from 12.4 to 14.4% for xanthan gum, respectively, with an additional rate of 2% of the biopolymer in both cases and a decrease in density and the volumetric weight. The increase in biopolymer concentration reduced the angle of friction and increased cohesion from 42 to 105 kPa after adding 2% xanthan gum and from 51 to 126% for guar gum. They reported that guar was more effective than xanthan gum in these types of treatments.

2.3 b-1,3/1,6-Glucan

Chang and Cho (2012) studied the effect of the biopolymer b-1.3/1.6-glucan in Korean residual soil (Asian countries). They obtained the following conclusions: (i) The polymer treatment of b-1.3/1.6-glucan served to strengthen the residual soil, and (ii) it also provided technical, economic and environmental advantages compared to conventional cement treatments (eco-friendly additive). The same authors Chang and Cho (2014) treated residual soil with beta-1,3/1,6-glucan. They found that beta-1,3/1,6-glucan treatment modified the mechanical properties of the matrix and appeared to be a useful application to improve soil compaction performance.

2.4 Chitin and Chitosan

Chitin and chitosan are highly basic polysaccharides used frequently. Chitosan is chemically derived from chitin, which is an N-acetylglucosamine polymer consisting of units linked together by bonds and can therefore be considered the amino equivalent of cellulose. Chitosan is obtained from partially diacetylated chitin. The main source of these polymers is the shell of crustaceans, which undergoes deproteinization followed by demineralization before use (Delphine and Philippe 2004).

Hataf et al. (2018) treated low plasticity clays from Shiraz Iran with a chitosan-like biopolymer in the form of powdered shrimp shell treated and mixed with clay soil. They concluded that incorporation of chitosan could increase the interparticle interaction of soil particles leading to the improvement of mechanical properties. Additionally, they highlighted that chitosan had the potential for short-term soil stability in wet soils and had no impact on the mechanical properties in the dry state.

2.5 Agar

Agar is a polysaccharide obtained from red seaweed. It has been used in Japan for more than 350 years in various fields: food; construction; industrial; and agar where it has been applied as a pore-creating agent (Niaounakis 2015).

Khatami and O'Kelly (2013), Chang and Cho (2014), and Hataf et al. (2018) found that agar gum was very effective in improving the strength of sandy soil; it could minimize environmental problems.

2.6 Welan Gum

Welan gum is an exopolysaccharide used as a modifier in industrial applications such as cement manufacturing. It is produced by fermentation of sugars by bacteria of the genus Alcaligenes. Niaounakis (2015) and Sonebi (2006) showed that welan gum was a viscosity-modifying agent; it allowed the increase of the pseudoplastic behavior and the thinning by shearing.

2.7 Gellan Gum

Gellan gum is a microbial polysaccharide. It is a gelling agent generally used in food processing for its hydrogel rheology. Chang et al. (2016) conducted a laboratory study on soil stabilization using gellan gum; they treated sandy soil with gellan gum at different concentrations (0.5, 1.0, 1.0, 1.5, 2.0 and 5.0% relative to soil mass). It was concluded that gellan gum improved strength, cohesion and friction angle of sandy soil even at low concentrations.

3 Conclusions

Recent research is interested in the use of biopolymers such as xanthan gum, guar gum, gellan gum, beta glucan and chitosan for soil improvement. The following conclusions can be drawn from all the studies previously carried out on soil stabilization by biopolymer:

- Xanthan gum is more effective on fine soils than on coarse-grained soils.
- Agar gum is a very effective way to improve the resistance of sandy soil.
- Guar gum is more efficient than xanthan gum in these types of treatments.
- Welan gum is a viscosity-modifying agent.
- Chitosan has the potential for short-term soil stability in humid soils and has no impact on the mechanical properties in the dry state.

Finally, the treatment efficiency varies according to the type of biopolymer used and the type of soil treated; therefore, a wide range of results could be obtained from these soil improvement tests.

References

Ayeldeen, M.K., Negm, A.M., EL Sawwaf, M.A.: Evaluating the physical characteristics of biopolymer/soil mixtures. Arab. J. Geosci. **9**(5), 371 (2016)

Ayeldeen, M.K., Negm, A.M., El Sawwaf, M.A., et al.: Enhancing mechanical behaviors of collapsible soil using two biopolymers. J. Rock Mechan. Geotech. Eng. **9**(2), 329–339 (2017)

Cabalar, A.F., Wiszniewski, M., Skutnik, Z.: Effects of xanthan gum biopolymer on the permeability, odometer, unconfined compressive and triaxial shear behavior of a sand. Soil Mech. Found. Eng. **54**(5), 356–361 (2017)

Chang, I., Cho, G.: Strengthening of Korean residual soil with β-1, 3/1, 6-glucan biopolymer. Constr. Build. Mater. **30**, 30–35 (2012)

Chang, I., Cho, G.: Geotechnical behavior of a beta-1, 3/1, 6-glucan biopolymer-treated residual soil. Geomech. Eng. **7**(6), 633–647 (2014)

Chang, I., Jooyoung, I.M., Cho, G.: Geotechnical engineering behaviors of gellan gum biopolymer treated sand. Can. Geotech. J. **53** (10), 1658–1670 (2016)

Chang, I., Jooyoung, I.M., Prasidhi, A.K., et al.: Effects of Xanthan gum biopolymer on soil strengthening. Construct. Build. Mater. **74**, 65–72 (2015)

Chen, R., Zhang, L., Budhu, M.: Biopolymer stabilization of mine tailings. J. Geotech. Geoenviron. Eng. **139**(10), 1802–1807 (2013)

Delphine, R., Philippe, D.: Les (bio) polymères biodégradables: l'enjeu de demain. Chim. Nouv. **22**(86), 10 (2004)

Guo, L.: Investigation of Soil Stabilization Using Biopolymers. Doctoral dissertation. Iowa State University (2014)

Hataf, N., Ghadir, P., Ranjbar, N.: Investigation of soil stabilization using chitosan biopolymer. J. Clean. Prod. **170**, 1493–1500 (2018)

Khatami, H.R., O'Kelly B.C.: Improving mechanical properties of sand using biopolymers. ASCE J. Geotech. Geoenviron. Eng. **139** (2013). https://doi.org/10.1061/(ASCE)GT.1943-5606.0000861

Niaounakis, M.: Biopolymers: applications and trends (2015)

Sonebi, M.: Rheological properties of grouts with viscosity modifying agents as diutan gum and welan gum incorporating pulverised fly ash. Cem. Concr. Res. **36**(9), 1609–1618 (2006)

Swain, K.: Stabilization of Soil Using Geopolymer and Biopolymer. Doctoral dissertation. National institute of technology, Rourkela (2015)

Examples of Geotechnical Problems Dealing with Temperature Change

Hamed Hoseinimighani and Janos Szendefy

Abstract

Nowadays, the development of technology and industry such as nuclear waste disposal, landfill and geothermal structures has given the temperature change in soils a high priority among research topics. It is essential to have suitable knowledge of thermal effects on soils to make efficient design of these structures. Indeed, the lack of this information can lead to malfunction of structures dealing with temperature change in soils. To this end, this paper introduces some engineering problems to have a better understanding of their nature. It also investigates the possible role of temperature during the design of such structures dealing with thermal issues.

Keywords

Temperature • Nuclear waste disposal • Landfill • Geothermal structures • Geo-environmental engineering

1 Introduction

Studies related to temperature change and its potential effect on soil properties and behavior started at mid-twentieth century with conducting consolidation tests at different temperatures as well as thermal tests under constant load (Gary 1936; Paaswell 1967). The range of temperature in early studies was usually between 10 and 50 °C. The reason for such limitation was related to the researcher's interest, which was the temperature difference between the laboratory and the field where the samples were being taken. Nowadays, however, the development of technology and industry has caused new and more complicated engineering applications to arise, encouraging the researchers to have a better understanding about the temperature effect in soil up to 100 °C and higher. During the design, it is important to determine the temperature effects on the chosen assumptions and parameters to make sure that the functionality and safety of the structure are not jeopardized.

2 Nuclear Waste Disposal

Clayey soils can be considered an ideal environment for nuclear waste disposal for its abilities, such as seepage control, pollution prevention, heat insulation and radiation protection (Bernier et al. 2004). However, chemical reactions of the waste can cause the soil to face temperature change up to 100 °C. In this respect, recent studies on high-level nuclear waste (HLW) disposal program have been carried out in France and Switzerland by the French National Radioactive Waste Management Agency (Andra) and the German Repository Safety Research Institution (GRS) (Ballarini et al. 2017)–(Zhang 2018). Also, a study on bentonite as a popular sealing material for HLW disposal in Finland (Akinwunmi et al. 2019) could serve as an illustrating example.

Insights on the effect of temperature on strength and volume change of the host soil are required since any excessive deformation and failure could cause leakage of hazardous substances. The effect of temperature on water flow and permeability is quite critical as well. Besides, it is essential to consider the behavior of the soil on the long term and how it could affect the environment for the next generations.

3 Waste Management

For waste management and design of landfills, geosynthetic clay liner (GCL) is often used as a mechanical and hydraulic barrier to ensure both the safety of the design (e.g., in slopes) and leakage prevention. GCL is a layer of bentonite captured between layers of geotextiles (and sometimes geomembrane)

H. Hoseinimighani (✉) · J. Szendefy
Budapest University of Technology and Economics, Budapest, 1111, Hungary

H. El-Askary et al. (eds.), *Research Developments in Geotechnics, Geo-Informatics and Remote Sensing*, Advances in Science, Technology & Innovation,
https://doi.org/10.1007/978-3-030-72896-0_19

(ASTM D4439 2018). The surrounding areas containing GCL can witness a rise in temperature caused by some chemical reactions of wastes and/or temperature fluctuation due to climate change (Tournassat et al. 2015). For instance, rise of temperature up to 50 °C in copper leach pads (Thiel and Smith 2004), 70 °C in nickel leach pads (Steemson and Smith 2009), 60 °C in municipal wastes (Yeşiller et al. 2015) and even more than 100 °C in aluminum waste (Stark et al. 2012) has been reported. Therefore, special attention needs to be directed toward the effects of temperature on the shear strength of GCL and the interface properties between GCL and soil to ensure the safety of the structure. Change in permeability by temperature is critical because it might lead to leakage of chemicals and dangerous substances and fluid to the environment.

4 Geothermal Structures

In recent years, pollution and global warming-related issues and effects of fossil energy on that have directed the attention toward finding a renewable and sustainable source of energy. Ground source heat pump (GSHP) is a common type of geothermal structure for space heating and cooling (Aresti et al. 2018). GSHPs are connected to a network of buried tubes, called ground heat exchanger (GHE), through which the water circulates. Nevertheless, due to high excavation costs, energy piles have become popular to make both mechanical and geothermal structures (Wang et al. 2017). Because of soil and ground being the source of energy in geothermal structures, it is of high importance to have sufficient knowledge about the ground temperature profile and thermal properties. Therefore, in situ, laboratory and numerical studies have focused on temperature profile and its thermal properties such as thermal conductivity and diffusivity (Seward and Prieto 2018). Heat pump function and circulation of fluid through the soil and foundation will cause the temperature fluctuation on pile–soil interface, pile and the surrounding soil. The first experiment regarding this issue was carried out by Morino and Oka (Morino and Oka Jan. 1994). During the few past years, temperature change and its effect on shear behavior of pile–soil interface, side and end bearing capacity of the pile, axial and radial deformation, etc., have gained popularity among researchers (Yavari et al. 2016).

5 Thermal Treatment of Contaminated Soils

The previously mentioned applications usually have a temperature range below 100 °C. However, the change in soil properties should be considered under extreme temperature, for instance, thermal treatment of contaminated soil (Chaudier 2013). Contaminated areas need remediation to prevent the migration of toxic material to the surrounding areas. Nowadays, various treatment methods can be applied. They could be physical (washing, flushing, thermal, vacuum extraction, solvent extraction), chemical (stabilization and solidification) or bioremediation techniques (Estabragh et al. 2016). Although thermal treatment offers fast and effective removal of the contamination, change in physical and mechanical properties of soil under such extreme temperature should be evaluated. Indeed, it provides critical information in the decision-making process that occurs at the beginning of a project (Chaudier 2013). Accordingly, it is also essential to evaluate the quality of the ground for geotechnical purposes after the remediation.

6 Conclusion

The roles of the temperature and its possible effects on physical and mechanical properties of soil have been discussed by a few examples of engineering applications mentioned above. This paper aimed to highlight the role of temperature change in geotechnical problems. Indeed, these insights can help ensure the quality and safety of designing the structures dealing with temperature change. Furthermore, this study tried to make up for the lack of focus on the long-term behavior of soils, which was noticeable in the literature.

References

Akinwunmi, B., Sun, L., Hirvi, J.T., Kasa, S., Pakkanen, T.A.: Influence of temperature on the swelling pressure of bentonite clay. Chem. Phys. **516**, 177–181 (2019)

Aresti, L., Christodoulides, P., Florides, G.: A review of the design aspects of ground heat exchangers. Renew. Sustain. Energy Rev. **92** (March), 757–773 (2018)

Armand, G., Bumbieler, F., Conil, N., de la Vaissière, R., Bosgiraud, J. M., Vu, M.N.: Main outcomes from in situ thermo-hydro-mechanical experiments programme to demonstrate feasibility of radioactive high-level waste disposal in the Callovo-Oxfordian claystone. J. Rock Mech. Geotech. Eng. **9**(3), 415–427 (2017)

ASTM D4439: Standard Terminology for Geosynthetics. Standard by ASTM International (2018)

Ballarini, E., Graupner, B., Bauer, S.: Thermal–hydraulic–mechanical behavior of bentonite and sand-bentonite materials as seal for a nuclear waste repository: Numerical simulation of column experiments. Appl. Clay Sci. **135**, 289–299 (2017)

Bernier, F., Demarche, M., Bel, J.: The Belgian demonstration programme related to the disposal of high level and long lived radioactive waste: achievements and future works. In: Symposium Proceedings. Scientific Basis for Nuclear Waste Management XXIX (2004)

Chaudier, S.: Thermal remediation alters soil properties: a review . Les Lettres Rom. **67**(1–2), 105–126 (2013)

Estabragh, A.R., Babalar, M., Javadi, A.A., Afsari, E.: Impacts of heating and surfactant treatments on the geotechnical properties of a cohesive soil. Int. J. Mech. Sci. **144**, 909–918 (2018)

Gary, H.: Progress report on the consolidation of fine-grained soils. In: First International Conference on Soil Mechanics and Foundation Engineering, pp. 138–141 (1936)

Morino, K., Oka, T.: Study on heat exchanged in soil by circulating water in a steel pile. Energy Build. **21**(1), 65–78 (1994)

Paaswell, R.E.: Temperature effects on clay soil consolidation. J. Soil Mech. Found. Div. **93**(3), 9–22 (1967)

Seward, A., Prieto, A.: Determining thermal rock properties of soils in Canterbury, New Zealand: comparisons between long-term in-situ temperature profiles and divided bar measurements. Renew. Energy **118**, 546–554 (2018)

Stark, T.D., Martin, J.W., Gerbasi, G.T., Thalhamer, T., Gortner, R.E.: Aluminum waste reaction indicators in a municipal solid waste landfill. J. Geotech. Geoenviron. Eng. (2012)

Steemson, M.L., Smith, M.E.: The development of nickel laterite heap leach projects. In: Proceedings of ALTA, pp. 1–22 (2009)

Thiel, R., Smith, M.E.: State of the practice review of heap leach pad design issues. Geotext. Geomembr. **22**(6), 555–568 (2004)

Tournassat, C., Bourg, I.C., Steefel, C.I., Bergaya, F.: Surface properties of clay minerals. In: Natural and Engineered Clay Barriers, pp. 5–31 (2015)

Wang, D., Lu, L., Cui, P.: Simulation of thermo-mechanical performance of pile geothermal heat exchanger (PGHE) considering temperature-depend interface behavior. Appl. Therm. Eng. **139**, 356–366 (2018)

Yavari, N., Tang, A.M., Pereira, J.-M., Hassen, G.: Effect of temperature on the shear strength of soils and the soil–structure interface. Can. Geotech. J. **53**(7), 1186–1194 (2016)

Yeşiller, N., Hanson, J.L., Yee, E.H.: Waste heat generation: A comprehensive review. Waste Manag. **42**, 166–179 (2015)

Zhang, C.L.: Thermo-hydro-mechanical behavior of clay rock for deep geological disposal of high-level radioactive waste. J. Rock Mech. Geotech. Eng. **10**(5), 992–1008 (2018)

Uniaxial and Static Compression Load Effect on Foam Glass Aggregate

Waleed Sulaiman Mustafa and Szendefy János

Abstract

Foam glass aggregate represents an excellent example of lightweight material made of waste glass as the primary material. It can be used in several geotechnical applications such as fill behind retaining walls, lightweight fill, and isolation floor layer. To evaluate the material strength, the uniaxial load was applied on small cubic shape samples with dimensions between 20 and 30 mm, prepared from big sized foam glass aggregate. Static compressional loads were conducted on foam glass aggregate samples to understand the bulk behaviour of the material. The analysis also considered important parameters such as Young's module (E) and oedometric compression module (E_{oed}) which have useful design extensions.

Keywords

Foam glass aggregate • Uniaxial and static compression test • Young's module • Oedometric compression module

1 Introduction

To consume more energy and get sustainable environment, foam glass became more popular than other traditional construction material. The material is produced from various types of waste materials such as recycled glass (Bernardo et al. 2007), fly ash (Bai et al. 2017), and waste mine. At first, foam glass was only used as an isolated panel. Recently, it has become more prominent as lightweight aggregate for its easier to handle compared to other construction materials (Scheffler and Colombo 2005). Foam glass has good resistance against bioactivity and bacteria attacks (Chen et al. 2006). As lightweight aggregate, foam glass can be used in many geotechnical applications such as fill behind retaining walls, lightweight fill, and isolation floor layer.

Due to lack of research, understanding foam glass aggregate behaviour under different loading conditions is still limited. Therefore, there is a need for more investigations in order to define material properties and assign traditional geotechnical parameters—such as compression module. In this work, uniaxial and static compressional load tests were conducted on foam glass aggregate material to provide information on important material parameters, namely Young's module (E) and oedometric compression module (E_{oed}) which will be helpful for future design purposes.

2 Materials and Methods

Figure 1a shows foam glass aggregates produced with various particle sizes. Particle size distribution curves (Fig. 1c) show that the material mainly contained gravel and sand sized particles. Depending on coefficient of uniformity (Cu = 1.66 and 1.58) and coefficient of curvature (Cc = 1.13 and 1.18) for both as received and 20% compacted foam glass aggregate respectively, the material is classified as poorly graded, with one major grain size range(16–64) mm.

For laboratory tests and to get a required particle size dimensions, small scale of cubic shape samples from larger foam glass aggregate particles could only be fabricated into roughly 20–30 mm small cubic sizes due to their irregular shape and were made to determine parameters such as grain density, uniaxial compressive strength for the material.

W. S. Mustafa (✉) · S. János
Department of Engineering Geology and Geotechnics, Budapest University of Technology and Economics, Budapest, Hungary
e-mail: wmustafa@edu.bme.hu; waleed.sulaiman@dpu.edu.krd

W. S. Mustafa
Duhok Polytechnic University, Duhok, Kurdistan Region, Iraq

H. El-Askary et al. (eds.), *Research Developments in Geotechnics, Geo-Informatics and Remote Sensing*, Advances in Science, Technology & Innovation,
https://doi.org/10.1007/978-3-030-72896-0_20

Fig. 1 **a** Foam glass aggregate, **b** foam glass prism sample, **c** particle size distribution

The small cubic samples were tested for uniaxial loading by applying constant displacement speed (0.01 mm/min) until failure or 20% deformation was reached. A number of test samples failed due to excessive deformations caused by of small imperfections in sample preparation.

For static compression tests of this material, a conventional geotechnical apparatus was impractical as the single particle could exceed the normal sample size (ranged between 16 and 64 mm in diameter). Therefore, a special test frame was constructed with 300 mm in diameter and 250 mm in high, which fulfilled the suggested 5D (largest particle diameter) size rule.

To achieve 20% compaction in the laboratory, a heavy load plate was used for compacting foam glass aggregate as done on-site compaction. Due to the size of the individual particles, a perfectly flat surface was unobtainable. This, in turn, caused the load plate to transfer loads on distinct points, rather than uniformly significantly altering test results. To decrease the effect of the rough surface, a gravel and sand mix was used as a load distribution layer on top (Fig. 2). First 4/8 mm gravel was used to "wedge" the foam glass particles' surface and create a more flat face. Then, it was covered with 0/1 mm sand to create the smooth surface on which the load plate could uniformly transfer loads.

3 Results and Discussions

3.1 Uniaxial Load

The final results of the uniaxial compression tests conducted on foam glass small cubic samples are shown in Fig. 3 (the solid blue line shows the average test results, while the two dashed lines show the scatter of the results). The two red lines are fitted on the elastic and plastic sections of the curve.

Two distinct sections are discernable on the graph. The first is a stiff and elastic section up to $\sim$ 1.2% relative deformation and 1200 kPa uniaxial stress. In other words, the material has a very stiff elastic behaviour, its Young's module is roughly 100 MPa. The second, plastic section, is significantly softer with only 4 MPa stiffness. According to Young's module value and previous research (Obrzud and Truty 2012), the material behaves as loosely compacted granular materials.

3.2 Static Compressional Load

The time deformation curves shown in Fig. 4 reveal that the material stiffness decreases as applying bigger amount of

Fig. 2 Foam glass aggregate laboratory sample preparation (left to right: foam glass aggregate, with gravel layer, sand load distribution layer)

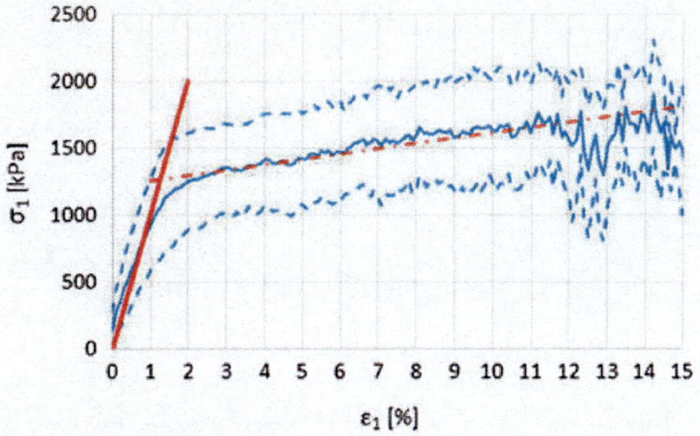

Fig. 3 Uniaxial loading test results of foam glass material sample

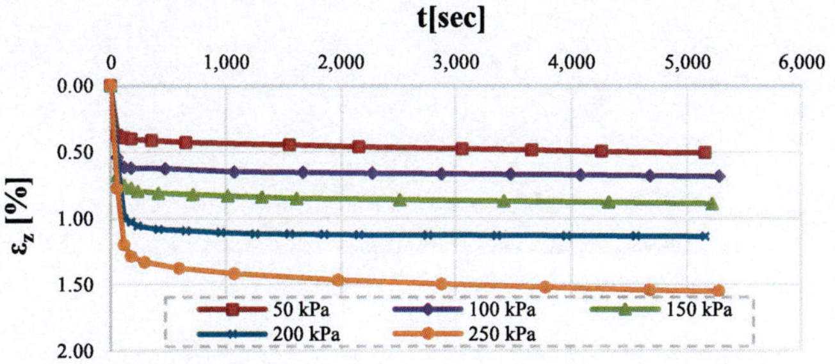

Fig. 4 Time deformation curves of foam glass aggregate

applied loads. Same increase in the amount of deformation occurs after applying extra (50) kPa on foam glass aggregate samples with a slight difference at (250) kPa.

The calculated oedometric modulus (E_{oed}) for all loading values (50, 100, 150, 200 and 250) kPa at approximately (5000) sec are about (9.8, 14.6, 16.8, 17.6 and 16.1) kPa, respectively. The results suggest that foam glass aggregate can be used in non- structural geotechnical applications, such as embankments, retaining wall backfill, and pipe bedding where the stresses should not exceed ultimate strength of the material.

4 Conclusions

The material testing results allow the following conclusions: The foam glass material itself is rigid (E = 100 MPa), but once it is broken down into aggregate form, the stiffness drops to 5–10%. This sort of behaviour is commonly observed in loosely compacted granular materials. Besides, as the stiffness decreases with loading, foam glass aggregate is not expected to take up major loads. We suggest that foam glass aggregate should not be subjected to stresses that exceed its ultimate strength far from extra deformation due to particle crushing at contact positions of the material.

References

Bai, J., Yang, X., Xu, S., Jing, W., Yang, J.: Preparation of foam glass from waste glass and fly ash. Mater. Lett. **136**(November 2017), 52–54 (2014)

Bernardo, E., Cedro, R., Florean, M., Hreglich, S.: Reutilization and stabilization of wastes by the production of glass foams. Ceram. Int. **33**(6), 963–968 (2007)

Chen, Q.Z., Thompson, I.D., Boccaccini, A.R.: 45S5 Bioglass®-derived glass-ceramic scaffolds for bone tissue engineering. Biomaterials **27**(11), 2414–2425 (2006)

Obrzud, R., Truty, A.: The Hardening Soil Model—A Practical Guidebook Z Soil. PC 100701 Report (2012)

Scheffler, P., Colombo, M.: Cellular Ceramics: Structure, Manufacturing, Properties and Applications. Wiley-VCH Verlag GmbH Co. KGaA, Weinheim, p. 669 (2005)

Bioengineering Technique for Protecting Submersible Roads in Haor Districts of Bangladesh

Mohammad Shariful Islam, Mumtahina Binte Latif, and Tanzila Islam

Abstract

A Haor is a wetland ecosystem in the north eastern part of Bangladesh. In Haor areas, submersible roads are constructed for transportation, so that biodiversity is not hampered. Roads and embankment failures are severe in this region as most of them become submerged during monsoon. But the conventional practices of embankment protection are expensive and sometimes ineffective. In many countries including Bangladesh, the bioengineering technique is being used for slope protection, where vetiver grass (*Vetiveria zizanioides*) is playing an efficient role. The main objective of this investigation is to determine the tolerance of vetiver system under submergence and the strength of rooted soil in wet condition. Among four constructed models, two models (VM) are filled up with Haor soils and covered with vegetation. Another two models (VGM) are filled with soil and covered with geo-jute and vegetation. To determine the effect of increasing moisture content on the strength of rooted soil, direct shear tests are conducted at moisture content 10% and 25% for samples reconstituted with 1% (w/w) root. Test results show that shear strength of vetiver-rooted soil is 1.05–1.32 times higher than that of bare soil depending on the moisture content. It is observed that vetiver tillers are sustainable under seven days of submergence. In other words, it can survive against flash flood.

Keywords

Bioengineering • Haor • Submersible road • Shear strength • Vegetation

1 Introduction

The north eastern part of Bangladesh is locally known as Haor, comprising Brahmanbaria, Habiganj, Kishoreganj, Netrokona, and Sunamganj districts. It physically consists of a bowl-shaped shallow depression, also known as a back-swamp (Alam et al. 2010). The pre-monsoon flash flood originating from Assam and Meghalaya causes immense damage every year. Roads and walkways in this region also go under water during monsoon. As maintaining biodiversity is necessary, submersible roads are constructed in these areas. Due to erosion and washing of soil particles from the embankment surface, the roads and slopes suffer damages frequently. Devastating flood, excessive rainfall, and tidal surge accelerate the failure process which results in immense damage to agriculture and infrastructures. Besides, reducing the effects of rain induced erosion, slope protection is important against wave action for the long-term stability. In such cases, bioengineering can be an effective solution for submersible roads, not to mention that it is an environment-friendly approach. Past research (Yong and Yu 2014; Tang et al. 2018) indicates that the use of vetiver grass (*Vetiveria zizanioides*) is a successful bioengineering method to protect the slopes which can sustain under water up to two months. In Bangladesh, vetiver is being used in coastal regions, hilly areas and Haor districts, and the investigations are being continued successfully (Islam and Badhon 2017; Islam et al. 2013). It is necessary to determine the effects on the growth of the tillers in case of the frequent change of water level during monsoon. The current study aims to evaluate the shear strength of rooted soil and effectiveness of vetiver in protecting submersible roads.

M. S. Islam (✉) · M. B. Latif · T. Islam
Department of Civil Engineering, Bangladesh University of Engineering and Technology (BUET), Dhaka, 1000, Bangladesh
e-mail: msharifulislam@ce.buet.ac.bd

H. El-Askary et al. (eds.), *Research Developments in Geotechnics, Geo-Informatics and Remote Sensing*, Advances in Science, Technology & Innovation,
https://doi.org/10.1007/978-3-030-72896-0_21

2 Methodology

2.1 Model Fabrication

To check the sustainability of vetiver under two conditions of submergence, four glass models were fabricated. Then, representative soils of Haor area (fine sand, according to USCS) were collected. Vetiver tillers were collected from a nursery prepared at BUET premises (GPS: 23.725° N, 90.38° E). The schematic diagram of the model is shown in Fig. 1a. Firstly, the bottom of the models was filled with crushed stone up to 10 cm. Secondly, geotextile (3.5 mm) was laid over the stones so that the sand would not pass with water. Afterward, the model was filled up with Haor soil up to 61 cm. All models were constructed based on two types of protection system, using vetiver plantation (VM) and vetiver plantation with geo-jute (VGM). In case of plantation with geo-jute (1.5 mm) models (VGM), geo-jute was laid over the filling sand. Vetiver tillers were planted with a center-to-center spacing of 10 cm (Fig. 1a). The models were put on the rooftop of the Civil Engineering Building of BUET (GPS: 23.729° N, 90.393° E). Vetiver grasses were allowed to grow under enough sunlight and appropriate moisture. The bottom tap was always open to pass the extra water. Two representative models of VM and VGM were filled up with water for submergence after full growth of the tillers for 130 days, and both taps were then closed (Fig. 1b). At first, they were submerged for seven days, the weather was fully sunny, and no rainfall occurred. After 7 days, both taps were opened to allow the water to pass. After few days, they were submerged again. The study is still going on. The remaining two models of VM and VGM were representative of frequent rise and fall of water in Haor region. For 7 days, they were continuously submerged under water for 3 h.

2.2 Direct Shear and Permeability Test Setup

Direct shear test (CD) was conducted on the prepared reconstituted soil samples according to ASTM D 3080. Tests were conducted for both bare soil and 1% (w/w) root-mixed composite soil samples at 10% and 25% moisture content, because over this moisture content bare soil could not hold the water and got drained out from soil. After the vetiver's growth for 130 and 7 days of submergence, soil was dug out from the models of VM and VGM. Geo-jute was already decomposed in the soil.

All types of soil, i.e., bare soil and soils of VM and VGM, were oven dried at first. Then, water was added to soils to attain 10% and 25% moisture content. Chopped vetiver roots of 1 cm length were mixed with the wet soil. For all the root-mixed soil samples, root content was 1% (w/w). The roots of the vegetation interacted with the soil to produce a composite material. In field conditions, the roots were spread in multi-direction, so the direct shear test was also simulated as such by root inclusion in soil. The prepared samples were kept in a desiccator to keep the moisture content unchanged. For fine sand, the falling head permeability test was performed according to ASTM D 5084.

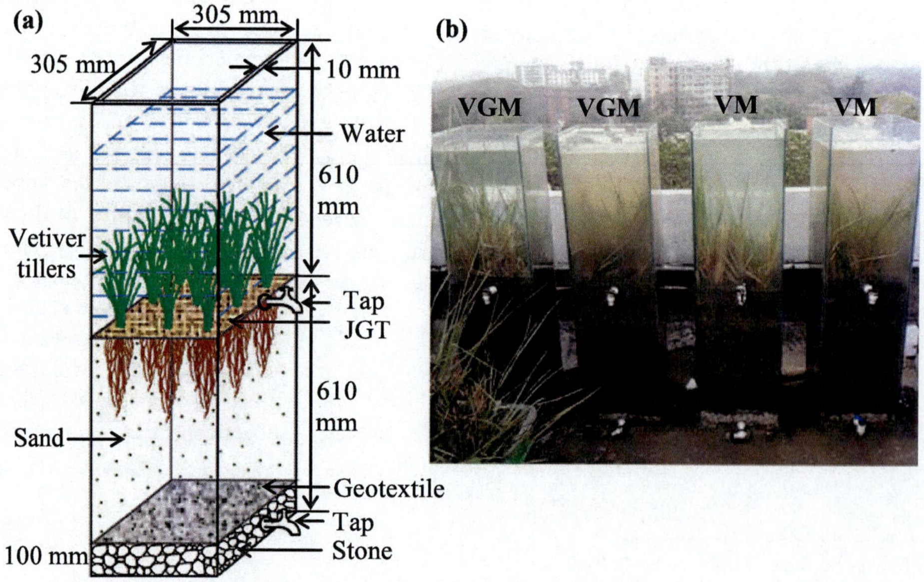

Fig. 1 **a** Schematic diagram of submerged glass model with protection of VGM and **b** vetiver tillers submerged under water after 130 days

3 Results and Discussions

3.1 Properties of Soil

The soil is found as fine sand where F.M = 0.82, D_{10} = 0.10 mm, and percentage fines = 4.8. Specific gravity (G_s), density (γ_{bulk}), and permeability (k) of soil are 2.66, 17.3 kN/m^3, and 8.36 × 10^{-3} m/s, respectively. These findings imply that the soil is of medium permeability.

3.2 Sustainability and Growth of Vetiver Under Submergence

From four glass models of VM and VGM, it is observed that some new shoots have grown and the old shoots have become greener than earlier. No bad odor is registered which indicates that no sapling has gotten rotten. From the outside view of the glass models, it is observed that the root of the tillers reached the bottom of the model. The tillers were not uprooted to continue further study. After draining out the water, vetiver saplings are noticed to be growing properly like before. Thus, it can be concluded that vetiver is sustainable under seven days of flash flood.

Tillers were pulled out from two fluctuating submergence models, VM and VGM to observe the physical conditions after 137 days. After that the shoot and root length, shoot and root matrix diameter, and the number of new shoots were measured (Table 1). The biodegradable geo-jute accelerated the growth rate of vetiver system getting mixed with the soil.

3.3 Strength-Deformation of Rooted Soils

Three types of reconstituted soil samples (bare, soil of VM, and VGM) were tested under different normal stresses (i.e., 12.5, 25, and 50 kPa). We tried to keep the dry unit weight of the soil sample as same as in the field condition. Moisture content (w_n) of the samples was 10% and 25%. For normal stress 12.5 kPa, shear strength of VM and VGM increased to 24% and 32%, respectively, than that of bare soil. When the normal stress was increased to 25 kPa, the shear strength of VM and VGM increased to 8% and 12%, respectively, than that of bare soil. In case of normal stress 50 kPa, shear strength of VM and VGM increased to 7% and 10%, respectively, than that of bare soil. This indicates that the application of vetiver root increases the shear strength of soil. In case of VGM, the shear strength was found greater than that of VM, as geo-jute was degraded with soil, and it accelerated the stabilization process. In case of 12.5 kPa, the increment of strength was greater than that of 25 and 50 kPa. As the submersible roads are usually of 1 m height (LGED, JICA 2005), the soil was subjected to low stress. As the increment rate of strength is higher in case of low normal stress, the bioengineering technique is appropriately applicable for submersible roads. Shear displacement of rooted soil taken from VM and VGM was found as 1.4–2.5 times and 1.4–1.8 times higher than that of bare soil, respectively. This suggests that the failure strain increases and the rooted soil is ductile. The test results indicate that the average angle of internal friction of vetiver root-mixed reconstitute soil samples is higher than those of the reconstitute bare soil,

Table 1 Performance study on growth of vetiver under submergence for VM and VGM

Parts of vetiver	Parameters	Value/condition	
		VM	VGM
Shoot	Length (cm)	28–82	22–64
	Width (cm)	1	1
	Diameter of the bush (cm)	8	10.5
	Color	Green	Green
Root	Length (cm)	60–75	60–70
	Single root diameter (cm)	0.2	0.2
	Root matrix diameter (cm)	8	13
Leaf	Color	Green, yellowish brown	Green, yellowish brown
Tiller	Inflorescence	No	No
	Number of tillers grown in one point	31	64

Table 2 Comparison of shear strength properties of the reconstituted bare and rooted soil

M.C. (%)	Bare soil		Rooted soil of VM		Rooted soil of VGM	
	ϕ'	c' (kPa)	ϕ'	c' (kPa)	ϕ'	c' (kPa)
10	33.3	9.26	34.3	13.1	34.2	14.4
25	37.3	8.86	37.4	11.11	37.9	12.91

ϕ' = Angle of friction, and c' = Apparent cohesion

which is shown in Table 2. Besides, despite the soil being sand, cohesion values were found that is termed as apparent cohesion (c'). It might be caused by negative capillary pressure or root cohesion.

4 Conclusions

The main findings of the study are as follows:

(1) The inclusion of vetiver root increased the shear strength and ductility of soil–root matrix. However, the increment rate of shear strength was higher in case of lower normal stress. In other words, vetiver grass plantation would be effective for protecting low-height submersible road embankments in Haor areas.
(2) A study of 137 days revealed that vetiver root grew in the range of 60–75 cm. The number of tillers per point grew from 2 to 31 and 64 in case of VM and VGM, respectively. From the model study, it was found that vetiver grass could survive under continuous submergence and fluctuating submergence for 7 days durations' flash flood.

This investigation suggested that plantation of vetiver could be a cost-effective, sustainable and eco-friendly method for submersible road protection.

References

Alam, M.S., Quayum, M.A., Islam, M.A.: Crop production in the Haor areas of Bangladesh: insights from farm level survey. Sci. J. Krishi Found. **8**(2), 88–97 (2010)

Islam, M.S., Badhon, F.F.: Sandy slope stabilization using vegetation. In: Proceedings, International Conference on Disaster Risk Mitigation, Dhaka, Bangladesh (2017)

Islam, M.S., Arifuzzaman, Hossain, M.S., Nasrin, S.: Effectiveness of vetiver root in embankment slope protection: Bangladesh perspective. Int. J. Geotech. Eng. **7**(2), 136–148 (2013)

LGED, JICA: Road design standards rural road **76** (2005)

Tang, V.T., Fu, D., Binh, T.N., Rene, E.R., Thanh Sang, T.T., Singh, R.P.: An investigation on performance and structure of ecological revetment in a sub-tropical area: a case study on Cuatien River Vinh City, Vietnam. MDPI **10**(5), 636 (2018)

Yong, W., Yu, T.X.: Ecological revetment reconstruction of Huang Jia River in Licha. Appl. Mech. Mater. **641–642**, 156–161 (2014)

Geophysical Approaches in Engineering Geology Studies

Geoelectrical Study of the Mides Dam Site (Southern Tunisia)

Moucz Gouasmia, Abdelkader Mhamdi, Ferid Dhahri, and Mohamed Soussi

Abstract

During the last years, facing the scarcity of water, the arid countries have moved toward the hydrogeology of rivers' sites. Dam construction along rivers is indeed a water management process at the national level. However, this complicated task requires a wide range of engineering disciplines. Geological surveys help recognize the planned construction area. The present study aims to recognize the lithology of the wadi Oudei dam site (Southern Tunisia) and to define and locate potential impermeable layers within the riverbed to ensure water retention upstream the dam wall. For this purpose, we applied the geoelectrical method to investigate the lithology of the study area. The survey highlighted several geoelectrical units encompassing a coarse deposits fluvial aquifer overlaying a very thick clayey level that could serve as an efficient impermeable underground.

Keywords

Geoelectrical investigation • Hydrogeology • Dam site • Mides

M. Gouasmia (✉) · A. Mhamdi · F. Dhahri
Faculty of Sciences of Gafsa, University of Gafsa, Sidi Ahmed Zarroug, 2112 Gafsa, Tunisia

M. Gouasmia · A. Mhamdi · M. Soussi
Faculty of Sciences of Tunis, LB18ES07: Sedimentary Basins and Petroleum Geology, University of Tunis El Manar, El Manar II, 2092 Tunis, Tunisia

F. Dhahri
Faculty of Sciences of Tunis, Laboratory 3G (Geodynamics, Geo-Digital and Geomaterials), University of Tunis El Manar, El Manar II, 2092 Tunis, Tunisia

1 Introduction

In arid and semi-arid regions, socioeconomic activities and sustainable regional development are limited by water resources scarcity. Water resources management is a global challenge integrating all water-related issues involving scientific knowledge, water management, treatment technologies, and institutional policies (Koundouri and Karousakis 2006). In the regions where water resources are insufficient, dams are constructed to intercept rainwater or to obstruct the natural flow of groundwater providing artificial storage aiming to water resources increase.

In west-southern Tunisia, the oasis of Mides (North of Tamerza) is nowadays poorly irrigated, especially during the summer season and dry period. In this region, water resources are limited; they engender a near-permanent deficit. However, the alluvial deposits along the Oudei wadi are of great hydrogeological interest. These deposits were investigated by a geoelectrical method to determine the thickness and the geometry of their lithological units. The studied site was also evaluated in view of maximizing the infiltration of floodwaters by the construction of superficial or underground water reservoirs.

2 Materials and Methods

The geoelectrical prospecting seemed to be an efficient method used to better evaluate the hydrogeological potential of the aquifer in the studied site. This tool helped to identify the facies and the thickness of the alluvium deposits within the Oudei wadi and to detect the impermeable substratum that could serve as a barrier preventing any loss (Gouasmia et al. 2007; Karastathis et al. 2002) from the proposed water-retention structure. Twenty Vertical Electrical Sounding (VES) surveys were carried out by Schlumberger configurations with AB varying from 140 to 400 m. VES was placed in the way to cover the entire area to be surveyed along with six profiles, five of which are almost in the E-W direction and one in the N–S direction (Fig. 1).

H. El-Askary et al. (eds.), *Research Developments in Geotechnics, Geo-Informatics and Remote Sensing*, Advances in Science, Technology & Innovation,
https://doi.org/10.1007/978-3-030-72896-0_22

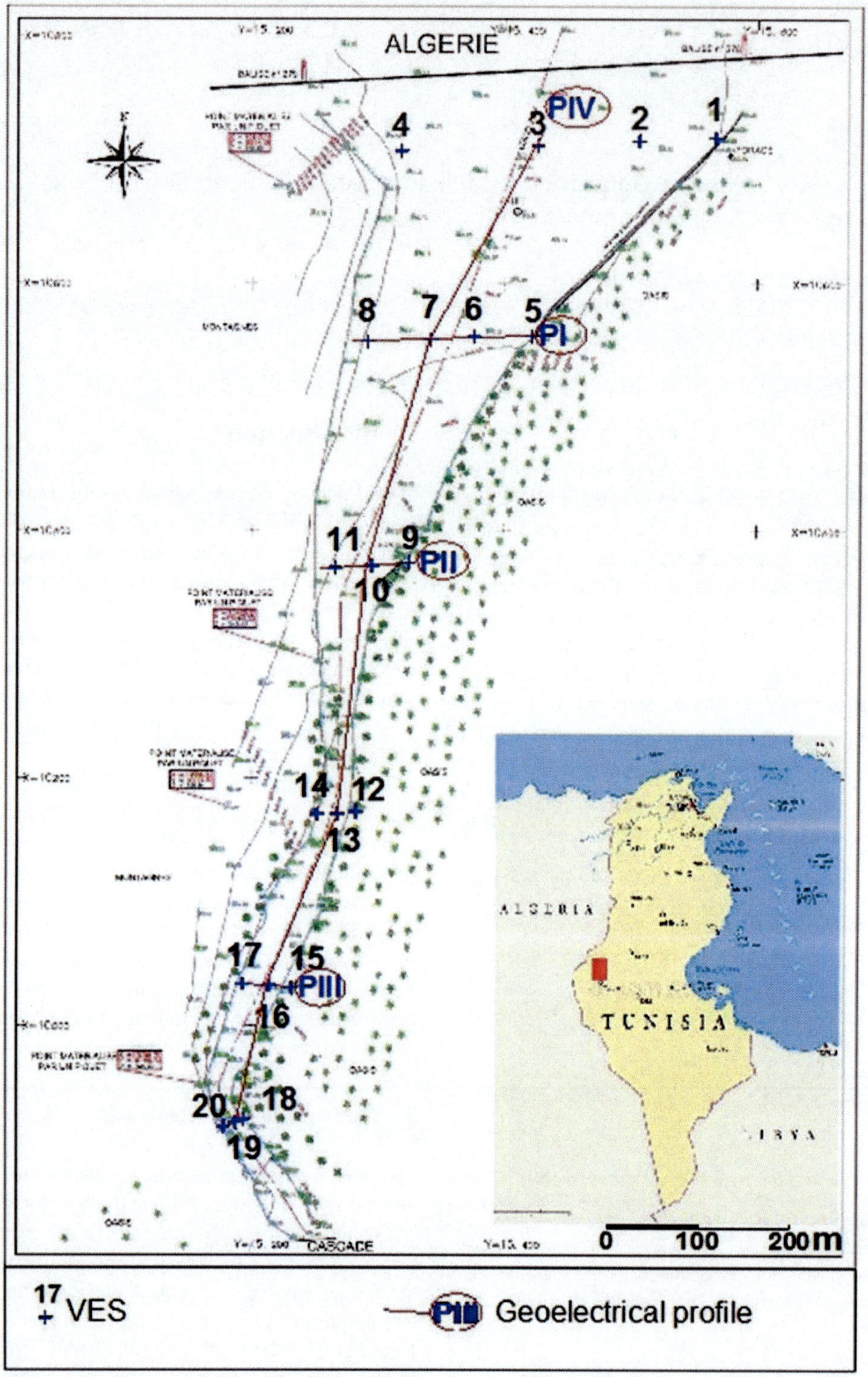

Fig. 1 Location map of the study area, VES, and geoelectrical profiles

3 Results and Discussion

3.1 Geological and Hydrogeological Setting

The area of the bed of Mides river investigated in this study lies 1 km west of the Mides village. The Oudei wadi draining from Algeria crosses J. Brikiss and cuts a gateway (Mides wadi) into the limestones of the Berda Formation. Upstream from this gateway, the wadi bed drains over the El Haria clays (Paleocene) and the limestone and phosphate series of the Eocene (Ben Mammou and Mellouli 1994).

The aquifer of Mides is confined in the alluvium of Oudei wadi. The detrital deposits encompass coarse sands and gravels on the first meters. Downward sands become fine to very fine with several levels of clayey sand. The alluvium aquifer of the wadi is recharged mainly by the infiltration of floodwaters and by the direct infiltration of occasional rainwater (Ben Mammou and Mellouli 1994). The underground flow of groundwater related to Oudei wadi appears to be from N to S.

3.2 Geophysical Investigation

VES was used to draw some geoelectrical profiles (Fig. 2) and an isopach map (Fig. 3) for the alluvium aquifer. The geoelectrical profiles highlight the existence of several lithological levels. These levels may be affected by geoelectrical discontinuities (faults) that are responsible for some uplifted and subsiding blocks.

The deposits in the northern part of Mides wadi are essentially resistant. They correspond to sands, gravel, and alluvium. These levels form an alluvial aquifer (12–20 Ωm). Toward the South, a conductive level related to El Haria clays (7–9 Ωm) exhibits a thickness and depth decreasing from N to S. These clays are overlaid by resistant levels representing to alluvium facies potentially considered an aquifer unit. The resistant substratum could be attributable to limestones (35–91 Ωm). These limestones exhibit a depth that decreases from N to S.

The isopach map (Fig. 3) shows that the thicknesses of the alluvium aquifer vary from less than 1 m to more than

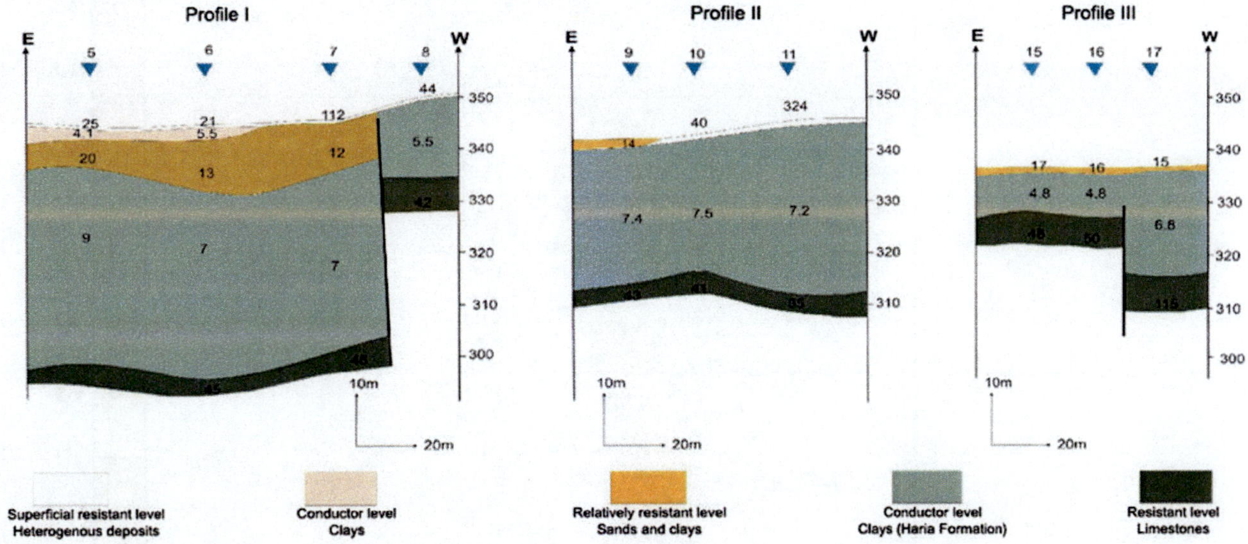

Fig. 2 Geoelectrical profiles

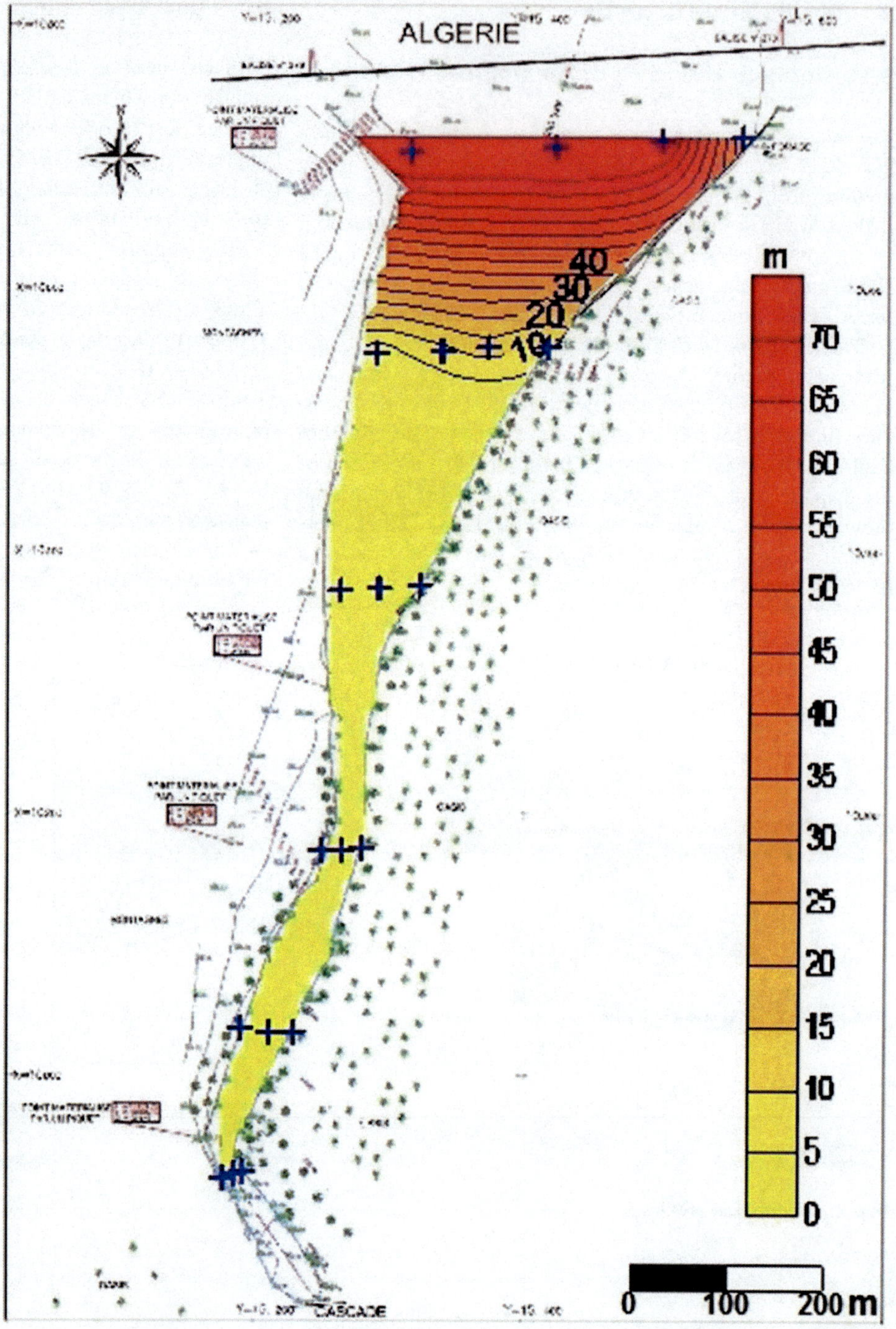

Fig. 3 Aquifer isopach map of the study area

50 m. The most important values are found toward the N (near the Algerian–Tunisian border), particularly in the northwestern part of the wadi, whereas the lowest values are detected at the south of the studied site. Thickness variations from N to S seem brutal; they do not happen gradually, which could argue for the existence of faults.

4 Conclusions

The geoelectrical investigation of the Oudei wadi in the oasis of Mides (west-southern Tunisia) for hydrogeological and geotechnical purposes highlights the existence of an alluvial unit with significant hydrogeological interest. This aquifer encompasses several lithological levels affected by some faults responsible for brutal lateral thickness variations. This investigation also permits to detect an impermeable clayey substratum that can serve for the construction of a water retention structure such an underground dam. The geoelectrical method seems to be an efficient tool for geotechnical evaluation of sites when a decision is needed to construct structures for water resources management, especially in arid and semi-arid regions.

References

Ben Mammou, A., Mellouli, M.: Création de retenues d'eaux souterraines au centre et au sud-ouest de la tunisie. Le projet de Tamerza. Secheresse **2**, 85–92 (1994)

Gouasmia, M., Mhamdi, A., Chekhma, H., Lahmadi, M., Amri, F., Ben Dhia, H.: Etude géophysique du site du barrage d'Oum Laksab (Tunisie centre Ouest). Revue Française de Géotechnique. N° 118, pp. 37–42 (2007)

Karastathis, V.K., Karmis, P.N., Drakatos, G., Stavrakakis, G.: Geophysical methods contributing to the testing of concrete dams. J. Appl. Geophys. **50**, 247–260 (2002)

Koundouri, P., Karousakis, K. (ed.): Water Management in Arid and Semi-arid Regions: Interdisciplinary Perspectives. Edward Elgar Publishing (2006)

Application of Electromagnetic Wave Computed Tomography in Deep Karst Caves Exploration

Sheng-gen Huang, Jie Lin, Qi-kun Huang, and Rong-zhu Liang

Abstract

Electromagnetic wave computed tomography technology is an effective way of detecting underground karst caves. This paper establishes a numerical model for electromagnetic wave CT detection using finite difference time domain algorithm to interpret the propagation mechanism of electromagnetic wave in karst cave detection. The propagation rules obtained from numerical simulation are drawn as follows: (1) When electromagnetic wave propagates inside the karst cave, refraction, reflection, diffraction, and obstacle gain behavior will occur. (2) The refraction has relatively more influence on the energy absorption of electromagnetic wave while the disturbance of the waveform is principally attributed to the reflection. Furthermore, the tomographic images gained from the experiment are interpreted according to the simulation conclusions. The analysis of tomographic images shows a general agreement with exploration results. The research presented herein aims to serve as a based theory for analyzing tomography results in karst caves exploration.

Keywords

Geophysical method • FDTD • Numerical model • Absorption coefficient

1 Introduction

At present, drilling and geophysical methods are still the most common means in karst exploration (Abdallatif et al. 2015). Conventional drilling method is more suitable for finding out the local state of the cave although some data is hard to obtain, such as the overall size, distribution and filling condition. As for the geophysical methods, they can rapidly detect the whole distribution of the cave (Martínez-Moreno et al. 2014). Electromagnetic wave CT (computed tomography) technology, as one of the geophysical methods, can provide an image of internal structure by inversion calculation (Xie and Xu 2015). Nowadays, the application of electromagnetic wave CT technology mainly focuses on grouting quality detection (Bai and Li 2013), concrete quality detection (Kyun 2008) and karst detection (Kašpar and Pecen 2010). Nevertheless, engineering practice has revealed that in the case of caves filled with various stuffing this technology fails to provide satisfying results, which highly restricts the promotion of this method.

When the electromagnetic wave CT technology is applied to explore karst caves, the interpretation and judgment of propagation mechanism and energy attenuation of electromagnetic wave are primarily based on medical research results and previous exploration. Thus, the theoretical basis is considered inadequate.

Based on the FDTD algorithm (Ahmadi et al. 2015), the model to detect underlying cavities using electromagnetic wave CT technology is established through MATLAB. The PML and the PEC are set at the boundary of the model to solve the problem of the electromagnetic wave getting back into the model after it comes out of it (Du et al. 2017). The propagation law of electromagnetic wave inside the cavern is obtained by simulation.

2 Materials and Methods

Based on the FDTD algorithm, the model measuring 20 m × 20 m is settled through MATLAB software. The cave is set to be a circle with a radius of 2 m, and the center coincides with the center of the model. The filling materials inside the karst cave are loose and full of water. The

S. Huang · J. Lin (✉) · R. Liang
Faculty of Engineering, China University of Geosciences, Wuhan, 430074, China
e-mail: 1201710398@cug.edu.cn

Q. Huang
Civil Engineering Faculty, University of Western Australia, Clawley, Perth, 6009, Australia

H. El-Askary et al. (eds.), *Research Developments in Geotechnics, Geo-Informatics and Remote Sensing*, Advances in Science, Technology & Innovation,
https://doi.org/10.1007/978-3-030-72896-0_23

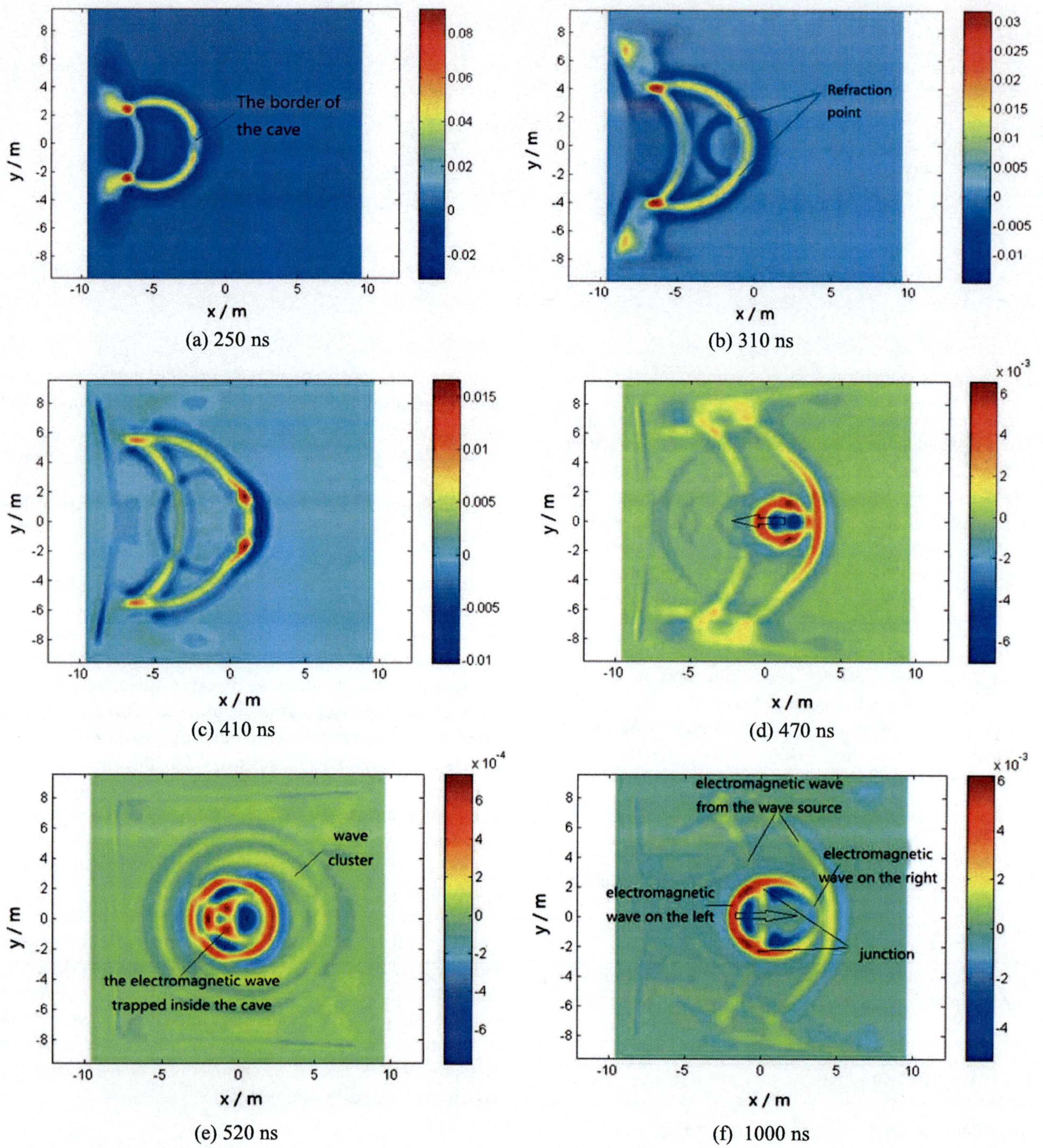

Fig. 1 Simulation images of electromagnetic wave after several time steps

saturated sands with relatively high porosity are adopted as the filling medium. The limestone is selected as the surrounding rock.

To ensure the accuracy of simulation results, the PML is set outside the model with a thickness of 2.5 m and a reflection coefficient of 10–8. Besides, the model is cut off from the external by PEC to prevent electromagnetic wave propagating outside the model from being introduced to the computational model again. The whole model is divided into 36,100 grids, where every grid measures 0.1 m × 0.1 m.

In the simulation, the current source is adopted to simulate the emission source of electromagnetic wave. The selected waveform of the excitation source is the normalized Gaussian derivative waveform with a signal frequency of 8 MHz. Considering the center of the cave as the origin of the coordinate axis, then the transmitting point is 10 m away from the receiving point.

3 Results

Figures 1 shows the images describing the transmission of electromagnetic wave in the model after several computational time steps when the transmitting point and the receiving point are located at points (− 5.0) and (5.0). To be specific, Fig. 1a–c delineates the simulation images of electromagnetic waves propagating from the left border of the cave to the right border, while Fig. 1d–f depicts the propagation images of electromagnetic wave trapped inside the cave.

4 Discussion

Figure 1a–b indicates that the electromagnetic wave is refracted when it passes through the cave border. It can be attributed to the dielectric coefficient and the permeability of the filling, which are significantly different from those of the intact limestone. As a result, strong refraction of electromagnetic wave occurs when it passes through the interface of two kinds of medium. Comparing the two images reveals that the transmission form of electromagnetic wave changes after the refraction at the cave border. Thus, the energy attenuation of electromagnetic wave is mainly dominated by the refraction phenomenon.

As presented in Fig. 1a–c, the electromagnetic wave reflects strongly at the right border of the cave. The intensity is stronger at the right border of the cave than the left. This observation is explained by the diffraction of the electromagnetic wave inside the cave. When the electromagnetic wave bypasses the saturated sand and gravel, it will be separated in the same phase. When the components of separated electromagnetic wave reach the right border, they will interact with each other, leading to the enhancement of intensity, referred to as the obstacle gain phenomenon.

Furthermore, comparing Fig. 1a, c reveals that the intensity of electromagnetic wave decreases from 0.015 v/m to 0.012 v/m after refraction. The amplitude is reduced by about 20%, which is smaller than that caused by refraction.

According to Fig. 1c, d, the electromagnetic wave could be divided into two parts when it reaches the right border. One part propagates out directly from the right border after refraction. The other part reflects at the right border and then travels back by the way it comes. Owing to the existence of filling material and pore water, this part of electromagnetic wave is trapped inside the cave. At the same time, the phenomena of reflection, diffraction, and barrier gain occur within the cave.

Figure 1e, f demonstrates that the trapped electromagnetic waves diffuse around and travel to the outside of the cave finally in the process of reciprocating propagation between the left and right borders. When the electromagnetic wave diffused from the cave encounters the wave propagating from the wave source, they will interfere with each other. Then, the intensity of the latter increases and the waveform becomes significantly disturbed at the meeting point.

Validating these results is the next step. To this end, the tomographic images gained from Nanhu section in Line Eight of Wuhan Metro are interpreted according to the simulation conclusions. The apparent absorption coefficient of each grid to electromagnetic wave is calculated through SIRT algorithm, and the tomography result obtained by Suffer software is shown in Fig. 2.

According to Fig. 2, the absorption coefficient of the electromagnetic wave in the middle area decreases from 0.52 to 0.42 dB/m, whereas the absorption coefficient of the outermost rock layer to electromagnetic wave is less than 0.4 dB/m. Therefore, it can be inferred that the region with a more significant absorption coefficient in the middle area is the strongly developed karst fissure area.

Near the area of − 25.6 to − 29 m and − 30.5 to − 34 m in YRK4128 borehole, the absorption coefficient of rock medium to electromagnetic wave decreases from 0.56 to 0.46 dB/m in the dark red region. Taking into consideration that the rock mass is fragmented in this region, it is possible to infer that there is a cave filled with loose stuffing. The simulation results show that electromagnetic wave is refracted, reflected and diffracted when it passes through the cave filled with stuffing. Also, some electromagnetic wave is retained in the cave. Consequently, the absorption coefficient will be more significant in this medium, while the absorption coefficient gradually decreases outside the cave. Hence, it can be speculated that the stratum distribution in this region is as follows: a strongly developed karst fissure area, a weakly developed karst fissure area and an intact limestone area, in a downward order.

Furthermore, the absorption coefficient of rock formation to electromagnetic wave decreases from 0.4 to 0.28 dB/m with no abrupt change from − 34 m to − 38 m. This area is, therefore, the stratum of weakly weathered limestone. The absorption coefficient of bedrock is less than 0.28 dB/m. It is deduced that the bedrock in this range is the intact limestone with weak weathering degree.

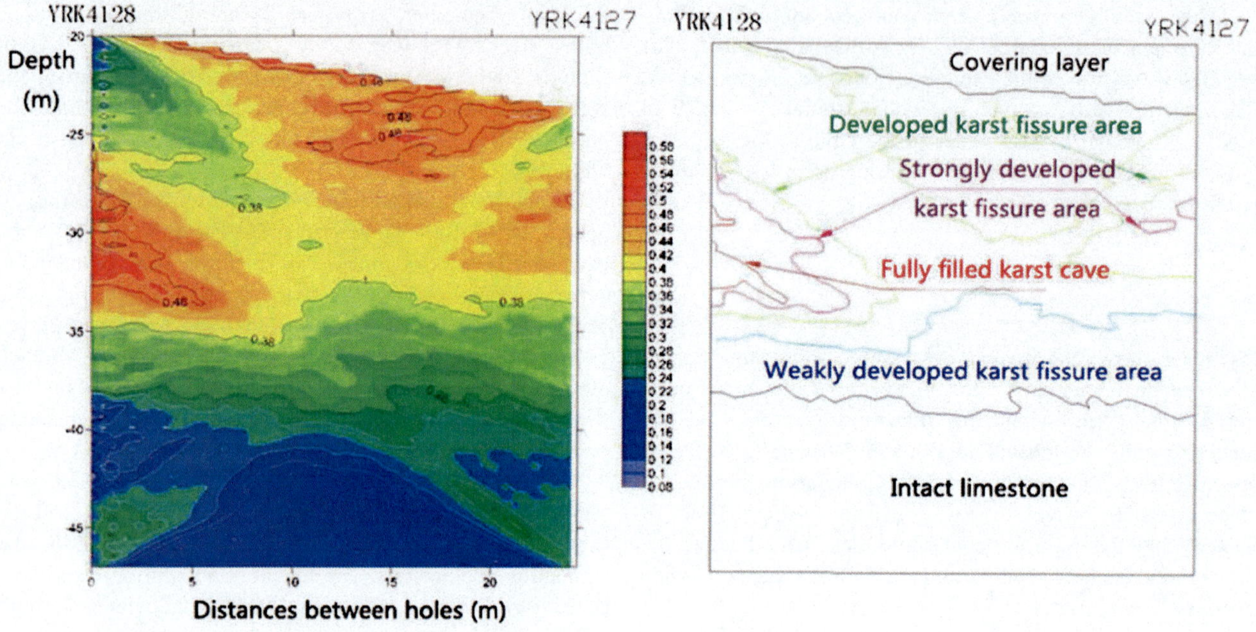

Fig. 2 Tomography result of profile YRK4128-YRK4127

5 Conclusions

The analysis above leads to the following conclusions: (1) The intensity of electromagnetic wave attenuates when it passes through intact limestone. If the influence of pores in limestone is to be ignored, the waveform of electromagnetic wave will hardly be disturbed. Also, the attenuation value of electromagnetic wave in intact limestone is proportional to the propagation distance. (2) Owing to the existence of karst cave and the filling material, the phenomena of refraction, diffraction and obstacle gain occur inside the cave. (3) With reflection present, a part of electromagnetic wave is trapped inside the cave. Due to the influence of the stuffing, the energy of electromagnetic wave will fade away when it passes through the pore water as well as the filling particles. (4) When the electromagnetic wave passes through the cave containing with filling material, the energy absorption of electromagnetic wave is influenced by the refraction phenomenon, rather than the reflection phenomenon.

References

Abdallatif, T., Khafagy, A.S.A.B., Khozym, A.: Geophysical investigation to delineate hazardous cavities in Al-Hassa Karstic Region, Kingdom of Saudi Arabia. Eng. Geol. Soc. Territory **5** (2015)

Ahmadi, R., Fathianpour, N., Norouzi, G.H.: Detecting physical and geometrical parameters of some common geotechnical targets through their effects on GPR responses. Arab. J. Geosci. **8**(7), 4843–4854 (2015)

Bai, H.R., Li, J.J.: Evaluation of grouting effect detection in Goaf. Appl. Mech. Mater. **438–439**, 1080–1083 (2013)

Du, Z., Liu, J., Liu, J., Xu, F., Li, Y.: High-precision acoustic modeling with second-order staggered difference. Arab. J. Geosci. **10**(21), 473 (2017)

Kašpar, M., Pecen, J.: Detection of caves in a karst formation by means of electromagnetic waves. Geophys. Prospect. **23**(4), 611–621 (2010)

Kyun, P.S.: Detecting defects in or behind tunnel lining concrete using electromagnetic wave radar with change of antenna type and frequency domain. Mod. Phys. Lett. B **22**(11), 989–994 (2008)

Martínez-Moreno, F.J., Galindo-Zaldívar, J., Pedrera, A., Teixido, T., Ruano, P.: Integrated geophysical methods for studying the karst system of Gruta de las maravillas (Aracena, Southwest Spain). J. Appl. Geophys. **107**, 149–162 (2014)

Xie, J.L., Xu, J.L.: Ground penetrating radar-based experimental simulation and signal interpretation on roadway roof separation detection. Arab. J. Geosci. **8**(3), 1273–1280 (2015)

The Use of Geophysical Data for the Valorization of High Silica Limestone Deposit Reserves

Mezza Safa, Ben M'Barek-Jemaï Moufida, Ben Salah Imed, Ben Saad Ayssar, and Tiss Houcine

Abstract

This study corresponds to a characterization of the lower Eocene limestone reserves in Bizerte region in the North of Tunisia. The valorization of this useful substance consists on the one hand in its integration in the raw mixture for the manufacture of cement and on the other hand in its industrial use to fabricate granulates and marble stones. The studied limestone in Bizerte region is taken from the abandoned limestone quarry of the company Bizerte Cements considered as sterile reserves due to its high silica content. The characterization of the siliceous limestone begins with the detection of the natural fracturing at the deposit. This step is realized through direct measurements on ground, seismic refraction allowing the exploration of the subsoil and delimitation of the available reserves. The alteration of the siliceous limestone is especially preponderant in the superficial beds where the velocities lie between 400 and 2000 m/s. However, this limestone seems to be sane and not cracked with velocities oscillating between 2200 and 3600 m/s. Hence, the more the difference of velocities is visible, the more the rock is altered. The integration of this raw material in the composition of the raw powder allows filling of the silica defect that knows the current exploitation of the limestone while maintaining the quality of the cement produced by the company Bizerte Cements. The analyses of samples taken for the fabrication of granulates prove that the chemical, physical and mechanical properties of the limestone having a content of $SiO_2 > 7\%$ allow it to be used as a gravel for hydraulic concrete. The fabrication essays of marble stones from this siliceous limestone indicate that this one is characterized by a content of $26.18\% < SiO_2 < 32.93\%$ and $28.30\% < CaO < 33.54\%$. The siliceous limestone taken from Bizerte region has a considerable density varying from 2.50 to 2.75 g/cm^3 and a low porosity that varies from 4.34 to 4.36%. This abandoned limestone of the lower Eocene has a compressive strength ranging from 89.58 to 151.39 MPa and a flexural strength that exceeds 14 MPa. The Ypresian siliceous limestone is a very resistant rock well matched with the standards requirements.

Keywords

Siliceous limestone • Seismic refraction • Cement • Granulates • Marble stones

M. Safa · B. M'Barek-Jemaï Moufida · B. S. Imed (✉) · B. S. Ayssar
Department of Geology, Bizerte Sciences University, 7021 Zarzouna, Tunisia

T. Houcine
Cement Bizerte Laboratory, Baie de Sebra, Bizerte, Tunisia

1 Introduction

The exploitation of useful substances has become a crucial activity for the industrial development of a country. In Tunisia, there is a large variety of potential deposits of useful substances suitable for producing building materials, mainly the limestone which represents a significant resource of granulates and marble stones. The understanding and management of this mineral wealth is one of the determining factors for socioeconomic development.

The high consumption of useful substances for a developing country like Tunisia leads to the depletion of certain natural resources and the overexploitation of reserves which often harms the environment. For sustainable development, it is necessary to use mineral resources as rationally as possible (UNIDO 2005). The management of the mineral heritage is entrusted to the Quarry and Explosives Management of the Ministry of Housing and Territory Arrangement (MEHAT). This service takes into account economic studies to set the country's strategic orientations in mineral matters. The qyarry and explosives management

H. El-Askary et al. (eds.), *Research Developments in Geotechnics, Geo-Informatics and Remote Sensing*, Advances in Science, Technology & Innovation,
https://doi.org/10.1007/978-3-030-72896-0_24

mainly concerns granulates, marble rocks and building stones. Limestone is also used in the cement industry as a rock carrying lime as well as in the chemical industry and even in the agro-alimentary depending on its properties (Achour 2007).

Besides, studying fracture behavior in depth is fundamental for the exploitation of a quarry. It allows the control of the blasting parameters and thus the optimization of the shooting plan. This information is obtained only from geophysical data processing, particularly by the seismic data. The underground exploration method allows the delimitation of the sedimentary layers and the knowledge of their degree of alteration and cracking from the surface observation of wave behavior along their deep propagation (Coulombez 2007).

In this respect, this paper proposes a characterization of the raw material exploitable by the company Bizerte Cements, mainly the siliceous limestone of the lower Eocene (Ypresian) considered as a sterile and abandoned reserve. The characterization of the siliceous limestone requires the detection of the natural fracturing at the deposit. This step is realized through direct measurements on the ground and seismic refraction. Further, the valorization of this useful substance is achieved by its integration in the raw mixture for the manufacture of cement, as well as in its industrial use to fabricate granulates and marble stones.

2 Materials and Methods

The studied limestone in Bizerte region is taken from the abandoned limestone quarry of the company Bizerte Cements that is considered as sterile reserves due to its high silica content. The characterization of the siliceous limestone begins with the study of the natural fracturing at the deposit that is done by direct measurements on ground and by the use of geophysical data, particularly seismic refraction in order to explore the sub-soil and to define the available reserves. The valorization of the Ypresian siliceous limestone is manifested by:

- Sampling for integration essay in the raw powder of cements;
- Sampling for fabrication essay of marble stones;
- Sampling for fabrication essay of granulates.

The geochemical analyses of the collected samples allow (i) the assessment of the quality of the deposit, (ii) the delimitation of the homogeneous content zones and (iii) the establishment of the geochemical model in order to manage the raw material depending on the qualitative and quantitative requirements. The digital mapping refers to the establishment of maps realized by gathering information and then by integrating data on the appropriate software. This mapping guides the exploitation of quarries and predicts the potential deposit reserves. The software used (ArcMap, ArcCatalog and ArcToolbox) allows the geo-referencing of maps (Triki et al. 2014).

The geochemical analysis by X-ray fluorescence is a non-destructive spectral technic that operates the photoluminescence of atoms in the X-rays domain to determine the elemental composition (Monnin et al. 2006). It assesses the geochemical composition in terms of percentages of oxides. The crystal structure of materials is carried out through X-ray mineralogical analysis (Voglis et al. 2005). The samples intended for the manufacturing essays of marble stones and granulate also undergo a physical and geotechnical analysis. The control of the rock's physical and mechanical properties permits the definition and the characterization of their industrial application domain (Milazzo and Blasi 2003).

3 Results and Discussions

3.1 Geophysical Data

The measurements of the fracturing in the studied quarry and the geophysical data represented by seismic refraction are complementary. This suggests that the alteration of the siliceous limestone is especially preponderant in the superficial beds. At this level, the velocities range between 400 and 2000 m/s. However, deeper velocities oscillate between 2200 and 3600 m/s. More the difference of velocities is noticed, more the rock is altered and cracked (Triki et al. 2014). These results promote the delimitation of the exploitable reserves estimated for about **8,178,000** tons of abandoned limestone and allow the orientation of the exploitation plan according to the natural fracturing detected by seismic refraction. The study of fracture behavior in depth is significant for the exploitation of a quarry allowing the control of the blasting operation in the quarry. Therefore, the geophysical data can be considered as an exploration method of the underground. Indeed, the data highlight the alteration level and the cracking degrees of the sedimentary layers which control the optimization of the shooting plan (quality and quantity of explosives to use) (Coulombez 2007).

3.2 Geochemical Data

The geochemical data collected are based on the results of the chemical analysis of the limestone samples taken from the studied deposit at different levels. They are reported in the geochemical maps obtained by a referencing compared to the UTM/Carthage Zone 32 N projection system (Fig. 1). The geochemical map of iso-concentration of CaO reflects the spatial distribution of lime and silica through the limestone deposit. The most dominant CaO content is between 40 and 45%, while the lowest value of CaO is between 25 and 30%. It coincides with the nearest limestone to the blackish marl reserves. The geochemical model shows that the limestone content of silica increases from the North to the South with a value of SiO_2 ranging from 8 to 14%. Going southwards, the silica concentration increases gradually to reach 32%, coinciding with the most siliceous limestone in the deposit adjoining the blackish marl of the Paleocene. This type of mapping helps to monitor the quality of the quarry during the phase of exploitation, as well as the exploration in order to monitor and manage the deposit reserves according to the qualitative requirements (Roy 1999).

3.3 Use of Siliceous Limestone in Cement Making

Raw materials from the study area are gathered in a calculation tool that takes into account all the available reserves in a quarry. It also allows the estimation of the geochemical composition of the raw powder and even the cement to produce (Rahhal and Talero 2008). The integration of the siliceous limestone in the composition of the raw powder allows filling of the silica defect that recognizes the current exploitation of the limestone while maintaining the quality of the cement produced by the company Bizerte Cements (Table 1). This mixture obtained is made with 67% of the regular limestone exploited added to that 33% of the siliceous limestone. Indeed, the respect for these mixing ratios between regular and siliceous limestone permits to obtain a limestone heap with a geochemical composition that respects the Tunisian cement manufactories specifications ($SiO_2 <$ 10% and a $CaO > 46\%$). These results serve to estimate the marl heap as well as the raw mix designing.

3.4 Use of Siliceous Limestone in Marble Field

The industrial essay of marble stones from this siliceous limestone reveals the composition of $26.18\% < SiO_2 < 32.93\%$ and $28.30\% < CaO < 33.54\%$. The siliceous limestone taken from Bizerte region has a considerable density varying from 2.50 to 2.75 g/cm^3 and a low porosity that varies from 4.34 to 4.36%. The porosity controls all the other physical parameters of the rock (density, permeability, electrical conductivity, etc.). Hence, the porosity increases with

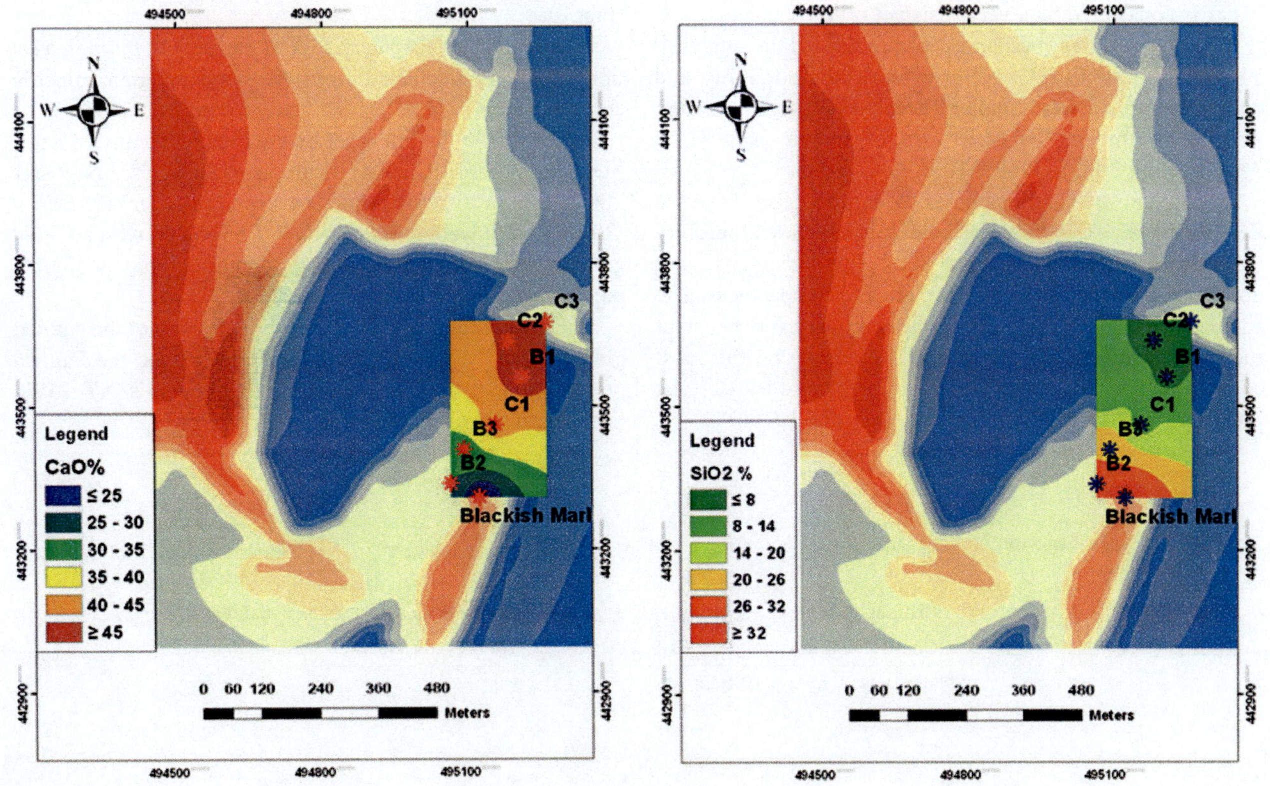

Fig. 1 Geochemical maps of CaO and SiO_2 along the siliceous limestone deposit

Table 1 Geochemical analyses of the raw mix before and after integration of the siliceous limestone

Oxides wt%	Current raw powder	Raw powder after integration of the siliceous limestone	Gap
SiO_2	13.21	13.12	0.09
Al_2O_3	3.26	2.98	0.28
Fe_2O_3	2.75	2.72	0.03
CaO	42.62	42.01	0.61
MgO	0.80	1.31	0.51
SO_3	0.36	0.62	0.26
K_2O	0.18	0.10	0.08
Na_2O	0.02	0.04	0.02
P_2O_5	0.36	0.40	0.04
LOI wt%	34.36	34.45	0.09

the decrease of the rock resistance (Torrenti and Bendboudjema 2005). This abandoned limestone of the lower Eocene has a compressive strength ranging from 89.58 to 151.39 MPa and a flexural strength that exceeds 14 MPa (Table 2).

From a comparison between the obtained results in this study and the classification of rocks set up by BRGM (El Amrani et al. 2009), the sample B2 admitting a compressive strength higher than 50 MPa (50 MPa $\leq$ 80.96 $\leq$ 100 MPa) is qualified as a compact and a resistant limestone. Concerning the other siliceous limestone, B3 which has the highest compressive strength (100 MPa $\leq$ 151.39 $\leq$ 250 MPa) is considered as a very resistant rock (Fig. 2).

The physical and mechanical parameters of the analyzed siliceous limestone are used to evaluate the durability and compatibility of every sample. These properties are of high importance for the industrial use of marble stones (El Amrani et al. 2009) (Table 3).

Table 3 shows that the samples B2 and B3, taken from the abandoned siliceous limestone, have the highest durability and compatibility index (DCI > 800). This indicates a very good quality of the rock with a remarkable compactness. The siliceous limestone of the lower Eocene is then a very resistant rock with considerable properties. It seems to be the most compatible for the industrial use as marble stones (CFMR 2004).

3.5 Use of Siliceous Limestone in Granulate Field

The chemical analyses of samples taken for the fabrication of granulates prove that the studied limestone is characterized by a content of SiO_2 = 11.19%, CaO = 46.07%, SO_3 = 0.03%, $CaCO_3$ = 82.26%, K_2O and Na_2O less than 0.1%.

The granularity of the analyzed limestone in this study is measured according to the European Standards EN 933–1 by a series of AFNOR sieve with square stitch (BRGM 2000) (Fig. 3).

This granulate characterization is carried out through a set of coefficients calculated from the obtained granulometric curve (Table 4).

According to the limits set by the French Comity of Rock Mechanics (CFMR 2004) (European Standard EN 933-1 2008), **the analyzed samples correspond to gravel** (Me = 20 mm) very well classified (S_0 = 1.58 < 2.5) with a varied granulometry implicating that this granulate has negligible porosity and permeability.

The identification of the application domain of the studied granulate is based on the evaluation of its mechanical properties according to the Tunisian Standards **NT 21.30** (Achour 2007) (Table 5).

Table 2 Mechanical analyses of the marble stones

Samples (MPa)	B2	B3	Ascertainment	Ultimate thresholds of European Standards (Bars)
Bending test	> 14	> 14	No break or crack	70–80
Compression test	89.58	151.39	No break or crack	500–600

NB= nota: 1 MPa = 10 Bars

Fig. 2 Marble samples from Bizerte cement quarry

Table 3 Durability and compatibility index and relative classification of the analyzed siliceous limestone

Samples	Durability and compatibility index	Relative classification
B2	887	**2**
B3	977	**1**

Bold is used to show the standards requirements

This abandoned limestone of the lower Eocene has a Los Angeles coefficient (L.A) value of 24.6% and a Micro-Deval coefficient (M.D) estimated for about 29.49%. This makes it more suitable as gravel for hydraulic concrete than as pavement and superficial coat.

4 Conclusion

This study is a characterization of the Ypresian limestone of the North East of Tunisia (Bizerte area) considered as a sterile material according to its silica content. The analysis of the fracturing with geophysical data allows the organization of a predictional operational plan. It further leads to the delimitation of the available reserves in the deposit. The integration of this raw material in the raw powder composition provides very encouraging results for the use of this raw material in cement industry according to the Tunisian and European standard requirements. The industrial essay of marble stones suggests that this siliceous limestone has similar properties to marble stone of the lower Eocene existing in Central Tunisia (Kesra marble). The industrial analysis of granulates proves that this limestone with $SiO_2 < 7\%$ can serve in making all types of hydraulic concrete, particularly the exceptional concrete adopted in civil engineering.

Fig. 3 Granulate samples from Bizerte cement quarry

Table 4 Characterization of the granulometry of the studied samples

Q_1 (mm)	Q_3 (mm)	D_{10} (mm)	D_{60} (mm)	Me (mm)	S_0	C_U
10	25	5	20	20	1.58	4

Table 5 Characterization of the mechanical properties and the application domain of the studied granulate

	Los Angeles coefficient (L.A)	Micro-Deval coefficient (M.D)
Granulate samples	24.6%	29.49%
Concrete	**$\leq$ 40**	**$\leq$ 35**
Pavement	**25 $\leq$ LA $\leq$ 30**	**20 $\leq$ MDE $\leq$ 25**

Bold is used to show the standards requirements

References

Achour, T.: Étude de l'influence de la nature des granulats sur les propriétés des bétons hydrauliques : cas des granulats calcaires tunisiens. Université Henri Poincaré, Nancy Université et l'École Nationale d'Ingénieurs de Tunis, Thèse de Doctorat en Génie Civil (2007)

BRGM: Bureau de Recherche Géologique et Minière. Guide méthodologique de sélection des pierres des monuments en termes de durabilité et compatibilité. Projet de recherche MET-D03 (2000)

CFMR: Comité Français de Mécanique des Roches. Manuel de mécanique des roches: Tome 2 Les applications. Ecole des mines de Paris (2004). ISBN: 2-911762-45-2

Coulombez, C.: Rapport d'option : Analyse et optimisation des pratiques d'abattage à l'explosif dans une carrière de granulats. Centre de Géosciences Paris Tech (2007)

El Amrani, E.L., Hassani, I., El Azhari, H.: Evaluation des propriétés physico-mécaniques des pierres de construction du Maroc à partir des vitesses des ondes P et de la résistance au choc. Bulletin De L'institut Scientifique, Rabat, Section Sciences De La Terre **31**, 41–54 (2009)

Milazzo, G., Blasi, P.: L'extraction des pierres ornementales dans les pays ACP. Centre pour le Développement de l'Entreprise Bruxelles (2003)

Monnin, Y., Dégrugilliers, P., Bulteel, D., Garcia-Diaz, E.: Petrography study of two siliceous limestone submitted to alkali-silica reaction. Cem. Concr. Res. **36**(8), pp 1460–1466 (2006). https://doi.org/10.1016/j.cemconres.2006.03.025

Norme Eropeenne EN 933-1: Caractéristiques géométriques des granulats : Détermination de la granularité. Analyse granulométrique par tamisage (2008)

Rahhal, V., Talero, R.: Calorimetry of Portland cement with metakaolins, quartz and gypsum additions. J. Therm. Anal. Calorim. **91**, 825–834 (2008). https://doi.org/10.1007/s10973-006-8250-6

Roy, D.M.: Alkali-activated cements opportunities and challenges. Cem. Concr. Res. **29**, 249–254 (1999). https://doi.org/10.1016/S0008-8846(98)00093-3

Torrenti, J.M., Bendboudjema, F.: Mechanical threshold of cementitious materials at early age. Materialsand Struct. **38**, 299–304 (2005). https://doi.org/10.1007/BF02479294

Triki, I., Hentati, I., Trabelsi, N., Zairi, M.: Evaluation de techniques d'interpolation spatiale à l'aide de l'extension Geostatistical Analyst d'ArcGIS : Cas du système aquifère phréatique de Sfax. Laboratoire Eau Energie Environnement. Ecole nationale d'ingénieurs de Sfax (2014)

UNIDO: Organisation des Nations Unies pour le Développement Industriel. Les pierres marbrières tunisiennes: Caractéristiques techniques et ressources disponibles. Unité pour la Promotion des investissements (2005)

Voglis, N., Kakali, G., Chaniotakis, E., Tsivilis, S.: Portland-limestone cement, their properties and hydration compared to those of other composite cement. Cem. Concr. Comp. **27**, 191–196 (2005). https://doi.org/10.1016/j.cemconcomp.2004.02.006

A Comparative Assessment of the Earthquake-Related Settlement in Embankment Dams Using Artificial Neural Networks and Multivariate Adaptive Regression Splines

Abdelatif Zeroual, Messaoud Djeddou, and Ali Fourar

Abstract

The need for new advanced methods in modeling the seismic behavior of earth dams and evaluating the resulting deformations is justified by the uncertainties surrounding the conventional methods. In this respect, the present study focuses on the prediction of crest settlement of embankment dams under variant earthquake loading. To this end, it applies two approaches: the feed-forward back-propagation neural network (FBNN) and multivariate adaptive regression splines (MARS). A total number of 152 datasets of historically documented cases of earthquake are gathered for models development and comparative performance between model forecasts. The obtained results indicated that both models could be used as reliable tools to predict the earthquake-related crest settlement in embankment dams.

Keywords

Earthquake-related crest settlement • Embankment dam • Simplified dynamic analysis • FBNN • MARS

A. Zeroual (✉)
Faculty of Science and Applied Science, Department of Hydraulic, University L'Arbi ben M'hidi, Oum El Bouaghi, Algeria
e-mail: zeroual.abdelatif@univ-oeb.dz

M. Djeddou
Research Laboratory in Subterranean and Surface Hydraulics (LARHYSS), Faculty of Sciences and Technology, Mohamed Khider University, Biskra, Algeria

A. Fourar
Faculty of Technology, Department of Hydraulic, University of Batna 2, Batna, Algeria

1 Introduction

Using artificial intelligence approaches (AIAs) to model the complex relationship between earthquake and embankment dam parameters may be considered the most precise method in predicting the earthquake-related deformation based on historical real cases data. Given the complexity of the seismic phenomenon, there are no accurate mathematical explicit equations that relate the independent to the dependent variables to assess the response of such dams. Therefore, semi-empirical formulas have been developed to obtain a preliminary estimate of the permanent, earthquake-induced deformation of embankment dams (Jibson 2007). These procedures based on the sliding block approach are generally not sufficiently accurate (Singh et al. 2007). Recently, researchers have used soft computing techniques (FBNN, RBFNN, SVM, MLR) for predicting deformations of embankment dams (Javdanian et al. 2018; Zeroual et al. 2018; Javdanian and Pradhan 2019).

MARS was developed as a nonparametric statistical approach with no assumptions about the underlying functional relationships between dependent and independent variables (Zhang and Goh 2013; Goh and Zhang 2014).

2 Materials and Methods

2.1 Database and Model Development

In this study, two soft computing models (FBNN and MARS) were developed to predict earthquake-induced settlement in embankment dams. Both predictive models were developed using a comprehensive datasets of 152 published cases of earthquake-related settlement in embankment dams. Respectively, 70% of the gathered data was used in the training phase, while 30% was used in the testing phase (Singh and Debasis 2009). Also, the ANN used was a back-propagation neural network with three layers. The

H. El-Askary et al. (eds.), *Research Developments in Geotechnics, Geo-Informatics and Remote Sensing*, Advances in Science, Technology & Innovation,
https://doi.org/10.1007/978-3-030-72896-0_25

input layer consisted of seven input neurons (DT, H, a_y, a_{max}, T_D, T_P, M_w), while the output layer consisted of a single neuron which represents the earthquake-induced crest settlement (CS_{EQ}) (Table 1).

3 Results

The trial-and-error approach was carried out to find the optimum models. As a result, the optimum architecture was found (7-10-1) for the developed FBNN model, whereas the optimum MARS model predicting (CS_{EQ}) was contained of 35 basis functions (BFs).

Table 2 depicts the comparison of the observed output values with the output values predicted from the developed models for training and testing sets, respectively.

4 Discussion

Table 2 indicates that both models performed acceptably, predicting results with low error. Also, it can be noted that predicted $(CS_{EQ})_{Pred}$ values are quite close to the observed $(CS_{EQ})_{Obs}$ values in both models, as their r values are in good agreement with the observed values ($r > 0.9$). Furthermore, it is rather difficult to select a more accurate model for predicting CS_{EQ}. Thus, in this study, the prediction performance indices have been combined and used together to compare the performances of developed models (Yesiloglu-Gultekin et al. 2013). Therefore, the new performance index (PI) is established for this purpose as follows:

$$PI = \left[\left(R^2 + r\right) - (RMSE + MSE)\right] \quad (1)$$

The obtained results show that the accuracy of the MARS model is slightly lower compared to that of the FBNN one (Table 2); even though, MARS is considered more computationally efficient and easier to be interpret than BPNN (Zhang and Goh 2016). Besides, the effect of the input variables was determined from the training datasets of the models as shown in Table 3. Despite the difference in results between the two predictive models, the most important influence was registered for the geotechnical characteristics (a_y and T_D), whereas M_w had the lowest important influence.

To verify and validate the developed models, a new case (Kaddara rockfill dam subjected to Boumerdas earthquake loading) is selected with input data that did not previously exist in the database. Table 4 shows the selected variables.

Table 1 Details of the input and output variables used in this study (all data)

Variables	Min.	Max.	Average	Median	Std. Dev.	Kurtosis	Skewness
DT*: dam type	1	13	6.388	7	2.487	0.124	− 0.332
H: height of dam	2.5	235	42.147	29	40.117	3.844	1.798
M_W: magnitude of the earthquake	4.5	8.3	7.019	7.1	0.823	1.018	− 1.049
a_{max}: maximum horizontal earthquake acceleraton	0.004	0.9	0.285	0.235	0.211	0.597	1.040
T_P: predominant earthquake period	0.25	0.96	0.377	0.32	0.131	2.172	1.524
T_D: fundamental period of dam	0.05	2.74	0.580	0.51	0.436	3.386	1.476
a_y: yield acceleration	0	0.55	0.163	0.15	0.110	1.026	0.811
CS_{EQ}: crest settlement	0.001	32	1.313	0.1045	3.746	37.772	5.652

*Dam types: 1: zone levee, 2: multi-zone levee, 3: zone earth dam, 4: zone embankment, 5: zone hydraulic fill dam; 6: Multi-zone hydraulic fill; 7: compacted multi-zone dam; 8: multi-zone rock fill dam; 9 Concrete faced rock fill dam (CFRD); 10: concrete faced decomposed granite or gravel dam; 11: natural slope; 12: upstream constructed tailings dam; 13: downstream constructed tailings dam

Table 2 Comparison between the predictive models

Model	Set	R^2	r	RMSE	MSE	ΣPI
FBNN (7–10-1)	Training	0.938	0.969	0.02960	0.00088	3.694
	Test	0.903	0.950	0.03431	0.00117	
MARS (35FBs)	Training	0.982	0.991	0.01595	0.00025	3.681
	Test	0.848	0.921	0.04324	0.00187	

Table 3 Relative importance of inputs variable

Model	Variable						
	a_y	T_D	T_P	a_{max}	H	DT	M_w
MARS	100	62.08	4.37	31.66	28.56	21.51	11.58
BPNN	64.65	100	9.05	16.43	92.89	37.21	3.46

Table 4 New input data (Kaddara dam)

Dam	Earthquake			Input variable						Reference
	Dis. Ep	Date	DT	H (m)	M_W	a_{max} (g)	T_P (s)	T_D (s)	a_y (g)	
Kaddara	30	21/05/2003	8	106	6.8	0.23	0.32	1.35	0.30	Louadj et al. (2009)

Table 5 Comparison of results between predictive models and simplified dynamic methods

Methods	Simplified dynamic method		BPNN	MARS	Note
	Bureau	Swaisgood			
$(CS_{EQ})_{Pred}$ (cm)	6.93	8.24	1.60	0.0579	$(CS_{EQ})_{Obs}$ = 1 cm
Relative error (%)	593	724	15	94.21	

Table 5 presents the results obtained from the Bureau method (2011), Swaisgood method (2003) and the developed models (FBNN and MARS). The comparison of the results indicates the ability of these models to provide accurate results.

5 Conclusions

In this study, the BPNN and MARS models were developed to assess the earthquake-induced crest settlement of embankment dams, CS_{EQ}. Assessing the accuracy of the developed models indicates that although both FBNN model (ΣPI = 3.694) and MARS model (ΣPI = 3.681) offered acceptable accuracy, the BPNN model has a little higher accuracy. Then, relative importance analysis was conducted to assess the effect of each of the input parameters on the crest settlement of the embankment dams (CS_{EQ}), where geotechnical characteristics (a_y and T_D) were found to have the most significant influence. Finally, the application of the developed models was compared to the simplified dynamic methods for assessment of crest settlement of Kaddara rockfill dam.

References

Bureau, G.: Quick assessment of potential seismic deformations of embankment dams. Risk Analysis, Dam Secur. Critical Infrastruct. Manage. **301** (2011)

Goh, A.T., Zhang, W.G.: An improvement to MLR model for predicting liquefaction-induced lateral spread using multivariate adaptive regression splines. Eng. Geol. **170**, 1–10 (2014)

Javdanian, H., Pradhan, B.: Assessment of earthquake-induced slope deformation of earth dams using soft computing techniques. Landslides **16**(1), 91–103 (2019)

Javdanian, H., Zarif Sanayei, H.R., and Shakarami, L.: A regression-based approach to predict crest settlement of embankment dams under earthquake shaking. Sci. Iranica (2018)

Jibson, R.W.: Regression models for estimating coseismic landslide displacement. Eng. Geol. **91**(2–4), 209–218 (2007)

Louadj, S., Vincens, E., Bahar, R., Laouami, N.: Identification of Keddara dam behavior during Boumerdes earthquake on May 21, 2003. Int. J. Geotech. Eng. **3**(1), 117–132 (2009)

Singh, R., Debasis, R.: Estimation of earthquake-induced crest settlements of embankments. Am. J. Eng. Appl. Sci. **2**(3), 515–525 (2009)

Singh, R., Debasis, R., Das, D.: A correlation for permanent earthquake-induced deformation of earth embankments. Eng. Geol. **90**(3–4), 174–185 (2007)

Swaisgood, J.R.: Embankment dam deformations caused by earthquakes. In: Pacific Conference on Earthquake Engineering, p. 014 (2003)

Yesiloglu-Gultekin, N., Gokceoglu, C., Sezer, E.A.: Prediction of uniaxial compressive strength of granitic rocks by various nonlinear tools and comparison of their performances. Int. J. Rock Mech. Min. Sci. **62**, 113–122 (2013)

Zeroual, A., Djeddou, M., Fourar, A.: Artificial neural network application for the Prediction of earthquake-induced crest settlement in rockfill dams. In: First International Conference on Dams, ICDBiskra (2018)

Zhang, W., Goh, A.T.: Multivariate adaptive regression splines and neural network models for prediction of pile drivability. Geosci. Front. **7**(1), 45–52 (2016)

Zhang, W.G., Goh, A.T.C.: Multivariate adaptive regression splines for analysis of geotechnical engineering systems. Comput. Geotech. **48**, 82–95 (2013)

Geotechnical Settings of Volcanic Residual Soils and Derived Engineering Problems in El Hierro Island (Spain)

Fabiola Fernández-Baniela, Daniel Arias-Prieto, and Álvaro Rubio-Ordóñez

Abstract

This article presents the geotechnical settings of the materials filling the bottom of an inactive volcanic crater on El Hierro Island (Canary Islands, Spain). This natural depression is selected to build the upper reservoir of El Hierro Hydro-Wind Plant (Gorona del Viento, S.A.). In this respect, and as per requirement of a geological study (geological mapping, geochemical analysis and petrographic study), an exhaustive geotechnical investigation (rotary boreholes, dynamic penetrations and trial pits) and geophysical surveys (seismic refraction and electrical tomography profiles) are carried out at this site. This article deals with residual soils generated from the weathering of volcanic materials, their physical properties and mechanical behavior. The deficient geotechnical characteristics of this type of soil imply a high risk of ground settlement, which must be foreseen in advance, carrying out a soil improvement by the most appropriate engineering technique.

Keywords

Geological engineering • Volcanic residual soils • Settlements

F. Fernández-Baniela (✉)
Faculty of Geology, University of Oviedo, Spain. IDOM Consulting, Engineering, Architecture, S.A.U., Avda. Monasterio de El Escorial 4, 28049 Madrid, Spain
e-mail: fbaniela@idom.com

D. Arias-Prieto · Á. Rubio-Ordóñez
Department of Geology, University of Oviedo, Jesús Arias de Velasco s/n, 33005 Oviedo, Spain

1 Introduction

On El Hierro Island (Canary Islands), a new energy model of fully renewable energy sources, combining wind and hydraulic energy, was implemented by the company Gorona del Viento El Hierro S.A. By the construction of El Hierro Hydro-Wind Plant, IDOM company carried out the detailed engineering design. One of the main elements of this plant is the upper water reservoir, whose geometry was designed to adapt to the topography of the selected site: a depression corresponding to an inactive volcanic crater. The geotechnical research campaign carried out by IDOM for the Geological-Geotechnical Study revealed the high thickness of soils that fill the crater (locally higher than 30 m), consisting mainly of soils generated by the intense weathering that affected the volcanic materials, characterized by their low bearing capacity and high deformability.

For the correct engineering design of any structure in volcanic soil areas, determining the existence and thickness of this type of soils is essential. To this end, a geotechnical research campaign suitable for this type of materials is recommended. Also, the soil improvement technique to be adopted should be the most suitable for the site conditions in order to guarantee the stability of the structure to be built.

2 Site Investigation Works

Three investigation campaigns were carried out by mechanical and geophysical techniques (Fig. 1): 16 boreholes (depths between 20 and 57 m), 9 dynamic penetrations of DPSH type (depths between 5 and 26 m), 10 trial pits, 9 seismic refraction profiles (total length of 540 m) and 2 electrical tomography profiles (total length of 465 m); standard penetration tests (SPTs) and pressuremeter tests were performed into the boreholes, and identification, resistance and deformability laboratory tests were carried out on numerous soil samples.

H. El-Askary et al. (eds.), *Research Developments in Geotechnics, Geo-Informatics and Remote Sensing*, Advances in Science, Technology & Innovation,
https://doi.org/10.1007/978-3-030-72896-0_26

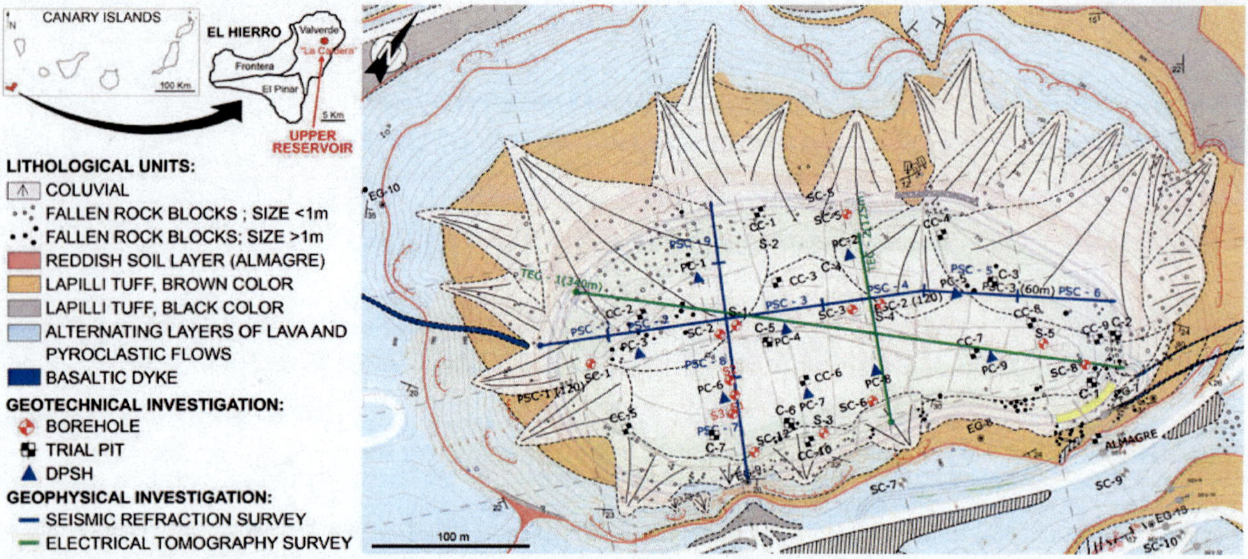

Fig. 1 Geological mapping of the upper reservoir showing the location of the investigations

3 Results

3.1 Geological Settings

The site selected for the upper reservoir is an inactive crater, named "La Caldera", with slopes comprising competent and resistant volcanic rocks ("aa" lava flows) and whose bottom is filled with soils derived from weathering of volcanic materials as well as colluvial soils on the surface.

The soils of volcanic origin are basically silty and clayey, with some sand; they are characterized by low dry density and very high void ratio (Wesley 2001; Shoji et al. 1993), which makes them highly deformable.

The results of the laboratory tests confirm that the structure of these soils corresponds to the typical structure generated by weathering of volcanic materials (Lizcano et al. 2006), synthesized in Fig. 2.

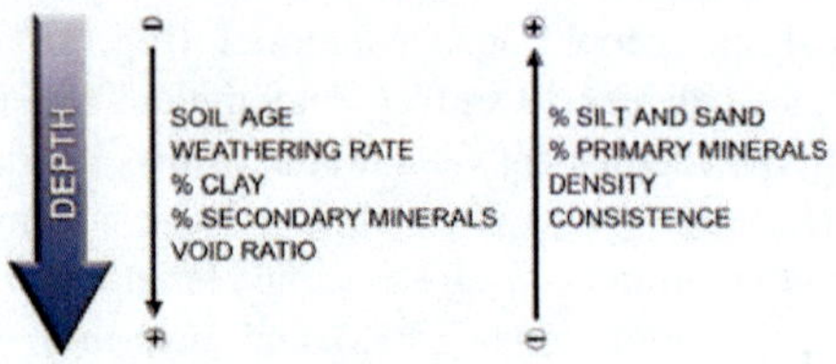

Fig. 2 Variation of soil features with depth

3.2 Geotechnical Settings

It was a complicated process to obtain undisturbed samples in these materials. Indeed, the drilling of the rotary boreholes produced their disintegration which required that much of the investigations of these materials be performed from "in situ" tests: pressuremeter tests to determine their deformability (Elastimeter-2 equipment, model 4181, from Oyo Corporation, Japan) and standard penetration tests (SPTs) inside the boreholes and dynamic penetrations (DPSHs) distributed inside the crater to determine the soil strength. The results of all these tests are noticeably worse in the southeast area of La Caldera (in black color in Fig. 3) than in the rest of the investigated area.

Most of the N_{SPT} values obtained in the southeast area were in the 5–15 range; values less than 5 were recorded between 28 and 38 m depth, where a layer of black plastic silts was identified (MH according to Unified Soil Classification System), characterized by a dry density of 0.90 g/cm^3 and a void ratio of 2.27. Values of N_{20} less than 5 were recorded up to 11 m deep. In most of the 46 pressuremeter tests performed, the value of the pressuremeter modulus obtained (E_p) was less than 20 MPa, and even less than 10 MPa in the tests carried out in the boreholes located in the southeast area. The low values of penetration resistance and pressuremeter modulus of these soils, the high thickness of soils detected (up 46.5 m) and the presence of the layer of black silts of low density and high porosity (not identified in other areas) led to interpret the southeast area of La Caldera as the probable location of the emission point of the crater.

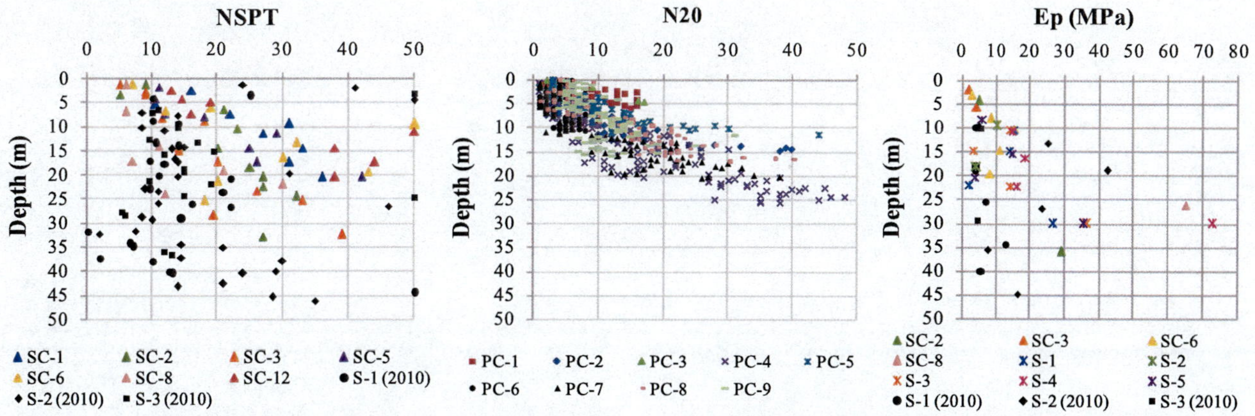

Fig. 3 Results of "in situ" tests carried out in La Caldera versus depth. N_{SPT}: values obtained in SPT tests (number of blows required to penetrate 300 mm into the soil); N_{20}: values obtained in DPSH investigations (number of blows required to penetrate 200 mm into the soil); E_p: pressuremeter modulus obtained in the pressuremeter tests carried out into the boreholes

3.3 Engineering Implications

The partial preloading was the technique selected to improve the soil at the bottom of the upper deposit, using the volume of excavated material. This election was conditioned in part by the singularities of El Hierro Island, declared a Biosphere Reserve in 2000 Network, and by constraints specific to this work (availability of machinery on the island and execution deadlines). Due to the deficient geotechnical characteristics of these materials, after the execution of the preloading, the value of the accumulated settlement generally varied between 0.1 and 0.4 m, reaching a maximum value of 1.5 m in the southeast area of La Caldera (Fernández-Baniela et al. 2018), where the most considerable thickness of soils was detected.

4 Concluding Remarks

Soils generated by weathering of volcanic materials seem to have deficient geotechnical characteristics that can be summarized as follows:

- Soil age, weathering rate, clay and secondary minerals content and void ratio increase with depth.
- Low values of dry density ($\leq$ 1 g/cm^3) and high values of void ratio (> 2) are usually characteristic of these soils.
- The mechanical behavior of these soils is inadequate: They exhibit low resistance (N_{spt} < 15 and N_{20} < 5) and high deformability (E_p < 10 MPa).
- From the engineering point of view, the geotechnical settings of this type of soils imply a high risk for settlements.

Acknowledgements We thank Gorona del Viento El Hierro, S.A., for granting IDOM the realization of the engineering design of El Hierro Hydro-Wind Plant. Special thanks to the IDOM staff and external consultants who were involved in the geotechnical research and the Geological-Geotechnical Study.

References

Fernández-Baniela, F., Arias, D., Rubio-Ordóñez, A.: Causes of ground settlements in upper reservoir of Hydro-wind Plant of El Hierro and method implemented for soil improvement. 2nd Int. Res. Conf. Sustain. Energy Eng. Mater. Environ. (IRCSEEME) Proc. **2**(23), 1468 (2018). https://doi.org/10.3390/proceedings2231468

Lizcano, A., Herrera, M.C., Santamarina, J.C.: Suelos derivados de cenizas volcánicas en Colombia. Rev. Int. de Desastres Naturales, Accidentes e Infraestructura Civil **6**(2), 167–198 (2006)

Shoji, S., Nanzyo, M., Dahlgren, R.A.: Volcanic Ash Soils: Genesis, Properties and Utilization. Elsevier Science, Amsterdam (1993)

Wesley, L.D.: Consolidation behavior of allophane clays. Géotechnique **51**(10), 901–904 (2001)

The Investigation of the Stress Field Evolution in Soil Samples During Vertical Loads with Computer Tomography

Viktória Mikita, Balázs Kovács, Imre Czinkota, Hasan Eteraf, and Zsolt Pinjung

Abstract

This paper presents the behavior of loose soils under vertical loads investigated by computed tomography. In order to determine the different bearing characteristics and to understand the compaction distributions in the soil samples, a static compression test series were carried out with the help of a cylindrical loading head. After the compression experiments, CT measurements were performed on the soil samples. By analyzing the CT data matrices, the relationship between density distribution and depth was described for rapid compaction processes in the subsurface environment by exponential approximation. In the soil samples with a secondary structure, a zone with higher compactness is formed directly under the load surface, which tends to decrease to a depth from the plane of the load, until it approaches the in situ density state.

Keywords

Soil compaction • Density distribution • Computer tomography • Loose soil • Stress field

1 Introduction

In soil mechanics, one of the most critical features of the soil is the resistance to vertical loads and the resulting deformations. The non-destructive exploration of deformations in the soil structure during the geotechnical tests provides additional information to the parameters and boundary conditions used in conventional measurement techniques (Cavé et al. 2009).

The soil skeleton and its deformations can be investigated in a non-destructive way by computed tomography (Petrovic et al. 1982). The results of the medical CT measurements are in Hounsfield units (HU), with a density-related parameter. Therefore, the evolution of the deformation field can be determined with computer tomography.

The measurements aim to (i) provide additional information on the estimation of optimal sample size in traditional geotechnical tests (as uniaxial or triaxial measurements) and (ii) determine the compaction rate distribution of soil samples in cohesive soils by a non-destructive way.

2 Materials and Methods

For the investigations, a very loose soil model was prepared. The optimal moisture content (for maximum compaction) of the soil was determined with Proctor tests. The cylindrical soil samples were made—according to the Proctor measurements—from semi dry ($w = 12.6\%$), medium wet ($w = 17.5\%$) and wet ($w = 26\%$) sandy loam soil with 105 mm diameter and 70 mm height. In order to achieve homogeneity, an automatic sample compaction device that maintains a constant pressure (0.2 bar) for every layer was applied to the soil sample built up from 5 layers. Furthermore, by measuring the thickness of each of the compressed layers, the apparatus allowed to determine the wet density and initial porosity of the soil column.

An X-ray-compatible plastic and wood-based soil sample holder allowed to continuously monitor the evolution of the deformation field in the soil sample using computer tomography measurements. The vertical load was applied at the motorized platform of the CT equipment in five loading steps by a 55 mm diameter and 30 mm height. The vertical

V. Mikita (✉) · H. Eteraf · Z. Pinjung
University of Miskolc, Miskolc, Hungary
e-mail: hgmv@uni-miskolc.hu

B. Kovács
Gáma-Geo Ltd., Miskolc, Hungary

I. Czinkota
Szent István University, Gödöllő, Hungary

H. El-Askary et al. (eds.), *Research Developments in Geotechnics, Geo-Informatics and Remote Sensing*, Advances in Science, Technology & Innovation,
https://doi.org/10.1007/978-3-030-72896-0_27

displacement (penetration) was controlled during the investigations.

After each loading steps (5 mm vertical displacement), CT measurements were performed with a medical CT (Siemens Somatom Definition AS+). The advantage of the test method was that the sample did not move from its initial position during the test series. Therefore, the CT matrices of the different load steps could be compared to each other: This method minimized the technical error in the analysis of the deformation field.

3 Results

Assuming a uniform load distribution at the surface of the soil sample, the pressure distribution in the soil was described by isobaric lines. The shape of the isobars varies with the load-bearing capacity of the soil; they can be characterized by lower depth and horizontally wider shape under higher loads. The resulted compaction field related to the applied pressure after the compression experiment (loading step series) is illustrated in Fig. 1.

After the compression experiments, the values of CT data matrixes were analyzed. Under natural conditions in loose soils and sediments, the slow compaction processes can be described by an exponential relationship (1) between the density (compaction) distribution and the depth (Athy 1930). With the modification of Athy's equation, the rapid compaction processes can be described as follows (Mikita 2013).

$$\overline{\mathrm{HU}} = b \cdot e^{(a_{\mathrm{HU}}(x_0 - x))} + c \qquad (1)$$

In Eq. (1), the a_{HU} is the HU change coefficient [−], which coefficient describes the degree of change in HU values as a function of depth. The parameter b is proportional to the absolute magnitude of the changes, while the factor c is the HU value for the initial compaction state of the soil.

Figure 2 shows that the compaction characteristics of soil samples at variant moisture contents are different. The compaction of the medium dry (w = 12.6%) soil sample is low and almost uniform along the depth. A higher compaction rate compared to the medium dry soil is reached at a higher moisture level (w = 26%), but the maximum compaction occurs at the medium wet (w = 17.5%) soil sample.

The largest change in compactness (density or HU change) occurs in the first two loading steps, which density changes are less and less caused by the additional loads. The effect of the first loading step can be detected up to 25 mm depth from the contact surface. In the second loading step, with 15 mm indentation, the whole depth of the sample is affected by the load. The bottom side effect resulting from the geometry of the sample starts to act from the second loading step.

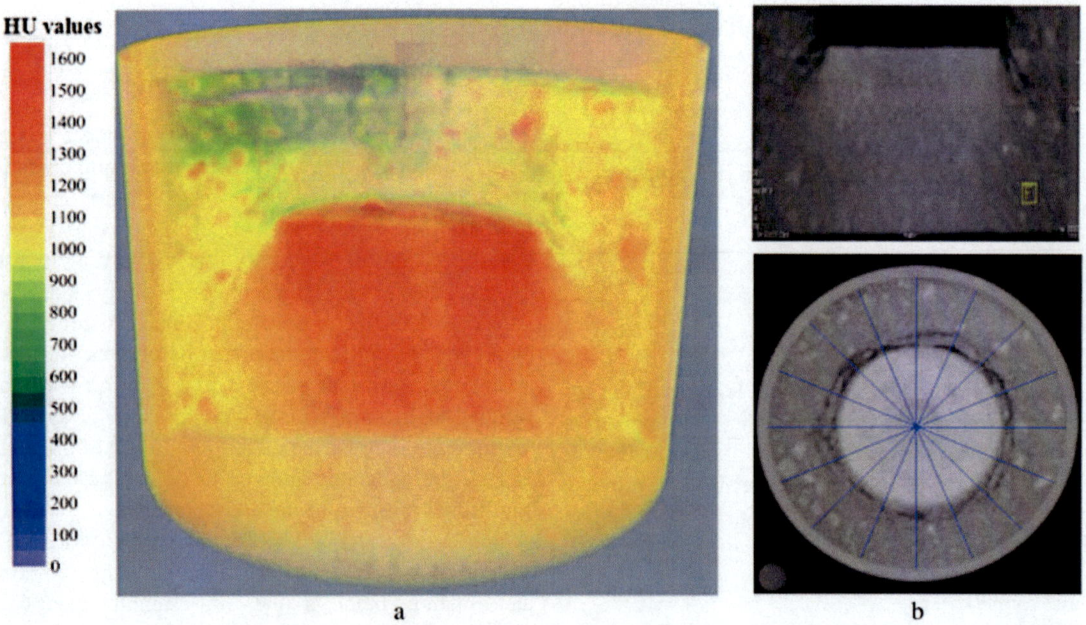

Fig. 1 Soil sample with the loose and compacted zones (**a**) reconstructed from CT images (**b**)

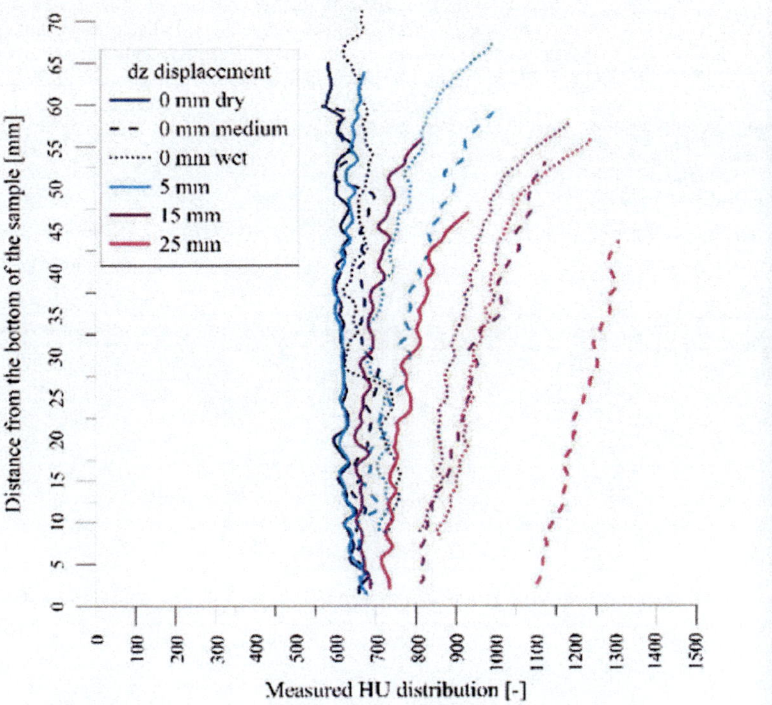

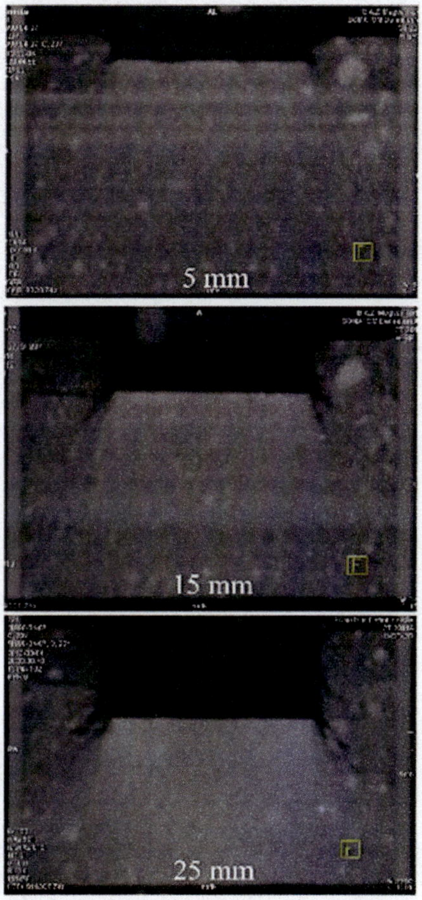

Soil	*am*			*b*		
Sandy loam	*depth of indentation [mm]*			*depth of indentation [mm]*		
state/moisture content	*5*	*15*	*25*	*5*	*15*	*25*
dry (12.6%)	0.1407	0.0651	0.0505	19.45	170.72	171.71
medium (17.5%)	0.0233	0.0064	0.00005	210.72	608.48	1031.79
wet (26%)	0.1079	0.0218	0.018	149.51	336.25	390.30

Fig. 2 Measured HU distributions of a sandy loam soil sample versus depth after each loading steps

4 Discussion

During the compaction experiments under the loading head, a highly compacted zone is developed, but on the edges of the head, cracks occurred. This phenomenon is consistent with the deformation theories, according to which the starting and the endpoints of the curves delineating the area of the plastic deformation under the horizontal loading surface are the edges of the loading surface (Szabó 1980).

In the soil samples with a secondary structure, a zone with higher compactness is formed directly under the load surface in the first loading steps, which tends to decrease to a depth (exponential relationship), until it approaches the in situ state.

For the determination of optimal sample size, it should be taken into account that in case of soil samples with secondary structure, the bottom effect occurs if the penetration depth reaches 20% of the total sample height.

5 Conclusions

The findings establish that the density change caused by the load can be determined with the help of CT images. Spatial HU distribution profiles provide essential information on the compression characteristics of the sample, helping to determine the depth of the loads' impact and to estimate the place and the magnitude of the occurring bottom effect.

Acknowledgements The described article/presentation/study was carried out as part of the EFOP-3.6.1-16-2016-00011 "Younger and Renewing University—Innovative Knowledge City—institutional development of the University of Miskolc aiming at intelligent specialisation" project and the GINOP-2.3.2-15-2016- 00031 "Innovative solutions for sustainable groundwater resource management" project implemented in the framework of the Szechenyi 2020 program. The realization of this project is supported by the European Union, co-financed by the European Social Fund and the European Structural and Investment Funds.

References

Athy, L.F.: Density, porosity and compaction of sedimentary rocks. Am. Assoc. Petrol. Geol. Bullet. **14**, 1–24 (1930)

Cavé, L., Al, T., Xiang, Y., Vilks, P.: A technique for estimating one-dimensional diffusion coefficients in low-permeability sedimentary rock using X-ray radiography: comparison with through-diffusion measurements. J. Cont. Hydrol. **103**, 1–12 (2009)

Mikita, V.: The application possibilities of CT in the field of hydrogeology and engineering geology. PhD thesis. Miskolc (2013)

Petrovic, A.M., Siebert, J.E., Rieke, P.E.: Soil bulk density analysis in three dimensions by computed tomographic scanning. Soil Sci. Soc. Am. J. **46**, 445–450 (1982)

Szabó, I.: Alapozás. Tankönyvkiadó, Budapest (1980)

Predictive Models in Geological Engineering

Strength, Disintegration and Self-sealing Characteristics of Mudrocks

Zeynal Abiddin Erguler and Guzide Kalyoncu Erguler

Abstract

Mudrocks are the most sensitive sedimentary rocks in terms of strength, deformation and slaking characteristics against change in environmental and climatic conditions. Various worldwide recorded engineering problems (e.g., landslide, slope instability, embankment failures, open-pit and underground excavation problems and undercutting related rockfalls) generally have direct or indirect relationships with these rocks. Clay minerals that are extremely sensitive to water content change constitute the most dominant mineralogical component of mudrocks. In addition to their problematic characteristics, mudrocks have promising features for modern engineering projects. For nearly three decades, mudrocks (e.g., Opalinus Clay, Boom Clay) have been investigated for safe storage of high-level radioactive waste by considering their low hydraulic conductivity and high swelling and self-sealing potential. Furthermore, the reduction of conventional fossil fuels and worldwide high energy demands have led scientists to do more investigation to achieve alternative energy sources from mudrocks. Considering low durability and strength-dependent mudrock-based engineering problems and the importance of these rocks for extraction shale gas and storage nuclear waste, the failure, slaking and self-sealing behaviors of these clay-bearing rocks are comprehensively evaluated in this study.

Keywords

Clay-bearing rocks • Disintegration • Mudrock • Self-sealing • Shale • Slaking

1 Introduction

Mudrocks (shale, claystone, mudstone, siltstone) form more than 60% of sedimentary rocks (Potter et al. 1980), include about two-thirds of the sedimentary section (Blatt 1982) and cover one-third of the total earth's surface (Franklin 1983). Average mineralogical compositions of these rocks were well documented by Pettijohn (1975) as clay minerals, quartz, feldspar, carbonates and iron oxide with percentages of 58%, 28%, 6%, 5% and 2%, respectively. Mudrocks commonly known as fine to very fine-grained siliciclastic sedimentary rocks are widely encountered problematic geological materials in many important engineering projects. These rocks are recognized as problematic rocks due to their low durability, low strength and high swelling potential when exposed to moisture (Sarman et al. 1994). Various landslides and slope instabilities that have occurred until now reveal the best example of such problematic features of mudrocks (Moon and Beattie 1995). In addition, based on literature review, Erguler and Ulusay (2009a) also emphasized that mudrocks were related to many foundation, slope stability, underground excavation and embankment failure problems. Furthermore, until present day, various rockfall problems which caused loss of life and property have been recorded in many countries, particularly in the USA, due to the time-dependent disintegration and erosion of mudrocks called as "undercutting." Also, Dick et al. (1994) stated that mudrock breakdown caused numerous problems in engineering projects such as slope instabilities, underground mining-related problems. Despite the engineering problems, mudrocks, particularly their clay minerals dominant constituents, have been known as raw material for pottery and

Z. A. Erguler (✉)
Department of Geological Engineering, Kütahya Dumlupınar University, 43270 Kütahya, Turkey
e-mail: zeynal.erguler@dpu.edu.tr

G. Kalyoncu Erguler
Mineral Research & Exploration General Directorate (MTA), 06800 Çankaya, Ankara, Turkey

H. El-Askary et al. (eds.), *Research Developments in Geotechnics, Geo-Informatics and Remote Sensing*, Advances in Science, Technology & Innovation,
https://doi.org/10.1007/978-3-030-72896-0_28

adobe for very long time. Nowadays, mudrocks are popular for their direct or indirect usage in many areas and modern projects such as absorbents, ceramic products, primary source of oil and gas, nuclear waste disposal, landfills, etc. (Kalyoncu Erguler 2020).

The steady increase in demand for energy with each day and the reduction of conventional fossil fuels have called for more research in order to achieve alternative energy sources. As a result, significant technological developments have been acquired in extraction of gas (shale gas) trapped in mudrocks. Particularly, present advances in drilling technology and hydraulic fracturing techniques provide shale gas to be alternative geo-energy sources. Recently, geological units have been considered as repositories for nuclear wastes in developed countries, and these clay-bearing rocks (e.g., Opalinus Clay, Boom Clay) have been selected as important research topics for their low hydraulic conductivity and self-sealing characteristics of induced fractures (Kalyoncu Erguler 2020). Considering the effect of mudrocks on many engineering problems and their importance for modern engineering projects, comprehensive database on strength, deformation, disintegration and self-sealing properties of these clay-bearing rocks is discussed in this study.

2 Classification and Mechanical Properties of Mudrocks

Based on several input parameters such as percentage of clay-size component, stratification thickness, mineralogical composition, anisotropy, strength, durability, hand-texture characteristics and induration, various classification systems were previously proposed for defining mudrocks and for understanding their physical properties and mechanical behavior. However, it was found that engineering geologists mostly preferred to use the classification system, which considered engineering performance of mudrocks, recommended by Potter et al. (1980). Mudrocks have a wide range of strength and deformability properties depending on geological age, mineralogical composition, lithological properties and degree of induration (Erguler, 2018). Previous studies (Erguler and Ulusay 2009a; Czerewko and Cripps 2012) stated that the uniaxial compressive strength (UCS) values of these rocks changed between 0.6 and 100 MPa. However, their strength and deformation properties ultimately decreased with the increase in water content. Erguler and Ulusay (2009a) measured more than 90% loss in the UCS, modulus of elasticity and tensile strength of saturated mudrocks in comparison with their oven-dried conditions. They recommended general empirical models for predicting these mechanical parameters of mudrocks at any water content by taking dry unit weight, porosity and water absorption by weight as independent variables.

3 Disintegration Behavior of Mudrocks

3.1 Durability Evaluation of Mudrocks Based on Laboratory Approaches

Mudrocks generally disintegrate rapidly when subjected to physical weathering processes, particularly "freezing–thawing" and "wetting–drying" (Kalyoncu Erguler 2020). Despite of responsibility of "freezing–thawing," "wetting–drying" and "heating–cooling" cycles (Hale and Shakoor 2003) on mudrock breakdown, previous studies (Franklin and Chandra 1972; Cripps and Taylor 1981; Moon and Beattie 1995; Dick et al. 1994) mainly developed laboratory-based approach to measure nondurable mudrock behavior under wetting–drying processes. The disintegration behavior of rocks is called "slaking"; the slaking phenomenon of rocks is determined by the slake durability (Id) (Franklin and Chandra 1972), the jar slake and the slake index tests (Kalyoncu Erguler 2020). Although Id is the most favored test to determine slaking behavior of rocks, several weakness and limitations of this method (e.g., mechanical breakdown during test, accepting fragments of various sizes retained in 2-mm mesh drum as durable) were emphasized by Erguler and Ulusay (2009c). In order to overcome the limitations related to slake durability test and to accurately simulate slaking behavior of mudrocks, Erguler and Shakoor (2009b) proposed a new quantitative method and a new durability classification system based on this method. This quantitative method quantifies the fragment size distribution of the slaked material of various sizes in terms of "disintegration ratio" (Fig. 1).

3.2 Field-Based Disintegration Behavior of Mudrocks

It is observed that the laboratory-based methodologies are not sufficient to accurately determine disintegration behavior of mudrocks at in situ conditions. Therefore, in order to incorporate the effect of all weathering processes on mudrocks and so to accurately assess the durability of these rocks, field behavior of these rocks should be considered. For this purpose, as shown in Fig. 2, the depth of undercutting at the road cuts provides a very useful database for understanding disintegration behavior of mudrocks at natural condition (Kalyoncu Erguler 2020). The direct measurement of undercutting rate of these rocks is critical for the safety of the related projects in terms of predetermination of undercutting-based instabilities and acquisition of the necessary measures in advance. It has been determined that the undercutting rates of Neogene-aged mudrocks situated in Turkey change between 1.0 and 4.6 cm/year. The total annual precipitations in this site have been recorded as a

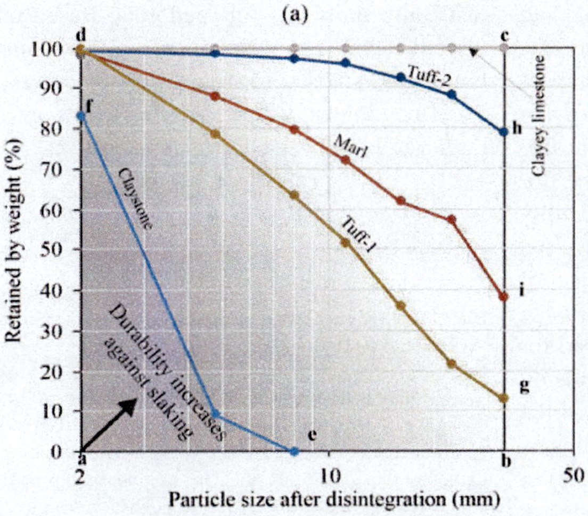

(b)

Rock type	Disintegration ratio (D_R)
Claystone	$D_R = \frac{Area\ (aefa)}{Area\ (abcd)} = 0.15$
Tuff-1	$D_R = \frac{Area\ (abgda)}{Area\ (abcd)} = 0.60$
Marl	$D_R = \frac{Area\ (abida)}{Area\ (abcd)} = 0.76$
Tuff-2	$D_R = \frac{Area\ (abhda)}{Area\ (abcd)} = 0.95$
Clayey limestone	$D_R = \frac{Area\ (aefa)}{Area\ (abcd)} = 1.00$

■ Entirely durable (D_R=0)
□ Entirely non-durable (D_R=1)

Fig. 1 **a** Disintegration size distribution curves of different rocks after the second cycle of slake durability test and **b** the disintegration ratio values for typical samples (Erguler and Shakoor 2009; Ozdemir and Erguler 2022)

Fig. 2 **a** Typical undercutting observed at road cut excavated in 2008, **b** inserting nails to measure time-dependent depth of undercutting on May 9, 2017 (Kalyoncu Erguler 2020), **c** the depth of undercutting on March 16, 2018 (Kalyoncu Erguler 2020; Ozdemir and Erguler 2022)

range of 36–56 cm (Ozdemir and Erguler 2022). Shakoor and Rogers (1992) and Niemann (2009) specified that differential weathering of mudrocks located in the Appalachian region, with precipitation ranging between 100 and 130 cm, was commonly recorded in the case of exposing beneath more resistant rocks. They found that the undercutting rates of these Devonian–Permian-aged rocks changed between 1.3 and 9.5 cm/year.

4 Self-sealing Characteristics of Mudrocks

The self-sealing behavior of mudrocks is geomechanical, hydro-mechanical and geochemical process-based mechanism. It is generally evaluated as a reduction in permeability (hydraulic conductivity) of rocks. Several self-sealing mechanisms are responsible for the decrease in permeability of mudrock. Bock et al. (2010) specified these significant factors as "compaction of the intact rock body," "mechanical closure of fractures," "physico-chemical factors (e.g., swelling, slaking)," "surface slaking, sedimentation and clogging of fractures" and "precipitation of minerals." The effect of self-sealing processes should be defined by considering in situ stress and temperature values as well as time parameter. Therefore, a deep scientific knowledge beyond the classical geomechanical perspective on hydro-mechanical properties of shales and modernized laboratory equipment is required. Although there are some numerical approaches to be used in self-sealing behavior, these approaches can only incorporate easy and precise processes such as stress increases dependent fracture closure (Kalyoncu Erguler 2020).

The hydraulic conductivity values of mudrocks generally change between 1×10^{-13} and 1×10^{-9} m/s at atmospheric pressure (Freeze and Cherry 1979). By performing permeameter and in situ packer tests, Marschall et al. (2005)

determined that this parameter ranged from 2×10^{-14} to 2×10^{-12} m/s for tectonically disturbed Opalinus Clay. Despite the low permeability values, technological advances in hydraulic fracturing and horizontal drilling provide required tools to acquire oil and gas from these impermeable geological units. In addition to mudrock importance for unconventional gas and oil production, the low permeability and good self-sealing characteristics of these rocks have led scientists to investigate their usability in safe storage of hazardous wastes such as nuclear waste disposal (Fisher et al. 2013) and sequestration of CO_2. The time required for the closure of the cracks in mudrocks especially depends on the mineralogical composition and geological history of the rock. It can be concluded that while self-sealing mechanisms cause fracture closure within a short time (a few months) in plastic clays, this process may be slower and take years for more indurated mudrocks (Bock et al. 2010).

5 Conclusions

Mudrocks—known as the widespread fine-grained siliciclastic sedimentary rocks—are critical for many engineering projects for their low hydraulic conductivity, self-sealing, slaking and water content dependent strength and deformation characteristics. Particularly, in the case of achieving sufficient scientific background, their low permeability and self-sealing features are thought to be of strategic importance in hosting nuclear waste disposal. However, to accurately determine the effect of self-sealing processes of fractures on production efficiency of shale gas by laboratory experiments is challenging. The effect of these processes should be defined by considering in situ stress and temperature values as well as the time parameter. Therefore, it can be concluded that current laboratory-based approaches are insufficient to include effective self-sealing mechanisms such as time-dependent creep deformations, slaking, micro-slaking, swelling, precipitation of new minerals.

The UCS values of these rocks ranged from 0.6 to 100 MPa, but the increase in water content dramatically decreased their strength and deformation characteristics. Statistically significant empirical models were recommended to estimate water content-dependent variations in these parameter based on their physical properties.

The disintegration behavior of mudrocks could not be accurately measured by performing slake durability index test due to several limitations. Thus, a new quantitative method called as "disintegration ratio" was proposed. Mudrock disintegration caused "undercutting"-dependent instability problems for road cuts. Based on measurements directly taken from Neogene and Devonian–Permian-aged mudrocks exposed to weathering processes in different climatic conditions, it was specified that mudrock undercutting rates varied between 1.0 and 9.5 cm/year.

References

Blatt, H.: Sedimentary Petrology. W.H. Freeman and Company, San Francisco (1982)

Bock, H., Dehandschutter, B., Martin, C.D., Mazurek, M., De Haller, A., Skoczylas, F., Davy, C.: Self-sealing of Fractures in Argillaceous Formations in the Context of Geological Disposal of Radioactive Waste. NEA No. 6184 (2010)

Cripps, J.C., Taylor, R.K.: The engineering properties of mudrocks. Q. J. Eng. Geol. **14**, 325–346 (1981)

Czerewko, M.A., Cripps, J.C.: Mudrocks, clays and pyrite. In: Burland, J.B., Chapman, T., Skinner, H.D., Brown, M.J.Z. (eds.) ICE Manual of Geotechnical Engineering. Volume 1 Geotechnical Engineering Principles, Problematic Soils and Site Investigation, pp. 481–516. ICE Publishing, London (2012)

Dick, J.C., Shakoor, A., Wells, N.: A geological approach toward developing a mudrock-durability classification system. Can. Geotech. J. **31**, 17–27 (1994)

Erguler, Z.A.: Encyclopedia of Engineering Geology. In: Shale, Bobrowsky, P.T., Marker, B. (eds.), pp. 829–830. Springer International Publishing AG, Switzerland (2018). ISBN: 978-3-319-73566-5

Ergüler, Z.A., Ulusay, R.: Water-Induced Variations in Mechanical Properties of Clay-Bearing Rocks (2009a)

Erguler, Z.A., Shakoor, A.: Quantification of fragment size distribution of clay-bearing rocks after slake durability testing. Environ. Eng. Geosci. **XV**(2) (2009b)

Erguler, Z.A., Ulusay, R.: Assessment of physical disintegration characteristics of clay-bearing rocks: disintegration index test and a new durability classification chart. Eng. Geol. **105**, 11–19 (2009)

Fisher, Q., Kets, F., Crook, A.: Self-sealing of faults and fractures in argillaceous formations: evidence from the petroleum industry. NAB 13-06 (2013)

Franklın, J.A.: Evaluation of Shales for Construction Projects: An Ontario Shale Rating System: Research and Development Branch, Ministry of Transportation and Research. Toronto, Report RR 229, 99 p. (1983)

Franklin, J.A., Chandra, A.: The slake-durability test. Int. J. Rock Mech. Min. Sci. **9**, 325–341 (1972)

Freeze, R.A., Cherry, J.S.: Groundwater, p. 604. Prentice-Hall Inc., Englewood Cliffs, New Jersey (1979)

Hale, P.A., Shakoor, A.: A laboratory investigation of the effect of cyclic heating and cooling, wetting and drying, and freezing and thawing on the comprehensive strength of selected sandstones. Environ. Eng. Geosci. **IX**(2), 117–130 (2003)

Kalyoncu Erguler, G.: The failure, slaking, and self-sealing behavior of clay-bearing rocks. Arab. J. Geosci. **13**, 857 (2020). https://doi.org/10.1007/s12517-020-05896-5

Marschall, P., Horseman, S., Gimmi, T.: Characterisation of gas transport properties of the opalinus clay, a potential host rock formation for radioactive waste disposal. Oil Gas Sci. Technol. **60** (1), 121–139 (2005)

Moon, V.G., Beattie, A.G.: Textural and microstructural influence on the durability of Waikato coal measures mudrocks. Q. J. Eng. Geol. **28**, 303–312 (1995)

Niemann, W.L.: Lessons learned from rates of mudrock undercutting measured over two time periods. Environ. Eng. Geosci. **XV**(3), 117–131 (2009)

Ozdemir, S., Erguler, Z.A.: Investigation of the slaking behavior of weak geological units in terms of undercutting rate. Bull Mineral Res Explor., 1–22 (2022). https://doi.org/10.19111/bulletinofmre.898013

Pettijohn, F.J.: Sedimentary Rocks, 3rd edn., p. 628. Harper and Row, New York (1975)

Potter, P.E., Maynard, J.B., Pryor, W.A.: Sedimentology of Shale, p. 553. Springer-Verlag, New York (1980)

Sarman, R., Shakoor, A., Palmer, D.F.: A multiply regression approach to predict swelling in mudrocks. Bullet. Assoc. Eng. Geol. **XXXI** (1), 102–121 (1994)

Shakoor, A., Rodgers, J.P.: Predicting the rate of shale undercutting along highway cuts. Bullet. Assoc. Eng. Geol. **XXIX**(1), 61–75 (1992)

Analysis of Auscultation Data of a Dam Case of the Algerian Dams

Samira Chaba, Marzzouk Cherrared, and Malek Bouhadef

Abstract

The analysis of dams auscultation data (leaks, under pressure) requires statistical methods that are more or less elaborate to separate the different sources of variation of the measured phenomenon. Methods based on statistical analyses such as the HST (seasonal hydrostatic) linear regression analysis model have been commonly used. However, they prove to be a powerless tool for interpreting the behavior of dams. A new approach is developed based on the use of artificial neural networks suitable for solving some particular problems. Indeed, the method of neural networks can model the nonlinearities. For this neuronal model, the actual data recorded at the Keddara dam is used.

Keywords

Auscultation • Neural networks • Perceptron • Piezometer • Drain

1 Introduction

Methods of statistical analysis of auscultation measurements allow to separate the respective influences of several explanatory factors (water coast, season) introduced in the model. The HST (seasonal hydrostatic) multiple linear regression analysis, which is used by EDF more than thirty years ago (Wilm and Beaujoint 1967), is among these statistical methods. It proves robust and simple for the interpretation of dam osculation measures. HST allows to determine the hydrostatic effects (due to the water load), the seasonal effects (due to the thermal effects) and the effect of the aging of the structure. Some limitations of HST have been identified: When the explanatory variables (water coast, season) are linked by dependence relationships, the separation of effects is imprecise. To overcome these shortcomings of classical analysis, models based on neural networks have been tested.

The Keddara dam oscillation measurements are tested and compared with those of the neural network and HST. The input variables in this model are the date, the season, the coast of the restraint, while the output variables are the piezometric coasts.

2 Methods of Measurement Analysis

2.1 HST Model

In the original model, the explanatory factors are taken into account, the first two reversible effects and the third irreversible effect: the level of the water body, the hydrostatic effect, noted '*H*', the date in the year, used for the seasonal effect (variation of temperature), noted '*S*'.

- The first effect is presented by the equation $H(z)$, which is a polynomial function of the fourth degree of the retaining coast:

$$H(z) = A_1Z + A_2Z^2 + A_3Z^3 + A_4Z^4 \tag{1}$$

The second effect is reversible, it presents the thermal state of the work; it would depend only on the season (S). The seasonal law that is substituted for the thermal law $S(\theta)$ is then a periodic function of time (period of one year).

In the absence of temperature measurements in the mass of the structures, it was decided to choose the explanatory variable. The variable season was assimilated to an angle of 0° on January 1 and 360° on December 31

S. Chaba (✉) · M. Cherrared · M. Bouhadef
Department of Geotechnics and Hydraulics, University of Science and Technology Houari Boumediene, Algiers, Algeria

H. El-Askary et al. (eds.), *Research Developments in Geotechnics, Geo-Informatics and Remote Sensing*, Advances in Science, Technology & Innovation,
https://doi.org/10.1007/978-3-030-72896-0_29

$$S(\theta) = A_5 \sin(s) + A_6 \cos(s) + A_7 \sin^2(s) + A_8 \sin(s).\cos(s) \quad (2)$$

The third effect accounts for the influence of time (effect of aging); it includes a negative exponential term representing the damped evolution, a positive exponential term representing the accelerated evolution and a linear term representing the persistent evolutions.

$$T(t) = A_9 e^{-t} + A_{10} e^{t} + A_{11} t \quad (3)$$

So the HST model is written as:

$$X(z, \theta, t) = H(z) + S(\theta) + T(t) \quad (4)$$

2.2 Model by Artificial Neural Networks (ANNs)

Artificial neural networks are an alternative for mathematical modeling; they are part of nonparametric and nonlinear statistical models, able to answer the problems of identification, prediction, etc. The construction of a RNA involves the determination of its architecture (number of layers and number of artificial neurons in each layer) and the choice of activation functions associated with the neurons. The neuron is characterized by its state, its connections with other neurons and its activation function. The state of each neuron is influenced by the injected input data and then transmitted throughout the neural network, layer after layer, to the output layer through the interconnections thus giving the NNA its overall behavior. The multilayer perceptron (MLP) is the most commonly used model for simulating nonlinear relationships. The ANN learning step consists of calculating and optimizing the informative weight of the nonlinear relationship between input and output. This step is iterative. It is followed by two verification steps 'test' and 'validation'. As in any modeling approach, these last steps are essential before operating an ANN. For our model, an MLP is used with the function 'ReLU' as an activation function for all neurons (Fig. 1). The algorithm adopted for learning is that of optimizing ADAM (A Method for Stochastic Optimization), because it is faster and it allows a better convergence to a minimum of the error.

3 Application of HST and NNA Models on the Dam from KEDDARA

The dam Keddara is located in the territory of the Wilaya Boumerdes, 8 km south of Boudouaou and 35 km east of the capital Algiers. It is part of the Isser-Keddara development intended to meet the drinking water needs of the Algiers agglomeration. Its reservoir is fed by inputs from

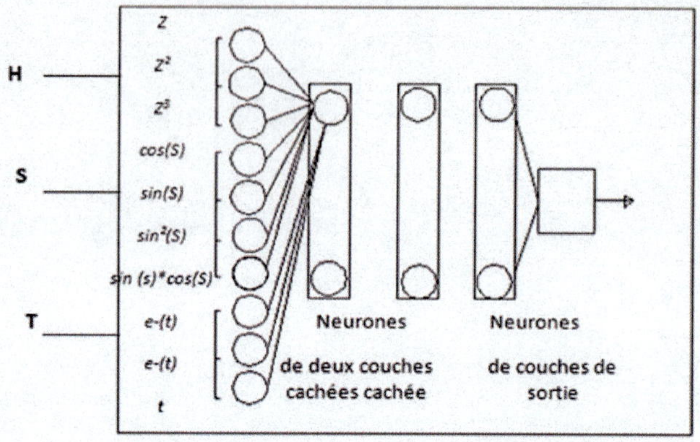

Fig. 1 Multilayer perceptron with two layers of neurons hidden and one layer of neurons output.

wadis Keddara, El Haad and the transfer gallery from the dam Hamiz. The Keddara dam is a riprap dike with a central core made of clay.

3.1 Neuron Network Application with Ten Explanatory Variables

The neural network requires ten explanatory variables in the input: the dimension of the water reservoir (Z, Z^2, Z^3), the season (cos(S), sin(S), sin2(S), sin(S) $\times$ cos(S)) and the time: $e-(t)$, $e(t)$, (t). The architecture consists of nine input-layer neurons, six in the first hidden layer and three in the hidden second layer and one output neuron. The results obtained by the neural networks are compared with those of the HST method. After 100 iterations, the neural network converges to a solution that is numerically more efficient than HST.

Surface or gallery piezometers show less variable variations, except in the case of exceptional rainfall where a sudden rise in the piezometric level occurs. The piezometric ratings estimated by the network tend to better follow the values observed compared to the HST model. The results (Figs. 3 and 4) obtained display the good performance of the model with neural networks which exceeds that of the HST model. This performance is well explained during the calibration and validation of the results. Consequently, it results in a better prediction.

3.2 Neuron Network Application with Three Explanatory Variables

The neural network (Figs. 5 and 6) requires three input variables: the water reservoir (Z), the season (S) and the time (t); the architecture is that of the 10 variables (Fig. 2).

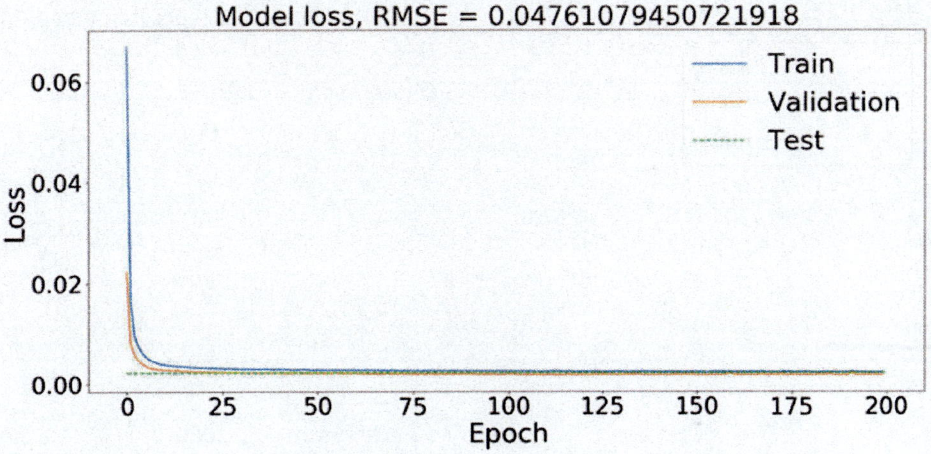

Fig. 2 Evolution of the quadratic error on the basis of learning and the test

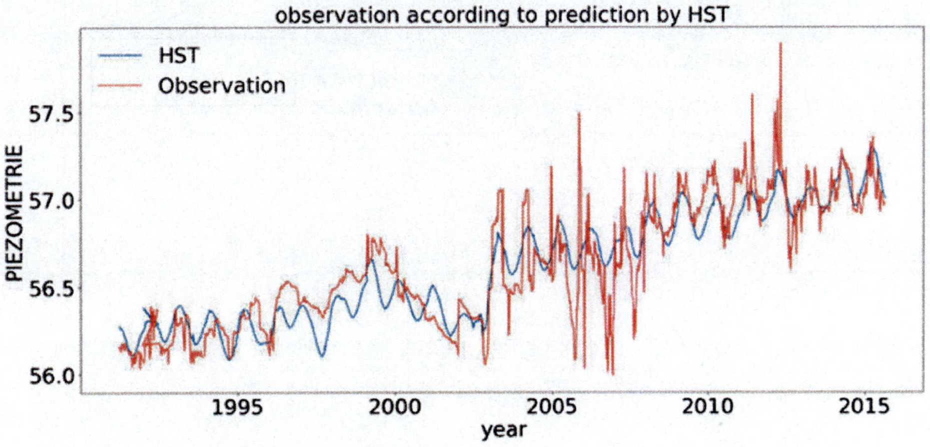

Fig. 3 Comparative evolution of HST and data observations

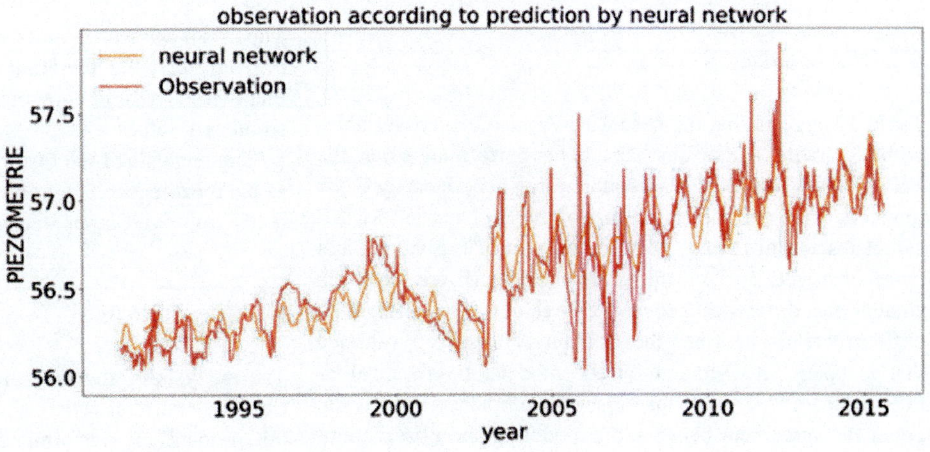

Fig. 4 Comparative evolution of HST and neurons

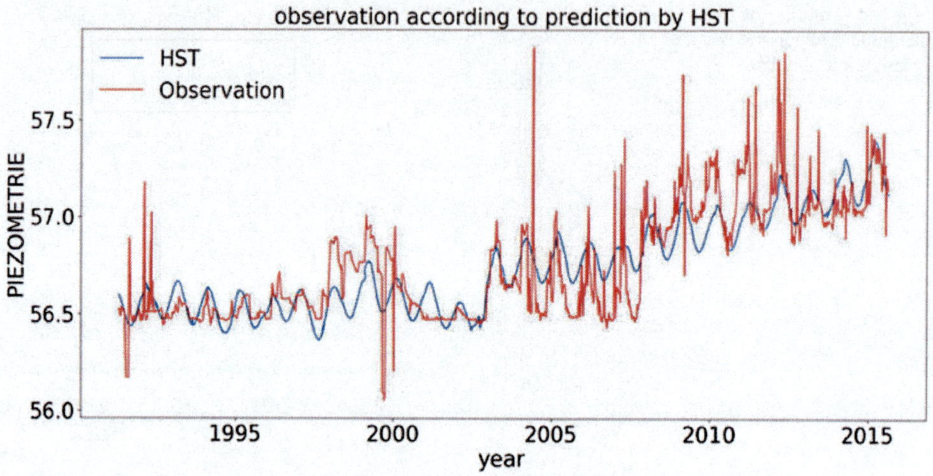

Fig. 5 Comparative evolution of HST and neurons

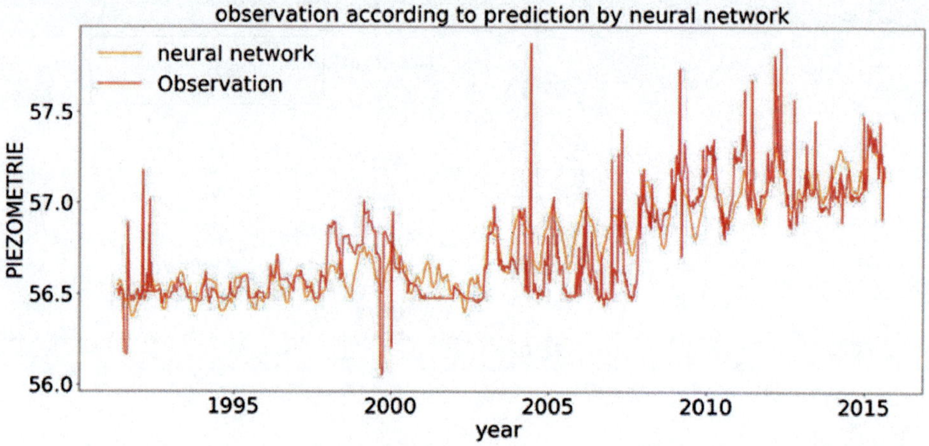

Fig. 6 Comparative evolution of HST and data observations

4 Discussion

Table 1 suggests that the model RNA with eleven variables with the gradient algorithm has a better performance than the HST model. The root mean square error generated by RNA model is small and inferior to that of the HST model; RMSE calculates the difference between the simulated values and those observed. In fact, the lower the RMSE criterion, the smaller the difference between the values. The Nash–Sutcliffe criterion quantifies the fraction of observed variance that is taken into account by the models. It takes values ranging from $-\infty$ to 1; the closer the efficiency is to 1, the better the agreement between the modeling and observation. The Nash values obtained during validation vary between 0.67 and 0.78 for the HST and NNA model, which implies that there is a good agreement between the observed and the simulated values.

The analysis of the results of the explanatory factors such as the piezometric height predicted by the neural networks is better improved compared to the HST model.

5 Conclusion

The analysis of the measurements of the auscultation data from the neural networks showed a new perspective. Beyond the numerical performance that was expected, it is

Table 1 RMSE and Nash–Sutcliffe coefficients calculated for HST and NNA

Piezometer	The root mean square error 'RMSE'	Nash–Sutcliffe
HST	0.05020790698	0.7403908026
NNA with ten variables	0.0473200579020	0.7844466841
NNA with three variables	0.0561379901899	0.67521001314

interesting to note that cross-effects between explanatory variables (water coast, season, age of the dam) are clearly highlighted. This observation may sometimes account for the difficulties of the HST model in correcting the data. Neurons should be used with caution: The calculated effects should be drawn close to the stall measures, and the exploration of all possible cross-effects should be examined to understand how neurons have improved modeling. The problem of over-learning (with measurement noise, for example) can be easily mastered by monitoring the error on a test basis that has not seen learning. In the case of over-learning, a decrease in the number of iterations or a simplification of the architecture of the neural network (decrease of the degrees of freedom) is generally sufficient to solve the problem. The results obtained by the neural networks are compared with those of the HST method. These results indicate that artificial neural networks (ten variables) can provide satisfactory answers for the prediction of osculation measurements compared to more classic methods. Therefore, it can be concluded that neural networks have better predictive power. These findings can be used and interpreted as such, or be further adopted for the development of simplified physical-based models.

References

Alexandre, G., Simon, Maxime, J., Royer, J., Florian, M., Jean-Paul, F.: Analysis of measurements of auscultation of dams with neural networks. Theme 3—innovation and auscultation (2012)

Bonelli, S., Félix, H., Torment, R.: Interpretation of dams auscultation measures by multiple linear regression HST. January 1998 with 294 Reads—2nd National Conference JN-FIAB98 (1998)

Penot, I., Daumas, B., Fabre, J.P.: Monitoring behavior. In: International Water Power & Dam Construction, pp 24–27 (Dec 2005)

Smail, N., Bekkouche, A., Rouissat, B.: Modeling and auscultation, tools for the analysis of the behavior of dams. In: International Conference Non-Saturated Soils and Environment "UNSATlemcen09" October 2009

Willm, G., Beaujoint, N.: The methods of monitoring dams in the service of the hydraulic production of Électricité de France. Old problems and new solutions. In: Proceedings of the Ninth International Congress on Large Dams, Istanbul, pp. 529–550 (1967)

Tailing Dam Forming Considering the Consolidation Properties of the Soils

Alan Lolaev, Aleksan Oganesyan, Alexander Badoev, and Emil Oganesyan

Abstract

Based on the operating conditions of alluvial storage in permafrost, it should be noted that the increase in the annual capacity of the layer may not provide its complete consolidation during the short spring–summer period before freezing. Thus, it will reduce the stability of the structure. In order to optimize the inwash technology of the tailing dam, it is necessary to determine the time of tail's consolidation. Based on laboratory tests, we developed an approach to coefficients definition of filtrational and secondary consolidation of tails during alluvium depending on a physical condition, density, water saturation degree. The mathematical model to calculate the time of their consolidation also has been developed.

Keywords

Tailing dam • Consolidation • Network planning • Tails • Alluvium

1 Introduction

As a rule, the changes in the tailing dams forming technology during its forming do not stop the technological process of alluvium. At the initial stage of operation, when the relief of the terrain allows storage wastes without the construction of the enclosing dam, issues of stability and technological schemes do not have such an acute significance. In this case, any technology is applicable. However, when the natural capacity of the dam is exhausted, and it is necessary to increase capacity during operation, special attention should be given to their stability. In this case, the technological scheme, the parameters of the alluvium, the individual sequence of the inwash, etc., may affect the stability of individual sections and the structure as a whole.

In case of changes in the volume of mineral raw materials, storage site conditions, the definition the optimization of the technological parameters of the tailing dams forming during its exploitation makes possible to predict and operatively manage the structure stability. The technological scheme of construction and operation, a whole complex of factors, should be considered, including climatic conditions. The influence of climatic conditions on the tailing dams' technological scheme forming process is of particular importance for objects located in the cryolithozone.

2 Methods and Results

A complex of experimental studies allowed to develop an algorithm for the tailing dam formation (see Fig. 1).

At the first stage of the work, initial data need to be prepared for further calculations: the geometric characteristics of the existing tailing dam with the definition of areas; the diameter of the main pipeline; the number of outlets and their diameters.

At the second stage—initial data formation, it is necessary to determine the volume of incoming tails and calculate the height of the layer of annual alluvium in accordance with the actual areas of alluvium.

Third stage—depending on the daily volume of tails and through put of one release, the required number of outlets and the width of the alluvium sector and the total number of sectors for which tailings will be stored must be determined.

Fourth stage—need to perform the alluvium of the tailings dam and indicate the total time of the inwash onto the beach (the number of days per year with a temperature above – 5 °C), which can be determined according climatic conditions of the region (Lolaev et al. 2016; Lolaev and Butygin 2005).

A. Lolaev (✉) · A. Oganesyan · A. Badoev · E. Oganesyan
NCIMM (STU), 44 Nikolaeva Str., Vladikavkaz, 362021, Russia
e-mail: proflolan@gmail.com

A. Lolaev · A. Oganesyan · A. Badoev
VSC RAS, 22 Markusa Str., Vladikavkaz, 362027, Russia

H. El-Askary et al. (eds.), *Research Developments in Geotechnics, Geo-Informatics and Remote Sensing*, Advances in Science, Technology & Innovation,
https://doi.org/10.1007/978-3-030-72896-0_30

Fig. 1 Algorithm of tailing dam formation

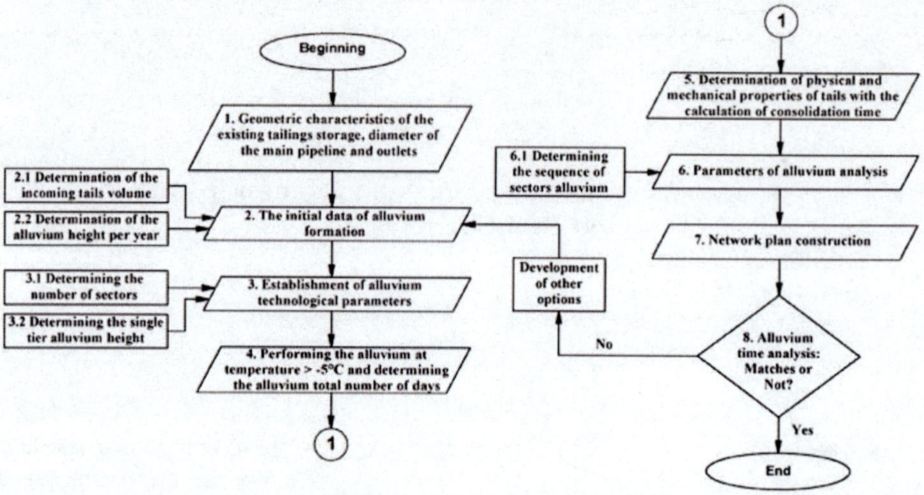

Besides, it is necessary to set a limit on the height of the annual inwash. This parameter is essential for dams located in the cryolithozone. The value of the maximum permissible height of alluvium per year should ensure the conditions of freezing of the alluvial massif and depends on the climatic characteristics of the area, chemical composition, physical and mechanical properties of the tails. In order to determine the value of the maximum permissible height of annual alluvium, separate studies should be conducted for each specific case.

The fifth stage—determination of tails physical and mechanical properties. To establish the dependence of the skeleton density of the soil on its moisture content, a series of individual soil tests for compaction must be carried out with a consistent increase in its moisture content. The test results should be made in the form of a graph. The number of individual tests for plotting should be at least six and also sufficient to identify the maximum value of the soil skeleton density (Lolaev et al. 2018, 2016).

Next, the plasticity limits of tails must be determined. The ratio of filtration and rheological phenomena in the process of soil consolidation varies depending on the density, humidity, characteristics of the soil structure and the magnitude of the load acting on them. The upper limit of humidity is moisture at the yield point. (Lolaev et al. 2018, 2016).

As it is known, the process of consolidation is conventionally divided into two phases, primary or seepage and secondary, due to creep of the soil skeleton. The time of completion of the filtrational consolidation stage (C_v) has to be determined using the consolidation curves constructed in the coordinates displacement (s)—logarithm of time (lg t) according to Casagrande method. The coefficient of secondary consolidation C_α also needs to be determined by the tangent of the angle between the linear segment of the curve in the area of secondary consolidation and a straight line parallel to the abscissa axis (Lolaev et al. 2018, 2016; Gosstroy et al. 1989).

The consolidation time can be calculated by changing the geometric characteristics of the samples under laboratory conditions (Lolaev et al. 2018, 2016):

$$t = \frac{F}{C_v} \times \frac{h}{h - s} \qquad (1)$$

where t is the consolidation time, min; F is the cross-sectional area of the ring, cm^2; C_v is the consolidation coefficient, cm^2/min; h is the initial height of the layer, cm; h_k is the final layer height, cm; s is the displacement, cm

The regression polynomial (polynomial) equation should be prepared (2) as an arranged in ascending powers of the studying factor and simultaneously linear to all coefficients. The time of tailings consolidation should be determined at known design values of humidity and density (Lolaev et al. 2018).

$$y = f(x) = b_0 + b_1x_1 + b_2x_2 + \cdots b_nx_n \qquad (2)$$

Alluvium at the storage area must be carried out in several stages. Moreover, between the end of the alluvium of the lower tier and the beginning of the alluvium of the subsequent tier, a technological break should be established.

The sixth stage—analysis of the structure alluvium parameters. Based on the operating conditions of alluvial tailings in the permafrost zone, it should be noted that the increase in the annual capacity of the layer may not provide its complete consolidation during the short spring–summer period before freezing. Thus, it will reduce the stability of the structure (Lolaev et al. 2018, 2016).

It is necessary to accurately determine the time of tailings consolidation, which is the most important factor in optimizing the technology of the alluvial in the conditions of the permafrost zone.

Furthermore, the sequence of all-in-sectors needs to be determined to ensure a sufficient amount of time during the process of tailings consolidation.

The seventh stage—in order to fulfill the regulatory requirements for optimizing the process of inwash, the network planning method has to be applied.

As a result of the construction and calculation of the network plan, the duration of the critical path in days is determined, which allows to conclude whether the entire volume of all the inwashed tails is laid in the required time frame (Lolaev et al. 2018, 2016).

At the last stage, the compliance of the obtained number of days required for tailings inwash needs to be checked. In case of non-compliance, it is necessary to return to the stage of determining the initial data.

3 Conclusions

This work proposes an algorithm of tailing dam forming considering the consolidation time, and network planning method has been developed. It aims to improve the accuracy of determining the technological parameters of the structure formation, which is the basis for reducing the cost of these facilities construction and operation.

The proposed algorithm was applied during the inwash of the tailing dam located on the territory of north of Siberia. It proved effective by providing completion of the consolidation process when laying tails in a certain period of time before freezing.

The algorithm also can be used by mining enterprises, research and development organizations, as well as in universities for educational purposes.

References

Lolaev, A.B., Badoev, A.S., Oganesyan, A.K.: Application of numerical methods for the optimization of the technological parameters of the tailing dam alluvium in permafrost region. In: Proceedings of 2-nd International Seminar "Numerical Analysis in Geotechnics" NAG2018, Ho Chi Minh City, Vietnam, (CD-edition) (2018)

Lolaev, A.B., Dzeboev, S.O., Badoev, A.S., Arutiunova, A.V., Ilaev, V.E.: Determination of the tails consolidation parameters of alluvial tailings dam in permafrost region. In: Proceedings of the 1st International Conference on Natural Hazards and Infrastructure, Chania, GREECE (2016)

Lolaev, A.B., Butygin, V.V.: Geological and ecological problems of industrial hydraulic engineering in cryolithozone. Nedra, Moscow (2005).(in Russian)

PNIIS Gosstroy USSR Recommendations for Determining the Parameters of Creep and Soil Consolidation by Laboratory Methods. Moscow, Stroyizdat, (1989) (in Russian)

Practical Criterion of Flow Stability in Unsaturated Soil

Mari Sato

Abstract

Flow in unsaturated soil cannot be accurately calculated using existing theories for two reasons: (i) The theories consider several assumptions, and (ii) the associate simulations are conducted under ideal conditions. Therefore, the aim of this study is to re-evaluate the existing theories and propose a simple method to determine the stability of vertical flow in actual unsaturated soil, based on the velocity term in Bernoulli's principle. As a result, a localized high hydraulic gradient might be imposed when there are local variations in water content under relatively wet conditions.

Keywords

Unsaturated • Judgment criteria • Bernoulli's principles • In situ • Water content

1 Introduction

Despite the numerical analyses conducted recently, the seepage situation inside the ground cannot be measured precisely in practical sites. Situations not satisfying the analytical assumptions should be also considered. To this end, this paper proposes a criterion that can be easily determined from in situ data by applying Bernoulli's principle, serving as an energy law, magnitude relations of common unsaturated geotechnical theories, and Darcy's law. Bernoulli's principle states that the velocity term $v^2 \approx 0$ is assumed under analytical conditions. However, this study considers a ground not satisfying these relations as a situation where the calculations are incomplete. This is performed under a one-dimensional simple case. The study reveals that there are several situations under the risk of high localized hydraulic head, where the flow cannot be calculated.

2 Preconditions and Equations

Vertical flow is estimated between two points: point 1 and point 2 in a small element. Pressure, velocity, and height for point n are denoted by P_n, V_n, and z_n, respectively. Bernoulli's principle is shown as follows:

$$\frac{1}{2}v_1^2 + \frac{P_1}{\rho} + gz_1 = \frac{1}{2}v_2^2 + \frac{P_2}{\rho} + gz_2 + gh_{12} \quad (1)$$

Here, h_{12} is a hydraulic head loss from point 1 to point 2, and $P/\rho g$ is ψ, which is the negative pressure head. The soils between points 1 and 2 are estimated as uniform materials with equal porosity, cross-sectional area, and fluid density ρ. Then, volumetric water content of each point is denoted by θ_n, and Darcy's law is applied.

$$v_n = \frac{k^s K_r(\theta_n)}{\theta_n}\frac{\mathrm{d}h}{\mathrm{d}z} \quad (2)$$

where k^s is the saturated hydraulic conductivity assumed to be constant among two points. $\mathrm{d}h/\mathrm{d}z$ is hydraulic gradient at point n. Equation (2) represents that Darcy velocity is translated to the true flow velocity in voids, v_n. $K_r(\theta)$ is the relative permeability coefficient and function of θ. Equation (1) is transformed to the following equation; note that point 1 is higher point than point 2 in height. Upward direction is defined as positive.

$$\frac{1}{2}\left(v_1^2 - v_2^2\right) = -\frac{\Delta P}{\rho} - g\Delta z + gh, h = h_{12} \quad (3a)$$

$$\Delta P = P_1 - P_2 \quad (3b)$$

$$\Delta z = z_1 - z_2 = \text{const.} > 0 \quad (3c)$$

M. Sato (✉)
Shimane University, 690-8504 Nishikawatsu, Matsue, Japan
e-mail: maris@life.shimane-u.ac.jp

H. El-Askary et al. (eds.), *Research Developments in Geotechnics, Geo-Informatics and Remote Sensing*, Advances in Science, Technology & Innovation,
https://doi.org/10.1007/978-3-030-72896-0_31

These concepts of equations and preconditions are common in geotechnical engineering, e.g., Bear (1972). As for the velocity term, there is no concern for calculations given that it is negligible comparing to other terms. This study uses the velocity term for judging the flow stability for concerning the situations out of precise calculations.

3 Evaluation of the Stability of Unsaturated Flow

3.1 Classification of Stable or Unstable Flow

For stable flows, (i) the magnitude relations between permeability, matric suction, and water content are established as shown in Fig. 1: Matric suction is reduced/permeability coefficient is increased owing to an increase in water content ($\partial\psi/\partial\theta < 0,\ \partial k/\partial\theta > 0$), (ii) localized hydraulic gradient and flow direction are equal to their averaged values ($dh/dz = \text{const.}$ in entire portion), (iii) the direction of the entire hydraulic head h corresponds to the direction of velocity ($v_1 + v_2 < 0$ if $h > 0$, downward), and (iv) Darcy's law, given in Eq. (2), is satisfied. When water content is relatively high, the following equations are assumed to be satisfied based on previous studies (vanGenuchten 1980; Childs and Collis-George 1950). It is assumed that the values of θ_1 and θ_2 are not massively different from each other.

$$\frac{\partial\left(\frac{K_r(\theta)}{\theta}\right)}{\partial\theta} \geq 0 \rightarrow \left(\frac{K_r(\theta_1)}{\theta_1} - \frac{K_r(\theta_2)}{\theta_2}\right)(\theta_1 - \theta_2) \geq 0 \quad (4)$$

The gradient of permeability is known to gradually increase following the increase in water content (Childs and Collis-George 1950), as shown in Fig. 1, and therefore, Eq. (4) is satisfied over certain water content. Please note that the case not considered under the equation is described in (3). In addition, $k(\theta_n)$ is related to $P_n = \psi_n \rho g$, and a few models and experiments (vanGenuchten 1980; Childs and Collis-George 1950) are already proposed considering the schematic tendency. This study comes to define the relationships among matric suction and water content as follows:

$$\frac{\partial P}{\partial\theta} \geq 0 \quad (5)$$

Finally, the four abovementioned conditions along with Eqs. (2)–(5) are all determined to be either satisfied or not in the relations of positive and negative. The case that satisfies all assumptions considered is called *stable*.

3.2 Simple Criterion of Boundaries of Stability for Unsaturated Flow

Only the cases satisfying the following equation are regarded as stable conditions (see Fig. 2).

$$|h - \Delta z| \geq \Delta\psi,\ \Delta\psi = \Delta P/\rho g \quad (6)$$

$\Delta\psi$ is the gap of the matric suction between point 1 and point 2. $|h - \Delta z| = \Delta\psi$ and $v^2 \approx 0$ is a known equation, being applied to various models (vanGenuchten 1980; Bear 1972). $\Delta\psi$ is supposed to be very close to $|h - \Delta z|$ in the practical ground in the stable condition, as shown in the estimated area in Fig. 2. It is considered the velocity and the locational variations are very small in the soil. When the cases do not satisfy Eq. (6), (i) $K_r(\theta_n)$ or (ii) dh/dz is mistakenly evaluated, or (iii) localized velocity v_n has a different

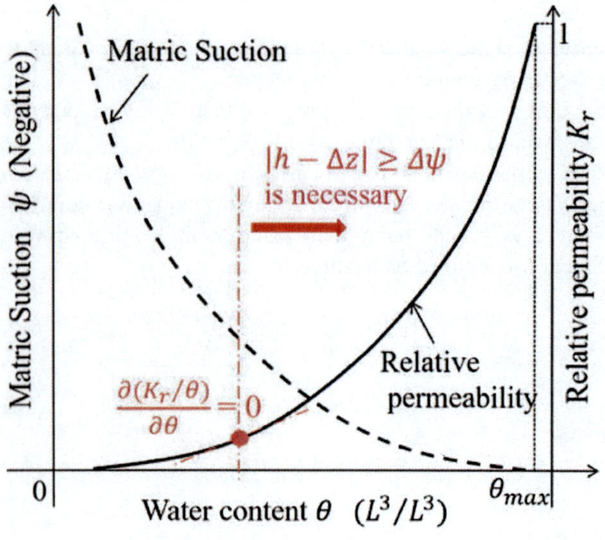

Fig. 1 Relationships between ψ, K_r, and θ and schematically gradient

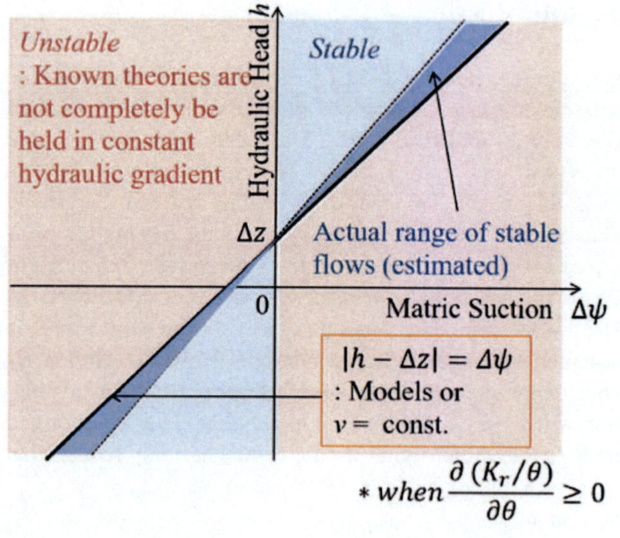

Fig. 2 Proposed criterion of boundary of flow stability among h, Δz and $\Delta\psi$

direction from that of h. It is inferred that the localized hydraulic gradient is not constant. There are two cases $\theta_1 > \theta_2$ or $\theta_1 < \theta_2$, and the former is introduced; $\frac{K_r(\theta_1)}{\theta_1}\frac{dh}{dz}\Big|_{n=1} - \frac{K_r(\theta_2)}{\theta_2}\frac{dh}{dz}\Big|_{n=2} < 0$, if $\theta_1 > \theta_2$, when the cases do not satisfy the stable conditions. The following equation is then obtained from Eq. (2):

$$1 \geq \frac{\theta_1 K_r(\theta_2)}{\theta_2 K_r(\theta_1)} > \frac{dh}{dz}\Big|_{n=1} \Big/ \frac{dh}{dz}\Big|_{n=2}, \text{ if } \theta_1 > \theta_2 \text{ and } |h - \Delta z| < \Delta\psi \tag{7}$$

Correspondingly, the dry portion requires a higher hydraulic gradient than the wet one under unstable conditions. Figure 3 depicts the schematically hydraulic gradient line of upward flow ($h < 0$).

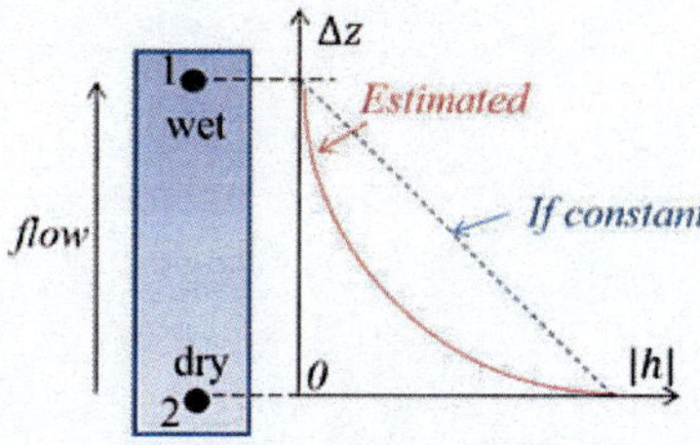

Fig. 3 Example of schematically hydraulic gradient line in unstable flow

3.3 Application Range of the Proposed Criterion on Water Content

This criterion is not completed in a lower water content zone, ranging $0 \geqslant \partial\left(\frac{K_r(\theta)}{\theta}\right)/\partial\theta$. Unstable flow condition could be stabilized because $(|v_1| - |v_2|)\ (\theta_1 - \theta_2) \leqslant 0$, in this range. The behavior of the flow of unsaturated soil is expected to be separated by the gradient of permeability, as shown below and illustrated in Fig. 1.

$$f_\theta = \frac{\partial\left(\frac{K_r(\theta)}{\theta}\right)}{\partial\theta}, f_{\theta_b} = 0 \tag{8}$$

The criterion in Eq. (6) should be considered, when $f_\theta \geq 0$ and $\theta > \theta_b$. θ_b is defined as the boundary of water content in this study. Dry conditions below θ_b indicate that water flows smoothly even under unstable conditions.

4 Discussion

This fundamental evaluation proposed a simple criterion applicable except when the water content is low: Under the conditions, $f_\theta \geq 0$ is satisfied. Generally, coarser materials exhibit a steeper gradient than finer ones (Childs and Collis-George 1950); it is indicated that coarser materials are governed primarily by this proposed criterion. Under unstable conditions, localized high hydraulic gradient might be imposed especially at the drier portion. Therefore, it is predicted that flow is not smooth in the unstable area and is concentrated in the stable one. $\Delta\psi$ can be measured using a tensiometer; water content can also be measured in situ. Unstable conditions might occur under heavy rainfalls and rise of the ground water level at the same time, for example. In contrast, flow can be controlled by varying water content distributions and flow direction in the ground.

5 Conclusions

A simple criterion under a one-dimensional condition is proposed in this study, in which water content boundary is $\partial\left(\frac{K_r(\theta)}{\theta}\right)/\partial\theta$ and $|h - \Delta z| \geqslant \Delta\psi$ is stable. Unstable conditions exhibit the risk of high localized hydraulic gradient under relatively wet conditions. This criterion can (i) evaluate flow under conditions other than analytical conditions and (ii) be applied under in situ urgent conditions.

Acknowledgements The grammars and expressions are partially proofed by Editage.

References

Bear, J.: Dynamics of Fluids in Porous Media. Elsevier, New York (1972)

Childs, E.C., Collis-George, N.: The permeability of porous materials. Proc. R. Soc. Lond. Ser. A Math. Phys. Sci. **201**(1066), 392–405 (1950)

van Genuchten, M.T.: A closed-form equation for predicting the hydraulic conductivity of unsaturated soils. Soil Sci. Soc. Am. J. SSSA, 892–898 (1980)

Optimization of the Calculation of the Piles of Fixed Offshore Platforms

Latif Aslanov

Abstract

For precast metal piles, an assessment of the load-bearing capacity according to the tables of building regulations does not correspond to reality and leads to significant errors. For depths of more than 35 m, it is, in fact, impossible. The dynamic method (building regulations) for predicting the load-carrying capacity of precast piles of a small length with the driving parameter for offshore waters using mechanical and pile hammers gives results with unacceptable errors. For long piles, this method is not applicable at all. In this work, the leading causes of these shortcomings are analyzed. Further, the corresponding scientific and methodological foundations of new methods for calculating the load-bearing capacity of pile foundations are developed.

Keywords

Pile • Pile foundation • Pile foundation calculation method • Rational design • Pile testing

1 Introduction

Theoretically, the method was based on solving the problem of one-dimensional dynamics problem of force propagation process in a hammer-pile-soil system by numerical methods (by finite elements method). If we consider a pile as a one-dimensional bar, we can broadly express the impact propagation process as follows:

$$\frac{\partial^2 u}{\partial t^2} = F(u, x, t) \tag{1}$$

Here $u = u(x, t)$ is the displacement of the cross section of pile, m; t is time, s;

F is an integro-differential operator expressing the stress state arising in the cross section of the pile with regard to viscous–elastic–plastic resistance of the soil.

To eliminate the analytic complexity created by the mathematical model of physical–mechanical properties of the soil, instead of this equation, a hammer-pile-soil system is replaced by different elements with characteristic relations (elastic, plastic, viscous) (Fig. 1).

In the results, the solution of the equations is reduced to the following successively calculated equations. For displacement of pile elements

$$u_{i,t} = u_{i,t-1} + \Delta t v_{i,t-1} \tag{2}$$

for elastic forces arising between pile elements, $F_{i,t} = k_i\left(u_{i,t} - u_{i-1,t}\right)$

for plastic displacement of pile elements

$$U^0_{i,t} = \begin{cases} u_{i,t} + Q_i & \text{for} \quad u_{i,t} + Q_i \prec u^0_{i,t-1}, \\ u^0_{i,t-1} & \text{for} \quad u_{i,t} + Q_i \geq u^0_{i,t-1} - Q_i, \\ u^0_{i,t} - Q_i & \text{for} \quad u^0_{i,t} - Q_i \succ u^0_{i,t-1}, \end{cases} \tag{3}$$

$v_{i,t} = v_{i,t-1} + \left(F_{i-1,t} - F_{i,t} - R_{i,t}\right) \cdot \frac{\Delta t}{m_i}$ for velocities of the elements.

$R_{i,t} = \left(u_{i,t} - u^0_{i,t}\right) \cdot P_i \cdot \left\{1 + \left(\tau_i v_{i,t-1}\right)^{\mu} \text{sign}\left[\left(u_i - u^0_i\right) v_i\right]\right\}$ for resistance of soil to elements. Here, $u_{i,t}$ is the general displacement of the i-th element in time, t, m; $u^0_{i,t}$ is plastic displacement of the i-th element at time t, m; $\partial_{i,t}$ is the velocity of the i-th element in time t, m/s;$F_{i,t}$ is the inner force influencing the i-th element, kN; k_i is the longitudinal rigidity of the elements, kN/sm; m_i is the mass of the elements, t; Δt is the time of the passing of the impact wave of the element, s and is chosen as $\Delta t \leqslant \min\left(\frac{m_i}{x_i}\right)^{\frac{1}{2}}$. Q_i is the

L. Aslanov (✉)
Azerbaijan University of Architecture and Construction, AynaSultanova St.5, AZ1073 Baku, Azerbaijan
e-mail: latif.aslanov@bk.ru

H. El-Askary et al. (eds.), *Research Developments in Geotechnics, Geo-Informatics and Remote Sensing*, Advances in Science, Technology & Innovation,
https://doi.org/10.1007/978-3-030-72896-0_32

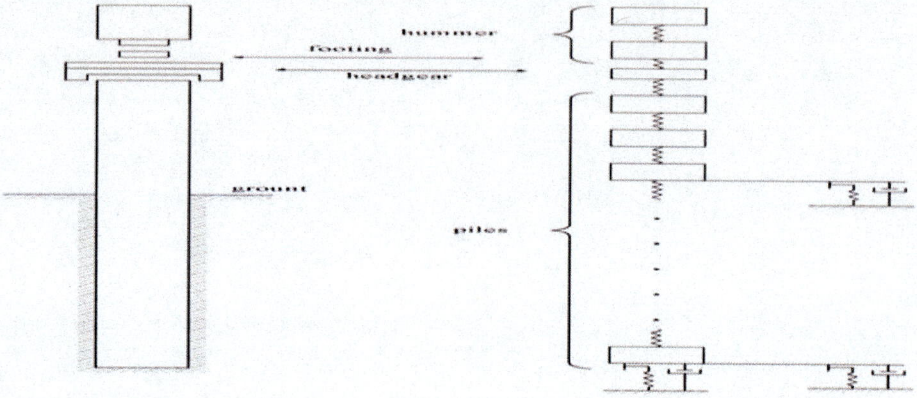

Fig. 1 Scheme of partition of the pile in finite elements

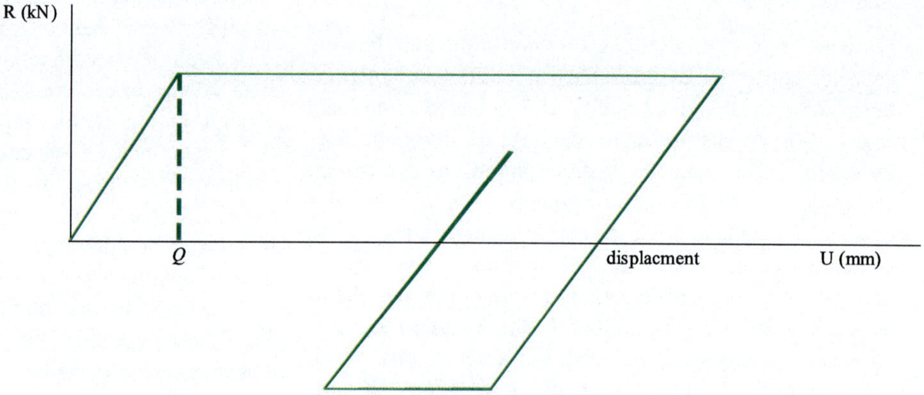

Fig. 2 Diagram of the dependence of resistance on total displacements of the pile and the soil

elastic deformation limit of the soil influencing the i-th pile element. For a great majority of soils, Q_i changes within $0,1 \div 0,4$ sm.

$R_{i,t}$ is the resistance of the soil to pile element, kN; P_i is the external friction rigidity of the soil to the pile element within the elastic deformation, kN/sm; τ_i is the viscosity factor of the soil, and for lateral surface resistance of the pile, it was accepted $0.001 \div 0.004$ s/cm, for longitudinal resistance $0.005 \div 0.005$ s/sm.

Currently, various literature sources on piles are available, some of them are given in [$1 \div 9$].

For clay soils at a depth to 100 m by the formula

$$f_i = k_f f_3 \tag{4}$$

where k_i is a coefficient accepted depending on the soil flow index I_L; if $I_L \leqslant 0,4$, then $k_1 = 0,5 + 0,0143\ h_i$; if $0,4 \prec I_L \leqslant 0,5$, then $k_i = 0,55 + 0,0125\ h_i$; if $I_L > 0,5$, then

$$k_f = 0,65 + 0,01\ h_i \tag{5}$$

Here h_i is the distance from the sea bottom surface to the middle of the i-th layer, $m; f_3$ is oil's design resistance on the lateral surface of the pile at a depth of 35 m depending on the soil flow index of the i-th layer.

Settings or methods or materials and methods or … etc.

A one-dimensional dynamic problem in the process of force propagation in the hammer-pile-soil system is solved by a numerical method (by the finite element method).

2 Results

2.1 Results (Subsection Sample/Only if Needed!)

See Fig. 2.

3 Discussion

Theoretically, the method aims to solve the problem of one-dimensional dynamics of force propagation process in a hammer-pile-soil system by numerical methods (by finite elements method). To eliminate the analytic complexity created by the mathematical model of physical–mechanical properties of the soil, instead of this equation, a hammer-pile-soil system is replaced by different elements with characteristic relations (elastic, plastic, viscous) (Fig. 1). Although the diagram of pile resistance to the soil in the general case is of elastic–plastic character (Fig. 2), it is possible to take into account the viscosity that weakly shows itself in impact process.

4 Conclusions

By means of the one-dimensional wave theory, a convenient scientific-methodical basis was developed for analyzing long-size pile driving process. It can estimate the stress–strain state in any stage of hammer or pile driving process as well as the total load-bearing capacity of the pile. The calculations can determine the dependence of the resistance on the pile construction with respect to soil on the number of impacts expended for driving its unique length.

References

Aslanov, L.F.: Wave interaction of offshore structure and shelf soil through large section piles with a 'hard core' on the half-space model. Neft. Khozyaystvo **2**, 78–81 (2015a)

Aslanov, L.F.: Interaction between large cross-sections bored piles with 'hard core' under dynamic loads and shelf soils. Sci. Bull. Natl. Min. Univ. **5**, 21–25 (2015b)

Subanantharaj Palammal, J., Senthilkumar, P.K.: Arab J. Geosci. **11**, 706 (2018). https://doi.org/10.1007/s12517-018-4032-2

Sharafi, H., Maleki, Y.S., Fard, M.K.: Arab J. Geosci. **9**, 5 (2016). https://doi.org/10.1007/s12517-015-2051-9

Wang, Y.X., Guo, P.P., Ren, W.X., Yuan, B.X., Yuan, H.P., Zhao, Y.L., Shan, S.B., Cao, P.: Laboratory investigation on strength characteristics of expansive soil treated with jute fiber reinforcement. Int. J. Geomech. 04017101 (2017). https://doi.org/10.1061/(ASCE)GM.1943-5622.0000998

Wang, Y.X., Guo, P.P., Dai, F., Li, X., Zhao, Y.L., Liu, Y.: Behaviour and modelling of fiber reinforced clay under triaxial compression by using the combining superposition method with the energy based homogenization technique. Int. J. Geomech. (2018). https://doi.org/10.1061/(ASCE)GM.1943-5622.0001313

Yuan, B., Xu, K., Wang, Y., Chen, R., Luo, Q.: Investigation of deflection of alaterally loaded pile and soil deformation using the PIV technique ASCE Int. J. Geomech. (2017a).https://doi.org/10.1061/(ASCE)GM.1943-5622.0000842

Yuan, B., Sun, M., Wang, Y., et al.: Full 3D displacement measuring system for 3D displacement field of soil around a laterally loaded pile in transparent soil. ASCE Int. J. Geomech. **19**(5), 04019028 (2019)

Yuan, B., Chen, R., Deng, G., Peng, T., Luo, Q.Z., Yang, X.: Accuracy of interpretation methods for deriving p–y curves from model pile tests in layered soils. ASTM J. Testing Eval. **45**(4), 1238–1246 (2017b)

Reliability Analysis of Shallow Foundation Using Response Surface Methodology and Multi-objective Genetic Algorithm

Brahim Lafifi and Ammar Rouaiguia

Abstract

This paper presents a reliability-based analysis of a shallow foundation subjected to a central vertical load, using the limit equilibrium theory. The reliability index is calculated by applying the first order reliability method (FORM). A deterministic model based on the Meyerhof equations is employed to compute the ultimate bearing capacity of the foundation. The concept of response surface method (RSM) is used for the assessment of the Hasofer–Lind reliability indexes by generating an approximate polynomial function for the performance function. In this study, unit weight and the soil shear strength parameters are considered as random variables. The optimization of the Hasofer-Lind reliability index is performed using a combination of response surface model and the multi-objective genetic algorithm.

Keywords

Reliability analysis • Shallow foundation • RSM • Optimization • Genetic algorithm

1 Introduction

A reliability-based approach for the analysis of foundations is more rational since it enables one to consider the inherent uncertainty of each input parameter. Nowadays, this is possible due to the improvement in the knowledge of the statistical properties of soil (Phoon and Kulhawy 1999). This paper presents a reliability-based analysis of a shallow foundation subjected to central vertical load. Previous investigations on the reliability analysis of foundations focused on either the ultimate or the serviceability limit state (Dalia et al. 2008; Sivakumar Babu and Srivastava 2007). This work examines the limit states in the analysis of foundations, using a deterministic formulation based on the Meyerhof equation for calculating the ultimate bearing capacity. The response surface methodology is applied to approximate the performance functions with the random input parameters, which are unit weight γ and the soil shear strength parameters c and φ.

2 Basic Reliability Index

Several methods such as first order reliability method (FORM), second order reliability method (SORM) and point estimate method (PEM) have been extensively applied to estimate the reliability index. For a linear performance function defined as $g(x) = C - D$, where C is the capacity and D is the demand, $g(x) > 0$ denotes safe region while, $g(x) < 0$ represents the failure region. Currently, the Hasofer-Lind reliability index β_{HL} is widely used and is defined as the shortest distance from the mean value point of the random parameters to the limit state surface; its matrix formulation is (Hamrouni et al. 2018):

$$\beta_{HL} = \min_{x \in F}(\sqrt{(x-\mu)^T C^{-1}(x-\mu)} \tag{1}$$

where x is the vector of n random parameters, μ is the vector of their mean values and C their covariance matrix. The solution of Eq. 1 is obtained by considering the constraint $g(x) \leq 0$, where the limit state surface $g(x) = 0$ separates the domain of random parameters into a failure region given by $g(x) < 0$ and a safe region represented by $g(x) > 0$ (Fig. 1).

B. Lafifi (✉) · A. Rouaiguia
Laboratory of Civil Engineering and Hydraulics, University 8 Mai 1945 Guelma, Guelma,, Algeria
e-mail: blafifi@gmx.fr

H. El-Askary et al. (eds.), *Research Developments in Geotechnics, Geo-Informatics and Remote Sensing*, Advances in Science, Technology & Innovation,
https://doi.org/10.1007/978-3-030-72896-0_33

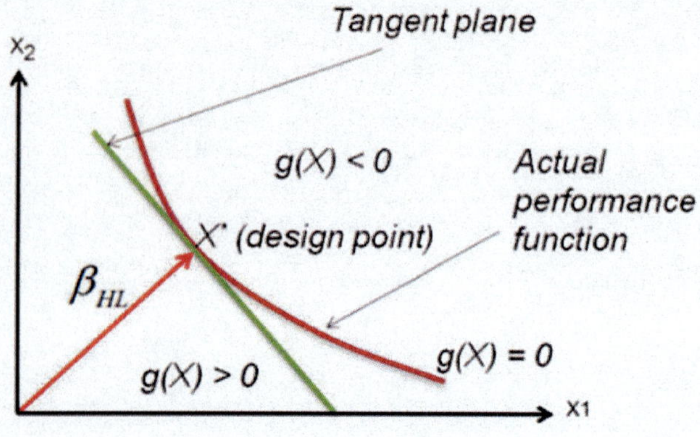

Fig. 1 Hasofer-Lind reliability index and design point for nonlinear performance function

The values of the design points are expected to give insight into the partial safety factors of each of the random parameters as shown below:

$$F_c = \frac{\mu_c}{c^*}; \quad F_\varphi = \frac{\tan(\mu_\varphi)}{\tan(\varphi^*)}; \quad F_\gamma = \frac{\mu_\gamma}{\gamma^*} \tag{2}$$

3 Deterministic Model of Bearing Capacity and Performance Function

Meyerhof's equation for the ultimate bearing capacity q_u for a shallow strip foundation of width B resting on a horizontal soil at a depth D_f and subjected to a central vertical load can be calculated as follows:

$$q_u = cN_c s_c d_c + \gamma D_f N_q s_q d_q + 0.5\gamma B N_\gamma s_\gamma d_\gamma \tag{3}$$

The bearing capacity factors, N_c, N_q and N_γ; (s_c, s_q, s_γ); and depth factors (d_c, d_q, d_γ) are obtained from equations available in the literature (Bowles 1996). The performance function is obtained with respect to the ultimate bearing capacity of the soil as follows:

$$g = \frac{q_u}{P_s} - 1 \tag{4}$$

where q_u is the ultimate bearing capacity and P_s is the applied footing load.

4 Response Surface Method

The main idea of the RSM is to approximate the performance function by an explicit function of the random variables. The expression of this approximation is given by:

$$g(x) = a_0 + \sum_{i=1}^{n} a_i.x_i + \sum_{i=1}^{n} a_i b_j.x_i x_j + \sum_{i=1}^{n} b_i x_i^2 \tag{5}$$

where x_i random variables; n number of the random variables; and a_i, b_i, $a_i b_j$ coefficients to be determined. In this work, three random variables were considered. They were characterized by their mean values μ_i and their coefficients of variation σ_i to evaluate the performance function $g(x)$ at the mean value point and the $2n$ points each at $\pm k$ where k = 1.65 in our case (see Table 1). Single replicate 3^n factorial design was used to fit a quadratic regression model. For this reason, a total number of 27 sample points was required.

5 Methodology

In the present study, a shallow foundation of width B = 1.0 m and depth D_f = 1.0 m was considered. The output response G was obtained from different values of the vertical applied load P_S, varying from small values up to the deterministic ultimate load corresponding to 27 sample points for three input variables (c, φ and γ). Results for the case of applied load P_S = *400* kN/m are presented in Table 2.

Using Design Expert V10 Software, a quadratic regression model is presented below by Eq. 6 with a coefficient of determination R^2 of 99.90%.

$$\begin{aligned} G = &-1.00075 + 0.21625 * C + 0.000053 * \varphi \\ &+ 0.3502 * \gamma - 0.016 * C * \varphi - 0.028 * \varphi * \gamma \\ &+ 0.00039 * C * \varphi^2 + 0.00062 * \gamma * \varphi^2 \end{aligned} \tag{6}$$

This regression model is an approximation of the response G based on the input design parameters (c, φ and γ).

6 Optimization of Reliability Index and Design Points Using Multi-Objective Genetic Algorithm

In this paper, RSM methodology and multi-objective genetic algorithm (NSGA-II) were coupled to solve the multi-objective optimization problems. The aim was to find conditions for the input parameters that lead to optimal values of the response outputs. For more information, please see reference (Meddour et al. 2018; Mezeh et al. 2019), about applying genetic algorithm optimization.

To minimize the Hasofer-Lind reliability index β_{HL} with the condition that $g(x) \leq 0$, constraints were used in this

Table 1 Soil parameters used in the analysis

Input parameters	Mean μ	Standard deviation (σ)	Cov (%)	$\mu-1.65\sigma$	$\mu+1.65\sigma$	Distribution
Unit weight γ(kN/m^3)	18.0	0.9	5	16.52	19.48	Normal
Cohesion C (kPa)	10.0	2	20	6.70	13.3	Normal
Friction angle φ(°)	30°	3	10	25.05	34.95	Normal

Table 2 Soil parameters used in the analysis

N°	Coded values			c	φ	γ	G
1	0	− 1	− 1	10	25.05	16.52	0.205
2	0	− 1	+ 1	10	25.05	19.48	0.309
3	− 1	+ 1	0	6.7	34.95	18.0	2.159
13	0	+ 1	0	10.0	34.95	18.0	2.613
14	+ 1	− 1	0	13.3	25.05	18.0	0.463
15	+ 1	− 1	− 1	13.3	25.05	16.52	0.411
25	− 1	− 1	0	6.7	25.05	18.0	0.051
26	− 1	0	− 1	6.7	30	16.52	0.672
27	0	+ 1	0	10	34.95	18.0	2.613

Table 3 Design points, reliability index and safety factors

P_S (kN/m)	F	c^* (kPa)	φ^* (°)	γ^* (kN/m^3)	β_{HL}	F_c	F_φ	F_γ
400	2.06	6.70	25.05	16.52	2.85	1.49	1.20	1.09
500	1.65	7.53	26.27	17.53	1.71	1.33	1.14	1.03
600	1.38	7.83	27.76	17.59	1.16	1.28	1.08	1.02
700	1.18	9.75	28.90	17.83	0.64	1.03	1.04	1.01
826	1.00	10.00	30.00	18.00	0.00	1.00	1.00	1.00

case. Table 3 presents the results of β_{HL} and the corresponding design points (c^*, φ^* and γ^*), for different values of the vertical applied load P_S. It's clear from the results above that the reliability index decreases with the increase of the applied load P_S.

7 Conclusions

In this work, the application of the response surface method in the assessment of allowable bearing pressure on shallow foundations is demonstrated by using Hasofer-Lind reliability index. The results show that reliability analysis may predict the range of allowable pressures on the footing. Furthermore, the values of c^*, φ^* and γ^* at the design point are found smaller than their respective mean values; they increase with the increase of the applied load P_S. Consequently, the partial safety factors of random variables decrease with the increase of the applied load.

References

Bowles, J.E.: Foundation Analysis and Sesign, 5th edn. McGraw-Hill Book Company (1996)

Dalia, S., Youssef, A., Soubra, A.: Reliability-based analysis of strip footings using response surface methodology. Int. J. Geomech. ASCE **8**(2), 134–143 (2008)

Hamrouni, A., Dias, D., Sbartai, B.: Reliability analysis of a mechanically stabilized earth wall using the surface response methodology optimized by a genetic algorithm. Geomech. Eng. **15** (4) (2018)

Meddour, I., Yallese, M.A., Bensouilah, H., Khellaf, A., Elbah, A.: Prediction of surface roughness and cutting forces using RSM, ANN, and NSGA-II in finish turning of AISI 4140 hardened steel with mixed ceramic tool. Int. J. Adv. Manuf. Technol. **97**, 1931–1949 (2018)

Mezeh, R., Mroueh, H., Housseingholian, M., Sadek, M.: New approach for the assessment of train/track/ground dynamics using in-situ measurements of high-speed train induced vibrations. Soil Dyn. Earthquake Eng. **116**, 50–59 (2019)

Phoon, K.-K., Kulhawy, F.H.: Evaluation of geotechnical property variability. Can. Geotech. J. **36**, 625–639 (1999)

Sivakumar Babu, G.L., Srivastava, A.: Reliability analysis of allowable pressure on shallow foundation using response surface method. Comput. Geotech. **34**, 187–194 (2007)

The Effect of Footing Shape on the Elastic Modulus of Soil

Lysandros Pantelidis

Abstract

In practice, geotechnical engineers often use triaxial elastic soil property values derived from the corresponding test or continuous probing test results (e.g. Cone Penetration Test), even for structures deforming under plane strain conditions, such as strip footings. Review of the available theoretical and experimental studies, however, led to the following observations: the triaxial elastic modulus differs from the respective plane strain one and in addition to the aspect ratio of footing, the soil modulus depends on the thickness of the soil medium. Based on experimental findings and on the elastic theory and using a sequential way of thinking, the elastic settlement at the centre of a $B \times L$ footing is suggested that be calculated considering the respective $B \times B$ footing along with the original distributed loading value.

Keywords

Plane strain conditions • Modulus of elasticity • Footing shape • Elastic settlement • Winkler's spring analysis • Shallow foundations

1 Introduction

As known, a solution for a plane stress problem can be applied to a plane strain problem and vice versa, provided that the correct elastic parameters for the problem considered are used. In this respect, for an ideal elastic isotropic material, the elastic constants of the medium in plane strain conditions are:

$$E_p = E_t/\left(1 - \nu_t^2\right) \quad \text{and} \quad \nu_p = \nu_t/(1 - \nu_t) \tag{1}$$

where, E is the elastic modulus and ν the Poisson's ratio, while the subscripts p and t denote plane strain and triaxial loading conditions respectively; see (Lee 1970; Poulos and Davis 1991). The effect of footing shape on the value of the elastic constants has also been shown experimentally (e.g. (Lee 1970; Elsamee 2013)). However, geotechnical engineers often use triaxial elastic soil property values even for structures deforming under plane strain conditions, such as retaining walls, strip foundations, trench excavations and tunnels.

In the present paper, the problem of determining the proper elastic modulus of soil for a general $B \times L$ footing is examined based on experimental results and the elastic theory. This value is of particular importance in calculating the modulus of subgrade reaction based on Winkler's spring hypothesis or the elastic settlement of footings. The analysis below follows a sequential way of thinking, revisiting experimental results and empirical expressions, interpreting results derived from the theory of elasticity, while it is driven by the findings of contemporary plate loading tests.

2 Derivation of the Proposed Relationship Between the $E_{B \times L}/E_{B \times B}$ Ratio and the Footing Shape

Experimental works referring to the effect of footing shape on the modulus of elasticity (E) or subgrade reaction (k_s) of soil are rather rare in the literature. In this respect, Lee (1970) found that the E_p/E_t ratio value is approximately equal to 1.4. For considering the effect of footing shape on the elastic settlement, Schmertmann et al. (1978) adopted Lee's E_p/E_t = 1.4 value suggesting the following empirical

L. Pantelidis (✉)
Cyprus University of Technology, 3036 Limassol, Cyprus
e-mail: lysandros.pantelidis@cut.ac.cy

H. El-Askary et al. (eds.), *Research Developments in Geotechnics, Geo-Informatics and Remote Sensing*, Advances in Science, Technology & Innovation,
https://doi.org/10.1007/978-3-030-72896-0_34

expressions for the elastic modulus of sands under triaxial and plane strain conditions respectively:

$$E_t = 2.5q_c \quad (2)$$

$$E_p = 1.4\,E_t = 3.5q_c \quad (3)$$

where q_c is the cone tip resistance of the CPT test (in same units as E). Terzaghi et al. (1996), in turn, adopting Lee's experimental results, suggested the following empirical interpolation function for use between the two extreme values of Eqs. 2 and 3:

$$E_{B\times L}/E_{B\times B} = 1 + 0.4\log(L/B) \quad (4)$$

where $E_{B \times B}$ and $E_{B \times L}$ are the moduli of elasticity of soil corresponding to a $B \times B$ and $B \times L$ footing respectively. For $1 \leq L/B \leq 10$, the $E_{B \times L}/E_{B \times B}$ ratio ranges between 1 and 1.4, while for $L/B > 10$ (assumed to correspond to plane strain conditions), it is equal to 1.4. What is not widely known, however, is that Lee (1970) in his publication entitled "Comparison of plane strain and triaxial tests on sand" used triaxial specimens of 3.6 cm in diameter for the axisymmetric case and 2.79 cm × 7.11 cm rectangular test specimens for what he called "plane strain" conditions. Thus, what Schmertmann et al. and later, Terzaghi et al. adopted for plane strain conditions, is actually for L/B ratio as low as 2.545 (= 7.11/2.79). Revisiting the $E_{B \times L}/E_{B \times B}$ versus L/B relationship, the $E_{B \times L}/E_{B \times B}$ ratio becomes equal to 1.4 for $L/B = 2.545$ if the "0.4" coefficient in Eq. 4 is replaced by unity, that is:

$$E_{B\times L}/E_{B\times B} = 1 + \log(L/B) \quad (5)$$

However, the empirical 1 + log(L/B) term is already well-known from the elastic settlement theory (Terzaghi et al. 1996), where the depth below the foundation level, up to which the contribution to elastic settlement is approximately equal to the 90% of the respective settlement corresponding to a semi-infinite mass is:

$$z_l = 2B[1 + \log(L/B)] \quad (6)$$

Elsamee (2013) provided a comparison table with k_s values for different rigid loading plates, relative densities of sand and embedment depths. He considered, among others, a square plate with edge 0.305 m ($L/B = 1$) and a 0.238 m × 0.307 m rectangular plate ($L/B = 1.29$). From the k_s values corresponding to these plates, the author found that $E_{B \times L}/E_{B \times B} = 0.986$. The derivation of the latter involved both Terzaghi et al.'s (1996) $k_{s2} = k_{s1}[(B_2 + B_1)/2B_2]^2$ empirical formula for plates on sands (because the two plates considered by Elsamee had different widths) and Steinbrenner's (1934) formula for elastic settlement analysis of $B \times L$ footings (because it contains the desired E parameter). For the same case ($L/B = 1.29$), Eqs. 5 and 4 gave $E_{B \times L}/E_{B \times B} = 1.11$ and 1.04 respectively. In this respect, it is mentioned that Terzaghi's empirical plate size transformation may have introduced an error, since $E_{B \times L}$ should be greater than $E_{B \times B}$. Roy and Deb (2015), in turn, provided k_s and settlement values for various plate sizes. Working with their results for the 7.5 × 7.5, 7.5 × 11.25 and 7.5 × 15 plates (in cm), the author found that, the $E_{B \times L}/E_{B \times B}$ ratio ranges from 1.094 to 1.121 for $L/B = 1.5$ (= 11.25/7.5) and from 1.134 to 1.168 for $L/B = 2$ (= 15/7.5). For $L/B = 1.5$ Eqs. 5 and 4 gave $E_{B \times L}/E_{B \times B} = 1.176$ and 1.070, while for $L/B = 2$, the same equations gave 1.301 and 1.120 respectively. It is noted that Roy and Deb's tests were conducted on clay.

The fact that Eq. 5 gave satisfactory results and that the empirical 1 + log(L/B) term is a well-established term derived from the elastic settlement theory gave rise to investigating the current problem from the elastic theory point of view. In this respect, the author noticed that working with Steinbrenner's (1934) shape factor (I_s), the $I_{s,B \times L}/I_{s,B \times B}$ ratio gave values comparable to the respective experimental $E_{s,B \times L}/E_{s,B \times B}$ ratio values (see Table 1) and to those obtained from Eq. 5; the Steinbrenner's shape factor can be found in any soil mechanics book [e.g. (Poulos and Davis 1991)]. As shown in Table 1, the derived $I_{s,B \times L}/I_{s,B \times B}$ values depend on soil thickness, H. Theoretically, the same should happen with the $E_{B \times L}/E_{B \times B}$ ratio, since, by definition, E equals to the ratio of the applied stress, q, over the corresponding strain, ε. Indicatively, from Hoek's law, it is inferred that two specimens with the same k_s value under unconfined compression have the same E value only if they have the same height, h:

$$E = q/\varepsilon = (q/\rho)h = k_s h \quad (7)$$

where ρ and h are the displacement and height of specimen respectively and $k_s = q/\rho$.

Another interesting point is that when the effect of footing shape on the elastic modulus of soil is considered while calculating the elastic settlement of a footing, the latter appears to be independent of the aspect ratio of footing (Pantelidis, 2020). That is, Steinbrenner's formula for the elastic settlement at the centre of footing becomes:

$$\begin{aligned}\rho_{(B\times L)} &= 4[P/(B\times L)](B/2)(1-\nu^2)\,I_{s,B\times L}/E_{B\times L}\\ &= 4(P/(B\times L))(B/2)(1-\nu^2)\,I_{s,B\times B}/E_{B\times B}\end{aligned} \quad (8)$$

Table 1 $I_{s,B \times L}/I_{s,B \times B}$ and $E_{B \times L}/E_{B \times B}$ values for various cases; the $E_{B \times L}/E_{B \times B}$ values refer to the experimental studies, while the $I_{s,B \times L}/I_{s,B \times B}$ values were obtained from the theory of elasticity

v	$I_{s,B \times L}/I_{s,B \times B}$ Thickness of compressible layer, H				$E_{B \times L}/E_{B \times B}$	$I_{s,B \times L}/I_{s,B \times B}$ Thickness of compressible layer, H				$E_{B \times L}/E_{B \times B}$
	B	z_1[a]	$2z_1$[a]	∞		B	z_1[a]	$2z_1$[a]	∞	
	L/B = 1.29 [Elsamee's(2013) case]					L/B = 2.545 [Lee's (1970) case]				
0	1.058	1.096	1.114	1.130	0.986[b]	1.138	1.333	1.415	1.498	1.40
0.25	1.048	1.092	1.112	1.130		1.096	1.313	1.405	1.498	
0.5	1.023	1.082	1.108	1.130		0.993	1.271	1.385	1.498	
	L/B = 1.5 (Roy and Deb 2015)					L/B = 2.0 (Roy and Deb 2015)				
0	1.085	1.152	1.182	1.210	1.094–1.121	1.121	1.254	1.310	1.365	1.134–1.168
0.25	1.068	1.145	1.178	1.210		1.090	1.240	1.303	1.365	
0.5	1.026	1.129	1.171	1.210		1.013	1.211	1.290	1.365	
	L/B = 10					L/B = 20				
0	1.156	1.625	1.882	2.267	–	1.156	1.699	2.011	2.660	–
0.25	1.084	1.564	1.841	2.267		1.082	1.623	1.948	2.660	
0.5	0.909	1.435	1.757	2.267		0.901	1.466	1.822	2.660	

[a] z_1 was calculated using Eq. 6
[b] Empirical transformation for plate size has been involved [see Terzaghi et al. (1996)]

where P is the axial load of the column. Because, however, a longer footing has greater influence depth, for the case of stratified mediums the $E_{B \times L}$ value should represent first an equivalent homogenous medium (see Pantelidis (2021)).

The area of applicability and limitations of the proposed approach depend on the applicability and limitations of Steinbrenner's formula.

3 Conclusions

Review of the available theoretical and experimental studies provided the following insights: the triaxial elastic modulus greatly differs from the respective plane strain one, the soil modulus depends on the aspect ratio of footing as well as the thickness of soil medium. These findings and the elastic theory suggest that the elastic settlement at the centre of a $B \times L$ footing should be calculated considering the respective $B \times B$ footing along with the original distributed loading value. Moreover, since the settlement of a square footing is approximately equal to the settlement of its equivalent circular footing, the problem could be reduced to calculating the settlement of the latter. Finally, for soil mediums thicker than the influence depth of footings, the $E_{B \times L} = [1 + \log(L/B)] \cdot E_{B \times B}$ formula could conveniently be used, where, $E_{B \times B} \approx E_t$.

References

Elsamee, W.N.A.: An experimental study on the effect of foundation depth, size and shape on subgrade reaction of cohessionless soil. Engineering **05**, 785–795 (2013)

Lee, K.L.: Comparison of plane strain and triaxial tests on sand. J. Soil Mech. Found. Div. **96**(3), 901–923 (1970)

Pantelidis, L.: Elastic settlement analysis for various footing cases based on strain influence areas. Geotech. Geol. Eng. **38**,4201–4225 (2020). https://doi.org/10.1007/s10706-020-01290-w

Pantelidis, L.: The equivalent modulus of elasticity of soil mediums for designing shallow foundations.Geotech. Geol. Eng. **39**(5), 3863–3873 (2021). https://doi.org/10.1007/s10706-021-01732-z

Poulos, H.G., Davis, E.H.: Elastic Solutions for Soil and Rock Mechanics. John Wiley, New York (1991)

Roy, S.S., Deb, K.: Effects of aspect ratio of rectangular foundations on subgrade modulus of soft soil. In: Proceedings of ICSGE2015 International Conference on Advances in Soft Ground Engineering, pp. 571–579. Research Publishing, Singapore (2015)

Schmertmann, J.H., Hartman, J.P., Brown, P.R.: Improved strain influence factor diagrams. J. Geotech. Eng. Div. **104**(8), 1131–1135 (1978)

Steinbrenner, S.W.: Tafelnzursetzungsberechnung. Die StraBe **1** (1934)

Terzaghi, K., Peck, R.B., Mesri, G.: Soil Mechanics in Engineering Practice. John Wiley & Sons (1996)

Effect of Water Content on Consolidation Parameters

Hasan Eteraf, Amir Mosallaei, Balázs Kovács, and Viktória Mikita

Abstract

Most practicing geotechnical engineers employ classical (saturated) soil mechanics to analyze geotechnical engineering problems even when unsaturated conditions are involved. This is misguided given that some of the fundamental features of unsaturated soil behavior cannot be properly represented without an understanding based on unsaturated soil mechanics. On a practical level, field applications such as the analysis of slope instabilities, landslides, earth dams, highways, and underground disposal of radioactive waste, all require a proper understanding of the behavior of unsaturated soils. Similarly, foundations as well as other geotechnical activities in regions where the natural soil is unsaturated to considerable depth necessitate significant insight into the behavior of these soils. The aim of this study is to investigate the effect of water content on the consolidation parameters. To this end, 4 samples with different water content were chosen, as a first step, for the investigation of consolidation behavior as representatives of different ranges of moisture content. Then, consolidation tests were applied to each sample to find and compare the consolidation behavior of the samples. Two consolidation parameters: volume compressibility coefficient and compression index were analyzed and compared. The results showed that consolidation behavior depends on moisture content.

Keywords

Unsaturated soil • Consolidation • Water content • Volume compressibility • Compression index

H. Eteraf (✉) · B. Kovács · V. Mikita
University of Miskolc, 3515 Miskolc, Hungary
e-mail: hghassan@uni-miskolc.hu

A. Mosallaei
Shahrood University of Technology, Shahrood, Semnan, Iran

1 Introduction

The soil behavior is mainly affected by the change in the initial moisture content, void ratio, scale, and overburden stress. Nevertheless, setbacks can be encountered with the sample scale and the initial moisture content effects on engineering applications in field conditions. There are some parameters that influence the consolidation settlement such as: the degree of saturation (Sr %) or initial moisture content (w %), the amount of overburden pressure (σ_p kPa), etc. (Rosine and Sabbagh 2016).

The investigations conducted heretofore, have demonstrated that fully saturated conditions do not always reflect conservative situations in terms of safety. In fact, important plastic deformation processes, such as collapse upon wetting, cannot be predicted using the effective stress principle of Terzaghi (Rampino et al. 2000).

In geotechnical engineering practice, conventional soil mechanics principles are applied for the design of the foundations, assuming that the soil is in a state of saturation. However, natural soils are more often found unsaturated as the ground-water table is at a great depth (Fattah et al. 2016).

Unsaturated soil is a three-phase material containing soil particles, water, and air. The mechanical behavior of unsaturated soil is strongly influenced by both pore air pressure (u_a) and pore water pressure (u_w) (Estabragh et al. 2017).

2 Materials and Methods

A comprehensive model of poroelasticity for consolidation in unsaturated soils has been recently developed. Still, it remains elusive how variations in soil texture and water content affect the consolidation behavior, and the underlying parameters deriving this behavior. Despite these long-standing advances, the effect of soil texture and initial

H. El-Askary et al. (eds.), *Research Developments in Geotechnics, Geo-Informatics and Remote Sensing*, Advances in Science, Technology & Innovation,
https://doi.org/10.1007/978-3-030-72896-0_35

water saturation on consolidation behavior in unsaturated soils receives little attention (Lo and Lee 2015).

The type of soil used in this research is SM according to the USCS classification system. Atterberg limit tests were carried out to determine the liquid limit (LL) and plasticity limit (PL). The properties of the soil are mentioned in Table 1.

In addition, we performed the Proctor test. The optimum moisture content and the maximum dry density were determined from the compaction curves which is shown in Table 2.

In this research, 4 samples with different water content were chosen for the investigation of consolidation behavior as representatives of different ranges of moisture content according to Table 3. After samples preparation, consolidation tests were applied on each sample, aiming to find and compare their consolidation behavior.

3 Results

Consolidation tests were performed in order to obtain the consolidation curves and parameters. Then, the consolidation parameters–volume compressibility coefficient (M_v) and compression index (C_c)—were calculated from the consolidation test data for each sample with its proper water content.

4 Discussion

For a better understanding, Table 4 shows the trend of void ratio difference (Δe) throughout the consolidation for different samples. According to Table 4., comparing the differences of void radio reveals that by increasing the normal stress the void ratio differences are more and more divergent. For instance, the difference of void ratio in 300 kPa is 0.0525, 0.0580, and 0.0883 for samples with 5, 10, and 15% moisture respectively. The differences between these values are also increasing compared with those at 600 kPa. In a range of 5–15% moisture content, the value of Δe is increasing with the increase of the moisture content. However, it is interesting to note that a sample with 20% moisture content does not follow this trend. This discrepancy could be attributed to optimum moisture content. Increasing moisture content up to the approximately optimum moisture content can lead to a clear increasing trend in value of void ratio difference. Whereas, after we pass optimum moisture content range, (sample with 20% moisture) the results are unpredictable, they do not follow the previous trend.

Figure 1 illustrates the comparison of the volume compressibility coefficient. Generally, with increasing the water content to 15%, the values of volume compressibility coefficient increased. Accordingly, it seems that this coefficient could be attributed to optimum moisture content range, considering the change of behavior after that point. For a range of 5–15% water content, with increasing loading during the consolidation, the intervals between the $\sigma - M_v$ curves increased. Figure 2 shows the comparison of compression index values. It reveals that by increasing the water content up to 15%, the compression index increased as well only to start decreasing after 15%.

5 Conclusions

It can be concluded that the consolidation behavior depends on moisture content. In fact, the values of Δe follow the increase of the moisture content up to the optimum moisture content. Afterward, the consolidation behavior becomes unpredictable, not following the previous trends.

By increasing the water content up to an approximate value of optimum moisture content, both the compression index and the coefficient of volume compressibility increase.

Table 1 Soil properties

LL	PL	PI	USCS	Type of soil
26.9	19.3	7.6	SM	Silty sand

Table 2 Results of standard proctor test

$\omega_{opt}(\%)$	$\rho_{D,Max}\left(\frac{g}{cm^3}\right)$
12.5	1.938

Table 3 Soil classification according to water content

Soil	Dry	Medium	Wet	Extra wet
Moisture content range	4–6.5	9–13	13–16	> 16
Chosen moisture content	5	10	15	20

Table 4 Values of Δe

σ (kPa) (%)	0	100	200	300	400	500	600
$\omega = 5$	0	0.0149	0.0258	0.0525	0.0575	0.0654	0.0762
$\omega = 10$	0	0.0257	0.0430	0.0580	0.0834	0.1057	0.1227
$\omega = 15$	0	0.0206	0.0412	0.0883	0.1050	0.1330	0.1542
$\omega = 20$	0	0.0173	0.0322	0.0639	0.0805	0.1011	0.1185

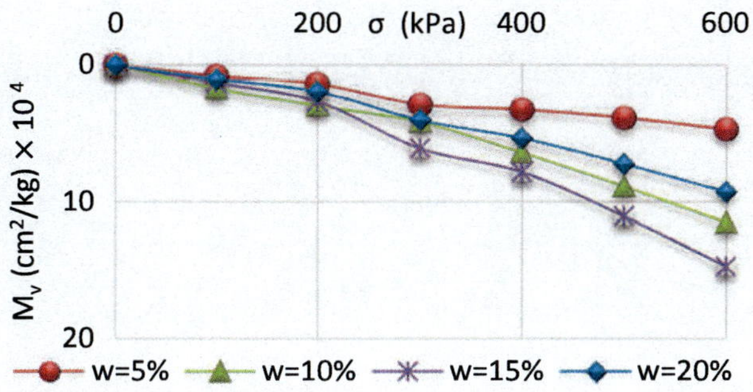

Fig. 1 Comparison of M_v

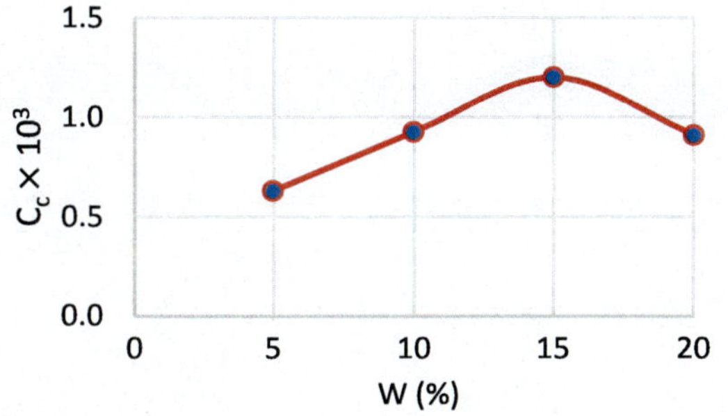

Fig. 2 Comparison of C_c

However, when the moisture content is higher than the optimum moisture content value, the compressibility rate is unpredictable and the compression index decreases. It's recommended to consider low permeability soil for a generalization of consolidation behavior investigation.

Acknowledgments The described article/presentation/ study was carried out as part of the EFOP-3.6.1-16-2016-00011 "Younger and Renewing University—Innovative Knowledge City—institutional development of the University of Miskolc aiming at intelligent specialisation" project and the GINOP-2.3.2-15-2016- 00031 "Innovative solutions for sustainable ground-water resource management" project implemented in the framework of the Szechenyi 2020 program. The realization of this project is supported by the European Union, co-financed by the European Social Fund and the European Structural and Investment Funds.

References

Estabragh, A.R., Moghadas, M., Moradi, M., Javadi, A.A.: Consolidation behavior of an unsaturated silty soil during drying and wetting. Soils Found **1**, **57**(2), 277–287

Fattah, M.Y., Salim, N.M., Irshayyid, E.J.: Experimental study on compressibility, volume changes, strength and permeability characteristics of unsaturated bentonite-sand mixtures. Eng. Technol. J. **34**(7 Part (A) Engineering), 1308–1323 (2016)

Lo, W.C., Lee, J.W.: Effect of water content and soil texture on consolidation in unsaturated soils. Adv. Water Resour. **1**(82), 51–69 (2015)

Rampino, C., Mancuso, C., Vinale, F.: Experimental behaviour and modelling of an unsaturated compacted soil. Can. Geotech. J. **37**(4), 748–763 (2000)

Rosine, T.N., Sabbagh, T.T.: Effect of moisture content and scale on fine-grained soil compressibility. GEO-CHINA 2016, p. 262, 2016 Jul 21

Desiccated Crack Behavior of Kaolin Clay Soil

Nor Shahidah Mohd Nazer and Charlene Sharmila Charles

Abstract

Crack desiccation is one of the challenging behavior of shrinkage soft soil. Existence of crack increases soft soil's permeability and reduces its effective stress under unsaturated conditions. This research is conducted on kaolin clay soil obtained from Sungai Pelek in Sepang, Malaysia. It aims to characterize the geometry of crack under desiccated conditions. To this end, three slurry samples with different thicknesses are placed in a circular aluminium pan and air-dried to allow cracking in a slow evaporation process. Crack parameters were recorded throughout the testing duration; including the length, angle, and width to acquire crack porosity, aperture, and intensity. The cracking analysis identifies four crack stages: the onset stage, the primary crack stage, the secondary crack stage and the steady-state stage. The onset crack stage is marked by the first appearance of crack after a rapid loss of water content due to drying. Thicker samples take longer to crack. The primary crack stage is indicated by the rapid and short development of primary cracks due to vertical and horizontal evaporation. The secondary crack stage, which is the longest stage for crack formation, is marked by the appearance of cracks that intersects with the primary cracks at intersecting angles between 60° and 139° to form a polygonal cracks pattern. The steady-state stage is indicated by complete drying where the formation of cracks is no longer observable. The relationships between crack parameters and crack stages are briefly analyzed and presented in this paper to gain insight into Kaolin's clay soil crack evolution.

Keywords

Desiccated • Crack • Kaolin clay • Soil

1 Introduction

Soft soil contains a high amount of clayey minerals, making it one of the most challenging geo-material to work with due to its low permeability characteristics. The poor performance of dried soft soil can be closely linked to major geohazards incidences such as landslide, embankment failure, and landfill leakage (Rahardjo 2014). This could be due to the presence of desiccated crack, which in turn changes the mechanical and physico-chemical of the ground (Kodikara and Costa 2013). Great research investigates the aspects of mechanical behavior of cracked soil and rock (field analysis, laboratory simulation, modeling, etc.) (Zhou et al. 2018; Wang et al. 2018). One of the vital aspects to look at is the geometry of crack patterns. The geometric of cracks is greatly influenced by its boundary conditions, surface area, and thickness (Ledesma et al. 2018). In this paper, the geometry analysis of desiccated crack is investigated in terms of crack intensity, porosity, and aperture to gain some insight into the evolution of desiccated crack of Kaolin's clay soil.

Soil Properties and Methods

Soil samples were extracted from a field in Sungai Pelek, Sepang in Malaysia. The samples were prepared into three slurry mixtures of different thicknesses then placed in a circular aluminium pan. The samples had an initial water content of 52% (Sample A), 69% (Sample B), and 81% (Sample C), respectively. The samples were then left in a room with uniform temperature and humidity for drying.

N. S. Mohd Nazer (✉) · C. S. Charles
Department of Earth Sciences and Environment, Universiti Kebangsaan Malaysia, 43600 Bangi, Malaysia
e-mail: shahidahnazer@ukm.edu.my

H. El-Askary et al. (eds.), *Research Developments in Geotechnics, Geo-Informatics and Remote Sensing*, Advances in Science, Technology & Innovation,
https://doi.org/10.1007/978-3-030-72896-0_36

Table 1 Characteristics of soil samples

Soil characteristics	Results
Bulk density (g/cm^3)	1.16
Moisture content (%)	17.94
Specific gravity	2.49
Clay percentage (%)	41.6
Liquid limit	44.64
Plastic limit	13.17
Plasticity index	31.47
pH	3.6
Hydraulic conductivity (cm/s)	5.63×10^{-9} to 9.22×10^{-10}
Mineral content	Quartz, Kaolinite
Element content	SiO_2, Al_2O_3
Optimum moisture content (%)	14.44
Maximum dry density (g/cm^3)	2.1
Dispersion	Non-dispersive

The weights of samples were observed on a daily basis for 57 days. Changes in crack geometry were captured by a digital camera, and the crack parameters were measured by Vernier caliper. The observation was carried out until the sample reached the steady-state stage; crack formation came to a complete halt. Table 1 summarizes the characteristics of these samples.

2 Results

Figure 1 indicates that the crack porosity and intensity increase with the decrease of soil thickness, whereas, the crack aperture decreases along with the soil thickness. These scenarios can be clearly seen in Fig. 2. Sample A being the thinnest has the highest crack intensity, resulting in a high porosity. Its aperture, however, appeared to be lower.

3 Discussion

The evolution of crack formation in Kaolin clay soil of Sungai Pelek shows four distinct stages of crack formations (Fig. 3). The onset crack stage is the transition point from saturated to unsaturated phase marked by the first breakthrough of crack. This point corresponds to the air-entry value in the soil-water characteristic curve (SWCC) (Fredlund et al. 2010). The primary crack stage is the second stage indicated by the rapid development of main cracks due to vertical and horizontal evaporation in a short period. The third stage is known as the secondary crack stage, where smaller cracks intersected with the primary cracks at intersecting angles of 60°–139° to form a hexagonal crack pattern (Ledesma et al. 2018). The fourth stage is the steady-state stage indicated by complete drying where the formation of cracks is no longer observable.

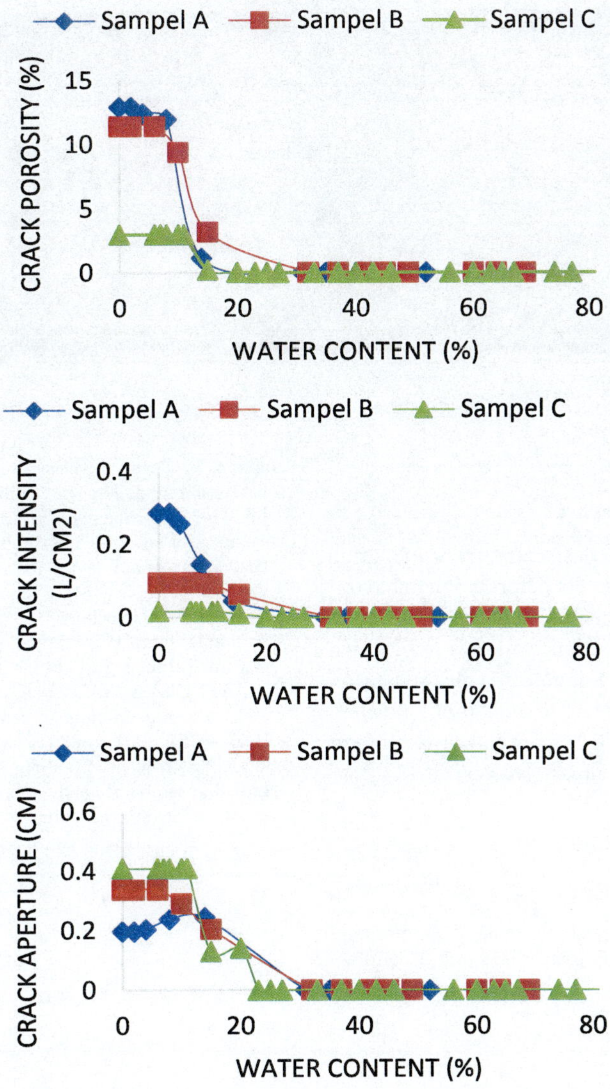

Fig. 1 Crack porosity, intensity, and aperture

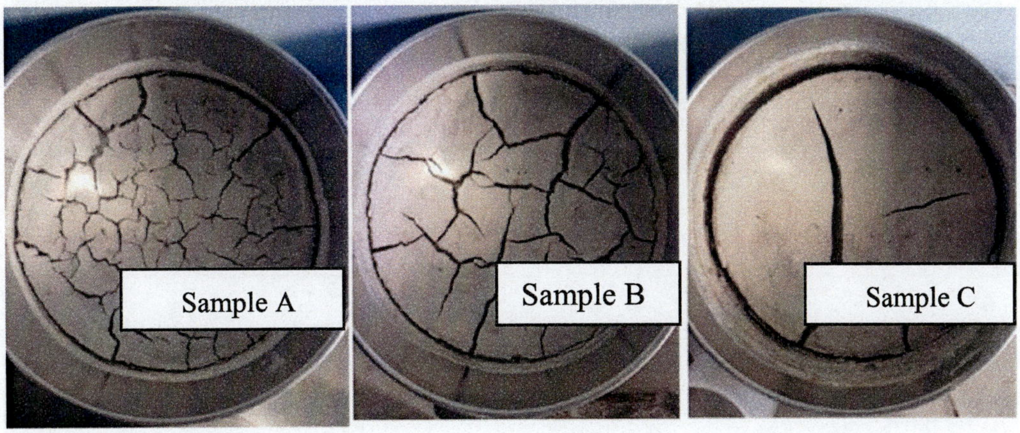

Fig. 2 Crack development in Sample A, B, and C

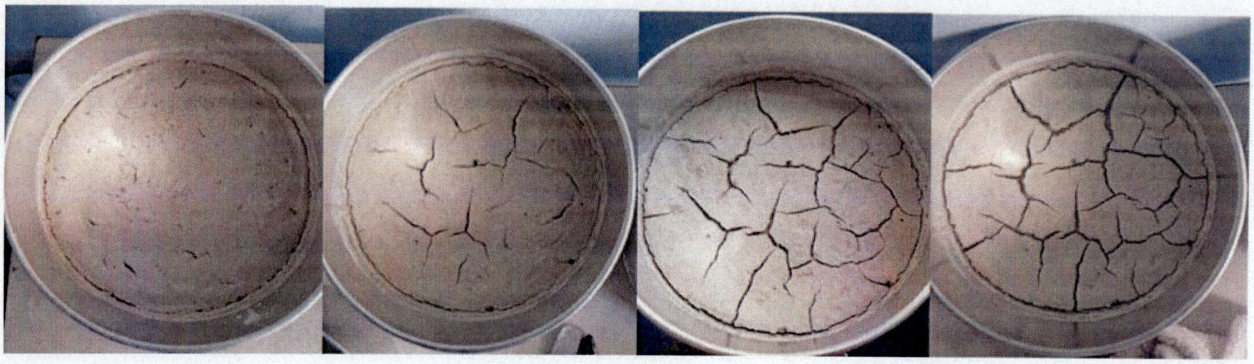

Fig. 3 4 stages of crack formation (left to right)

4 Conclusions

Preliminary understanding of the desiccated crack evolution in clay soil can be useful to predict its deformation behavior. Physical parameters shown by unique crack patterns can be attributed to the hydraulic performance of soft soil which directly affects its permeability. Future research focuses on soft soil's permeability under cracked condition. Further, different boundary conditions will be carried out to support the initial findings of these preliminary studies.

References

Fredlund, D.G., Houston, S.L., Nguyen, Q., Fredlund, M.D.: Moisture movement through cracked clay soil profiles. Geotech. Geol. Eng. **28**(6), 865–888 (2010)

Kodikara, J., Costa, S.: Desiccation cracking in clayey soils: mechanisms and modelling. In: Multiphysical Testing of Soils and Shales, pp. 21–32. Springer, Berlin (2013)

Ledesma, A., Prat, P.C., Lakshmikantha, M.R.: Boundary effects in the desiccation of soil layers with controlled environmental conditions. Geotech. Test. J. **41**(4), 675–697 (2018)

Rahardjo, P.P.: Geotechnical failures case histories of construction on soft soils, forensic investigations and counter measures in Indonesia. Int. J. Integr. Eng. **6**(2) (2014)

Wang, Z.F., Cheng, W.C., Shen, J.S., Ni, J.C.: Modeling of permeation and fracturing grouting in sand: laboratory investigations. J. Test. Eval. **46**(5), 2067–2082 (2018)

Zhou, X.P., Zhang, J.Z., Wong, L.N.Y.: Experimental study on the growth, coalescence and wrapping behaviors of 3D cross-embedded flaws under uniaxial compression. Rock Mech. Rock Eng. **51**(5), 1379–1400 (2018)

An Attempt to Apply the Kinematic Method of Rigid Solids in the Study of Bearing Capacity of Shallow Foundations

Boutahir Born Bencheikh and Belabed Lazhar

Abstract

In the geotechnical engineering field, shallow foundations are frequently needed to ensure good fieldwork stability. They are also intended to permanently and uniformly transmit all solicitation requests on the seating floor. Bearing capacity of foundations, durability, stability, design of shallow foundations, lead unfortunately to a serious problem where finding an adequate solution presents the main goal for several scholars. Indeed, the corresponding drawback is due only to the high number of reported damage that occurred in the structure of foundations and also the punching failure. Furthermore, stability analysis of shallow foundations against puncture has recently been carried out, in which a kinematic rigid body break based-method has been applied. In this respect, this paper proposes two mechanical models, as a contribution to the research field. These models are used for dwell computing the failure along the sliding surface based on the kinematic method of rigid solids.

Keywords

Shallow foundations • Bearing capacity • Punching • Kinematic method of rigid solids

1 Introduction

The kinematic method of rigid fracture solids is considered an effective tool widely used in the study of slope stability, slopes and retaining walls, different types of foundations and so on. Its principle passed on the limit analysis, which is given by Culmann in (1866) and Salençon in (1976) (Costet and Sanglerat 1969). It allows to compute the potentialities of a system's resistance in a given geometry. This can be achieved mainly by writing the compatibility between the balance of this system and the resistance of its constituent material. To this end, two mechanical models of possible rupture of the shallow foundations by punching are proposed in the following subsection.

B. B. Bencheikh (✉) · B. Lazhar
Laboratory of Civil Engineering and Hydraulics, University 8 Mai 1945, Guelma, Algeria

2 Kinematic Method of Rigid Fracture Solids and Principle of Minimum Safety

The type of plastic limit state that reaches a system of the soil mass after large enough deformations depends on the static and kinematic boundary conditions. The Experience has shown that the shallow foundations in most cases prefer to break along sliding surfaces. The calculation of this limit state of failure can be carried out according to the kinematic method.

2.1 Mechanical Model with Five Rigid Solids (Model I)

To study the total equilibrium of model I, the five solids are separated.

The cohesion forces C_i which act along the sliding surfaces are calculated as follows (Fig. 1) (Belabed 1999):

$$C_i = c * l_i.$$

l_i: length of the corresponding sliding surface. Cohesion forces also act against the direction of motion of rigid solids.

After the isolation of each body separately and the stabilization study of the solids N_γ and N_c are obtained by the case of a sliding footing (Salençon 1983), by the application of the kinematic method of, the load bearing factors rigid fracture solids; the calculations are done for different types of soils and for different slip angle (Chatzigogos 2007) θ_1, θ_2, θ_3, θ_4, θ_5, θ_6. To reduce the number of unknowns θ_1, θ_2,

H. El-Askary et al. (eds.), *Research Developments in Geotechnics, Geo-Informatics and Remote Sensing*, Advances in Science, Technology & Innovation,
https://doi.org/10.1007/978-3-030-72896-0_37

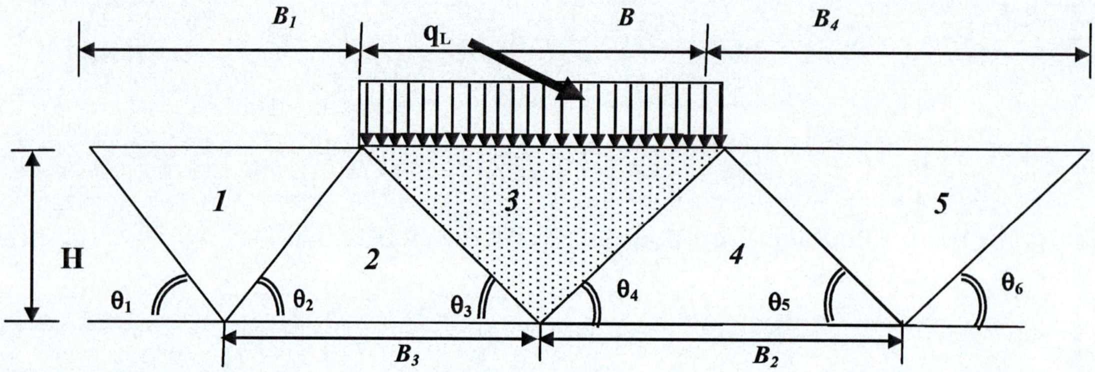

Fig. 1 Model of rupture to five rigid solid

θ_3, θ_4, θ_5 and θ_6, we take: $\theta_3 = \theta_4 = [(\pi/4) + (\phi/2)]$ and deduce according to the other angles θ_3, θ_4 (Zhou and Cheng 2014; Zhou et al. 2016).

$$\begin{aligned} N_\gamma = & -\frac{\sin(\theta_1+\theta_2)\sin^2\theta_3\sin^2\theta_4\sin(\theta_1-\phi)\cos(\theta_2-\phi)\sin(\theta_2-2\phi)}{\sin\theta_2\sin\theta_1\sin(\theta_3-2\phi)\sin(\theta_1+\theta_2-2\phi)\sin^2(\theta_3+\theta_4)} \\ & -\frac{\sin\theta_3\sin\theta_4}{\sin(\theta_3+\theta_4)}+\frac{\sin(\theta_2+\theta_3)\cos(\theta_3-\phi)\sin\phi\sin^2\theta_4\sin\theta_3}{\sin\theta_2\sin(\theta_3-2\phi)\sin^2(\theta_3+\theta_4)} \\ & +\frac{\sin(\theta_4+\theta_5)\cos(\theta_4-\phi)\sin\phi\sin^2\theta_3}{\sin\theta_5\sin(\theta_4-2\phi)\sin^2(\theta_3+\theta_4)}\times\sin\theta_4 \\ & -\frac{\sin(\theta_5+\theta_6)\sin^2\theta_3\sin^2\theta_4\sin(\theta_6-\phi)\cos(\theta_4-\phi)\sin(\theta_5-2\phi)}{\sin\theta_5\sin\theta_6\sin(\theta_4-2\phi)\sin(\theta_5+\theta_6-2\phi)\sin^2(\theta_3+\theta_4)} \end{aligned} \tag{1}$$

and:

$$\begin{aligned} N_C = & \frac{\sin\theta_4\begin{bmatrix}2\cos\phi\sin\theta_1\sin(\theta_2+\theta_3)\sin(\theta_1+\theta_2-2\phi)+\\+\sin\theta_2\sin\theta_3[\cos(2\theta_1-\phi)\sin(\theta_1-2\phi)-\sin\theta_1\cos(\theta_2-\theta_1+\phi)]\end{bmatrix}}{\sin(\theta_3+\theta_4)\sin(\theta_1+\theta_2-2\phi)\sin(\theta_3-2\phi)\sin\theta_1\sin\theta_2} \\ & +\frac{\sin\theta_3\begin{bmatrix}2\cos\phi\sin\theta_6\sin(\theta_4+\theta_5)\sin(\theta_5+\theta_6-2\phi)+\\+\sin\theta_2\sin\theta_3[\cos(2\theta_1-\phi)\sin(\theta_1-2\phi)-\sin\theta_6\cos(\theta_5-\theta_6+\phi)]\end{bmatrix}}{\sin(\theta_3+\theta_4)\sin(\theta_5+\theta_6-2\phi)\sin(\theta_4-2\phi)\sin\theta_6\sin\theta_5}+2\frac{\sin\theta_3\sin\theta_4}{\sin(\theta_3+\theta_4)} \end{aligned} \tag{2}$$

2.2 Mechanical Model with Seven Rigid Solids (Model II)

See Fig. 2.

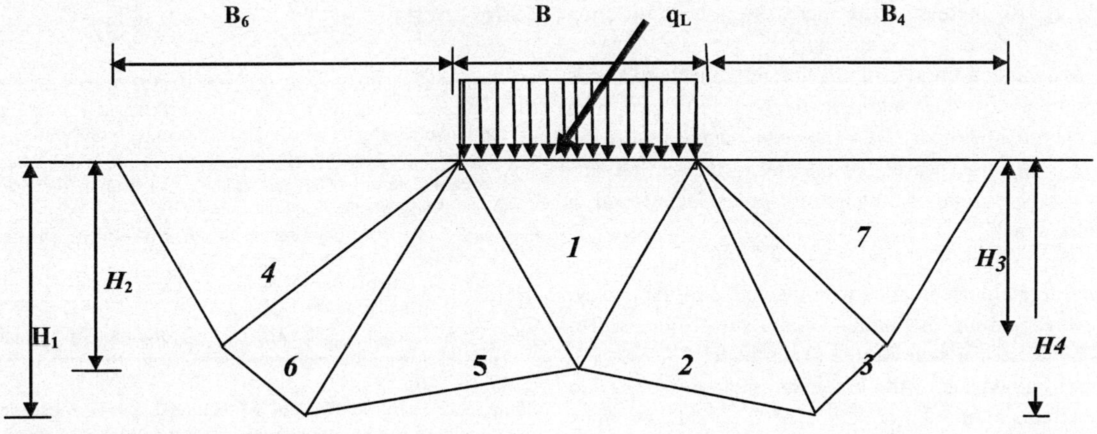

Fig. 2 Model of failure at seven rigid solids

3 Parametric Study

A programme under Mathcad has been developed to compare the proposed model (Model I, II). The results of the calculations are summarized in Tables 1 and 2.

Abbreviations used in the tables:

N_γ calculated: N_γ cal, $K_{P\gamma}$ calculated: $K_{P\gamma}$ cal, N_c calculated: N_c cal, K_{Pc} calculated: K_{Pc} cal, N_γ from coulomb and Caquot/Kerisel: $N_{\gamma K}$, N_c from coulomb and Caquot/Kerisel: N_{cK}, $K_{P\gamma}$ from coulomb and Caquot/Keérisel: $K_{P\gamma K}$, K_{Pc} from coulomb and Caquot/Kerisel: K_{PcK}.

For our case $\beta = 0$ (horizontal ground surface). $\delta_p = -\phi$ (since it is about the friction of the earth on itself).

4 Conclusion

This paper proposes a new approach to apply to the stability of the shallow foundations. The findings provide the following conclusions:

1. The failure always occurs when the angle of failure θ is greater than the internal friction angle ϕ which is physically justified ($\phi > \theta$).

Table 1 Coefficient the ratio K_p according to the 5 solid model and the Caquot/Kerisel model

ϕ (°)	θ_1 (°)	θ_2 (°)	θ_5 (°)	θ_6 (°)	λ (°)	$K_{P\gamma}$ cal	K_{Pc} cal	$K_{P\gamma K}$	K_{PcK}
10	50	50	50	50	−40	1.39	2.28	1.43	1.27
15	52.5	52.5	52.5	52.5	−37.5	1.69	1.52	1.75	1.58
20	55	55	55	55	−35	2.20	1.12	2.19	1.19
25	57.5	57.5	57.5	57.5	−32.5	2.82	2.54	2.84	2.57
30	60	60	60	60	−30	3.32	3.42	3.39	3.46
35	62.5	62.5	62.5	62.5	−27.5	5.69	4.97	5.75	5.01
40	65	65	65	65	25	9.75	8.31	9.71	8.18

N_γ calculated: N_γ cal, $K_{P\gamma}$ calculated: $K_{P\gamma}$ cal, N_c calculated: N_c cal, K_{Pc} calculated: K_{Pc} cal, N_γ from coulomb and Caquot/Kerisel: $N_{\gamma K}$, N_c from coulomb and Caquot/Kerisel: N_{cK}, $K_{P\gamma}$ from coulomb and Caquot/Keérisel: $K_{P\gamma K}$, K_{Pc} from coulomb and Caquot/Kerisel: K_{PcK}

Table 2 Bearing capacity factors N_c and N_γ according to the five-solid model and the model of Caquot/Kerisel

ϕ (°)	θ_1 (°)	θ_2 (°)	θ_5 (°)	θ_6 (°)	λ (°)	N_γ cal	N_c cal	$N_{\gamma K}$	N_{cK}
10	50	50	50	50	−40	0.98	8.22	1.00	8.30
15	52.5	52.5	52.5	52.5	−37.5	2.22	10.65	2.30	11.00
20	55	55	55	55	−35	4.80	14.38	5.00	14.80
25	57.5	57.5	57.5	57.5	−32.5	9.89	20.64	10.4	20.70
30	60	60	60	60	−30	21.64	30.68	21.8	30.10
35	62.5	62.5	62.5	62.5	−27.5	47.85	46.08	47.9	46.10
40	65	65	65	65	25	113.18	75.23	113	75.30

2. The limit load increases with increasing internal friction angle ϕ and decreasing cohesion c.
3. The modelling accuracy of the fracture mechanisms is ensured only by the kinematic method compared with other classical methods. This allows to accurately use the resistance against shearing of the soil in the sliding surfaces as well as against the cohesion forces developed in the same areas.
4. When the inclinations of the extreme solids (dimensions) are well fixed in accordance with the active thrust theory. The corresponding inclination of the stop force is also fixed in its critical state on the same theory.
5. The main drawback of the kinematic method, applied to rigid solids, is the introduction of different functions whose complexity increases with the increase in the number of rigid solids (to be taken into account). It also depends on the shapes of previous functions such as triangular form and curve form.

References

Belabed, L.: Application of the probabilistic safety concept on the verification of the overall stability of anchored retaining walls with the kinematics of rigid solids. Fr. J. Geotech. **89**, 49–54 (1999)

Chatzigogos, C.: Seismic behavior of superficial foundations, towards the consideration of a performance criterion in the design. PhD thesis, Solid Mechanics Laboratory (2007)

Costet, J., Sanglerat, G.: Practical Course in Soil Mechanics, vol. 2. Dunod Edition, Paris (1969)

Salençon, J.: Calculation at Break and Limit Analysis. ENPC Press, Paris (1983)

Zhou, X.P., Cheng, H.: Stability analysis of three-dimensional seismic landslides using the rigorous limit equilibrium method. Eng. Geol. **174**, 87–102 (2014)

Zhou, X.-P., Gu, X.-B., Yu, M.-H., Qian, Q.-H.: Seismic bearing capacity of shallow foundations resting on rock masses subjected to seismic loads. KSCE J. Civ. Eng. **20**(1), 216–228 (2016)

Geomechanical Studies Based on Numerical and Analytical Methods (T12)

Application of Classification Systems for the Assessment of Rock Mass Stability—Case of National Road 43, Jijel, Algeria

Chahra Yellas, Riad Benzaid, and Mustapha Tekkouk

Abstract

In this work, we present the classification of the rock masses in the "*Grottes Merveilleuses*" and "*Falaises*" areas along the national road no. 43. In various rock mass stability studies, rock mass rating (RMR), rock mass quality (*Q* System), and slope mass rating (SMR) are used as empirical methods for characterization of rock mass based on geological and geotechnical site data as well as physical and strength properties of rock samples. The study shows that the rock mass of the treated area corresponds to a partially stable class III normal rock with the possibility of block fall.

Keywords

Rock mass classification • RQD • RMR • *Q* System • SMR

1 Introduction

The risks of rock slope failure present considerable challenges with respect to motorway traffic, railway, and civil engineering infrastructure located in the immediate environment of the rock massifs. In order to minimize the risks associated with this natural hazard, studies and classification concepts for rock masses help to design the techniques for approaching and carrying out development projects. Several rock mass classification systems were developed from civil and mining engineering case studies conducted by several researchers to assess rock mass quality and rock slope stability, including rock quality designation (RQD) (Deere 1963), the rock mass quality system (*Q*) (Barton et al. 1974; Barton 2002), the rock mass rating (RMR) (Bieniawski 1973, 1989; Davood et al. 2015), the Geological Strength Index (GSI) (Hoek et al. 1995), and the slope mass rating (SMR) (Romana 1985; Sujatha and Thirukumaran 2018; Arie Noor et al. 2018; Sharma et al. 2019). The RMR, Q, and SMR classification systems were used in this study. The studied area concerns the RN 43 national road at the places known as "*Grottes Merveilleuses*" and "*Falaises*" (Fig. 1). It is mainly composed of rocks with carbonate and vacuolar facies, and it also presents numerous problems related to the slope instability revealed in recent years during road widening works.

2 Geological Context

The chain of Babors is limited to the east by the endogenous massif of El Aouana and the Mauritanian nappe Guerrouche, by the Bibans in the south-west, in the north by the Mediterranean Sea, and in the west, by the Western Babors. Our study is located in the eastern Babors massif. The chain of oriental Babors consists of three large tectonic units arranged from north to south (Fig. 1) such as the Breck-Gouraya unit, the Erraguene-Barbacha unit, and the Jebel Babors unit.

3 Description and Morphology of the Studied Massif

The first site called "*Les Grottes Merveilleuses*" is composed of massive Lias limestones cut into black-gray, irregularly dolomitized black-gray strands. The second site called "*Les Falaises*" is represented by black, oolitic limestones with a micritic texture. The dolomitization is intense giving a vacuolar appearance to the rock mass (Figs. 2 and 3).

C. Yellas (✉) · R. Benzaid · M. Tekkouk
Laboratoire de Génie Géologique (LGG), University of Jijel, BP. 98, Ouled Aissa, 18000 Jijel, Algeria
e-mail: chahra.yellas@univ-jijel.dz

H. El-Askary et al. (eds.), *Research Developments in Geotechnics, Geo-Informatics and Remote Sensing*, Advances in Science, Technology & Innovation,
https://doi.org/10.1007/978-3-030-72896-0_38

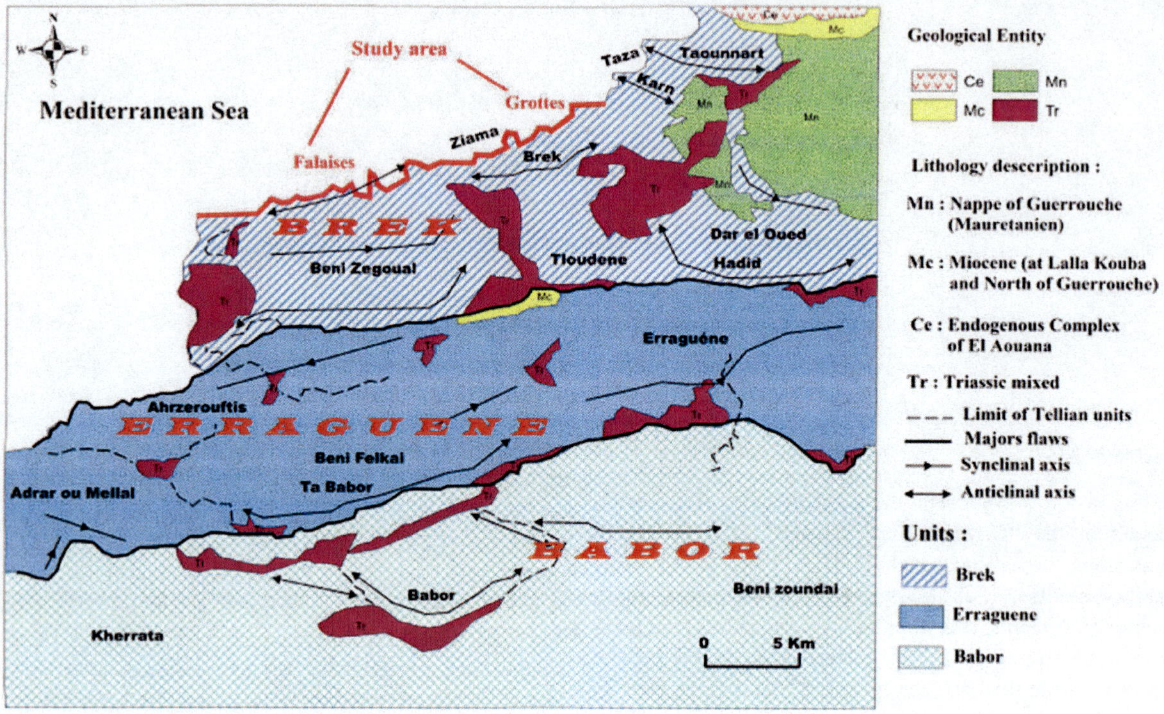

Fig. 1 Units and sub-units of oriental Babors (from Obert (1981), slightly modified)

Fig. 2 Field photographs show **a** carbonated mass in verticalized metric banks and presence of open mega fractures of Grottes; **b** highly tectonized carbonate cliff of Falaises; **c** example of karstic cavities (Grottes Merveilleuse)

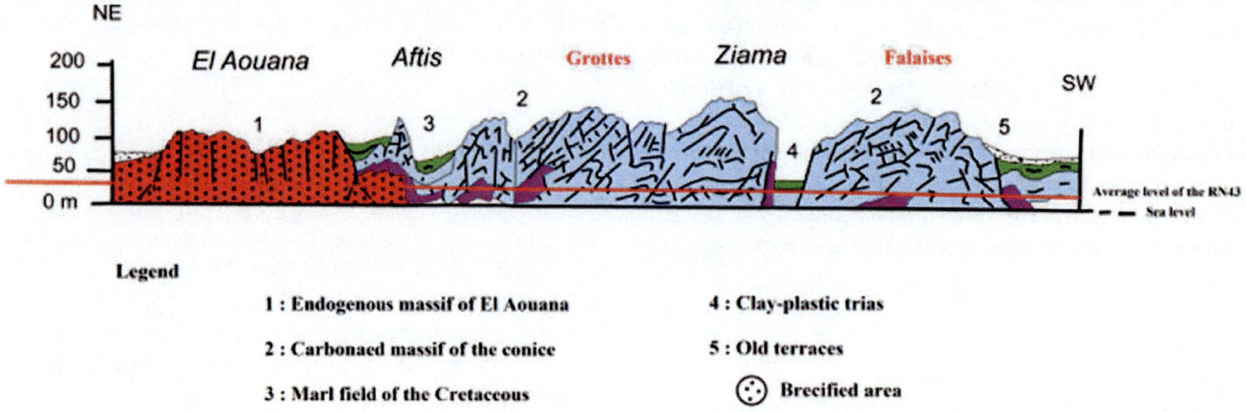

Fig. 3 Geological section showing the state of fracturing of the massif along the ledge (RN 43) Jijel-Bejaia

4 Results and Discussion

There are three stations (Grottes Merveilleuses, Dar El Oued, and Falaises) (Fig. 1), and for each station, we studied seven rock slopes. In this article, we present the results of one station (Falaises site). From the measurements made on the "*Falaises*" site, we present the results of the RMR and System Q calculations which allowed us to establish the correlation between these two parameters of the rock mass classification system. The values obtained related to the seven rock slopes studied are shown in Fig. 4. The calculated RMR values show that the minimum (RMR = 36) is related to rock mass 6, and the maximum (RMR = 54) is linked to rock mass 2 with an RMR average of 45.28, indicating that the rock mass is a good class III rock. The values calculated from the System Q show that the rock mass 4 has a minimum value (Q = 0.40), and the rock mass 7 presents a maximum value (Q = 3.03) with an average of 1.49 which indicates a bad class D rock. The processing of the initial data taken on site offers the possibility of correlating Bieniawski's RMR quality index with Barton's Q quality index using different types of mathematical expressions (Barton 2002; Bieniawski 1989). Our experimental results represented in Fig. 4, shows that the relation between RMR and Q can be estimated by Eq. 1 (Bieniawski 1984):

$$\text{RMR} = 9 \ln Q + 44 \quad (1)$$

According to Fig. 4, the relationship between RMR and Q System has a correlation coefficient (R^2 = 0.70) which shows that our experimental results are satisfactory. We also calculated the SMR, which corresponds to the modified RMR parameter by Romana (1985) and which makes it possible to evaluate the stability of the different treated rock

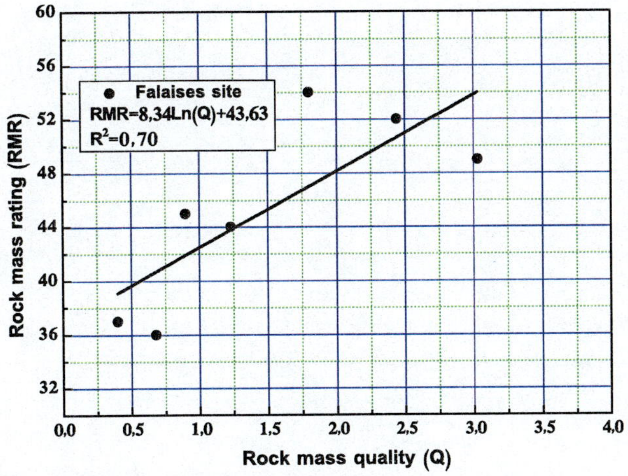

Fig. 4 Correlation between RMR and Q for seven rock slopes (in the *Falaises* site)

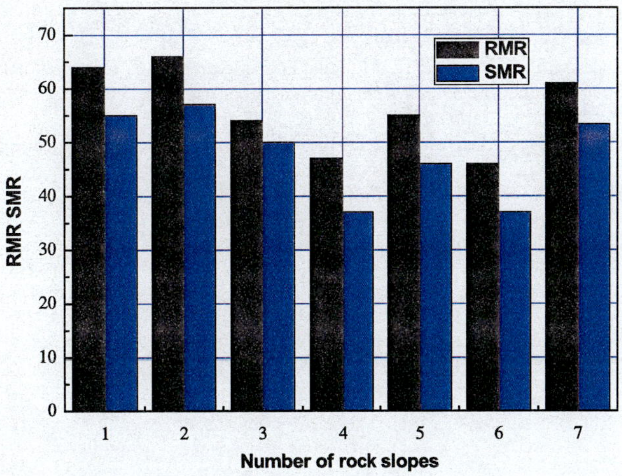

Fig. 5 Comparison between RMR and SMR for seven rock slopes (in the *Falaises* site)

slopes. Figure 5 shows the different RMR and SMR values calculated taking into account the seven rock masses. It is clear that the difference between the two parameters is not very important and that in some rock masses (namely 4 and 6) the reading of the SMR shows an unstable area which could be the potential site of planar failure or big wedges in the future due to the fact that the dip of the failure surface is oriented in the opposite direction of the road section.

5 Conclusion

The results of the classifications obtained show that the slopes of the study area are classified in the partially stable normal class (III) with a probability of falling from the block. The large number of tectonic discontinuities related to the deformation of the Baborien massif has severely cut the calcaro-dolomitic rock mass in heterogeneous block patterns requiring the application of the more advanced methods of rock engineering.

References

Arie Noor, R., Zufialdi, Z., Dicky, M., Iyan, H., Sri, M.: Modification of rock mass classification in the rock slope platy jointed Andesite at Seloharjo Area. Int. J. GEOMATE **16**(53), 163–170 (2018)

Barton, N.: Some new Q-value correlations to assist site characteristics and tunnel design. Int. J. Rock Mech. Min. Sci. **39**, 185–216 (2002)

Barton, N.R., Lien, R., Lunde, J.: Engineering classification of rock masses for the design of tunnel support. Rock Mech. Rock Eng. **6** (4), 189–239 (1974)

Bieniawski, Z.T.: Engineering classification of rock masses. Trans. South Afr. Inst. Civ. Eng. **15**, 335–344 (1973)

Bieniawski, Z.T.: Rock Mechanics Design in Mining and Tunneling, p. 272. Balkema, Rotterdam (1984)

Bieniawski, Z.T.: Engineering Rock Mass Classifications, p. 251. Wiley, New York (1989)

Davood, F., Gholam, R.K., Mojtaba, H.: Assessment of a modified rock mass classification system for rock slope stability analysis in the Q-system. Earth Sci. Res. J. **19**(2), 147–152 (2015)

Deere, D.U.: Technical description of rock cores for engineering purposes. Rock Mech. Eng. Geol. **1**, 17–22 (1963)

Hoek, E., Kaiser, P.K., Bawden, W.F.: Support of Underground Excavations in Hard Rock. Balkema, Rotterdam (1995)

Obert, D.: Etude géologique des Babors orientaux (Domaine tellien, Algérie). Ph.D. thesis, University of Paris 6, Paris, France (1981)

Romana, M.: New adjustment ratings for application of Bieniawski classification to slopes. In: Proceedings of International Symposium on Role of Rock Mechanics, ISRM, Zacatecas, Mexico, pp. 49–53 (1985)

Sharma, M., Sharma, S., Kumar, M., Singh, S.K.: Analysis of slope stability of road cut slopes of Srinagar, Uttarakhand, India. Int. J. Appl. Eng. Res. **14**(3), 609–615 (2019)

Sujatha, E.R., Thirukumaran, V.: Rock slope stability assessment using geomechanical classification and its application for specific slopes along Kodaikkanal-Palani Hill Road, Western Ghats, India. J. Geol. Soc. India **91**(April), 489–495 (2018)

A New Classification Method for Predicting Performance of Continuous Miners

Kaushik Dey and C. Uday Kumar

Abstract

The performance of any machine should be predicted prior to its deployment in a mine. The performance of excavation machines largely depends on the geomaterial and machine parameters. Indian coal mines are heading towards increased coal production from underground mines and, thus, would like to deploy continuous miner. Continuous miner is a machine similar to road header with a little modification in the cutter design. Therefore, it is essential to develop a predictive model for the performance estimation of continuous miner considering the geotechnical variability of Indian coal mines. This paper reports the development of such a predictive model. This new model is based on a geotechnical classification system, which gives the continuous miner a cuttability index (RMCI). RMCI is basically governed by the geotechnical parameters and machine parameters, and thus, it represents the machine–rock interaction. The RMCI is correlated with the machine performance expressed as an hourly cutting rate.

Keywords

Continuous miner • Geotechnical classification • Cuttability index

1 Introduction

The largest coal-producing countries like China, USA, and Australia are producing 95%, 33%, and 20% of their total coal production from underground (UG) mines, respectively. With just 15% share of coal production from UG mines, India needs a quantum jump in production from these mines due to the depletion of near surface deposits (Anon 2014a, b). Thus, there is an urgent need for implementation of bulk production technologies for deep-seated deposits (300–600 m depth). Currently, contribution from highly mechanized faces, namely longwall, blasting gallery, continuous miner, etc., has remained at a low level. Phased replacement of small un-economic mines with larger mines using modern technologies is absolutely necessary if underground operations are to be made profitable and competitive. Therefore, Indian underground coal mines are quickly moving towards the adaptation of mechanized mining, one of which is continuous miner (Uday Kumar 2015; www.cil.com).

2 Continuous Miner

Continuous miner (CM) (Fig. 1) is a mining machine that produces a constant flow of ore from the working face of the mine. The machine continuously excavates coal using a cutting steel drum and simultaneously gathers and loads the excavated quantities through a conveyor system. There are many variations available in the design of continuous miners, depending on the manufacturer; however, it mainly consists of (Ramesh and Ravi 2010):

(A) A central body to carry all other components mounted on some type of drive mechanism to provide mobility (most commonly crawler tracks).
(B) A cutting head, usually rotating drum(s) and/or chains with cutting picks attached with slewing arrangements.
(C) A gathering mechanism to pick up dislodged cut coal and deliver it into the loading system.
(D) A conveying system (usually a chain conveyor running in a steel trough from front to rear of the miner) as the loading/discharging system.

Currently, 19 underground coal mines are planned for continuous miner deployment to achieve a target production of 11.69 million tonnes per year (www.cil.com).

K. Dey (✉) · C. Uday Kumar
Department of Mining Engineering, Indian Institute of Technology, Kharagpur, India

H. El-Askary et al. (eds.), *Research Developments in Geotechnics, Geo-Informatics and Remote Sensing*, Advances in Science, Technology & Innovation,
https://doi.org/10.1007/978-3-030-72896-0_39

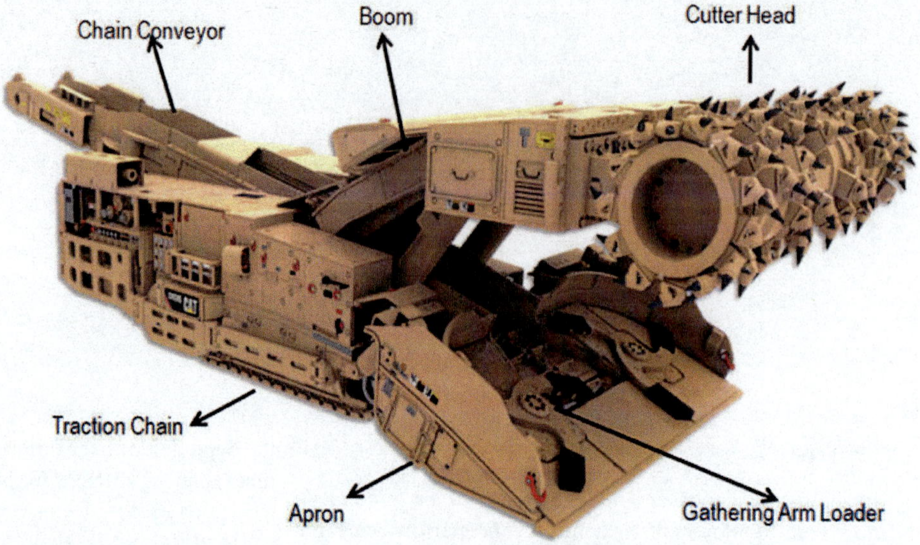

Fig. 1 Main elements of a continuous miner machine (www.mining.cat.com)

Table 1 Parameters influencing the performance of surface miner (Dey and Ghose 2011; Pradhan and Dey 2009; Dey and Sen 2012)

Geotechnical parameters	Machine parameters	Application
Density, brittleness, compressive strength, fracture energy, toughness index, tensile strength, sonic velocity, abrasivity (Schimazek-F, CERCHAR) volumetric joint count, stickiness of material, specific energy	Cutting tool configuration pick lacing, type of pick number of picks, tip material, drum weight, engine power, coolant for tips	Dimension of operating area, operator skill, specific requirements (lighting, dust suppression, etc.)

3 Parameters Influencing Performance of Continuous Miner

The performance of any cutting machine depends on a number of influencing parameters (Sen et al. 2000). These parameters may be classified into three groups as given in Table 1.

Previously, a number of researchers, namely Farmer (1986), Roxborough (1987), Bilgin et al. (1988), Gehring (1992), and Dey and Ghosh (2008, 2009), worked on the predictive models for rock cutting machines (like road header, surface miner, etc.) similar to continuous miner. However, performance prediction models for continuous miner for Indian coal mines are not available.

4 Predictive Model

4.1 Field Investigation

Performance of the continuous miner is observed for 19 different faces. The instantaneous cutting rates measured for a machine with a cutter-head power of 372 kW and a weight of 74.8 tonne, at different cutting cases with their properties at respective cutting rates. The rock properties monitored are uniaxial compressive strength (UCS), Brazilian tensile strength (BTS), RQD for each continuous miner face. All the faces observed are Bord and Pillar development faces. The performance of the continuous miner at each face is recorded as machine instantaneous cutting rate (MICR), in tonne/h.

4.2 Development of Model

On bivariate regression analysis, it has been found that UCS, BTS, and RQD exhibit an inverse power correlation with the MICR, whereas the power and weight of the machine show power correlation with the MICR. Considering these, a new geotechnical classification system is developed and proposed based on the measured data for prediction of the continuous miner performance.

$$\mathrm{CMCI} = \frac{P \times W}{\left[\mathrm{BI} \times \left(\frac{\mathrm{RQD}}{100}\right)^{\frac{1}{3}}\right]} \quad (1)$$

where

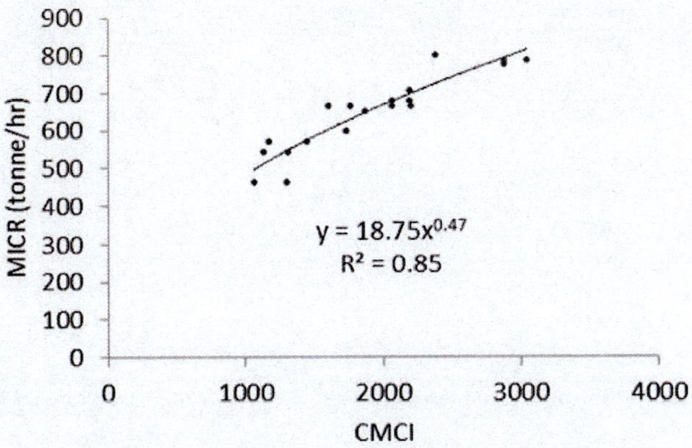

Fig. 2 Relation between CMCI and production rate (MICR)

CMCI	Continuous Miner Cuttability Index,
P	Power of machine (kW),
Brittleness Index (BI) (Altindag)	(UCS × BTS)/2,
W	Machine weight (tonne),
RQD	Rock quality designation (%),
BTS	Tensile strength (MPa),
UCS	Uniaxial compressive strength (MPa).

The CMCI is calculated and correlated with the measured instantaneous cutting rates for the measured data. The Continuous Miner Cuttability Index developed based on BI, RQD and machine properties (power and weight) shows a good correlation ($R^2 = 0.848$) with actual measured performance (Fig. 2). Hence, the proposed equation for predicted instantaneous cutting rate ($MICR_P$), in tonne/h, based on the CMCI, is established as:

$$MICR_P = 18.75 \times CMCI^{0.47}. \quad (2)$$

5 Conclusions

The CM performance is assessed based on some influencing geotechnical and machine parameters, namely UCS, BTS, RQD, machine power, and machine weight. A classification system is adopted to model the machines' ability to cut the rock. Thus, Continuous Miner Cuttability Index (CMCI) was developed. The developed model is used for 19 coal mining faces where the continuous miner is working to generate a model for performance prediction of continuous miner. The developed model is named as predicted instantaneous cutting rate ($MICR_P$). The developed PICR model has been found applicable within ±17% accuracy for Indian coal mining conditions. The statistical constants can be improved with varied data sets in the future.

References

Altindag, R.: Correlation of specific energy with rock brittleness concepts on rock cutting. J. South Afr. Inst. Min. Metall. 163–171 (2003)

Anon: Performance Statistics of Coal in India (2014a). www.idatainsights.com

Anon: Coal Summit Expo 2014. Preamble of 5th Coal Summit Conference, New-Delhi, India, 23rd and 24th September, p. 2 (2014b)

Bilgin, N., Seyrek, T., Shahria, K.: Golden horn clean up contributes valuable data, tunnels and tunneling. Tunnels Tunn. 41–44 (1988)

Dey, K., Ghose, A.K.: Predicting cuttability with surface miners—a rockmass classification approach. J. Mines Metals Fuels **56**(5 & 6), 85–92 (2008)

Dey, K., Ghose, A.K.: Selecting a surface miner—an algorithm. J. Mines Metals Fuels **57**(9), 282–287 (2009)

Dey, K., Ghose, A.K.: Review of cuttability indices and a new rockmass classification approach for selection of surface miners. Rock Mech. Rock Eng. **44**(5), 601–611 (2011)

Dey, K., Sen, P.: Planning and design with surface miners—some issues. J. Mines Metals Fuels **60**(11 & 12), 294–309 (2012)

Farmer, I.W.: Energy based rock characterization. In: Karamis, M. (ed.) International Proceedings of Symposium on Application of Rock Characterization Techniques in Mine Design, pp. 17–23. AIME, Littleton (1986)

Gehring, K.H.: Evaluation of Cutting Performance BBV, 1992-08-04, P102 (1992)

Pradhan, P., Dey, K.: Productivity improvement through selection of operating mode of surface miner—a computational approach. J. Mines Metals Fuels **57**(3 & 4), 67–75 (2009)

Ramesh, K.B., Ravi, S.K.: Under ground coal mining in India—technological options and challenges ahead. National Seminar, 28 August, Hyderabad, pp. 17–19 (2010)

Roxborough, F.F.: The role of some basic rock properties in assessing cuttability. Seminar on Tunnels, Wholly Engineered Structures, Sydney, AFCC, April, P122 (1987)

Sen, P., Pathak, K., Dey, K.: Performance analysis of continuous surface miner. Surf. Min. (Braunkohle and other material) **52**(4), 371–376 (2000)

Uday Kumar, C.: Performance analysis of continuous miner—a case study. Unpublished M.Tech. dissertation, IIT Kharagpur, pp. 2–40 (2015)

www.cil.com

www.mining.cat.com

Support Vector Regression of Los Angeles Abrasion on Mechanical Properties in Igneous Rocks

Güneş Ertunç and Yılmaz Özçelik

Abstract

In order to use rocks as an engineering material, the mechanical properties of rocks should be evaluated and characterized in detail. The Los Angeles abrasion (LAA) rate is considered one of the most important features of the rocks in engineering structures. The aim of this study is to estimate the Los Angeles abrasion of igneous rocks from various mechanical properties by using the unsupervised statistical learning method, support vector regression. To investigate the performance of the LAA predictions, regression statistics are drawn. As a result, a relationship between the LAA and the mechanical properties of igneous rocks was revealed and the regression characteristics were very well-reflected via the support vector regression method.

Keywords

Igneous rocks • Los Angeles abrasion • Regression • Support vector regression

1 Introduction

Earth scientists and rock engineers mostly use mechanical properties of rocks such as uniaxial compressive strength (UCS), tensile strength (TES), point load strength (PLS), and modulus of elasticity (YOU) in the design of surface and underground mining structures. These mechanical properties are measures of the different aspects of rock strength as standardized by international bodies (Brown 1981; ASTM, C.C.M. 2006).

The aim of this study is to reveal regression characteristics between the mechanical properties of igneous rocks with the Los Angeles abrasion rate. In order to do so, the support vector machine methodology is used. Support vector machines solve binary classification problems by formulating them as convex optimization problems (Vapnik 1995, 1999). This optimization problem consists in determining the maximum distance that distinguishes the hyperplane and the correct classification of as many training points as possible. SVMs represent this optimal hyperplane with support vectors. The support vector regression is a generalization of the classification problem where the model returns a continuously weighted output, as opposed to a finite set output (Khanna and Awad 2015).

The study area is an andesite quarry located 60 km NE of Ankara (Fig. 1), the capital city of Turkey. Volcanic units such as andesite, dacite, riolitic lava, and tuff occurred in the Tertiary Mamak Formation.

All of the mechanical properties of the samples were tested based on the guidelines (Ulusay et al. 2007). The sample list and descriptive statistics of each sample are given in Table 1.

2 Methodology

Support vector machines are based on a quadratic optimization problem and do not require the distribution function of the data. In the support vector classification, the aim is to determine the limit that separates the target data into two categories. The linear and nonlinear distinctions can be summarized as whether the separation line is linear or not. In the differentiation of more than two categorical variables, support vector machines produce solutions with one-vs-all (Rifkin and Klautau 2004) and one-to-one methods (Bishop 2006). SVM can also be used as a regression tool by applying the same principles as the SVM for classification, with few minor differences. In simple regression, the aim is

G. Ertunç (✉) · Y. Özçelik
Mining Engineering Department, Hacettepe University, 06800 Ankara, Turkey
e-mail: gertunc@hacettepe.edu.tr

H. El-Askary et al. (eds.), *Research Developments in Geotechnics, Geo-Informatics and Remote Sensing*, Advances in Science, Technology & Innovation,
https://doi.org/10.1007/978-3-030-72896-0_40

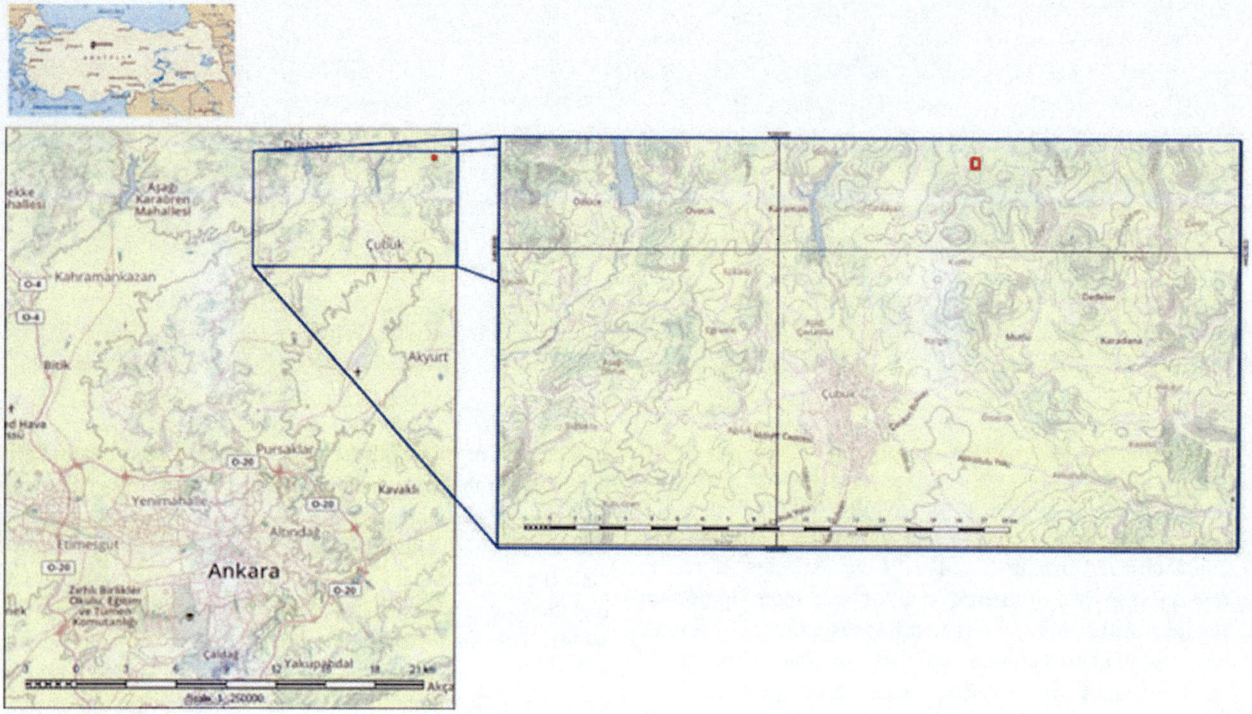

Fig. 1 Location of the study area

Table 1 Descriptive statistics

	Minimum	Maximum	Mean	Median	Standard deviation
UCS	25	131	79.15	79.38	24.26
PLS	3.50	14	13.45	28.7	16.2
TES	5.09	13.45	8.76	8.98	2.3
YOU	7.80	28.70	14.20	12.90	14.20
LAA	11.20	16.20	14.31	14.20	1.22

to minimize the error rate. As an alternative approach, SVR tries to fit the error within a certain threshold. This threshold's width relies on defining the loss function that ignores the errors placed at a certain distance of the actual value. This function is usually called epsilon intensive function. One of the key ideas in SVM classification and regression is a solution that performs a small part of a subset of the training which offers tremendous benefits by means of computational advantages. The existence of the global minimum and the reliable generalization limit is also optimized and ensured with the help of the epsilon intensive loss function. In addition, SVR allows nonlinear fitting problems by deploying a nonlinear kernel which is known as the kernel trick. The regression model is, thus, represented by combinations of the training points rather than a function of the features and some weights.

3 Results

The support vector regression algorithm has been used in order to predict the LAA from UCS, PLI, TES, and YOU properties. All of the mechanical properties versus the Los Angeles abrasion regression is modeled with a support vector regression using the Gaussian radial basis kernel (Fig. 2).

Most of the previous studies in the literature describe relationships in an exponential formula by means of a trend. This study also reveals this trend in an exponential way by an exponential term which is embedded in the Gaussian kernel. It can be seen from the regression statistics that the most correlated mechanical property with Los Angeles abrasion is the tensile strength (Table 2). Also, it was found that there is a strong correlation between the uniaxial

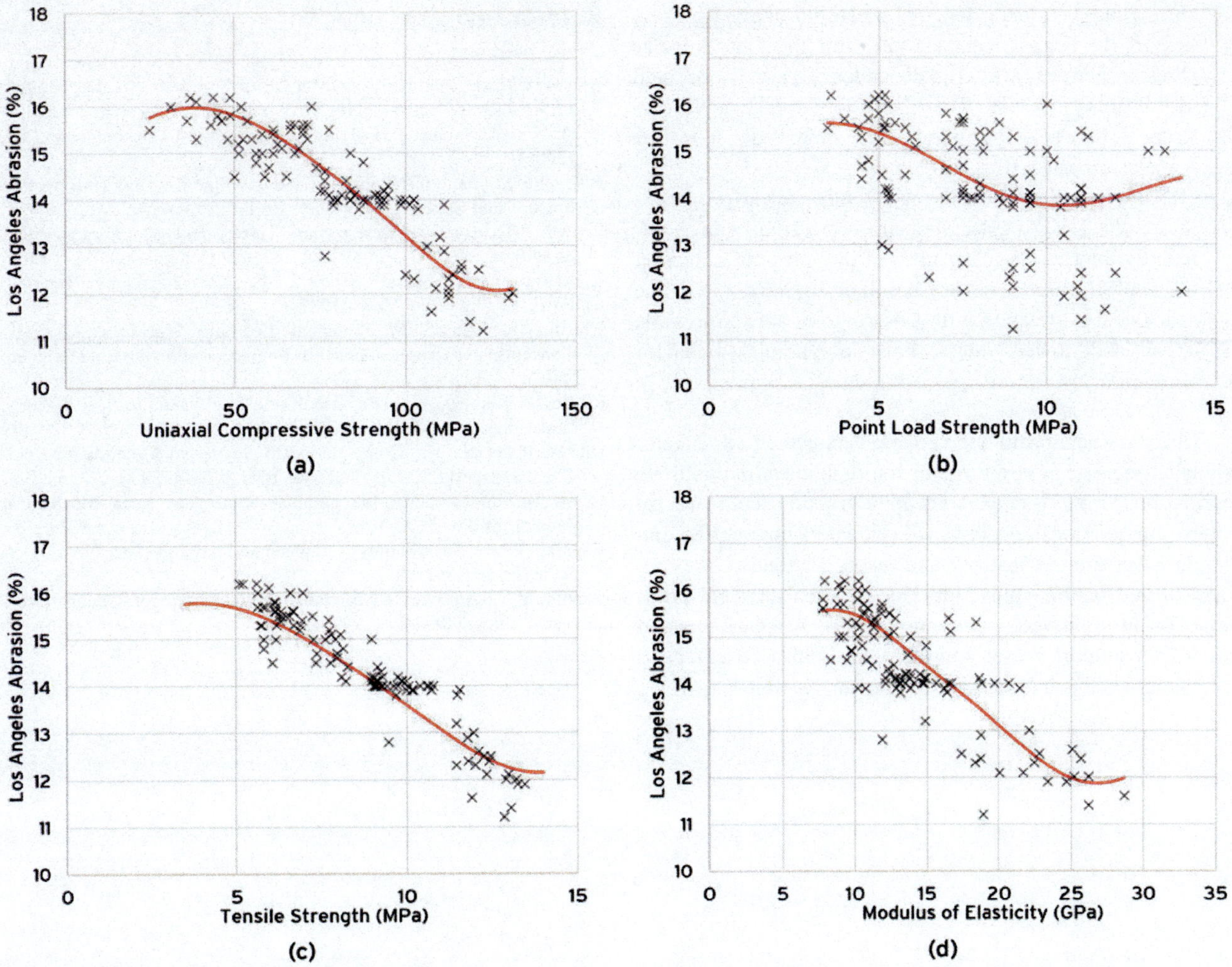

Fig. 2 Regression results (red curve). **a** UCS versus LAA, **b** PLS versus LAA, **c** TES versus LAA, and **d** YOU versus LAA

Table 2 Regression statistics

Independent variable	R^2	Sum of squares		F	Significance F
		Regression	Residual		
UCS	0.7982	126.28	31.92	419.33	0.0000
PLS	0.2221	35.14	123.06	30.27	0.0000
TES	0.8791	139.07	19.12	770.86	0.0000
YOU	0.6652	105.23	52.97	210.59	0.0000

compressive strength and the Los Angeles abrasion. The modulus of elasticity versus the Los Angeles abrasion regression is very similar to the uniaxial compressive strength versus the Los Angeles abrasion, mainly because of the uniaxial compressive strength and the modulus of elasticity of rocks have a strong correlation among them.

Predicting the Los Angeles abrasion from the point load strength would be less reliable since the correlation between them is fairly low when compared to the other mechanical properties. It can be clearly seen even in the scatter plot itself.

4 Conclusions

In this study, the Los Angeles abrasion characteristics of igneous rocks and correlations between this attribute and the mechanical properties of rocks were investigated. The conclusions are as follows:

1. Rocks with high uniaxial compressive strength, tensile strength, and modulus of elasticity are more resistant to

abrasiveness, as expected. All these mechanical properties are in negative correlation with the Los Angeles abrasion percentages. This dependency is well observed in this study's results. There is a very weak, or no correlation between the point load index and the Los Angeles abrasion (Fig. 2).

2. The lower and upper tails of the mechanical properties versus the Los Angeles abrasion regression behave differently than in the mid-zones. In intermediate ranges of mechanical properties, they are likely to have a negative near linear correlation with Los Angeles abrasion, while upper and lower values have polynomial behavior regression fit.

The study deals with one variable versus the Los Angeles abrasion regression at a time. In future studies, multivariate regression, i.e., 2D support vector regression, could reveal interesting relationships. Physical characteristics such as unit weight, apparent porosity, and shore hardness are not investigated in this study. Introducing such physical properties could improve the prediction of Los Angeles abrasion rate. This support vector approach could also be extended into sedimentary and metamorphic rocks as well.

References

ASTM, C.C.M.: Standard Test Method for Resistance to Degradation of Small-Size Coarse Aggregate by Abrasion and Impact in the Los Angeles Machine. ASTM International, West Conshohocken, PA (2006)

Bishop, C.M.: Pattern Recognition and Machine Learning (Information Science and Statistics). Springer, Berlin (2006)

Brown, E.T.: Rock Characterization, Testing and; Monitoring: ISRM Suggested Methods. Published for the Commission on Testing Methods, International Society for Rock Mechanics by Pergamon Press, Oxford, New York (1981)

Khanna, R., Awad, M.: Efficient Learning Machines: Theories, Concepts, and Applications for Engineers and System Designers (2015)

Rifkin, R., Klautau, A.: In defense of one-vs-all classification. J. Mach. Learn. Res. **5**, 101–141 (2004)

Ulusay, R., et al.: The Complete ISRM Suggested Methods for Rock Characterization, Testing and Monitoring: 1974–2006. Commission on Testing Methods, International Society of Rock Mechanics, Ankara, Turkey (2007)

Vapnik, V.N.: The Nature of Statistical Learning Theory, p. 188. Springer, Berlin (1995)

Vapnik, V.N.: An overview of statistical learning theory. Trans. Neur. Netw. **10**(5), 988–999 (1999)

Hasdrubal Field [Tunisia]: Geomechanical Integrity Study

Kais Ben Abdallah, Salma Souissi, Essaib Hamdi, and Sihem Ghomari

Abstract

Hasdrubal is a naturally fractured carbonate reservoir field located in offshore Tunisia. Over the last decade, the field has manifested serious integrity and water coning production issues. The extended network of discontinuities in addition to the impact of reservoir depletion renders the stability of the reservoir a major concern for the field operator. An integrity study of the field was necessary to accurately determine the ever-changing stress settings (magnitudes and orientation of stress tensors) in order to, accordingly, adjust the design of the well trajectory while properly assimilate their impact on the production trend. An elaborated workflow was initiated to fine-tune the mechanical proprieties of rock basing on geological information, drilling data, and field logs. The One-dimensional Mechanical Earth Model (MEM) was then built to validate wellbore failure and explain drilling incidences reported. The stress regime was also confirmed. It matches the geological history of the area with an alteration between Normal Faulting and Strike slip regimes (Albukhari et al. in First international conference on advances in rock mechanics. Hammamet, Tunisia 2018; Wendt et al. in Geomechanical study—Hasdrubal field 2005). The MEM was in a position to explain the drilling events for all the vertical wells, thus, providing a reasonable guide design for the wells under development. The shear stability of faults encountered by the horizontal well HAS-A1 was also appraised using the Fracture Stability Advisor software. The results show that at initial reservoir pressures the fractures in Hasdrubal field are stable and do not exceed the failure criterion. The previous interpretation is applicable to any major discrete discontinuity that is present in the same stress field and depicts similar rock-related strength properties.

Keywords

Fractured • Stability • Faults • MEM • Stability • Geomechanics

1 Introduction

The basic approach to tackle integrity-related issues starts with disclosing the complicated interaction between the rock behavior and the stress setting present over the different stages of the field lifespan (e.g., drilling, production). A recent study conducted on Hasdrubal field has reported long term production issues (Wendt et al. 2005). Despite the significant remaining reserves, a higher cut in water and a major drop in gas and oil deliverability have been observed. Pressure depletion of the reservoir and other geomechanical factors may have changed the properties of the rock matrix and the discontinuities network, thus, compromising the integrity and economical profitability of the field. To overcome previous limitations, the operator initiated a drilling campaign, in the vicinity of the horizontal well HAS-A1, within the western part of the field. The anticipated wells are intended to intercept the major faults system aiming to improve production through the network's fractures. A strong need for an investigation study has, therefore, been raised first to ensure wellbore stability during drilling and also to establish a more efficient development strategy that takes into account the geomechanical mechanisms involved and their considerable impact on the overall sustainability of the field.

K. B. Abdallah (✉) · S. Souissi · E. Hamdi
National Engineering School of Tunis, Tunis, Tunisia
e-mail: benabdal@ualberta.ca

S. Ghomari
APO-Shell Tunisia, Tunis, Tunisia

H. El-Askary et al. (eds.), *Research Developments in Geotechnics, Geo-Informatics and Remote Sensing*, Advances in Science, Technology & Innovation,
https://doi.org/10.1007/978-3-030-72896-0_41

2 Geomechanical Modeling of Hasdrubal Field: Geological Setting, Framework, and Methodology

The Hasdrubal field is a major oil and gas offshore asset that has been operated by APO-Shell Tunisia since 2015. From the geological perspective, Hasdrubal's structure represents an NNW-SSE trending horst block created between a system of oblique slip faults that are typically associated with parallel to off-parallel fractures over the reservoir section of El-Geira formation (Wendt et al. 2005). The structure and evolution of the Hasdrubal field's fractures reflect the main tectonic events that have affected the Gabes-Tripoli Basin through its geological burial history. Initial compaction and dewatering caused grain micro-fracturing in grain-stone fabrics that were followed by lithification and forming of stylolite. Stylolite growth was accompanied by stylolite-associated fractures that developed perpendicular to the main extension direction (NW–SW). The active extension in the basin was accommodated by NW–SE extensional fault growth that initiated its own fracture associations. During the successive phases of extension, fault-related fractures were periodically reactivated and extended. The high sensitivity of those fractures to geomechanical settings may result in ongoing reactivation during field production and impact the type of fluid produced by shifting the flow path within the reservoir and across different sections of the basin.

The Hasdrubal field is challenging to study for several reasons among which are important variations in lithology and complex geological structural configuration with many major open faults and fractures. To adequately address previous challenges, a proven workflow was implemented to generate the geomechanical model associated with our field (Al-Ajmi et al. 2006; Wendt et al. 2005): (a) Initially, information from the analysis of logs and formations integrity measurements related to our reference wells (Has1, Has2, Has3, Has4) were collected and their quality was checked for accuracy then (b) Correlations that infer rock mechanics from field logs (i.e., shear and compressional waves velocities V_s and V_p) had been elaborated before the dynamic-to-static conversion was generated then calibrated by core tests results. Once all the required properties, including pore pressure (P_p), were measured; (c) a stress profile (magnitude and orientation of stress tensors) was calculated through Eqs. 1 and 2 then calibrated with field logs, laboratory measurements, and drilling reports.

$$\sigma_H = \frac{v}{1-v}\sigma_V - \frac{v}{1-v}\alpha P_p + \alpha P_p + \frac{vE}{1-v^2}\varepsilon_x + \frac{E}{1-v^2}\varepsilon_y \quad (1)$$

$$\sigma_h = \frac{v}{1-v}\sigma_V - \frac{v}{1-v}\alpha P_p + \alpha P_p + \frac{E}{1-v^2}\varepsilon_x + \frac{vE}{1-v^2}\varepsilon_y \quad (2)$$

3 Results and Discussion

3.1 Mechanical Earth Model: Wellbore Integrity and Faults Stability Analysis

Subsequent sections describe the methodology adopted to build a reliable Mechanical Earth Model for our field, mainly implying the calculation of the rock-mechanics properties in order to eventually estimate the associated stress tensors.

Rock deformation and strength properties: a theoretical model based on isotropic, homogeneous, linearly elastic medium correlations has been considered for the determination of the dynamic Young Modulus and Poisson Ratio of the different formations of the field. Clark Sydney and Press (1966) developed several empirical correlations to relate the sonic wave velocities to the elastics modulus of the material (Al-Ajmi et al. 2006; Knöll 2016; Wendt et al. 2005). The same correlations were adopted by the current study to generate dynamic Young Modulus and Poisson Ratio (Knöll 2016; Song 2012). Morals and Marcinew (1993) correlation, which proved to give a very good estimation of static moduli from dynamic moduli, was applied to obtain the static Young Modulus and Poisson Ratios. Static moduli were eventually deployed to generate the rock strength parameters of our model (UCS, TS, and Friction Angle) (Knöll 2016; Song 2012). In previous calculations, the Morales parameters were systematically adjusted with the lithology of the present rock (limestone or argillaceous sections) to ensure authenticity to real mechanical properties associated with the different geological layers of our field.

Stress state components (S_v, S_{Hmax}, S_{hmin}): The magnitude of Vertical stress (S_v) was computed by integrating the available density logs with depth along the well paths of the reference wells. For well sections with inconsistent density and quality compromised by an enlarged borehole, the density log is interpolated by interpreting the pseudo-density from the acoustic log using Gardner's relationship (Gardner et al. 1974). Regarding horizontal stresses; the poro-elastic model was adopted to interpret the initial estimation of minimum and maximum horizontal stress magnitudes for the four reference wells. By assuming a zero lateral tectonic strain (ε_x, ε_y), an initial assessment of stresses magnitude was obtained through Eqs. 1 and 2. These lateral strains are then

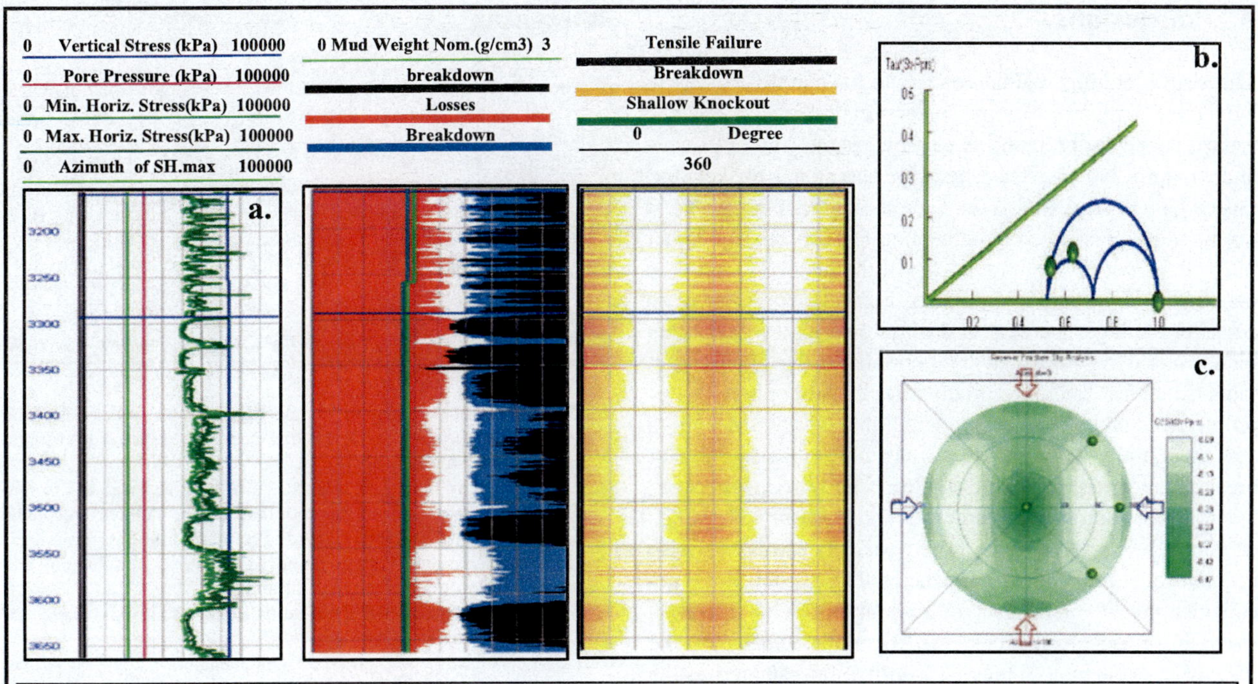

Formation	Top depth TVD (m)	Vertical stress MPa	Max. Horizontal Stress MPa	Min. Horizontal Stress MPa	Pore Pressure MPa	UCS MPa	Friction angle	Stress regime
El- Garia	3003	69.3	48.5	46.2	30	60-113	20-32	Normal/strike Slip

Fig. 1 Geomechanical assessment of HAS-A1 well: **a** MEM, **b** WSA, **c** FSA results, and **d** MEM summary for El-Geira formation (reservoir section)

adjusted so that the minimum horizontal stress is constrained and calibrated to the reported Leak Off Tests (LOTs) (Knöll 2016; Song 2012). The stresses orientations were inferred using the FMI logs available for HAS-3 and HAS-2 wells with plausible interpretation that the mean direction of minimum and maximum horizontal stresses are reported to be N80E and N170E; respectively (Barton 1997 approach (Al-Ajmi et al. 2006; Song 2012)). Based on the MEM results, the faulting stress regime of our field is found to be in alteration between normal faulting and strike slip regimes over the reservoir's interval with a difference in the horizontal stresses of the order of 10%, a result that showed great consistency with previous geological studies' findings (Albukhari et al. 2018; Wendt et al. 2005). Subsequently, the stresses and the estimated rock properties previously determined for the reference wells are reached to build the Mechanical Earth Model of the HAS-1 using Petrel platform. The MEM generated was then used to perform Wellbore Stability Analysis (WSA) and Fault Stability Analysis (FSA) of HAS-A1 (refer to Fig. 1b, c). Both the MEM and WSA predict severe breakouts and borehole enlargement during the drilling of the horizontal section of HAS-A1 well. The previous prediction did not match the filed observations, as the Photo Electric Factor (PEF) image of the borehole does not indicate any breakout events. Divergence between model prediction and field observations may be explained by the variability of the stress tensors along the horizontal section of HAS-A that was presumed constant in our MEM. Finally, the assessment of faults' shear stability was carried out using the Fracture Stability Advisor software. The interpretation of our study results (Fig. 1b), even while considering conservative strength estimates and worst-case stresses settings, shows that the fractures in Hasdrubal field are stable at virgin conditions and do not exceed the failure criterion. The same analyses also indicate that any major discrete discontinuity in the same stress field, and with similarly conservative estimates of strength, would also be well below the condition of being critically stressed.

4 Conclusions

The stress faulting regime concerned by our MEM was in agreement with the tectonic history of the region and the typical shape of the faults as cast by seismic surveys of the area (mainly Normal Faulting with occasional Strike Slip) (Albukhari et al. 2018; Wendt et al. 2005). From a good productivity perspective, intercepting the maximum number of fractures within the bottom part of the reservoir would be beneficial. However, this strategy needs to be revisited to consider the high potential of drilling mud losses. Suspected losses might be mitigated by minimizing the exposure to faults/fractures and potentially by cement plugging of problematic faults once encountered. It is worthy to mention that the simple 1D-MEM does not replicate the absence of breakout recorded in HAS-A1. Stress partitioning between layers and/or across the field is the likely cause of this discrepancy. Plasticity effects in the weaker intervals may also account for some of the discrepancies. The 1D-MEM did not consider the adverse effect of discontinuities on the rock strength of our formations. As the development of the 1D-MEM exclusively relies on information gathered at the well level. The Model is not fully capable of achieving a genuine simulation of in-situ stresses environment present at our reservoir. An accurate prediction can be achieved through a 3D-MEM that grasps the stresses and mechanical states at the full scale of our field, and, thus, it is highly recommended as prospective work.

References

Al-Ajmi, A.M., et al.: Wellbore stability analysis based on a new true triaxial failure criterion. Dissertation, University of KTH, Oklahoma, pp. 1–5 (2006)

Albukhari, T., Bishish, G., Abouzbeda, M., Madi, A.: Geomechanical wellbore stability analysis for the reservoir section in J-NC186 oil field. In: First International Conference on Advances in Rock Mechanics. Hammamet, Tunisia, Mar 2018

Clark Sydney, P., Press, F.: Seismic velocities. Handbook of physical-constants. In: Clark, S.G. (ed.) Geological Society of America, pp. 195–121 (1966)

Gardner, G.H.F., Gardner, L.W., Gregory, A.R.: Formation velocity and density the diagnostic basics for stratigraphic traps. Geophysics **39**, 770–780 (1974)

Knöll, L.: The process of building a mechanical earth model using well data. Master thesis, Department of Petroleum Engineering, Montanuniversität, Austria (2016)

Morals, R.H., Marcinew, R.P.: Fracturing of high-permeability formations: mechanical properties correlations. In: Annual Technical Conference and Exhibition, USA, SPE 26561 (1993)

Song, L.: Measurement of minimum horizontal stress from logging and drilling data in unconventional oil and gas. Master thesis, Department of Chemical and Petroleum Engineering, University of Calgary, Alberta (2012)

Wendt, A., Fuller, J., Merad, M.: Geomechanical study—Hasdrubal field. Schlumberger report to BG Group (Hasdrubal team, Tunisia), Dec 2005

Initial Layering Ground Model from Seismic Data Based on Hierarchical Ascendant Classification

Alexandre Bolève, Rod Eddies, and Yassine Benboudiaf

Abstract

De-risking an infrastructure project, such as nuclear power plants, requires a clear geoscientific understanding of the sub-surface. Depth to bedrock, the presence of weak zones, cavities, and faults represent fundamental information that is necessary to anticipate construction difficulties and protect the project from risks of cost and delay. Geophysical investigations are, therefore, essential to bring such information prior to any further decisions related to the project feasibility. For major infrastructure projects, many types of geophysical methods can be deployed to assess specific information (microgravity for cavities, resistivity for geology/hydrogeology, seismic for delineating stratigraphy, structure, and geotechnical properties). To help provide a representative ground model, data integration tools can be used to analyze an entire geophysical dataset. Two principal techniques can be applied to address this objective: clustering and learning. In this paper, we use a clustering technique called Hierarchical Ascendant Classification (HAC) to create an initial layering model from seismic data for a nuclear waste storage site. The HAC approach, derived from initial models, is defined by three units (soils, transitional material, and bedrock) and the depths of corresponding geological interfaces. A HAC approach can provide an objective means to build robust, initial 2D/3D ground models from geophysical data.

Keywords

Ground model • Geophysics • Principal component analysis • HAC

A. Bolève (✉) · R. Eddies · Y. Benboudiaf
Fugro France, 34 Allée du Lac d'Aiguebletté, 73375 Le Bourget-du-Lac, France
e-mail: a.boleve@fugro.com

1 Introduction

A ground model is an approximation of reality created for the purpose of solving a problem (Parry et al. 2014). For major infrastructure projects, the ground model will help to anticipate ground conditions that could affect project cost and delay. The key objectives of a geophysical project are to evaluate the ground conditions (lithology and thickness of the near surface formations, definition of the main geological/geotechnical units) and to assess any potential geo-hazards (identification of faulting with orientation, extension, and thickness; mapping of altered, weathered and fractured zones), all captured by a geophysical ground model (Eddies and Wood 2014). Frequently, geophysical investigations require the acquisition of a significant volume of data that requires a holistic approach for early-phase evaluation of the site as a whole. In this study, we demonstrate the use of Hierarchical Ascendant Classification (HAC) on seismic refraction and surface wave data in the context of a nuclear waste repository site located in the North-East of France. The field work focuses on the seismic data acquisition for the feasibility of low-level (sub-surface) nuclear waste storage (low activity—long term).

2 Application of HAC on 2D Geophysical Profiles

Clustering is the task of grouping a set of data in such a way that data in the same group (called a 'cluster') are more similar to each other than to those in other groups. HAC, a statistical procedure that uses an orthogonal transformation, is one clustering technique that can be used to analyze and classify geophysical data.

The clustering HAC approach has already been used in several geophysical domains as for example data filtering, seismic data stacking, enhanced data interpretation/characterization, and of course classification (Baba et al.

H. El-Askary et al. (eds.), *Research Developments in Geotechnics, Geo-Informatics and Remote Sensing*, Advances in Science, Technology & Innovation,
https://doi.org/10.1007/978-3-030-72896-0_42

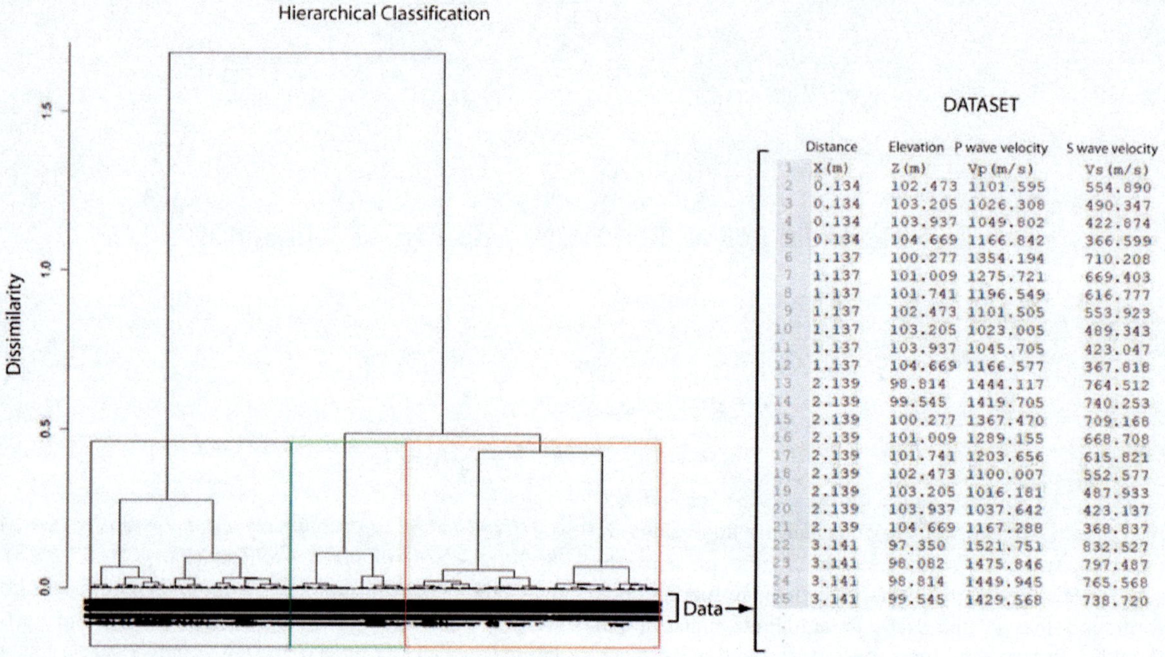

Fig. 1 Hierarchical Clustering tree. In this example, the tree has been cut off into three clusters (green rectangle, black and red). The vertical axis of the tree (dendrogram) represents the distance or dissimilarity between clusters. The horizontal axis represents the objects (data) and clusters. Each joining (fusion) of two clusters is represented on the graph by the splitting of a vertical line into two vertical lines. The vertical position of the split, shown by a short horizontal bar, gives the distance (dissimilarity) between the two clusters

2014; Davydenko and Grayver 2014; Wu and Bai 2018). In the present study, HAC was used to identify geological units based on analyzing seismic refraction and surface wave's 2D profiles.

The input variables are spatial (distance "*X*" and depth "*Z*" coordinates of data points) and geophysical parameters (compressional wave velocity "V_p" and shear wave velocity "V_s"). Because vertical variation of V_p and V_s can be relatively smooth, in some cases, interface identification can be difficult and need additional information. To overcome this difficulty, we propose to add first derivative data (vertical gradient V_p and V_s) to enhance the interface detection capacity of the HAC algorithm. The number of input variables, therefore, increases from 4 to 6 variables. Classification analysis is then performed under an Ascendant Hierarchical Classification (Fig. 1) to identify groups of similar geophysical signature (Kassambara 2017).

2.1 Results

Figure 2 presents the automatic Hierarchical Ascendant Classification results based on "*X*", "*Z*", "V_p", "V_s", "Grad V_p" and "Grad V_s" variables with an arbitrary tree cut off of eight clusters.

We observed that the addition of vertical gradient both for V_p and V_s increases the level of detail in the resulting layering model. The HAC is then independently performed for each 2D profile which can lead to classification inconsistency between 2D profiles. In order to make the classification consistent, we propose an interactive reclassification process based on a global statistic to reach clear cluster partitioning and, therefore, clear parameterization for each cluster (V_p and V_s velocities). Each ground unit of each profile that does not match with the global project statistic classification has, as a result, been reclassified. Global statistics are then recomputed to be the starting point of the next iteration. The results of the iterative reclassification process are shown in Fig. 3. After iteration n°3, an improvement in class definition is apparent, with much less overlap between clusters.

3 Discussion

An initial ground model was created through an iterative HAC analysis process. The technique has some limitations. Interfaces defined by cluster values remain constant through the iterative process as reclassification is based on profile cluster's dissimilarity.

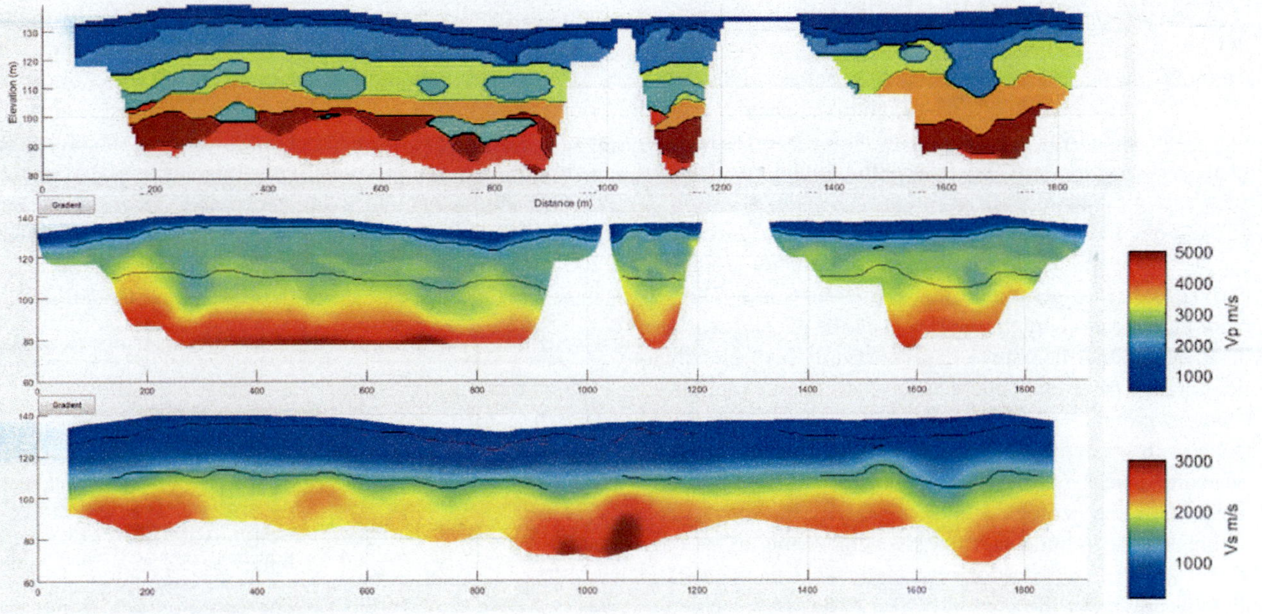

Fig. 2 Top: automatic ground model of eight clusters. Middle: V_p profile. Bottom: V_s profile

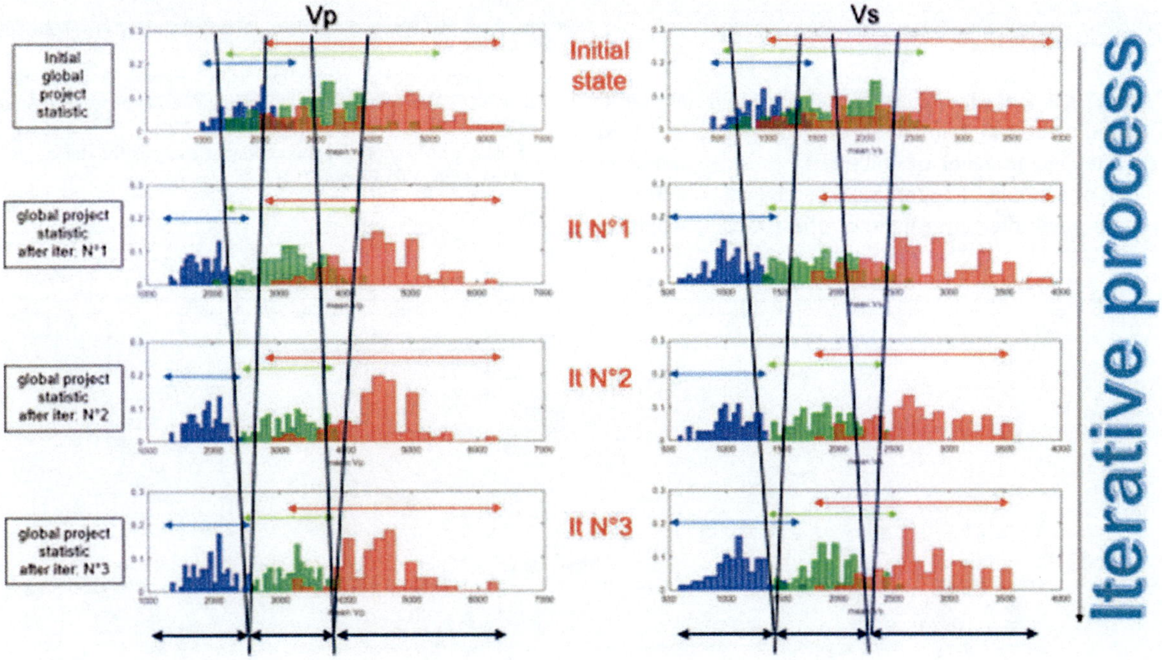

Fig. 3 Iterative reclassification process statistic for V_p and V_s. Left and right parts of the figure represent, respectively, *P* and *S* wave mean velocity distribution for each ground unit (surface: blue; transitional: green; deep-bedrock: red) from the initial state (top) to the end of the iterative reclassification process (bottom). The black lines point out the decrease of the ground unit overlap and consequently the efficiency of the iterative reclassification process

In the present work, both classifications with and without derived data have been realized, which led to the identification of "primary" ground unit (from geophysical data itself and a priori related to the geology) and "secondary" ground unit (from geophysical/derived data and a priori related to geology but could also be related to ground particularity such as a weak zone, cavities, etc.).

To combine geophysical and derived data during the same classification process, we need to properly weigh the derived data (not performed in the present work). The primary classification criteria must remain the geophysical data itself. Additional derivative data must only help classification in the case of subtle cluster dissimilarity and/or to enhance ground specificities that need to be highlighted (such as for example a rapid vertical wave velocity variation related to a weak zone, i.e., strong decrease of wave velocity inside the same geological unit).

Finally, the ground unit classification seems to be consistent with field observations especially when the bedrock is directly present at the ground surface. The HAC technique seems to provide an interesting tool to help build an initial ground model that could help optimize further investigations (geotechnical, etc.).

4 Conclusions

HAC analysis is a clustering technique that could help build initial ground models from multiple geophysical datasets. However, the identification of unit interfaces is dependent on the degree of contrast of geophysical parameters which is in some cases not sufficient to build a robust model. Addition of first derivative data seems to enhance the sensitivity of the method.

In all cases, complementary borehole or probe information is essential to calibrate an automatic classification. If available, existing borehole information can, therefore, be incorporated as an additional constraint for the HAC analysis. Finally, other clustering techniques (K-medoid) and learning techniques (machine, deep learning, Hastie et al. (2008)) based on a geophysical georeferenced database could also be implemented to build an initial model.

References

Baba, K., Bahi, L., Ouadif, L., Cherradi, C.: Multivariate statistical analysis tool for the interpretation of geoelectrical data: application to sterile bodies in the sidi chennane phosphatic deposit (Morocco). J. Mater. Environ. Sci. **5**(4), 10005–1012 (2014)

Davydenko, A.Y., Grayver, A.V.: Principal component analysis for filtering and levelling of geophysical data. J. Appl. Geophys. (2014)

Eddies, R., Wood, R.: Making the earth transparent: how engineering geophysics and an integrated geotechnical programme can help manage uncertainty and reduce project risk and cost. Ground Eng. J. (2014)

Hastie, T., Tibshirani, R., Friedman, J.: The Elements of Statistical Learning, Data Mining, Inference, and Prediction. Springer, Berlin (2008)

Kassambara, A.: Practical Guide to Principal Component Methods in R (2017)

Parry, S., Baynes, F.J., Culshaw, M.G., Eggers, M., Keaton, J.F., Lentfer, K., Novotny, J., Paul, D.: Engineering geological models—an introduction. IAEG Commission 25 (2014)

Wu, J., Bai, M.: Fast principal component analysis for stacking seismic data. J. Geophys. Eng. (2018)

Estimation of Long-Term Longitudinal Deformation Profile for a Tunnel in Squeezing Ground

Eugie Kabwe and Murat Karakus

Abstract

Tunnel excavation in squeezing ground is very challenging due to the difficulty in reliable predictions at the preliminary design stage. Employing creep constitutive models which consider long-term deformation helps tunnel response to predict realistically in squeezing ground. In this paper, tunnel closure was simulated using a proposed elasto-viscoplastic constitutive (EVPD) model that included fractal damage. The numerical simulation was performed in order to estimate the tunnel long-term longitudinal deformation and consider the rate of excavation advancement in squeezing ground. This was achieved by a tunnel model construction within FLAC^{3D} which portrayed a real tunnel case. The tunnel was simulated in steps of 5 m advance using both the proposed creep model and in-built models for verification purposes. It was observed that the proposed model estimates long-term longitudinal deformation within a tunnel excavated in squeezing ground. This may lead to a reliable design of tunnel support structure in squeezing ground.

Keywords

Creep • FLAC^{3D} • Long-term deformation • Squeezing ground

1 Introduction

The creep behaviour of rocks and rock masses in response to stress has a significant impact on the stability of underground tunnels. The behaviour might occur instantaneously after tunnel excavation and continue over time and it is often referred to as squeezing. Hence, the estimation and numerical modelling of the creep behaviour of rocks and rock masses is of great importance in tunnel design and stability analysis (Fahimifar et al. 2015). In order to describe this phenomenon, several researchers proposed models which include but not limited to the classic Burgers model connected in series with a Bingham component (Barla et al. 2008; Malan 1999; Pellet 2009; Sterpi and Gioda 2009; Sharifzadeh et al. 2013), CVISC model (Itasca Consulting Group, Inc. 2017), VIPLA model (Lemaitre 1996). These are employed to predict the long-term deformation responsible for squeezing. However, these models fail to explicitly estimate the squeezing related long-term deformation (Paraskevopoulou and Diederichs 2018). Hence, it is imperative that a creep model (Kabwe et al. 2020a, b) be proposed for the explicit estimation of long-term deformation in tunnels. In this paper, the long-term longitudinal deformation within a tunnel was estimated by a fractal derivative model (EVPD) that considers damage.

2 Fractal Derivative Creep Model

Creep behaviour can be accurately described by employing the fractal derivative constitutive equations, which employ scaling transform and can be applied in numerical techniques to estimate a variable range of applications (Cai et al. 2016; Yang et al. 2018).

E. Kabwe · M. Karakus (✉)
School of Civil, Environmental and Mining Engineering, The University of Adelaide, Adelaide, 5005, Australia
e-mail: murat.karakus@adelaide.edu.au

E. Kabwe
e-mail: eugie.kabwe@adelaide.edu.au

H. El-Askary et al. (eds.), *Research Developments in Geotechnics, Geo-Informatics and Remote Sensing*, Advances in Science, Technology & Innovation,
https://doi.org/10.1007/978-3-030-72896-0_43

2.1 Derivation of the Fractal Derivatives with Damage

Fractal derivative coincides with the classical integer order derivative employed in most formulations by using a metric transform $t = t^{\gamma}$ (Kilbas et al. 2006). The fractal derivative application is observed in the non-variable dashpot expressed by the following relation:

$$\sigma(t) = \eta \frac{\mathrm{d}\varepsilon(t)}{\mathrm{d}t^{\gamma}} \quad (0 \le \gamma \le 1) \tag{1}$$

where γ is the fractal derivative coefficient, t is the time, σ is the stress, η is the viscosity coefficient and ε is the strain. When $\gamma = 0$ the fractal dashpot acts as a spring which represents an ideal solid and when $\gamma = 1$ it acts as a Newtonian dashpot which represents an ideal fluid. If we consider constant stress which occurs during creep and applies scaling transformation, $t = t^{\gamma}$ we obtain an expression for strain as:

$$\varepsilon(t) = \frac{\sigma}{\eta} t^{\gamma} \quad (0 \le \gamma \le 1) \tag{2}$$

Equation 2 shows that the strain rate obeys the power law and incorporating damage, the fractal derivative formulation can be expressed in the form of fractal time. Applying the scaling transformation $t = t^{\gamma}$ and expressing fractal damage variable (D_{FD}) by Eq. 3, the influence of the fractal coefficient, γ and the fractal damage factor D_{FD} is illustrated by Fig. 1.

$$D_{\mathrm{FD}} = 1 - \mathrm{e}^{-\omega t^{\gamma}} \quad (0 \le \gamma \le 1) \tag{3}$$

The increase in ω has a progression effect on the material's rate of D_{FD} (Fig. 1). It shows that using $\omega = 1.4$, damage of the material reaches 90% within 3 h while at $\omega = 0.9$ it reaches 90% after 5 h. Figure 1b shows that the γ increment influences the rate of the D_{FD}. There is more damage in the material when higher-order values of ω and γ are used. The constitutive that incorporates D_{FD} is given by:

$$\sigma(t) = \eta_0 \mathrm{e}^{-\omega t^{\gamma}} \frac{\mathrm{d}\varepsilon(t)}{\mathrm{d}t^{\gamma}} \tag{4}$$

consider a stress level $\sigma(t) = \sigma$ and initial strain as $\varepsilon(0) = 0$ the constitutive expression for strain in a fractal dashpot incorporating D_{FD} is:

$$\varepsilon(t) = \frac{\sigma}{\eta_0 \omega}\left(\mathrm{e}^{\omega t^{\gamma}} - 1\right) \tag{5}$$

In the next section (Sect. 2.2), the derived fractal derivative with damage equation (Eq. 5) for the non-variable dashpot was integrated into the creep constitutive equations to build the EVPD model.

2.2 The Creep Constitutive Equations

The creep constitutive model employed is an extension of the Burgers model characterised by an EVP deviatoric behaviour and an EVP volumetric behaviour. The constitutive equations of this model are expressed as:

$$\begin{cases} \varepsilon = \frac{\sigma}{G_{\mathrm{M}}} + \frac{\sigma t^{\beta}}{\eta_1^{\beta}} + \frac{\sigma}{G_{\mathrm{K}}}\left[1 - \exp\left(-\frac{t^{\beta}}{T_{\mathrm{K}}}\right)\right], \sigma < \sigma_Y & \\ \varepsilon = \frac{\sigma}{G_{\mathrm{M}}} + \frac{\sigma t^{\beta}}{\eta_1^{\beta}} + \frac{\sigma}{G_{\mathrm{K}}}\left[1 - \exp\left(-\frac{t^{\beta}}{T_{\mathrm{K}}}\right)\right] + \left(\frac{\sigma - \sigma_Y}{\eta_2^{\gamma}} t^{\gamma}\right), \sigma \ge \sigma_Y & 0 \le t \le t_{\mathrm{a}} \\ \varepsilon = \frac{\sigma}{G_{\mathrm{M}}} + \frac{\sigma t^{\beta}}{\eta_1^{\beta}} + \frac{\sigma}{G_{\mathrm{K}}}\left[1 - \exp\left(-\frac{t^{\beta}}{T_{\mathrm{K}}}\right)\right] + \frac{\sigma - \sigma_Y}{\eta_2^{\gamma}\omega}\left(\mathrm{e}^{\omega t^{\gamma}} - 1\right), \sigma \ge \sigma_Y & t > t_{\mathrm{a}} \end{cases} \tag{6}$$

where t_{a} is the accelerated creep phase start time, β is the fractal coefficient, η_1^{β} and G_{M} are the Maxwell viscosity coefficient and shear modulus, η_2^{γ} is the VPD component

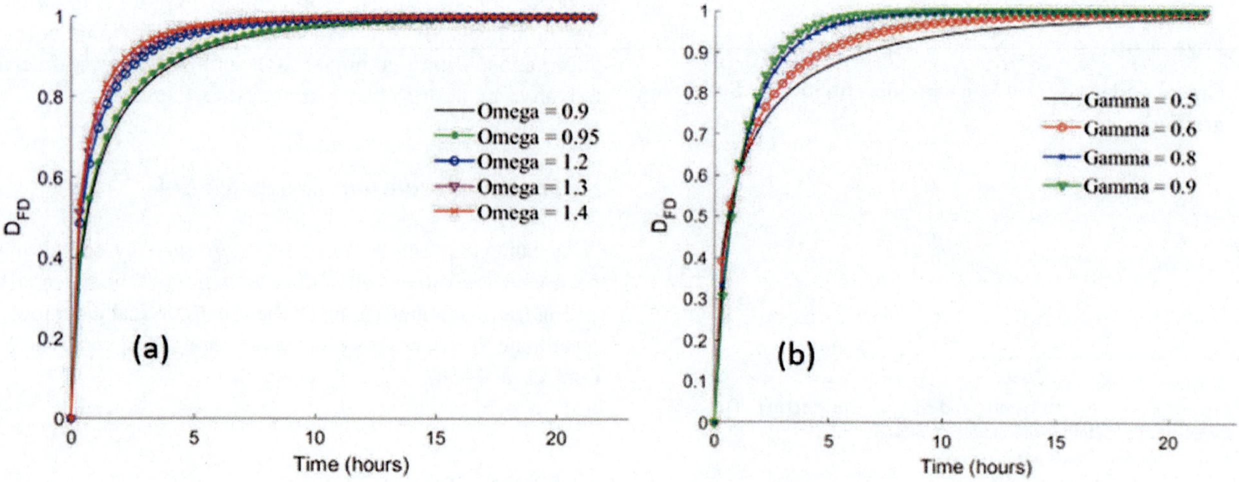

Fig. 1 Damage evolution with different values of **a** damage factor and **b** fractal order coefficient

viscosity coefficient and $T_K = \eta_K / G_K$. G_K and η_K are the Kelvin shear modulus and viscosity coefficient. Damage effect is a significant factor and should be considered in estimating the long-term deformations characterised by vis-coplastic (VP) strains. Hence, the addition of the VPD unit to the Burgers model improves its capability to estimate long-term deformations. The EVPD constitutive equations are expressed in a three-dimension (3D) stress state based on the equations (Eq. 6). In this stress state, the rock stress tensor (σ_{ij}) consists of the spherical stress tensor (σ_m) and deviatoric stress tensor (S_{ij}) which are expressed as:

$$\sigma_m = \frac{1}{3}(\sigma_1 + \sigma_2 + \sigma_3) = \frac{\sigma_{kk}}{3} \tag{7}$$

$$S_{ij} = \sigma_{ij} - \sigma_m \delta_{ij} = \sigma_{ij} - \frac{\sigma_{kk}}{3}\delta_{ij} \tag{8}$$

$$\sigma_{ij} = S_{ij} + \frac{\sigma_{kk}}{3}\delta_{ij} \tag{9}$$

where σ_1, σ_2 and σ_3 are the major, intermediate and minor principal stresses respectively. δ_{ij} is the Kronecker delta, σ_m is responsible for volume change of a rock mass while S_{ij} is responsible for the change in the shape of the rock mass, not the volume. The strain tensor (ε_{ij}) can also be decomposed into spherical strain tensor (ε_m) and deviatoric strain tensor $(\dot{e}_{ij})$:

$$\varepsilon_m = \frac{\varepsilon_{kk}}{3} \tag{10}$$

$$e_{ij} = \varepsilon_{ij} - \varepsilon_m \delta_{ij} \tag{11}$$

$$\varepsilon_{ij} = e_{ij} + \varepsilon_m \delta_{ij} \tag{12}$$

Based on Hooke's law the σ_m and S_{ij} in 3D stress state is computed by:

$$\sigma_m = 3K\varepsilon_m \tag{13}$$

$$S_{ij} = 2Ge_{ij} \tag{14}$$

Substituting Eqs. 13 and 14 into Eq. 6 and assuming S_{ij} is less than the yield strength S_Y the 3D creep constitutive equation for the model is expressed as:

$$\varepsilon_{ij} = \frac{S_{ij}}{2G_M} + \frac{\sigma_m \delta_{ij}}{3K} + \frac{S_{ij} t^\beta}{2\eta_1^\beta} + \frac{S_{ij}}{2G_K}\left[1 - \exp\left(-\frac{t^\beta}{T_K}\right)\right] \quad S_{ij} < S_Y \tag{15}$$

When $S_{ij} \geq S_Y$ and $t \leq t_a$ VP deformation terminates, the failure yield criterion (F) and VP potential function (Q_{vp}) are introduced and the fractal order VP strain rate using (Perzyna 1966) and is given by:

$$\varepsilon_{ij}^{VP} = \frac{\langle \phi(F) \rangle}{2\eta_2^\gamma} \frac{\partial Q_{vp}}{\partial \sigma_{ij}} t^\gamma \tag{16}$$

where $\langle \phi(F) \rangle$ is the VP nucleus that controls the magnitude of the VP strains given by:

$$\langle \phi(F) \rangle = \begin{cases} 0, & F < 0 \\ \phi(F), & F \geq 0 \end{cases} \tag{17}$$

When an associated flow rule $(F = Q_{vp})$ is adopted, Eq. (16) is reduced to:

$$\varepsilon_{ij}^{VP} = \left(\frac{f}{2\eta_2^\gamma}\right) \frac{\partial f}{\partial \sigma_{ij}} t^\gamma \quad (f \geq 0) \tag{18}$$

The Mohr-Coulomb (MC) yield function (f) is expressed as:

$$f = \sigma_1 - \sigma_3 N_\vartheta + 2c\sqrt{N_\vartheta} \tag{19}$$

where c is the cohesion, $N_\vartheta = \frac{1+\sin\vartheta}{1-\sin\vartheta}$, ϑ is the friction angle and the 3D creep constitutive equation is:

$$\varepsilon_{ij} = \frac{S_{ij}}{2G_M} + \frac{\sigma_m \delta_{ij}}{3K} + \frac{S_{ij} t^\beta}{2\eta_1^\beta} + \frac{S_{ij}}{2G_K}\left[1 - \exp\left(-\frac{t^\beta}{T_K}\right)\right] + \left(\frac{f}{2\eta_2^\gamma}\right)\frac{\partial f}{\partial \sigma_{ij}} t^\gamma \quad (f \geq 0) \tag{20}$$

when $S_{ij} \geq S_Y$ and $t > t_a$ damage terminates and the VP strains in the VPD component can be computed from the modified Perzyna (1966) incorporating the fractal damage as:

$$\varepsilon_{ij}^{VPD} = \frac{\langle \phi(F) \rangle}{2\eta_2^\gamma \omega}\left(e^{\omega t^\gamma} - 1\right)\frac{\partial f}{\partial \sigma_{ij}} \tag{21}$$

The creep constitutive equation under 3D stress state is further expressed as:

$$\varepsilon_{ij} = \frac{S_{ij}}{2G_M} + \frac{\sigma_m \delta_{ij}}{3K} + \frac{S_{ij} t^\beta}{2\eta_1^\beta} + \frac{S_{ij}}{2G_K}\left[1 - \exp\left(-\frac{t^\beta}{T_K}\right)\right] + \left(\frac{f}{2\eta_2^\gamma \omega}\right)\left(e^{\omega t^\gamma} - 1\right)\frac{\partial f}{\partial \sigma_{ij}} \quad (f \geq 0) \tag{22}$$

The constitutive models are coded using the object-oriented C++ as a user-defined constitutive model source code and implemented within $FLAC^{3D}$.

3 Numerical Simulation

The functionality of the implemented model was compared with an in-built CVISC model, which can compute the instantaneous strains and to some extent delayed deformations characterised by squeezing (Boidy et al. 2002). The instantaneous strains in CVISC are computed by the elastoplastic part which employs MC yield surface. The creep strain by the Burger's component while delayed deformation responsible for squeezing is not explicitly estimated (dependent on stress not on time). Hence, employing the EVPD model for an accurate estimation of the long-term deformation in squeezing ground is needed. The creep parameters used in the numerical analysis correspond closely to the field scale values to account for the rock mass discontinuities and heterogeneities (Fabre and Pellet 2006) (Table 1).

The numerical simulation was conducted within FLAC^{3D} code, Fig. 2 shows the half geometric meshed model employed which depicts a horseshoe tunnel excavated in a 100 m × 80 m × 50 m block. The boundary and initial conditions of the geometric model are set as; zero normal displacement at all faces as well as the symmetrical plane and initial stress of 8.4 MPa.

During the simulation, the excavation of the tunnel was achieved in 5 steps of 5 m advances. The simulation of the tunnel advance starts with an EP analysis and when the model equilibrium is reached at 200 steps, then the creep model was activated. The creep time span employed in the creep simulation is equivalent to 4 days (4.32E5 s) and the results are illustrated in Fig. 3.

Figure 3 shows that the long-term deformation is more within the EVPD model simulated tunnel as compared to the CVISC model. The total deformation estimated is approximately 26% and accounts for 2% more than the CVISC estimation (24%), this variance is attributed to the long-term deformation in the accelerated creep phase. However, the total deformation in the MC modelled tunnel is about 2.7% (Table 2). The CVISC model underestimates the squeezing related delayed deformations as compared to the EVPD model.

4 Conclusions

In this paper, it was shown that the long-term deformation within a tunnel excavated in squeezing ground can be properly simulated and estimated using the EVPD constitutive equations. In the tunnel simulation, actual staged excavation was considered with an advance of 5 m in 5

Table 1 FD creep model parameters (Barla et al. 2010)

η_K (MPa y)	G_K (MPa)	G_M (MPa)	η_1^β (MPa y)	η_2^α (MPa y)	β	α	c (MPa)	ϕ	ν	K (MPa)	σ^t (MPa)
4.26	498.1	566	27.98	0.64	0.9	0.65	0.61	28	0.2	942	0.0085

CVISC creep model parameters (Barla et al. 2010)

η_K (MPa y)	G_K (MPa)	G_M (MPa)	η_M (MPa y)	σ^t (MPa)	c (MPa)	ϕ	ν	K (MPa)
4.26	498.1	566	27.98	0.0085	0.61	28	0.2	942

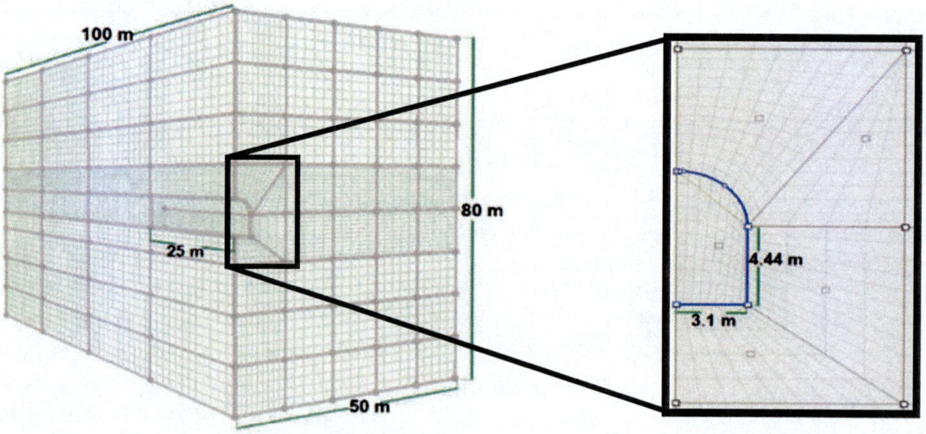

Fig. 2 Model geometry

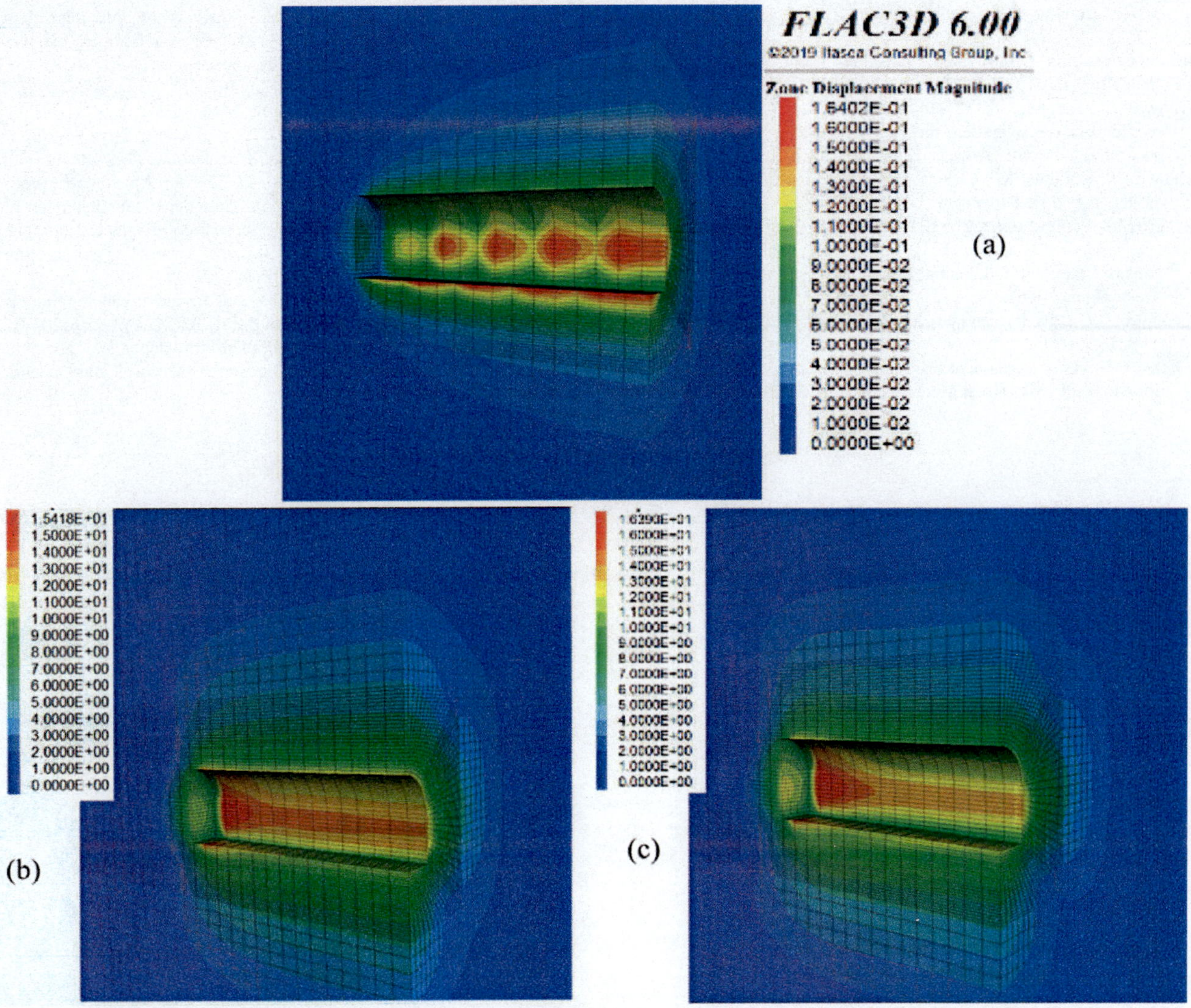

Fig. 3 Estimation of the total deformation: **a** MC model, **b** CVISC model and **c** EVPD model

Table 2 Total deformation within the tunnel

Boundary deformation	MC model	CVISC model	EVPD model
Total deformation (mm)	164	1540	1640

steps. The overall analysis proved that the EVPD model is a suitable tool for tunnel simulation and appropriate support system design in the long term.

References

Barla, G., Bonini, M., Debernardi, D.: Time dependent deformations in squeezing tunnels. In: The 12th International Conference of International Association for Computer Methods and Advances in Geomechanics (IACMAG), pp. 1–6 (2008)

Barla, G., Bonini, M., Debernardi, D.: Time dependent deformations in squeezing tunnels. ISSMGE Int. J. Geoeng. Case Histories **2**(1), 40–65 (2010)

Boidy, E., Bouvard, A., Pellet, F.: Back analysis of time-dependent behaviour of a test gallery in claystone. Tunn. Undergr. Space Technol. **17**(4), 415–424 (2002)

Cai, W., Chen, W., Xu, W.: Characterizing the creep of viscoelastic materials by fractal derivative models. Int. J. Non-Linear Mech. **87**, 58–63 (2016)

Fabre, G., Pellet, F.: Creep and time-dependent damage in argillaceous rocks. Int. J. Rock Mech. Min. Sci. **43**(6), 950–960 (2006)

Fahimifar, A., Karami, M., Fahimifar, A.: Modifications to an elasto-visco-plastic constitutive model for prediction of creep deformation of rock samples. Soils Found. **55**(6), 1364–1371 (2015)

Itasca Consulting Group, Inc.: FLAC^{3D} Version 6.0. User manual. Consulting Group, Inc., Minneapolis (2017)

Kabwe, E., Karakus, M., Chanda, E.K.: Creep constitutive model considering the overstress theory with an associative viscoplastic flow rule. Comput. Geotech. **124**, 103629 (2020a)

Kabwe, E., Karakus, M., Chanda, E.K.: Isotropic damage constitutive model for time-dependent behaviour of tunnels in squeezing ground. Comput. Geotech. **127**, 103738 (2020b)

Kilbas, A.A.A., Srivastava, H.M., Trujillo, J.J.: Theory and Applications of Fractional Differential Equations. Elsevier Science Limited, Amsterdam (2006)

Lemaitre, J.C.: JL Mécanique des matériaux solides. Dunod, Paris (1996)

Malan, D.F.: Time-dependent behaviour of deep level tabular excavations in hard rock. Rock Mech. Rock Eng. **32**(2), 123–155 (1999)

Paraskevopoulou, C., Diederichs, M.: Analysis of time-dependent deformation in tunnels using the convergence-confinement method. Tunn. Undergr. Space Technol. **71**(June 2017), 62–80 (2018)

Pellet, F.L.: Contact between a tunnel lining and a damage-susceptible viscoplastic medium. CMES Comput. Model. Eng. Sci. **52**(3), 279–295 (2009)

Perzyna, P.: Fundamental problems in viscoplasticity. In: Advances in Applied Mechanics, vol. 9, pp. 243–377. Elsevier, Amsterdam (1966)

Sharifzadeh, M., Tarifard, A., Ali, M.: Time-dependent behavior of tunnel lining in weak rock mass based on displacement back analysis method. Tunn. Undergr. Space Technol. Incorporating Trenchless Technol. Res. **38**, 348–356 (2013)

Sterpi, D., Gioda, G.: Visco-plastic behaviour around advancing tunnels in squeezing rock. Rock Mech. Rock Eng. **42**(2), 319–339 (2009)

Yang, X., Chen, W., Sun, H.: Fractional time-dependent apparent viscosity model for semisolid foodstuffs. Mech. Time-Depend. Mater. **22**(4), 447–456 (2018)

Principal Component Analysis of Geological and Tunnel Boring Machine Parameters in Hard Rock (Thailand)

Kannipa Motanated, Nalin Eardkeaw, and Premanan Photee

Abstract

Tunnel boring machine (TBM) performance prediction is essential for feasibility evaluation and planning purposes. Prediction models involve multiple geological and machine factors from field and laboratory studies. This paper presented an alternative method of feature selection for a less complex yet predictive model. The principal component analysis (PCA) was applied to a multivariate dataset of geological and tunnel boring machine parameters including thrust force, torque, the amount of small-, medium- and large-sized rock chips, quartz content, TBM penetration distance per revolution, and disc cutter wear of Mae Tang to Mae Ngat diversion tunnel project, Chiang Mai, Thailand. Variables with high loading factors from PC1 and PC2 are thrust force, torque, the amount of small- and medium-sized rock chips, and penetration rate. These five features have high contributions to the trends in data variation and should be considered as the most influential parameters when developing TBM performance models.

Keywords

TBM • Disc cutter • Rock chip • Quartz content • PCA

1 Introduction

During the excavation, multiple sets of disc cutters are grinding hard rock surfaces and creating rock chips. The ease with which a rock stratum is penetrated depends on TBM specifications, TBM operation parameters, and geological properties (Gong et al. 2007; Ren et al. 2018). The penetration rate and disc cutter consumption are also influenced by geological properties such as fracture spacing (Yin et al. 2016), mica and quartz contents, and uniaxial compressive strength (Armetti et al. 2018; Villeneuve 2016). Thus, the aforementioned factors are interrelated and make TBM performance prediction rather complex.

The objective of this research was to extract factors that are statistically significant for constructing a predictive model of TBM performance. The principal component analysis (PCA) was applied to the multivariate dataset of machines and geological parameters. The dataset consists of TBM operation parameters including thrust force and torque, the amount of small-, medium- and large-sized rock chips, quartz content, TBM penetration distance per revolution, and disc cutter wear.

2 Methods

This study gathered TBM thrust force (kN), torque (kN m) and penetration rate (mm/rev), disc cutter wear (mm), the percentage amount of small-, medium- and large-sized rock chips, and quartz content (%) from Mae Tang to Mae Ngat diversion tunnel project, Chiang Mai, Thailand in 2018. The TBM excavates through massive Carboniferous-Permian sandstone with few thin layers of shale, and chert, and massive, unweathered Triassic granite and granodiorite. Small-, medium-, and large-sized chips have diameters of <20 mm, between 20–60 mm, and ≥ 60 mm, respectively, based on the disc cutter tip width and average disc cutter spacing. The amount of each chip size was determined by analyzing the photographs from muck piles, collected at the end of each operating day. This study used Fiji which is a Java-based image processing program. Quartz content was calculated by point counting method from rock thin-section when quartz grains are distinguishable under a microscope, diameter ≥ 0.0625 mm. Otherwise, quartz content was analyzed by X-ray diffraction (XRD). Rock samples

K. Motanated (✉) · N. Eardkeaw · P. Photee
Department of Geological Sciences, Chiang Mai University, Chiang Mai, 50200, Thailand
e-mail: kannipa.motanated@cmu.ac.th

H. El-Askary et al. (eds.), *Research Developments in Geotechnics, Geo-Informatics and Remote Sensing*, Advances in Science, Technology & Innovation,
https://doi.org/10.1007/978-3-030-72896-0_44

collected for thin-section and XRD analysis were from the muck pile of each operating day. Disc cutter consumption was an average of the wear of all 32 disc cutters and measured at the end of each operating day. All factors except quartz content have been logged covering the tunnel length of 1484.67 m. This study divided the data into two sets—one has seven variables from 79 operating days and the other has all eight variables from 13 operating days. Only data from days with quartz content analysis are shown in Table 1.

The PCA is a dimensionality reduction technique. This technique was applied to the dataset in order to determine variables that have high correlations with the principal component and explain the multivariate data. The PCA was computed on "R" statistical computing and graphic software. Principal components (PC) are the eigenvectors of a correlation matrix which represent the strength of the associated variables.

3 Results and Discussion

The PCA was applied to the two datasets and the results are shown in Table 2. PCs with greater Eigenvalues than one were retained based on the Kaiser criterion (Kaiser 1960).

From Table 2 the first three and four principal components explain 80.24% and 89.15% of the variance of the first and the second dataset, respectively. Each principal component is a variation of a combination of variables; squared loading factor equals the percentage of the variance in an original variable that is reflected by a particular factor. PC1 of the two datasets share the same variables with high loading factors—the amount of small- and medium-sized chips. Variables with high loading factors of PC2 are thrust force, torque, and penetration rate. Therefore, trends in the TBM dataset are strongly controlled by these five variables, similar to those from Gong et al. (2007) and Heydari et al. (2019). The analysis shows that penetration rate is negatively correlated with thrust force and torque. However, the penetration rate is dominated by geological parameters such as rock uniaxial compressive strength and the volumetric joint count (Gong and Zhao 2009). Comparing the TBM performances from different fields having different rock types is ineffective.

When modeling the relationship between TBM operating, TBM performance, and geological parameters, a pair of parameters or composite parameters are typically compared at a time. This research presented an alternative statistical approach for TBM performance analysis. Because the TBM performance prediction depends on a combination of factors (Gong and Zhao 2009; Hassanpour et al. 2009), the PCA allows a simplified visualization of the relatedness of a set of correlated variables and is able to eliminate composite parameters.

4 Conclusions

Instead of analyzing the relationship between two parameters or composite parameters at a time, this study recommended applying the PCA to the TBM operating and performance parameters and geological data. PC1 and PC2 of the two PCA analyses explain more than 57% of the variance. Variables with high loading factors are the amount of small- and medium-sized chips, thrust force, cutterhead

Table 1 TBM operation parameters and penetration rate, disc cutter wear, amount of each rock chip size, and quartz content of an underground diversion tunnel project in Thailand

Thrust (kN)	Torque (kN m)	Penetration (mm/rev)	Disc cutter wear (mm)	Small chip (%)	Medium chip (%)	Large chip (%)	Quartz content (%)
2954.01	413.92	9.37	0.27	25.70	67.70	6.60	77.97
3261.14	560.04	11.17	0.36	44.56	52.74	2.70	87.41
3621.57	584.27	8.04	0.30	47.50	51.40	1.10	81.03
3465.81	444.12	10.60	0.22	48.30	50.94	0.76	75.06
3250.98	552.71	10.37	1.08	54.20	45.48	0.32	89.00
3990.71	462.20	8.33	0.06	55.10	44.48	0.42	88.25
3760.16	445.88	8.80	0.28	55.19	44.81	0.00	91.25
3257.24	471.87	10.11	0.39	81.87	18.13	0.00	76.79
3624.47	497.74	10.24	0.52	76.96	22.83	0.22	78.91
3623.87	488.82	11.44	0.25	82.26	17.74	0.00	88.54
3521.00	477.41	9.08	0.98	73.58	26.43	0.00	58.94
4165.48	446.10	8.80	0.45	36.67	62.50	0.83	78.33
3619.81	522.58	8.68	0.61	62.44	37.56	0.00	65.07

Table 2 Principal component analysis of (1) TBM data with seven variables from 79 operating days and (2) TBM data with eight variables, including quartz content, from 13 operating days

	First dataset (79 days)			Second dataset (13 days)			
	PC1	PC2	PC3	PC1	PC2	PC3	PC4
Eigenvalues	2.81	1.73	1.08	2.87	1.73	1.41	1.11
Cumulative %	40.15	64.81	80.24	35.92	57.57	75.21	89.15
Thrust	−0.183	**−0.581**	−0.139	0.093	**−0.699**	0.095	−0.131
Torque	0.298	**−0.546**	−0.157	0.177	0.201	−0.082	**−0.800**
Penetration rate	−0.351	0.465	−0.251	0.154	**0.541**	0.410	0.410
Cutter wear	−0.082	−0.052	**−0.910**	0.251	0.252	**−0.558**	−0.236
Small chip	**0.563**	0.095	−0.061	**0.561**	0.023	0.130	0.130
Medium chip	**−0.564**	−0.071	0.047	**−0.549**	−0.063	−0.134	−0.134
Large chip	−0.336	0.364	0.241	−0.489	0.333	−0.056	0.177
Quartz content	n/a	n/a	n/a	−0.134	0.013	**0.683**	−0.398

Factor loadings with absolute values $\geq$0.500 are in bold

torque, and penetration rate and strongly control trends in the TBM and geological data. However, the data are from Carboniferous-Permian sandstone, shale, and chert and Triassic granite and granodiorite of Chiang Mai, Thailand, additional data from other tunneling operations with similar lithology should be analyzed in order to verify the accuracy of the approach. Additionally, the data should be divided into training and testing datasets in future PCA.

References

Armetti, G., Migliazza, M., Ferrari, F., Berti, A., Padoveser, P.: Geological and mechanical rock mass conditions for TBM performance prediction-the case of "La Maddalena" exploratory tunnel, Chiomonte (Italy). Tunn. Undergr. Space Technol. **77**, 115–126 (2018)

Gong, Q., Zhao, J.: Development of a rock mass characteristics model for TBM penetration rate prediction. Int. J. Rock Mech. Min. Sci. **46**(1), 8–18 (2009)

Gong, Q., Zhao, J., Jiang, Y.: In situ TBM penetration tests and rock mass boreability analysis in hard rock tunnels. Tunn. Undergr. Space Technol. **22**, 303–316 (2007)

Hassanpour, J., Rostami, J., Khamehchiyan, M., Bruland, A.: Developing new equations for TBM performance prediction in carbonate-argillaceous rocks: a case history of Nowsood water conveyance tunnel. Geomech. Geoeng. Int. J. **4**(4), 287–297 (2009)

Heydari, S., Hamidi, J., Monjezi, M., Eftekhari, A.: An investigation of the relationship between muck geometry, TBM performance, and operational parameters: a case study in Golab II water transfer tunnel. Tunn. Undergr. Space Technol. **88**, 73–86 (2019)

Kaiser, H.: The application of electronic computers to factor analysis. Educ. Psychol. Measur. **20**(1), 141–151 (1960)

Ren, D.J., Shen, S.L., Arulrajah, A., Cheng, W.C.: Prediction model of TBM disc cutter wear during tunnelling in heterogeneous ground. Rock Mech. Rock Eng. **51**, 3599–3611 (2018)

Villeneuve, M.: Hard rock tunnel boring machine penetration test as an indicator of chipping process efficiency. J. Rock Mech. Geotech. Eng. **9**(4), 611–622 (2016)

Yin, L., Miao, C., He, G., Dai, F., Gong, Q.: Study on the influence of joint spacing on rock fragmentation under TBM cutter by linear cutting test. Tunn. Undergr. Space Technol. **57**, 137–144 (2016)

Geo-Informatics and Remote Sensing (T6): Recent Developments in Geo-Informatic Techniques

Time Series Forecasting Using Long Short-Term Memory Neural Networks: A Case Study of Seismogram

Hilal Nuha, Mohamed Mohandes, Bo Liu, and Ali Al-Shaikhi

Abstract

This paper presents an investigation of the long short-term memory (LSTM) neural networks for seismogram time series prediction. LSTM has been widely used for time series prediction problems and achieved excellent results therefore it is interesting to utilize this technique for seismology. A seismogram from Albuquerque, New Mexico (Anmo), USA provided by the IRIS website is used for the experiment. The seismogram is recorded to observe passive seismic activities like Earthquake. The data is recorded with a 16 kHz sampling frequency by a station in Albuquerque. The network is with 200 hidden units and trained with 250 maximum iterations. The LSTM is able to achieve 0.00306 of root mean squared error (RMSE).

Keywords

Seismogram • Time series forecasting • Long short-term memory

1 Introduction

In this world, there is a large number of sensors for geophysical data acquisition (IRIS at a Glance 2018). For example, the Incorporated Research Institutions for Seismology (IRIS) has a high-performance network of more than 150 permanent stations for global studies of earthquakes and deep Earth structures and more than 4000 portable instruments, including magnetotelluric systems. Each station is located at a unique geospatial position, continuously recording time series readings. Besides observatory purposes, there is a rising demand for geophysical time series forecasting (Liang et al. 2018; Nuha et al. 2019). This paper presents seismogram time series forecasting problem using Long short-term Memory (LSTM) neural networks. LSTM has been widely used for time series prediction problems and achieved excellent results. Recurrent neural networks (RNNs) with LSTM have become a scalable and effective model for many learning problems for time series analysis. This includes language modeling, speech synthesis, speech acoustic modeling, audio, and video analysis, etc. (Greff et al. 2017).

2 Methodologies

An LSTM network with sequence-to-sequence regression for time series prediction is utilized. Given each time instance of the input sequence, the network is trained to forecast the future time instance value (https://www.mathworks.com/help/deeplearning/ug/long-short-term-memory-networks.html; Hochreiter and Schmidhuber 1997).

The process of carrying memory forward is mathematically described as follows:

$$h_n = \phi(\boldsymbol{W}x_n + \boldsymbol{U}h_{n-1}) \tag{1}$$

where h_n is the hidden unit at time step n which represents a function of the input for the same time step x_n. The hidden unit of the previous time step h_{n-1} is multiplied by its own hidden unit using unit matrix $\boldsymbol{U}$. The input layer is transformed to the hidden layer using A weight matrix $\boldsymbol{W}$. Figure 1 illustrates an LSTM cell with output units $\boldsymbol{O}$ and output layer weights $\boldsymbol{V}$. LSTM employs Backpropagation Through time (BPTT) to train the networks. LSTM is able to preserve the error that can be back-propagated through layers and time. By preserving a more constant error, the

H. Nuha · M. Mohandes (✉) · B. Liu · A. Al-Shaikhi
King Fahd University of Petroleum and Minerals (KFUPM), Dhahran, 31261, Saudi Arabia
e-mail: mohandes@kfupm.edu.sa

H. Nuha
School of Computing, Telkom University, Bandung, Indonesia

H. El-Askary et al. (eds.), *Research Developments in Geotechnics, Geo-Informatics and Remote Sensing*, Advances in Science, Technology & Innovation,
https://doi.org/10.1007/978-3-030-72896-0_45

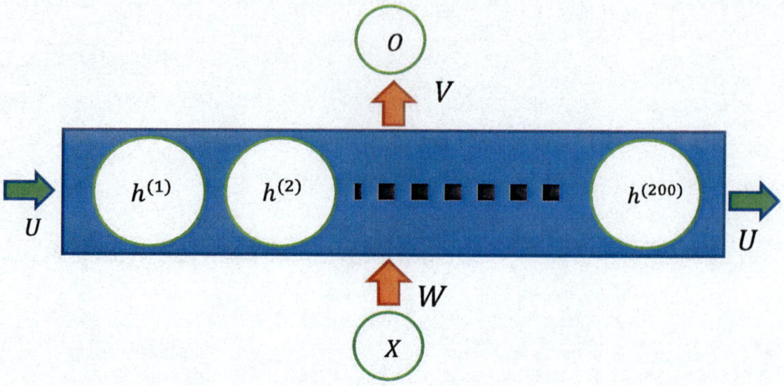

Fig. 1 An LSTM cell

Table 1 Comparison of LSTM and AR model

Algorithms	LSTM	AR, $P = 1$	AR, $P = 3$	AR, $P = 5$	AR, $P = 10$
RMSE	**0.00306**	0.0077	0.0090	0.0086	0.0089

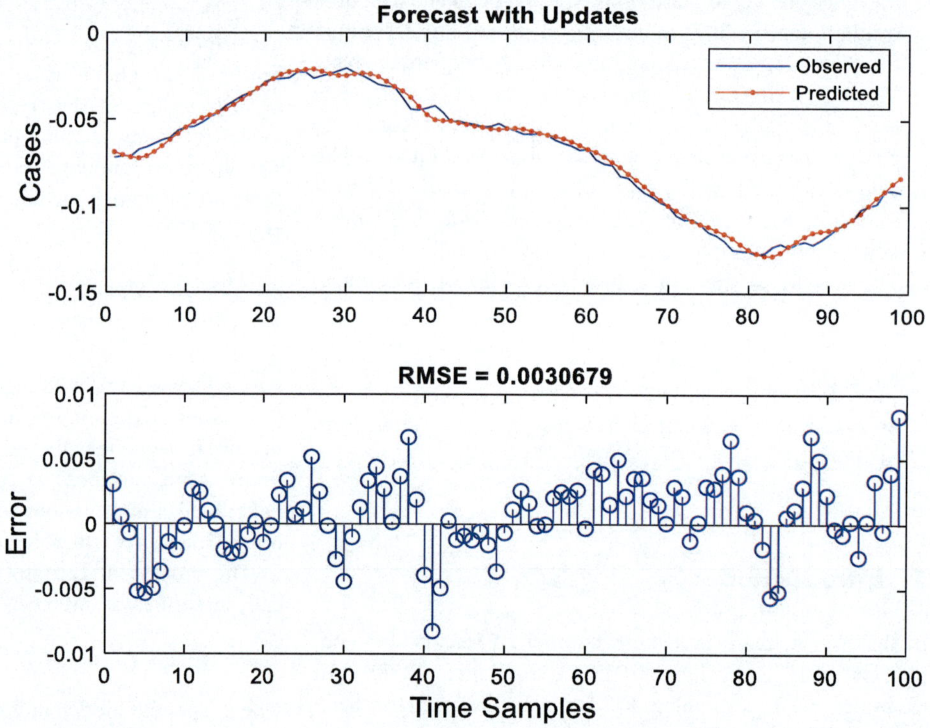

Fig. 2 Sample of the predicted seismogram

networks can continue to learn over many time steps (Nicholson 2019; Psychoula et al. 2018).

In this problem, we use a seismogram from Albuquerque, New Mexico (Anmo), USA provided by the IRIS website (IRIS Website 2018; Mousa and Al-Shuhail 2011). The seismogram is recorded to observe passive seismic activities like Earthquake. The data was recorded with a 16 kHz sampling frequency by a station in Albuquerque in January 2015. 99% of the data was used for training and the remaining data for testing. The network is illustrated by Fig. 1. The network is with 200 hidden units. Having 200 units in the hidden layer means that the layer produces a vector with 200 elements to be passed to the output layer (Tilk 2016). This number of hidden units allows the network to achieve more accurate predictions. The training is set to 250 maximum iterations to prevent overfitting.

For comparison, an Autoregressive (AR) model is adopted to perform prediction using different numbers of past values (*P*). The prediction error is 0.00306 of root mean squared error (RMSE). Table 1 summarizes the results of the experiments. It can be seen that LSTM outperforms the AR model for all sizes of past values. A seismograph sample used in the experiments is shown in Fig. 2. It can be deduced that the prediction error is much smaller compared with the original waveform amplitude.

3 Conclusions

This paper investigates the LSTM networks for seismogram time series prediction. The LSTM network outperforms the autoregressive model for prediction for different numbers of past values. The network was shown to be able to achieve 0.00306 of RMSE with a relatively small error compared to the waveform.

Acknowledgments This work is supported by the Center for Energy and Geo Processing (CeGP) at King Fahd University of Petroleum and Minerals (KFUPM), Dhahran, Saudi Arabia, under Project GTEC1801.

References

IRIS at a Glance 2018, Incorporated Research Institutions for Seismology, New York (2018)

IRIS Website, 1 June 2018 [Online]. Available: http://service.iris.edu/irisws/timeseries/1/query?net=IU&sta=ANMO

Liang, Y., Ke, S., Zhang, J., Yi, X., Zheng, Y.: GeoMAN: multi-level attention networks for geo-sensory time series prediction. In: IJCAI, Stockholm (2018)

Long Short-Term Memory Networks [Online]. Available: https://www.mathworks.com/help/deeplearning/ug/long-short-term-memory-networks.html

Greff, K., Srivastava, R.K., Koutník, J., Steunebrink, B.R., Schmidhuber, J.: LSTM: a search space odyssey. IEEE Trans. Neural Networks Learn. Syst. **28**(10), 2222–2232 (2017)

Hochreiter, S., Schmidhuber, J.: Long short-term memory. Neural Comput. **9**(8), 1735–1780 (1997)

Mousa, W., Al-Shuhail, A.: Processing of Seismic Reflection Data Using MATLAB™. Morgan Claypool (2011)

Nicholson, C.: A Beginner's Guide to LSTMs and Recurrent Neural Networks. Skymind [Online]. Available: https://skymind.ai/wiki/lstm. Accessed 2 Oct 2019

Nuha, H., Balghonaim, A., Liu, B., Mohandes, M., Deriche, M., Fekri, F.: Deep neural networks with extreme learning machine for seismic data compression. Arab. J. Sci. Eng. 1–11 (2019)

Psychoula, I., Merdivan, E., Singh, D., Chen, L., Chen, F., Hanke, S., Kropf, J., Holzinger, A., Geist, M.: A deep learning approach for privacy preservation in assisted living. In: IEEE International Conference on Pervasive Computing, Athens (2018)

Tilk, O.: Long short term memory. Quora **30**, 4 (2016) [Online]. Available: https://www.quora.com/Deep-Learning-What-does-it-mean-for-an-RNN-to-have-512-hidden-units. Accessed 2 Oct 2019

Interpolation Problem on Outlier Contaminated Seismogram Using Extreme Learning Machine

Hilal Nuha, Bo Liu, Mohamed Mohandes, and Ali Al-Shaikhi

Abstract

In this work, we present a weighted l_1 norm-based Extreme Learning Machine (ELM), namely, enhanced Regularized ELM (eRELM) for regression problems with outliers contaminated data. The enhancement includes the introduction of a weighting scheme based on l_1 norm to a regularization term in the optimization problem. Seismogram is typically recorded with high frequency. Therefore, it requires a large amount of storage. Interpolation allows the seismogram to be stored with the same quality with a smaller number of samples. Our proposed method shows its robustness on interpolation of outlier contaminated seismogram outperforming other ELM based regression techniques by achieving 0.001245 of root mean squared error.

Keywords

Outlier • Seismogram interpolation • Extreme learning machine • l_1 norm

1 Introduction

It is reported that almost 2.5 quintillions of bytes are created every day (Jacobson 2013) making large amounts of data more accessible for a machine learning task. However, the machine learning performance is reduced when the quality of data is poor. In fact, some outliers may exist in the data either due to environmental conditions or to machine error during data acquisition processes. The main problem when dealing with outliers is that the training algorithm has a tendency to favor outliers and eventually, results in a significant performance decline (Zhang and Luo 2015). A new type of learning model for single-hidden layer feed-forward neural networks (SLFNs), namely Extreme Learning Machine (ELM), has drawn a lot of attention from the machine learning communities (Huang et al. 2006; Miche et al. 2010; Barreto and Barros 2016; Zhang and Luo 2015). The ELM has been extensively implemented in many real world scientific and engineering applications, like wind power modeling (Wan et al. 2014), face recognition (Mohammed et al. 2011), and seismic data compression (Nuha et al. 2019). Motivated by work in Zhang and Luo (2015) on using l_1 norm for outlier robust ELM regressor, in this work, we propose a learning algorithm, referred to as the enhanced Regularized ELM (eRELM), with two learning phases. To evaluate its robustness, our proposed algorithm is then experimentally tested on a real dataset for regression. Methods in regression are able to facilitate the estimation of non-linear function and are tested for interpolation and prediction.

2 Methodologies

We consider a regression problem with N training samples $\{(\boldsymbol{x}_i, y_i)\}_{i=1}^{N}$, with input vector $\boldsymbol{x}_i \in \mathbb{R}^n$ and the corresponding desired output $y_i \in \mathbb{R}$. Assuming that the model perfectly satisfies the relationship between $\boldsymbol{x}_i$ and y_i, the SLFN with L hidden units is modeled as the following summation:

$$\sum_{i=1}^{L} \beta_i \sigma(\boldsymbol{w}_i . \boldsymbol{x}_j + b_i) = y_j, \quad j = 1, 2, \ldots, N \tag{1}$$

where $h = \sigma(x) = \frac{1}{1+\exp(-x)}$ denotes the activation function and $\boldsymbol{w}_i = [w_{i_1}, w_{i_2}, \ldots, w_{i_n}]^{\mathrm{T}}$ is randomly initialized input weight vector that connects n input units and the i-th hidden unit. b_i is the randomly initialized bias of the i-th hidden

H. Nuha · B. Liu · M. Mohandes (✉) · A. Al-Shaikhi
King Fahd University of Petroleum and Minerals (KFUPM), Dhahran, 31261, Saudi Arabia
e-mail: mohandes@kfupm.edu.sa

H. Nuha
School of Computing, Telkom University, Bandung, Indonesia

H. El-Askary et al. (eds.), *Research Developments in Geotechnics, Geo-Informatics and Remote Sensing*, Advances in Science, Technology & Innovation,
https://doi.org/10.1007/978-3-030-72896-0_46

Table 1 Performance of the algorithms on the regression tasks using real dataset

Tasks	Outlier rate (%)	RMSE				
		ELM	RELM	WRELM	ORELM	eRELM
Seismogram interpolation	0%	0.001239	0.001247	0.001247	0.001241	0.001241
	10%	0.005663	0.004746	0.001259	0.001246	0.001242
	40%	0.026107	0.024776	0.003337	0.001251	0.001245

unit. β_i represents the weight that connects the i-th hidden unit and the output unit. The output unit uses linear activation function without any bias. The hidden layer output values matrix $\mathbf{H}$ is defined as

$$\mathbf{H}(\mathbf{w}_1,\ldots,\mathbf{w}_L,\boldsymbol{x}_1,\ldots\boldsymbol{x}_N,b_1,\ldots,b_L) = \begin{bmatrix} h_{1,1} & \cdots & h_{L,1} \\ \vdots & \cdots & \vdots \\ h_{1,N} & \cdots & h_{L,N} \end{bmatrix}_{N\times L}. \tag{2}$$

We present a two-phase scheme to enhance the robustness of the ELM to outliers. First, an initial output weights $\boldsymbol{\beta}$, representing the initial model, and the corresponding initial error $\boldsymbol{e}$ are obtained using Regularized ELM (RELM) (Huang et al. 2012; Deng et al. 2009; Bartlett 1998) as

$$\boldsymbol{e}_0 = \boldsymbol{y} - \mathbf{H}\boldsymbol{\beta}_0 \tag{3}$$

$$\boldsymbol{\beta}_0 = \left(\boldsymbol{H}^{\mathrm{T}}\boldsymbol{H} + \frac{\boldsymbol{I}}{D}\right)^{-1}\boldsymbol{H}^{\mathrm{T}}\boldsymbol{y} \tag{4}$$

During this phase, there is no mechanism to alleviate the impact of outliers, which negatively leads the initial model away from the true one. In the second phase, an optimization consisting of an l_1 norm of the weighted error along with a regularization term of the output weights $\boldsymbol{\beta}$ is applied. This optimization problem can be iteratively solved using the augmented *Lagrange* Multiplier (ALM) method (Hestenes 1969; Powell 1969). The iterative solution of $\boldsymbol{\beta}_{k+1}$ is then given by

$$\beta_{k+1} = \boldsymbol{P}_v\left(\boldsymbol{y} - \boldsymbol{e}_k + \frac{\lambda_k}{\mu}\right), \tag{5}$$

where $\boldsymbol{P}_v = \left(\boldsymbol{H}^{\mathrm{T}}\boldsymbol{v}^2\boldsymbol{H} + \frac{2}{D\mu}\boldsymbol{I}\right)^{-1}\boldsymbol{H}^{\mathrm{T}}\boldsymbol{v}^2$, which is independent of iteration k. This enables us to reduce the computational cost by pre-calculating $\boldsymbol{P}_v$ as proposed in Zhang and Luo (2015). The penalty parameter $\mu = \frac{2N}{|y|_1}$ is typically applied, as recommended in Zhang and Luo (2015), Zhang et al. (2011), Yang and Zhang (2011). Error $\boldsymbol{e}_{k+1}$ is updated as follows

$$\boldsymbol{e}_{k+1} = \max\left(\left|\boldsymbol{y} - \boldsymbol{e}_k + \frac{\lambda_k}{\mu}\right| - \frac{\lambda_k}{\mu}, 0\right) \circ \operatorname{sign}\left(\left|\boldsymbol{y} - \boldsymbol{e}_k + \frac{\lambda_k}{\mu}\right| - \frac{\lambda_k}{\mu}\right) \tag{6}$$

where the operator $\circ$ denotes the element-wise vector multiplication. Several weighting schemes have been proposed on robust estimate of $\mathbf{v}$, as introduced by Suykens et al. (2002). In this work, we use the following error weighting scheme:

$$v_i = \begin{cases} \Lambda, & \left|\frac{e_i}{\hat{s}}\right| \le c \\ 1, & \text{otherwise} \end{cases} \tag{7}$$

where $\Lambda > 1$ and $\hat{s} = \frac{p_{75} - p_{25}}{2\times 0.6745}$ denoting a robust estimate of the standard deviation of the initial error variables $\boldsymbol{e}_0$. p_γ is the γth percentile of the initial error and the constant c is set to 1. The algorithm is then terminated when the maximum number of iterations is reached.

3 Experimental Results

In this section, we compare the performance of the developed algorithm with four variants of ELMs, namely: the original ELM, RELM, Weighted RELM (WRELM), and outlier robust ELM (ORELM) for interpolation problem. In this respect, we adopt a seismogram from Albuquerque, New Mexico (Anmo), USA provided by the IRIS website (2018). The seismogram is recorded to observe passive seismic activities like earthquakes. The data was recorded with a 16 kHz sampling frequency by a station in Albuquerque in January 2015. In this experiment, we cut the waveform to only 10,000 time instances where the outputs are in the interval $[-0.31, 0.25]$.

The results of our experiments are summarized in Table 1. The best results are typed in bold. It can be noticed that our proposed method performances are very close to the others when the data is not contaminated with any outliers. However, all l_2 norm-based methods performances decline heavily when the data is contaminated with outliers. The l_1 norm-based methods show their robustness against outlier contaminated data. However, the best performance is achieved by the eRELM.

4 Conclusions

In this work, we presented a new learning algorithm namely enhanced Regularized ELM (eRELM) using l_1 based optimization for regression problems with outlier contaminated data. An l_1 based cost function is able to minimize the effect of outliers inside the training data. Our proposed algorithm was experimentally tested on seismogram interpolation problem and achieved 0.001245 of RMSE outperforming other methods.

Acknowledgments This work is supported by the Center for Energy and Geo Processing (CeGP) at King Fahd University of Petroleum and Minerals (KFUPM), Dhahran, Saudi Arabia, under Project GTEC1801.

References

Barreto, G.A., Barros, A.L.B.: A robust extreme learning machine for pattern classification with outliers. Neurocomputing, 3–13 (2016)

Bartlett, P.L.: The sample complexity of pattern classification with neural networks: the size of the weights is more important than the size of the network. IEEE Trans. Inf. Theory 525–536 (1998)

Deng, W., Zheng, Q., Chen, L.: Regularized extreme learning machine. In: IEEE Symposium on Computational Intelligence and Data Mining, CIDM'09 (2009)

Hestenes, M.: Multiplier and gradient methods. J. Optim. Theory Appl. **4**, 303–320 (1969)

Huang, G.-B., Zhu, Q.-Y., Siew, C.-K.: Extreme learning machine: theory and applications. Neurocomputing **70**, 489–501 (2006)

Huang, G., Zhou, H., Ding, X., Zhang, R.: Extreme learning machine for regression and multiclass classification. IEEE Trans. Syst. Man Cybern. Part B (Cybern.) **42**(2), 513–529 (2012)

IRIS Website, 1 June 2018 [Online]. Available: http://service.iris.edu/irisws/timeseries/1/query?net=IU&sta=ANMO

Jacobson, R.: 2.5 quintillion bytes of data created every day. How Does CPG & Retail Manage It? 24, 4 (2013) [Online]. Available: https://www.ibm.com/blogs/insights-on-business/consumer-products/2-5-quintillion-bytes-of-data-created-every-day-how-does-cpg-retail-manage-it/

Miche, Y., Sorjamaa, A., Bas, P., Simula, O., Jutten, C., Lendasse, A.: OP-ELM: optimally pruned extreme learning machine. IEEE Trans. Neural Networks 158–162 (2010)

Mohammed, A.A., Minhas, R., Wu, Q.M.J., Sid-Ahmed, M.A.: Human face recognition based on multidimensional PCA and extreme learning machine. Pattern Recogn. **44**, 2588–2597 (2011)

Nuha, H., Balghonaim, A., Liu, B., Mohandes, M., Deriche, M., Fekri, F.: Deep neural networks with extreme learning machine for seismic data compression. Arab. J. Sci. Eng. 1–11 (2019)

Powell, M.: A method for nonlinear constraints in minimization problems. In: Fletcher, R. (ed.) Optimization, pp. 283–298. Academic Press, New York, NY (1969)

Suykens, J.A., Brabanter, J.D., Lukas, L., Vandewalle, J.: Weighted least squares support vector machines. Neurocomputing 85–105 (2002)

Wan, C., Xu, Z., Pinson, P., Dong, Z.Y., Wong, K.P.: Probabilistic forecasting of wind power generation using extreme learning machine. IEEE Trans. Power Syst. **29**, 1033–1044 (2014)

Yang, J., Zhang, Y.: Alternating direction algorithms for l-1 problems in compressive sensing. SIAM J. Sci. Comput. 250–278 (2011)

Zhang, K., Luo, M.: Outlier-robust extreme learning machine for regression problems. Neurocomputing **151**, 1519–1527 (2015)

Zhang, L., Yang, M., Feng, X.: Sparse representation or collaborative representation: which helps face recognition? In: IEEE International Conference on Computer Vision (ICCV), Barcelona (2011)

Geo-information Technologies for Monitoring Natural-Technical Systems to Ensure the Environmental Safety of the Coal Mining Process

Julia Ozaryan, Vitaliy Usikov, and Marina Bubnova

Abstract

The increasing demand for coal products leads to the increase in the mining activities. Consequently, in coal deposits areas, the impact on the environment will increase. Modern geo-information methods for monitoring changes in natural-technical systems allowed the authors to analyze the ecological situation in the area of the Bureinsky coal basin. It is an information system of constellation-Vega providing options of operational distant work with the satellite observation data. As a result of the study, the area of dust exposure at the site of open-cast mining of coal deposits has been determined. The concentrations of SO_2, CO_2, and NO_2 in the surface layers of the atmosphere have been revealed, and the dust concentration of in the adjacent territory to the village has been established. According to the satellite imagery, the dust exposure spatial and temporal dynamics and vegetation condition were calculated. The analysis of the NDVI dynamics for the period from 2001 to 2018 led to the conclusion that the vegetation quality indicators in the damaged territory are dependent on the complex production activities.

Keywords

Hard coal and brown coal deposits • Coal mines • Disturbed lands • Dust pollution • Remote sensing • Surface miners • Coal transportation

J. Ozaryan (✉) · V. Usikov · M. Bubnova
Mining Institute of the Far Eastern Branch of Russian Academy of Sciences, Khabarovsk, 680000, Russia

1 Introduction

Coal resources potential of Russia are over 274 billion tons of reserves. About 11% of the country's reserves (29.7 billion tons of coal) are concentrated in the Far East (State Report 2018). The development of technological solutions which make it possible to reduce the impact on the atmospheric air during open-cast coal mining is a pressing issue (Bubnova and Ozaryan 2015; Shapiroa and Bolsen 2019; Han and Chen 2019). A lot of research has been devoted to the study of the impact of coal mining and processing on the environment and population. However, the issue of the conditions of the intermittent permafrost zone remains unresolved. The goal of the study was to analyze the current state of the environment in the mining area of the Burein coal basin and develop technological recommendations to ensure environmental safety. In order to achieve this goal, the following objectives have been defined: (1) the identification of the principal sources of airborne coal dust, area calculation of the contaminated territory; (2) influence on the state of vegetation by NDVI and the possibility of natural restoration.

2 Materials and Methods

The processing of the data was fulfilled by QGIS complete with GRASS GIS. Software products as the Microdem, 3Dem, Landserf were used to solve some particular problems. The results were entered into a database developed by PostgreSQL/PostGIS for calculating a pollution indicator. Calculation and analysis of the NDVI vegetation index (Normalized Difference Vegetation Index) were carried out using QGIS and Vega-Science, a scientific analysis tool for satellite observation data. The landfill area was from 2 to 5 ha. Data on NDVI and climatic characteristics were analyzed for the period from 2001 to 2015 during the 29 week of the year, which is the middle of the growing season for the

H. El-Askary et al. (eds.), *Research Developments in Geotechnics, Geo-Informatics and Remote Sensing*, Advances in Science, Technology & Innovation,
https://doi.org/10.1007/978-3-030-72896-0_47

study area, located in the intermittent permafrost zone. For Landsat 8, when adding the invisible channels 5—near infrared NIR and 6—SWIR1 to obtain a color image to channels 2 and 4, we get almost identical pictures. Part of the spectrum of channel 5 is especially important for identifying the presence of water in the leaves of a healthy vegetation; therefore, various interpretations were carried out with the addition of channel 5. It has been established that the best combinations are channels 5-6-2 and 5-6-4. The satellite imagery obtained by MODIS and ASTER (Terra satellite), TM and ETM + (Landsat 5, 7, 8) systems with various spatial resolutions, digital elevation models: GTOPO30, SRTM30, SRTM3, topographic maps at a scale of 1:10,000–1:100,000 were the main sources of information in our study.

3 Results

Urgal coal deposit is the largest and most studied deposit of the Bureins coal basin. An area of more than 140–150 km^2 experiences anthropogenic impact during the exploitation of deposits underground and open-pit mining. The field of Urgal coal mine, which is located between rivers Urgal and Chegdomyn, is exploited to a depth of 500 m. On the right bank of river Chegdomyn, there are two villages—Lower and Upper Chegdomyn, and on the left bank of the river Urgal, there is *v*. Average Urgal, the recreation. Within the existing areas of open work (Bureinsky, Urgalsky open-cast mine), overburden rocks are stored in dumps, which are located at the edge of the goaf. The completed quarries are gradually flooded. Down the slope in the streams beds, there are sedimentations of quarry waters. The pollution of the surface layers of the atmosphere is: SO_2 = 2705; CO_2 = 162.924; NO_2 = 106,802 (t/year). The dustiness of the surface atmosphere in the village Chegdomyn ranges from 0.36 to 2.78 mg/m, with an average of 1.4 mg/m or 0.7 MPC (daily average values). Dust is a mixture of ash particles, soot (emissions from boiler rooms), and technogenic dust. It is generated by the blowing of small fractions from dumps and coal depots and during the drilling and blasting process. The satellite image processing and analysis data made it possible to identify the forms of technogenic relief and calculate its area: rock dumps occupy 1.35 km^2, coal cuts cover 7.01 km^2, and the underground mining area is 5.78 km^2. The total area of disturbed land is 14.14 km^2, which is 27% of the area of the mining site (51.4 km^2) (Fig. 1). A plume of dust streams spreads from the Urgalsky coal deposit and the loading station northeast of the quarry. The area of the dust plume distribution is identified by the gray contour. It is established that the transfer radius of coal dust ranges from 2 to 4 km in different directions.

We established remote monitoring sites in the places of greatest impact from the south-east and north-west sides of the quarry for the observation of vegetation quality using the vegetation index NDVI (Fig. 2). The studied sites are located on coal dumps which are subject to natural revegetation. Plant communities are represented by grass and bush communities. The index indicators of the background territory suggest the existence of dense vegetation, and the NDVI of the site located on coal dumps corresponds to sparse vegetation.

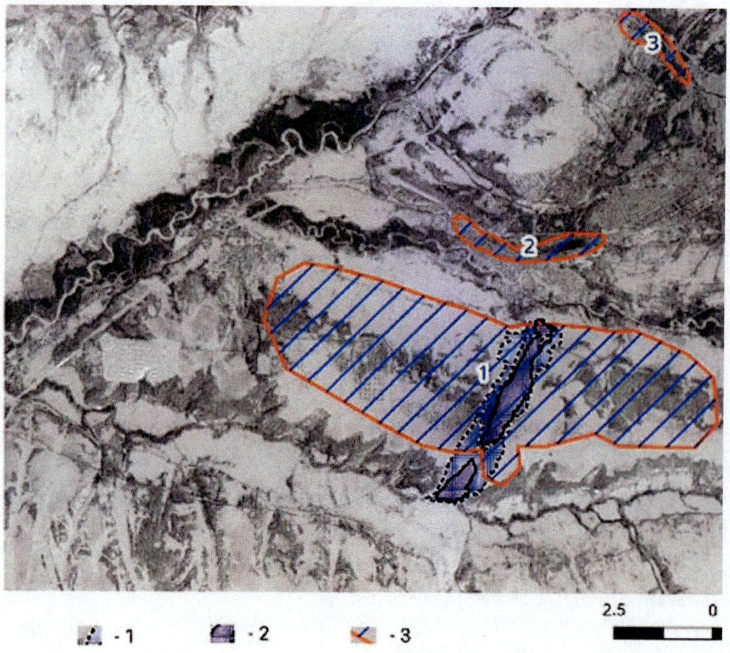

Fig. 1 Spread of coal dust, indicated by the pollution of snow cover of the Urgalsky coal basin 1—contour of a coal mine; 2—contour of waste dumps; 3—contour of dust distribution areas: 1—from the section; 2—from the train loading station; 3—from the mine

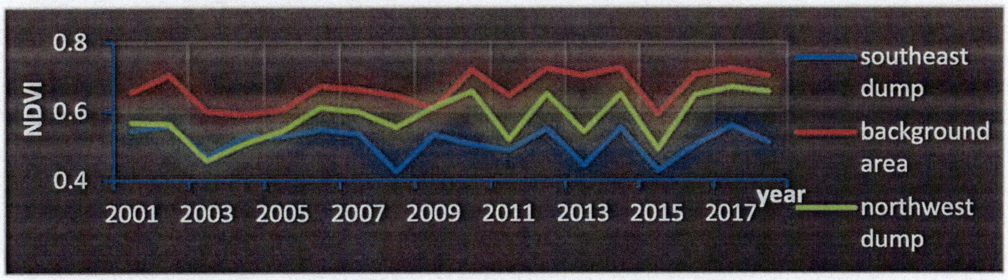

Fig. 2 NDVI dynamics in the zone of coal dust impact (2001–2018)

4 Discussion

The research shows that with an increase in the demand for coal products and an increase in the load on natural systems, there is the possibility of natural recovery, but the potential of the natural system is limited. While developing some environmental measures, it is necessary to determine rational action for fixing the dusty surface of dumps by reclamation or applying a protective layer. Dedusting the technological processes should be carried out regularly. The fact that Chegdomyn is located outside the limits of dust exposure was unexpected for the authors, but the level of dust pollution needs to be controlled in the future in the different seasons of the year. Previous studies (Borzenko and Shevtsov 2011) indicate a violation of the earth's surface and the formation of wastewater during coal mining at this site. Researchers have not explored the issue of natural potential of the territory revegetation yet. The authors are sure that without human intervention, disturbed territories will be restored after hundreds years, but the use of data on natural processes will reduce time.

5 Conclusions

As a conclusion, the spatial and temporal dynamics of dust exposure were identified. According to sensing satellite data, it was revealed that the village of Chegdomin is located outside the zone of local dust exposure. According to the satellite imagery, periods of anthropogenic impact and commissioning of sections of the deposit are determined. The main sources of emissions of pollutants and the radius of their distribution have been found. Vegetation quality indicators established using NDVI indicate that the overall condition of the plant communities was assessed as satisfactory. The imagery analysis for the summer period indicates that the secondary phytocenosis form on the waste sites, which means that the territory can be restored using the self-healing potential of the ecosystem. To reduce the impact on vegetation and restore the disturbed areas, it is necessary to make arrangements for dust elimination during drilling and blasting, excavation, loading, crushing, and screening operations. Moreover, it is necessary to reclaim coal dumps, the surface of which is eroding and experiencing pollution from the territory of the mining allotment. The work was performed in accordance with R&D AAAA-A18-118020590026-4.

References

Borzenko, E.A., Shevtsov, M.N.: Assessment of the impact of the objects of Urgalugol llc on the change of the environmental environment and the development of proposals for their reduction. New Ideas of the New Century, vol. 2.2 (2011)

Bubnova, M.B., Ozaryan, Yu.A.: Environmental monitoring of natural-mining systems based on remote sensing data. Ecol. Syst. Devices **11**, 15–22 (2015)

Han, S., Chen, H.: Interactions between organisational roles and environmental hazards: the case of safety in the Chinese coal industry. Resour. Policy **60**, 36–46 (2019)

Shapiroa, M.A., Bolsen, T.: Korean perceptions of transboundary air pollution and domestic coal development: two framing experiments. Energy Policy **126**, 333–342 (2019)

State report "On the state and environmental protection of the Russian Federation in 2017." M.- The Ministry of Natural Resources of Russia; NPP "Cadastre", 888 p (2018)

Managing Data Characterization Challenges—Unifying Geoinformatics Framework and Knowledge Management

Shastri L. Nimmagadda, Andrew Ochan, Neel Mani, and Torsten Reiners

Abstract

Data heterogeneity and multidimensionality are added challenges of upstream business research that can preclude the data integration process in diverse domains. The research purpose was to build data constructs and models that can articulate a unified approach and address the challenges of the data integration processes. Multidimensional data warehouse repository was proposed with an ontology-based data warehousing and mining approach, describing how the data constructs are brought together in multiple domains of the petroleum industry to integrate their domain knowledge in large size exploratory warehouse depositories. We envisage metadata with reservoir networks in favourable petroleum geology areas. The synergistic informatics approach can extract useful geological knowledge of prospective locales in large size sedimentary basins.

Keywords

Data modelling • Geosciences • Data integration • Knowledge management • Middle Eastern Basins

1 Introduction

Data are characterized differently in diverse application scenarios and domains. Many functional entities and dimensions exist in the oil and gas data sources, in particular in exploration, drilling, production, technical, logistics and marketing entities (Nimmagadda and Dreher 2012; Nimmagadda et al. 2017). Primarily, the data described as operational, archived, and external are either structured and/or unstructured. In dimensional modelling, these entities are considered as dimensions (Connolly and Begg 2015; Nimmagadda et al. 2018). Our objective was to interpret and envisage semantic-based geological and geophysical (G&G) attribute dimensions, test and validate them in dimensional models and methods. Connecting geological data events associated with various onshore and offshore exploration entities is the most challenging part of exploration and field development. Understanding of G&G data relationships and their characteristics needs a great deal of attention and knowledge of basin geology, which is not achievable by traditional methods of data management techniques (Nimmagadda et al. 2018). Data characterization guided novel data integration processes are needed that can facilitate making up geoinformatics solutions.

2 Data Characterization and Types

We presented a couple of critical data characteristic scenarios, *heterogeneity* and *multidimensionality*. *Heterogeneity* refers to the data that comes from many sources, mostly unknown and unlimited in many adaptable formats. Though the data related to upstream petroleum systems are meant to have been inherently interconnected, they are often misconstrued. The data relationships in the context of petroleum ecosystems' research comprehend with a concept, digital petroleum ecosystem as termed in Nimmagadda and Dreher (2012). The relationship is extended to explore the connectivity between elements and processes of diverse petroleum systems of various basins. We addressed the heterogeneity through integrated methodologies and data modelling approaches, interpreting transitional attribute dimensions as conceptualized and contextualized chain of events. They

S. L. Nimmagadda (✉) · T. Reiners
School of Management, CBS, Curtin University, Perth, WA, Australia
e-mail: shastri.nimmagadda@curtin.edu.au

A. Ochan
National Petroleum Authority, Entebbe, Uganda

N. Mani
Department of IT, Amity University, Noida, India

H. El-Askary et al. (eds.), *Research Developments in Geotechnics, Geo-Informatics and Remote Sensing*, Advances in Science, Technology & Innovation,
https://doi.org/10.1007/978-3-030-72896-0_48

constantly occur and interact through digital media in spatial dimensions (Han et al. 2001). *Multidimensionality*, however, characterizes datasets that have one or more dimensions. They are either derived from external data sources and/or from geographically and periodically varying dimensions with unknown boundary conditions, attributable in unknown areal extents of petroleum systems and data formats. The multidimensionality managed with cognizance is gently derived from heterogeneity. To analyze the phenomena, we deduced data models from data sources in super-basin contexts, computing fine-grained data views and representing them into meta-knowledge.

For conceptual data models, we needed to describe the data characteristics and business rules more precisely that represent complex geological environments. In this context, we acknowledge the support of an ontology-based data warehousing and mining approach, bringing multidimensional data into a single repository, transforming the data into coherent and standard formats and ensuring it is acceptable in any software and hardware platforms. The motivation of research came from geoscientific data characterization and description of entities, dimensions and objects (Connolly and Begg 2015). An entity is a discernible existence, which can also be abstract. An object is visible, and another is tangible but endures with encapsulation and inheritance properties. Objects influence onset of instructions or methods about what to do with data records. In the case of a dimension, it is a measure of identifiable existence or even an emerged conceptualized entity or dimension. The notion of dimension provides semantic information, especially amongst hierarchies and relationships between elements and processes of petroleum systems. An element is either an individual existence or a composite entity. Entities, objects and dimensions, used in the ontology modelling process are reusable in multiple domains and applications. For example, a reservoir object (Nimmagadda et al. 2017) that exists in a petroleum system domain can be used or reused in multiple applications of petroleum production and reservoir engineering projects. Similar is the case with reservoir dimension, which quantifies or qualifies within a generic petroleum system domain is reusable in other domain applications. In the context of digital petroleum ecosystems, entities, objects and dimensions and their attributes play significant roles in connecting multiple domains, where they are often associative (Connolly and Begg 2015). They are based on the types of data, adhered from kinds of factual instances, linked with elements and processes of petroleum systems, oil and gas fields and sedimentary basins in different tectonics. Each hierarchy can exhibit different data characterization challenges for which different data modelling techniques are needed.

3 Data Modelling Methodologies

The conceptual, logical and physical levels of data modelling approaches are proposed in the upstream business research. The conceptual model investigates the highest level of data relationships, either amongst entities, objects or dimensions. More emphasis is on dimensions, for organizing and modelling multiple attributes from heterogeneous datasets. In logical data modelling, the data are described with in-depth knowledge of associations, without any concern on physical organization, but at times conceptualizing and contextualizing the data attributes, based on types of oil and gas fields and their linked basins. Multidimensional modelling approach with schema architectures, used in petroleum industries are discussed at length in Nimmagadda et al. (2018). Data modelling workflow, as described in Fig. 1a, is designed to manage the variety of data characteristics and constraints that met in the upstream businesses. It starts from the selection of data and arrival of data modelling approach that best fits the basin types. Different schemes, as depicted in Fig. 1b (a–f), are deployed in petroleum data management that match a variety of upstream business scenarios and their data characteristics.

4 Results

In addition to data modelling, to exploit the untapped reservoirs in Middle Eastern basins, various data mining, visualization and interpretation artefacts were used. The data constructs and models associated with the integrated framework or workflow can interconnect geological structures and reservoirs in diverse petroleum systems of a basin or groups of basins. It included implementation models and linked integrated workflows that characterize the entire life cycle of exploration and field development project. As described in Nimmagadda and Dreher (2012), Nimmagadda et al. (2018), the ontologies developed for petroleum data sources were used and reused by multiple users for storing, processing and delivering quality information, in addition to integrating data science of geology in multiple basins. We interrogated the framework with metadata, use and reuse of data structures, including interpretation of data views in multiple application domains. A map view is presented in an implementation framework, as shown in Fig. 2, for visualizing and interpreting the reservoir connections, qualities and their distributions. As illustrated in Fig. 2, potential reservoir prone areas were interpreted, where favourable exploration and production extents were construed with prospective elements and processes of petroleum systems.

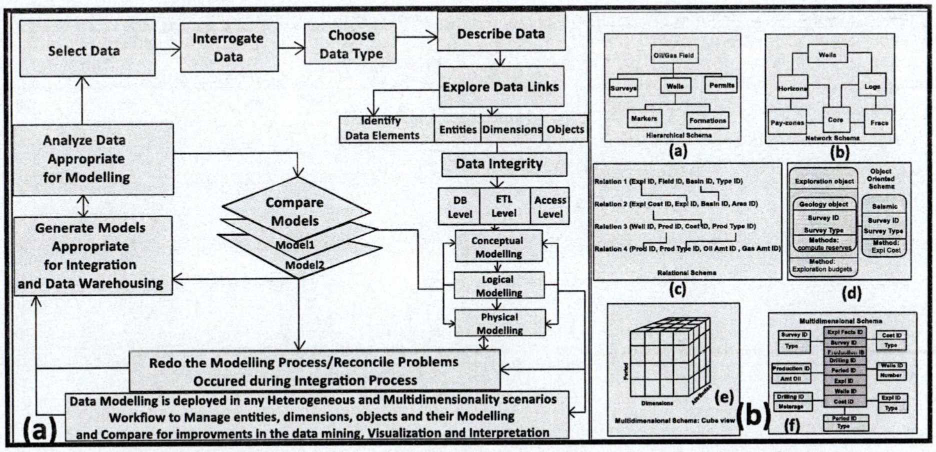

Fig. 1 Data modelling methodologies **a** data modelling workflow **b** various schemas

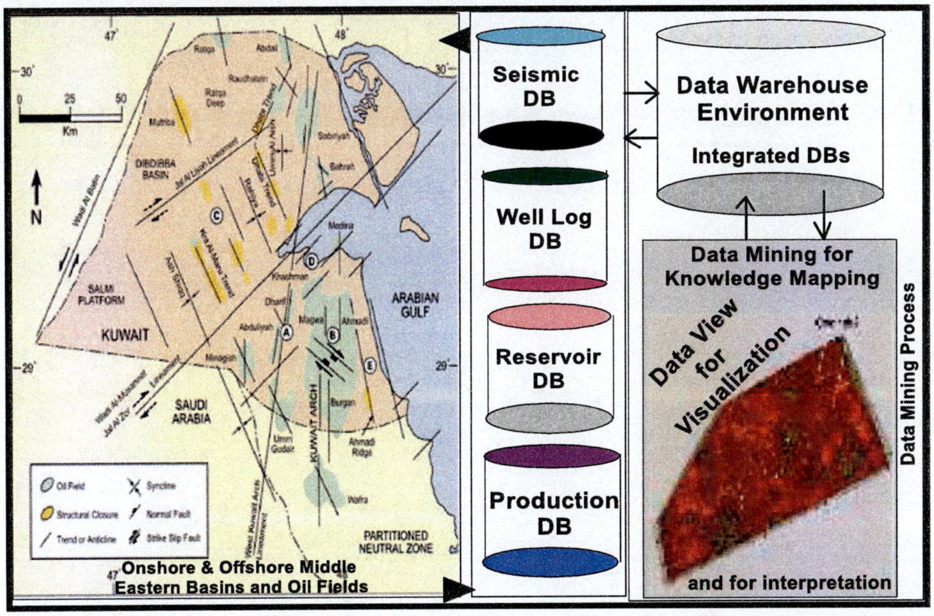

Fig. 2 An implementation framework, exploring reservoir connections from metadata

5 Conclusions and Recommendations

The scientific contribution describes challenges related to data science complexities that needed to resolve for data management and integration processes. Data modelling methods and their linked integrated frameworks have successfully been implemented to address the issues of data connectivity and integration in the upstream petroleum business contexts. Unified metadata is made available to explorers to detail exploration and field development plans with enhanced production activities in the investigating areas. The metadata can further facilitate contextual data processing including qualitative and quantitative interpretation of drillable reservoirs. New exploration areas are identified based on the interpretation of several data views drawn from knowledge-based, structured and unified metadata.

References

Connolly, T., Begg, C.: Database Systems, a Practical Approach to Design, Implementation and Management, 4th edn. Pearson, Sydney, Australia (2015)

Han, J., Kmber, M., Tung, A.K.H.: Spatial clustering methods in data mining: a survey. In: Miller, H., Han, J. (eds.) Geographic Data Mining and Knowledge Discovery, pp. 33–50. Taylor and Francis, London, New York (2001)

Nimmagadda, S.L., Dreher, H.: On new emerging concepts of petroleum digital ecosystem. J. Wiley Interdisc. Rev. Data Min. Knowl. Discov. **2**(6), 457–475 (2012). https://doi.org/10.1002/widm.1070

Nimmagadda, S.L., Reiners, T., Rudra, A.: An upstream business data science in a big data perspective. Procedia Comput. Sci. **112**, 1881–1890 (2017). https://doi.org/10.1016/j.procs.2017.08.236

Nimmagadda, S.L., Reiners, T., Wood, L.C.: On big data guided upstream business research and its knowledge management. J. Bus. Res. **89**, 143–158 (2018)

Spatiotemporal Changes in Date Palm Oases of Algeria Over the Last Century

Ali Mihi, Nacer Tarai, Abdelkrim Benaradj, and Haroun Chenchouni

Abstract

Date palm (*Phoenix dactylifera* L.) is one of the most important income-generating crops cultivated in hot desert regions worldwide. This study aims at monitoring the change in numbers of date palm trees (NDPT) over the past 100 years (1913–2013) in desert oases of the Ziban region (NE Algeria). Natural breaks classification (Jenks method) was used to classify the 17 municipalities studied in the region into four classes: low, medium, high, and very high based on the NDPT for 1913 and 2013. NDPT had doubled six times between 1913 and 2013, rising to ~3 million of date palm trees. An irregular spatial distribution in NDPT was noted in the study area, where the municipalities of Sidi-Okba and Tolga recorded very high NDPT during the last ten decades. The present paper reports the potential application of natural breaks classification method for data discretization with good adaptability and high accuracy.

Keywords

Date palm grove • Ziban region • Natural breaks classification • Data discretization • Spatial analysis • Oasis • Sahara Desert

A. Mihi (✉) · H. Chenchouni
Department of Natural and Life Sciences, FSESNV, Larbi Tebessi University, 12002 Tebessa, Algeria
e-mail: ali.mihi@univ-tebessa.dz

N. Tarai
Department of Agricultural Sciences, FSNV, University of Mohamed Khidder, 07000 Biskra, Algeria

A. Benaradj
Department of Natural and Life Sciences, University Center of Naama, Naama, Algeria

H. Chenchouni
Laboratory of Natural Resources and Management of Sensitive Environments 'RNAMS', University of Oum-El-Bouaghi, 04000 Oum El Bouaghi, Algeria

1 Introduction

Distributed between 10° N and 39° N of latitudes, the date palm (*Phoenix dactylifera* L.) is one of the most permanent domesticated fruit trees in the world. Date palm cultivation occurs mainly across hot desert regions of the Middle East and North Africa (Al-Khayri et al. 2015). Due to its social and economic importance and environmental-related benefit, date palm is the most important permanent crop in Algeria, with approximately 18 million date palms cultivated on an area stretching over 169,380 ha. In fact, estimates of Algerian date exports were about 27,660 million USD, and Algeria ranked fourth worldwide in dates production in 2012 (Bouguedoura et al. 2015). Recently, the Ziban region has been a representative and an excellent example of date palm cultivation in Algeria (Mihi et al. 2019).

To enhance the planning process and the management of the natural resources and habitats, it is necessary to understand the spatial and temporal changes of date palm land use. Visual representations of statistical data on a map represent useful tools for better understanding of spatial data characteristics of a given region (Mihi et al. 2020). Several methods can be used for spatial data classification including quantile, equal intervals, standard deviation, and natural breaks (Jenks method) (Ayalew et al. 2004). The latter is a classification method developed in 1967s and based on Jenks optimization algorithm. In recent years, natural breaks classification has been the most prevalent method for spatial data classification (Jenks 1967). The pillar of this method is to minimize the differences within the same class and maximize those among the different classes (Baz et al. 2009). The present work aims to monitor spatial and temporal changes in numbers of date palm trees (NDPT) cultivated at the region of Ziban using geographic information system (GIS) tools and Jenks classification. Furthermore, this study highlights the potential of natural breaks classification in the discretization with high accuracy of geostatistical data.

H. El-Askary et al. (eds.), *Research Developments in Geotechnics, Geo-Informatics and Remote Sensing*, Advances in Science, Technology & Innovation,
https://doi.org/10.1007/978-3-030-72896-0_49

2 Materials and Methods

The study area (Ziban region) is located at northeastern of Algeria and bordered from the South by the Aurès Mountains. In Algeria, Ziban region at the Wilaya 'Province' of Biskra appears as a transition space between the northern landscape systems (high plains and Atlas mountain ranges) and the Sahara Desert in the South. It covers an area around 21,671 km^2, which represents about 0.91% of national territory. The climate is hot and arid (De Martonne aridity index = 4). Agriculture is the main source of income for the local rural population, where date palm is the dominant crop in this region (Mihi 2021). The study area covers 17 municipalities (Fig. 1), where the date palm groves are confined at two subregions, viz. Zab Gharbi and Zab Chergui.

The sources involved in data collection process include some statistics from the Directorate of Agricultural Services of Biskra for the years 1913 and 2013 (Direction des Services Agricoles (DSA) 2013), field surveys and questionnaires filled by local farmers, some actors of date palm sector, and exporters of date palm fruits. Two different approaches were applied, comparative and analytic, based on the NDPT as a single classification criterion. To classify the 17 municipalities studied based on NDPT of each municipality, the Jenks optimization method was applied to generate two maps representing the NDPT distribution for the two years mentioned above. Accordingly, the municipalities were ranked and grouped into four classes: low, medium, high, and very high. All maps were created following the Universal Transverse Mercator (UTM) projection zone 31 North with the datum World Geodetic analysis (WGS) 1984. MapInfo software (version 9.0) was used in the cartography of all geostatistical data.

3 Results

Figure 2 shows the development of NDPT during the past 100 years, i.e., from 1913 to 2013. In total, NDPT increased from about 0.5 million trees (586,336) in 1913, to reach more than three million trees (3,233,706) in 2013, which corresponds to an increase of 82% in NDPT. In terms of cultivated areas, date palm surface area expanded about 69% between 1913 (84.61 km^2) and 2013 with 269.48 km^2 (according to Bouguedoura et al. (2015), 1 ha = 120 date palm trees).

The spatiotemporal distributions of NDPT were unevenly distributed in the study area. Date palm trees were more concentrated in two municipalities, viz. Sidi-Okba and Tolga, where NDPT recorded very high levels during the last one hundred years (Fig. 2). In 1913, the municipalities of

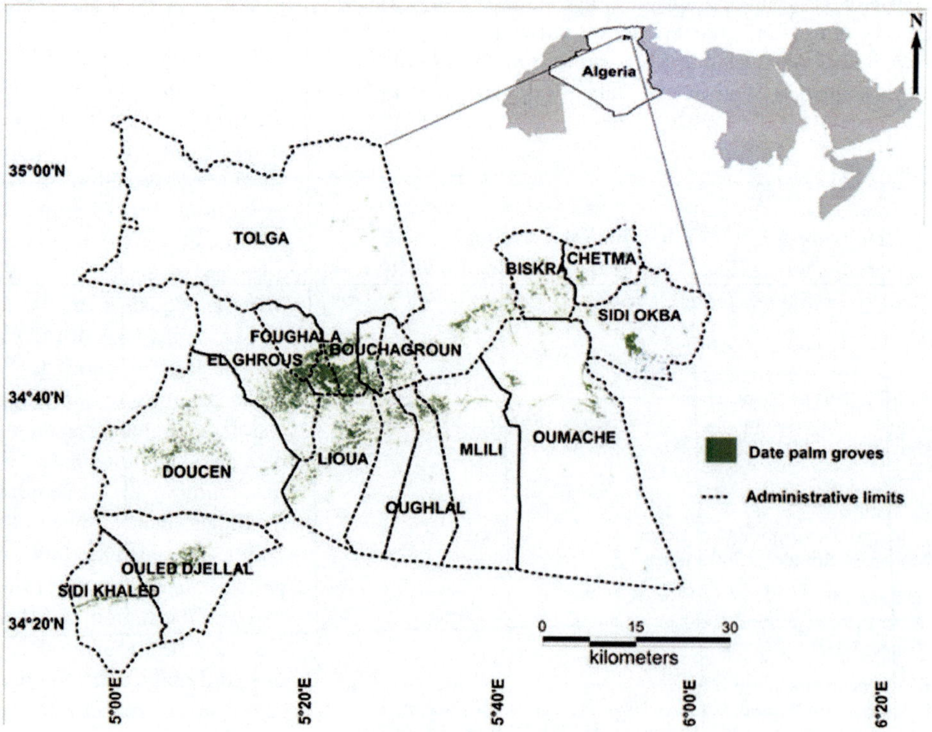

Fig. 1 Location of the study area in the region of Ziban, northeastern Algeria

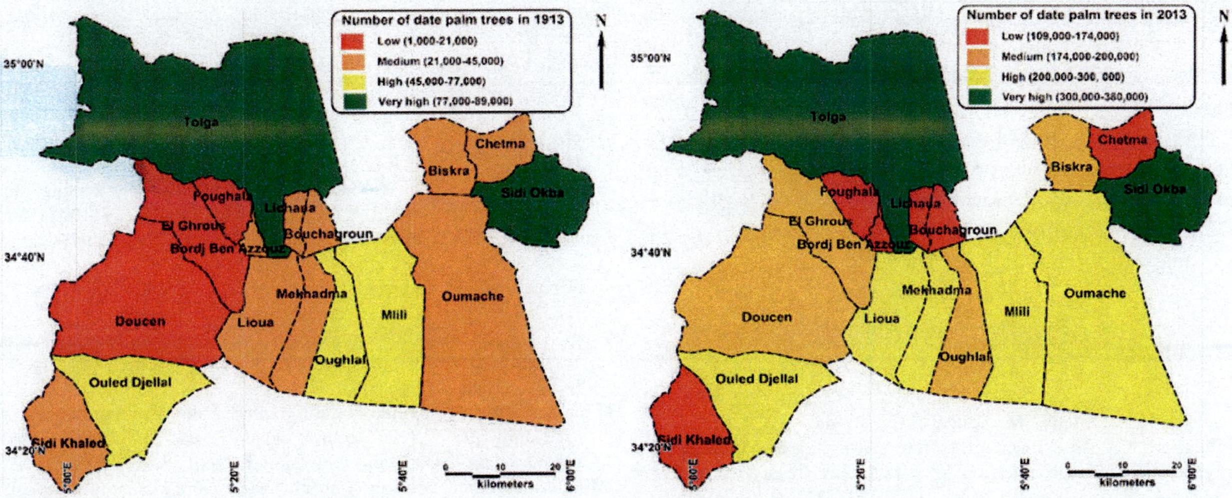

Fig. 2 Spatiotemporal changes of NDPT at 17 municipalities of Biskra (Ziban region, Sahara Desert of Algeria) for the years 1913 (left) and 2013 (right)

Doucen, Fhoughala, and El Ghrous showed low level of NDPT. These were followed by almost the rest of municipalities in the study area including Bordj Ben Azzouz, Bouchagroun, Biskra, Chetma, Lichana, Lioua, Mekhadma, Oumache, and Sidi-Khaled, which were classified within the medium class. The remaining three municipalities, namely Oughlal, Ouled Djellal, and Mlili, had a high level as shown in Fig. 2. In 2013, the spatial–temporal distribution of NDPT using natural breaks classification revealed that class of low level was represented by Bordj Ben Azzouz, Bouchagroun, Chetma, Foughala, Lichana, and Sidi-Khaled. The medium class level included four municipalities: Biskra, Doucen, El Ghrous, and Oughlal. While the high level of NDPT was distributed over the following municipalities: Ouled Djellal, Lioua, Mekhadma, Oumache, and Mlili.

4 Discussion

The current situation regarding the distribution and abundance of date palm trees depends on several variables that define the suitability of Saharan agricultural lands to the cultivation and establishment of date palms. However, other environmental factors and situations such as unfavorable soil conditions (mainly salinity and sodicity), long and intense droughts, and the choice of unsuitable varieties are among the limiting factors of date palm cultivation (Afrasinei et al. 2017; Lejars et al. 2017; Lamqadem et al. 2019; Mihi et al. 2018, 2019). In fact, since the implementation of different agricultural development programs such as the Accession to the Agricultural Land Ownership Program (APFA) in 1983 and National Plan of Agricultural and Rural Development (PNDA) launched in 2000, desert oases of Algeria have experienced high expansion of agro-ecosystem lands and farming areas. This extension in cultivated areas and NDPT was favored by the use of new technologies and sustainable agricultural practices with efficient water irrigation management. In Ziban region, drip irrigation system is currently the most water-saving method used in date palm plantation. Water is pumped from the Continental Intercalaire aquifer, which represents an enormous reserve with 91,900 km^3 in Ziban region (Lejars et al. 2017). In addition to the groundwater, two dams (Foum El Gherza in Seriana and Fountain of Gazelle in El Outaya) were intended to irrigate about 650,000 date palm trees in the Ziban region (Jaradat 2016). According to Bouguedoura et al. (2015), Deglet-Noor variety dominated in the study region with 49% of date palm cultivars cultivated. In reality, this variety has become the most commercial desert crop for its high socioeconomic benefits for local populations (Jaradat 2016). Besides, date palm oasis plays a vital role as farmland shelterbelt against desertification in hot desert regions. Another role it has among its ecosystem services is sustaining the balance between farmland/desert landscapes and the groundwater shortages (Mihi et al. 2018). Moreover, date palm production is an income-generating activity that offers add-value products, foreign exchange earnings, and ensures food self-sufficiency for national economy (Mihi 2018).

5 Conclusion

The spatiotemporal distribution of NDPT during the last ten decades from 1913 to 2013 showed an increase of 82%. Sidi-Okba and Tolga have very high levels of NDPT compared to the remaining15 municipalities in the region of

Ziban. The aim is to enhance the planning process and the management of the natural resources and habitats. To this end, decision makers, scholar groups, and farmers should make use of the thematic maps obtained from our research in order to assist in developing successful strategies and sustainable agro-economic policy in oasis ecosystems. Furthermore, this study demonstrated the potential application of Jenks optimization algorithm for data discretization with good adaptability and high accuracy.

References

Afrasinei, G.M., Melis, M., Buttau, C., Bradd, J.M., Arras, C., Ghiglieri, G.: Assessment of remote sensing-based classification methods for change detection of salt-affected areas (Biskra area, Algeria). J. Appl. Remote Sens. **11**(1), 1931–3195 (2017). https://doi.org/10.1117/1.JRS.11.016025

Al-Khayri, J.M., Jain, S.M., Johnson, D.V.: Date Palm Genetic Resources and Utilization. Africa and the Americas. Springer, The Netherlands (2015). https://doi.org/10.1007/978-94-017-9694-1

Ayalew, L., Yamagishi, H., Ugawa, N.: Landslide susceptibility mapping using GIS-based weighted linear combination, the case in Tsugawa area of Agano River, Niigata Prefecture, Japan. Landslides **1**(1), 73–81 (2004). https://doi.org/10.1007/s10346-003-0006-9

Baz, I., Geymen, A., Nogay, S.: Development and application of GIS-based analysis/synthesis modeling techniques for urban planning of Istanbul Metropolitan Area. Adv. Eng. Softw. **40**(2), 128–140 (2009). https://doi.org/10.1016/j.advengsoft.2008.03.016

Bouguedoura, N., Bennaceur, M., Babahani, S., Benziouche, S.E.: Date palm status and perspective in Algeria. In: Al-Khayri, J.M., Jain, S. M., Johnson, D.V. (eds.) Date Palm Genetic Resources and Utilization, pp. 125–168. Springer, The Netherlands (2015). https://doi.org/10.1007/978-94-017-9694-1_4

Direction des Services Agricoles (DSA): Annuaire des statistiques agricoles de la wilaya de Biskra. Biskra, Algeria (2013)

Jaradat, A.A.: Genetic erosion of *Phoenix dactylifera* L.: perceptible, probable, or possible. In: Ahuja, M.R., Jain, S.M. (eds.) Genetic Diversity and Erosion in Plants. Springer International Publishing, pp. 131–213. Springer, The Netherlands (2016). https://doi.org/10.1007/978-3-319-25954-3_4

Jenks, G.F.: The data model concept in statistical mapping. In: International Yearbook of Cartography, 7th edn., pp. 186–190 (1967)

Lamqadem, A.A., Afrasinei, G.M., Saber, H.: Analysis of Landsat-derived multitemporal vegetation cover to understand drivers of oasis agroecosystems change. J. Appl. Remote Sens. **13** (1), 014517 (2019). https://doi.org/10.1117/1.JRS.13.014517

Lejars, C., Daoudi, A., Amichi, H.: The key role of supply chain actors in groundwater irrigation development in North Africa. Hydrogeol. J. **25**(6), 1593–1606 (2017)

Mihi, A.: Etude écologique et cartographique de l'écosystème oasien par l'outil SIG et Télédétection: cas de l'oasis de Biskra, Algérie sud. Doctoral dissertation, University of Setif, Algeria (2018)

Mihi, A.: Dynamic simulation of future date palm plantation (*Phoenix dactylifera* L.) growth using CA–Markov model and FAO-LCCS data in Algerian dryland oases desert. Model. Earth Syst. Environ. (2021). https://doi.org/10.1007/s40808-021-01289-zC

Mihi, A., Nacer, T., Chenchouni, H.: Monitoring dynamics of date palm plantations from 1984 to 2013 using landsat time-series in Sahara Desert Oases of Algeria. In: El-Askary, H.M., et al. (eds.) Advances in Remote Sensing and Geo Informatics Applications, pp. 225–228. Springer, The Netherlands (2018). https://doi.org/10.1007/978-3-030-01440-7_52

Mihi, A., Tarai, N., Chenchouni, H.: Can palm date plantations and oasification be used as a proxy to fight sustainably against desertification and sand encroachment in hot drylands? Ecol. Ind. **105**, 365–375 (2019). https://doi.org/10.1016/j.ecolind.2017.11.027

Mihi, A., Benarfa, N., Arar, A.: Assessing and mapping water erosion-prone areas in northeastern Algeria using analytic hierarchy process USLE/RUSLE equation GIS and remote sensing. Appl. Geomat. **12**(2), 179–191 (2020) https://doi.org/10.1007/s12518-019-00289-0

Close-Up Imaging Telemetry Simulation of Low-Altitude ENA Emission Sources

Li Lu, Qing-long Yu, and Qi Lu

Abstract

The low-altitude ENA emission is located at magnetic latitude 50°–60°, 1500–3000 km altitude. Low orbit Sun-synchronous satellites are thus close to the ENA emission sources. Utilizing a 2-D neutral atom coding modulation imager onboard of Sun-synchronous orbit satellite was simulated, which can realize continuous monitoring of low-altitude ENA emission sources. Simulation studies show that (1) The low orbit Sun-synchronous satellite can approach ENA emission source both inside and outside of the radiation belt and telemetry ENA images with the temporal resolution about 5 s, which helps to analyze precipitation/injection processes of energetic ions of the ring current, as well as the causal sequence with the environment field changes; (2) The propagation distance of energetic ions can be calculated by using time difference of ENA energy spectrum response of different energy channels, identify if "the energetic ions come from the solar wind"; (3) Due to proximity to ENA emission source, the ENA imager will respond to medium- and small-magnitude signals of geomagnetic activity, which will help realize continuous collection and statistical analysis of massive data of geomagnetic activity without missing, and provide important data support for new space-based space weather forecasts.

Keywords

Magnetosphere • ENA imaging • Radiation belts • Space weather forecast

1 Introduction

Energetic ions are the main carriers of energy transport in geomagnetic activity, and most of the t energetic ions distribute in the ring current area. Therefore, real-time monitoring of the ion flux distribution in the ring current area is an effective way to study geomagnetic activity. The transport process of ring current ions during geomagnetic activity is controversial and inconclusive due to lack of effective measurement data.

Previous space explorations have been affected by the uncertainty of spatial or temporal variations of the data detected. The 2-D telemetry function of the neutral atom imager has solved the above problems. However, considering the time-varying scale of geomagnetic activity, the temporal and spatial resolution under the certain statistical requirements becomes the main obstacle to the function of this instrument.In other words, the ENA imaging probes can only obtain the global plasma distribution morphology currently (Brandt et al. 2002a, b; De Majistre et al. 2004; Perez et al. 2004). The following work attempts to apply this imaging model to the spatio-temporal evolution processes of plasma (Lu et al. 2016, 2019).

2 Low-Altitude ENA Emission Sources

According to TC-2/NUADU's ENA measurements and inversions during the geomagnetic activity on November 7, 2004 (Lu et al. 2008), 3-D ENA emission sources for different magnetic shells are shown in Fig. 1. The results show that ENA emission sources are located in low-altitude and high-latitude polar regions.

L. Lu (✉) · Q. Yu · Q. Lu
Laboratory of Space Environment Exploration, National Space Science Center of the Chinese Academy of Sciences, Beijing, 100190, China
e-mail: luli@nssc.ac.cn

L. Lu · Q. Yu · Q. Lu
Beijing Key Laboratory of Space Environment Exploration, Beijing, 100190, China

L. Lu · Q. Yu · Q. Lu
Key Laboratory of Science and Technology on Space Environment Situational Awareness, CAS, Beijing, 100190, China

H. El-Askary et al. (eds.), *Research Developments in Geotechnics, Geo-Informatics and Remote Sensing*, Advances in Science, Technology & Innovation,
https://doi.org/10.1007/978-3-030-72896-0_50

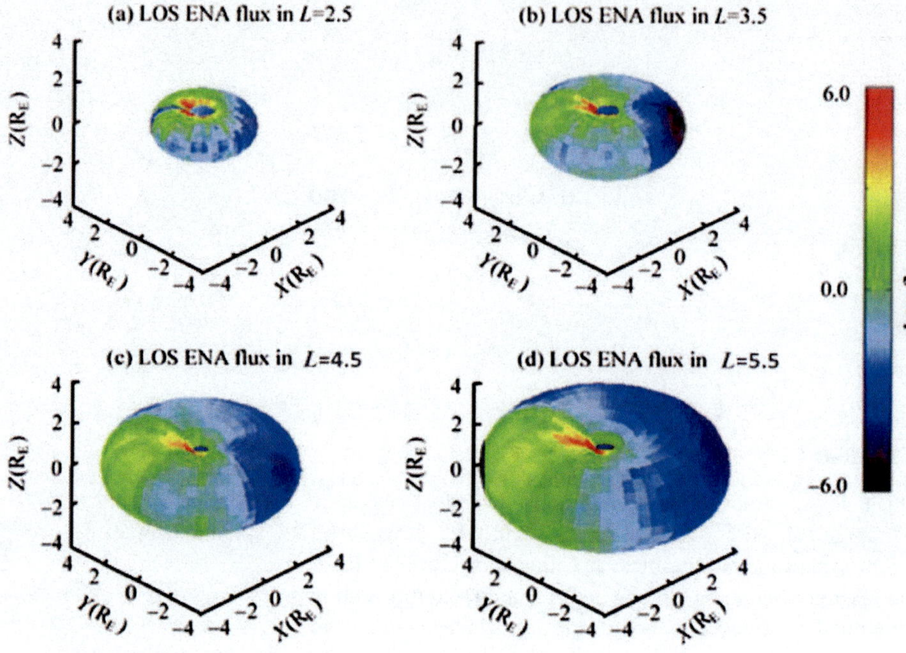

Fig. 1 3-D ENA emission source on magnetic shells with $L = 2.5$, 3.5, 4.5, 5.5, and on the right is the color scale of ENA flux (cited from Li et al. 2008)

3 Imaging Simulation of Low-Altitude ENA Emissions

3.1 Simulated Orbit Design

Sun-synchronous elliptical orbit has 1500 km apogee, 500 km perigee, and 98° inclination. A 2-D neutral atom imager is placed on the Sun-synchronous satellite, as shown in Fig. 2. Set instrument space angular resolution is 6° × 6°, the FOV of the instrument is 4π solid Angle. The simulation feature positions are given in Table 1.

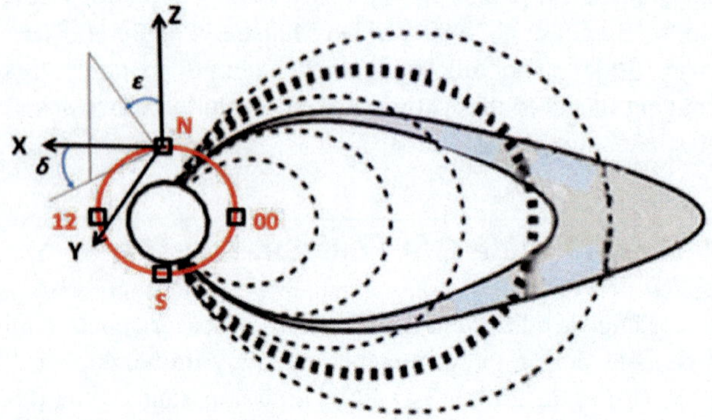

Fig. 2 Schematic diagram of Sun-synchronous orbit simulation test. The red curve in the figure is the Sun-synchronous orbit, and the black boxes are marked as imaging simulation feature positions

3.2 Simulation Equation

The counts recorded in each pixel of an ENA image, $C(\delta, \varepsilon)$ with elevation δ and azimuth ε, is represented in the simulation equation

$$C(\delta,\varepsilon) = \int \Delta E \Delta T \Delta \Omega j_{\text{ion}}(L,\varphi,\theta,E,\alpha) A(\delta,\varepsilon)\sigma(E) n(r,\varphi,\theta)\mathrm{d}V \quad (1)$$

where ΔE is the energy range; ΔT is the integral time for the pixel; $\Delta\Omega$ is the solid angle of the volume element pointing to the δ, ε pixel; j_{ion} is ion differential flux at the integral volume element; $A(\delta, \varepsilon)$ is the response function of a detector (Li et al. 2014); $\sigma(E)$ is the charge exchange cross sections (Smith and Bewtra 1987);. $n(r, \varphi, \theta)$ is the exospheric neutral atomic density, where r is geocentric distance, φ is longitude, and θ is latitude; $\mathrm{d}V$ is the volume element integral along the line of sight of the detector.

3.3 Simulation Results

By simulation, the 5 s integral 2-D ENA image at four different positions on the orbit (Table 1) was obtained, as shown in Fig. 3.

The polar angle of 90° (center of the circle in each plane in Fig. 3) is oriented toward the sky. Local time and north and south poles are marked with red characters on each plane. The simulation model adopts pixel settings of

Table 1 Parameters of four simulation positions

Position#	Latitude	Longitude	*X* (km)	*Y* (km)	*Z* (km)
1	0°	0°	7354.22217	0	500.00061
2	90°	0°	0	0	7871.19922
3	0°	180°	−7354.22217	0	500.00015
4	−90°	0°	0.00111	0	−6871.1997

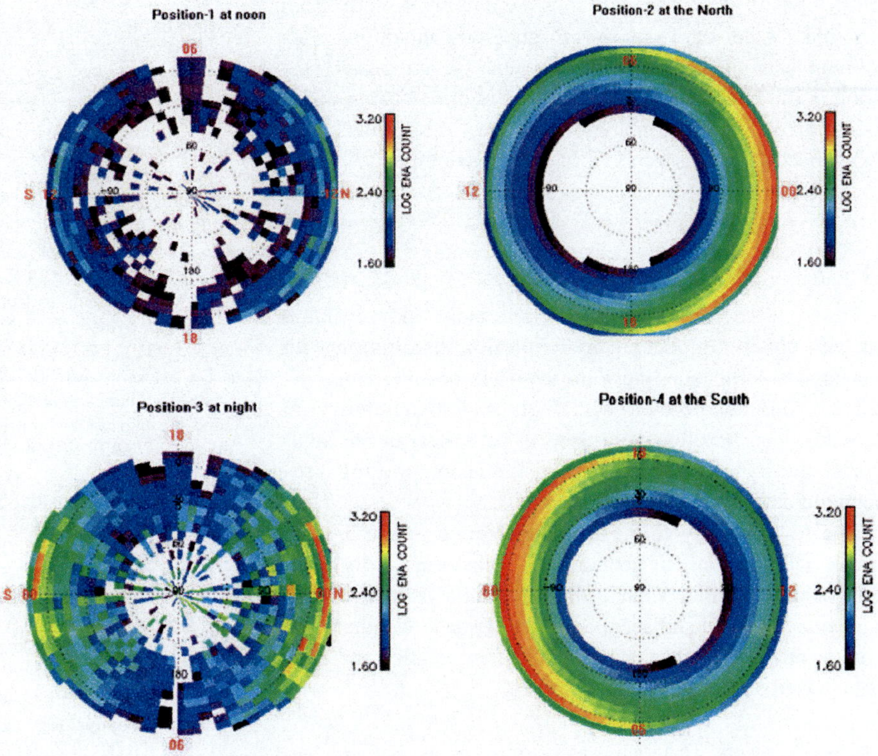

Fig. 3 ENA image for 5 s integral, 10–30 keV

6° × 6°. In order to compare the intensity of ENA counts at different locations, the same logarithmic color scale was used for the simulation of the each plane in Fig. 3.

The simulation results of position 1 and 3 (top left and bottom left of Fig. 3) show that the satellite can realize the north/south conjugate observation near the equator on the inside side of the radiation belt, and the ENA flux signals generated by the radiation belt in the north and south hemispheres are labeled with red letters S and N, respectively. Because the *Z* coordinate of the characteristic position 1 and 3 is 500 km north (see Table 1), the ENA counts at northern (red N) are larger than the southern (red S) in Fig. 3. The range of ENA counts is $-12° < \theta < 42°$ polar angle; $0° < \varphi < 360°$ Azimuth. The polar angle from 42° to 90° faces to the inner radiation belt.

According to the distribution of ENA counts at position 2 and 4 (top right and bottom right of Fig. 3), the location of ENA emission source is calculated as altitude, 1500–2800 km; latitude, 54°–60°.

4 Discussion

The neutral atom imaging includes the distribution information of ring current ions (Roelof 1987). Therefore, neutral atom imaging measurement is an effective means to monitor geomagnetic activity, especially ring current.

The ENA images detected by the neutral atom imager are always related to geomagnetic activity, and the 2-D coding modulation design is used to obtain high time resolution energy spectrum data. Different energy ions transport at different speeds. For the same transport distance, the shorter the transport time, the higher the energy. The emission source location of energetic ions can be judged according to the time response differences of energy spectrum with different energy channels: the smaller the time difference of energy spectrum response is, the shorter the transport distance of causing charge exchange ions is. It can be used to identify whether the energetic ions are coming directly from the solar wind or from the magnetotail injection. This problem was not solved before because the temporal resolution of the ENA images detected was poor. By using of 2D coding modulation technology make the time resolution of ENA spectral data as good as in situ measurements.

5 Conclusions

The low-altitude ENA source measurement and inversion not only obtain a global energetic ion flux distribution with measured significance, more important is that the ring current ion flux distribution with high spatial and temporal resolution can be effectively used in causal sequence analysis of the transport process of energetic ions during geomagnetic activity (Lu et al. 2016).

Due to proximity to the ENA source, the neutral atom imager will respond to geomagnetic substorms with moderate or lower order, which will be conducive to the mass data collection and statistical analysis of space weather, and provide important data support for the creation of new space-based space weather forecasts.

Acknowledgements This study was supported by the National Natural Science Foundation of China (Grant Nos. 41574152) and the Strategic Priority Program (SPP) on Space Science Advanced Research of Space Science Missions and Payloads (No. XDA15017100).

References

Brandt, P.C., De Majistre, R., Roelof, E.C., Ohtani, S., Mitchell, D.G.: IMAGE/high-energy energetic neutral atom: global energetic neutral atom imaging of the plasma sheet and ring current during substorms. J. Gephys. Res. **107**(A12), 1454 (2002a). https://doi.org/10.1029/2002JA009307

Brandt, P.C., Roelof, E.C., Ohtani, S., Mitchell, D.G., Anderson, B.: IMAGE/HENA: pressure and current distributions during the 1 October 2002 storm. Adv. Space Res. **33**, 719–722 (2002b)

De Majistre, R., Roelof, E.C., Brandt, P.C., Mitchell, D.G.: Retrieval of global magnetospheric ion distributions from high-energy neutral atom measurements made by the IMAGE/HENA instrument. J. Geophys. Res. **109**, A04214 (2004). https://doi.org/10.1029/2003JA010322

Li, L., McKenna-Lawlor, S., Balaz, J., Jiankui, S., Chuibai, Y., Jing, L.: Technical configuration and simulation of the NAIS-H for the MIT mission. Chin. J. Space Sci. **34**(3), 341–351 (2014) (in Chinese). https://doi.org/10.11728/cjss2014.03.341

Lu, L., Mckenna-Lawlor, S., Barabash, S., Balaz, J., Xing, L.Z., Chao, S., Bin, C.J., Ling, T.C.: Iterative inversion of global magnetospheric information from energy neutral atom (ENA) images recorded by the TC-2/NUADU instrument. Sci. China Technol. Sci. **51**(10), 1731–1744, 2008/10 (2008)

Lu, L., McKenna-Lawlor, S., Cao, J.B., Kudela, K., Balaz, J.: The causal sequence investigation of the ring current ion-flux increasing and the magnetotail ion injection during a major storm. Sci. China Earth Sci. **59**, 129–144 (2016). https://doi.org/10.1007/s11430-015-5121-7

Lu, L., McKenna-Lawlor, S., Balaz, J.: Close up observation and inversion of low-altitude ENA emissions during a substorm event. Sci. China Earth Sci. **62**, 1024–1032 (2019). https://doi.org/10.1007/s11430-018-9307-x

Perez, J.D., Zhang, X.X., Brandt, P.C., Mitchell, D.G., Jahn, J.- M., Pollock, C.J.: Dynamics of ring current ions as obtained from IMAGE HENA and MENA ENA images. J. Geophys. Res. **109**, A05208 (2004). https://doi.org/10.1029/2003JA010164

Roelof, E.C.: Energetic neutral atom image of a storm-time ring current. Geophys. Res. Lett. **14**(6), 652–655 (1987). https://doi.org/10.1029/GL014i006p00652

Smith, P.H., Bewtra, N.K.: Charge exchange lifetimes for ring current ions. Space Sci. Rev. **22**, 301–318 (1987)

ENA Coding Aperture Imager for STEM Mission at L5

Li Lu, Qing-Long Yu, Dan Wang, and Qi Lu

Abstract

The ENA coding aperture imager is designed for Solar Terrestrial Environment Monitor mission (STEM) at L5 which collect the energy neutral atoms generated by solar coronal mass ejection (CME) in 5–300 keV energy range. We can extend the energy spectrum of a single particle event to 10 meV. The original detection data of each ENA event generated by interplanetary CME, including the energy and position information of CME ions at the time of charge exchange carried by ENA, $P_i(t, x, y, E)$, were recorded with a two-dimensional detector. The ENA energy spectrum image, $P(x, y, E)$, was obtained by integrating the modulation data. The image reconstruction algorithm is applied to reconstruct the target image, that is, the ENA telemetry image generated by CME driven shock particle acceleration. The total field of view is 20° × 20°, and the center of the field of view is 10.27° away from the sun along the heliospheric line. The spatial resolution is 1° × 1°.

Keywords

Energetic neutral atom (ENA) • Coronal mass ejection (CME) • Solar–terrestrial relationship • Heliosphere

1 Introduction

The ENA coding aperture imager is designed for a Chinese mission: Solar Terrestrial Environment Monitor mission (STEM) at L5 which collects the energy neutral atoms generated by solar coronal mass ejection (CME) in 5–300 keV energy range. This coding aperture design with a large geometric factor for very low ENA flux environments allows the instrument to accumulate ENA's azimuth and energy spectrum information that entry detectors over a long period of time for reconstruction of ENA images. The instrument aims at the corona from L5 point, avoiding strong ultraviolet radiation from the sun's surface and collecting information on the spatial distribution of ion acceleration carrying by ENAs which is generated by EMC. It provides a new measurement for the study of mass ejection process of solar storm. In Sect. 2, we introduce the exploration environment; in Sect. 3, the payload scheme was expressed; and key indicators are shown in Sect. 4. The last, Sect. 5, is the conclusion.

2 Exploring Environment

At L5 (~1AU), ENAs (5–5000 keV) are produced by the acceleration of shock particles driven by a coronal mass ejection (CME). Assess the ENA flux generated by the CME-driven shock particle acceleration at 1AU, Fig. 1 (Wang et al. 2014).

The time scale of a coronal mass ejection (CME) is on the order of an hour. We intend to select the low-energy end with large ENA flux, energy (5–300 keV). At L5, if the

L. Lu (✉) · Q.-L. Yu · Q. Lu
Laboratory of Space Environment Exploration, National Space Science Center of the Chinese Academy of Sciences, Beijing, 100190, China
e-mail: luli@nssc.ac.cn

L. Lu · Q.-L. Yu · Q. Lu
Beijing Key Laboratory of Space Environment Exploration, Beijing, 100190, China

L. Lu · Q.-L. Yu · Q. Lu
Key Laboratory of Science and Technology On Space Environment Situational Awareness, CAS, Beijing, 100190, China

D. Wang
Institute of Spacecraft System Engineering, China Academy of Space Technology, Beijing, China

H. El-Askary et al. (eds.), *Research Developments in Geotechnics, Geo-Informatics and Remote Sensing*, Advances in Science, Technology & Innovation,
https://doi.org/10.1007/978-3-030-72896-0_51

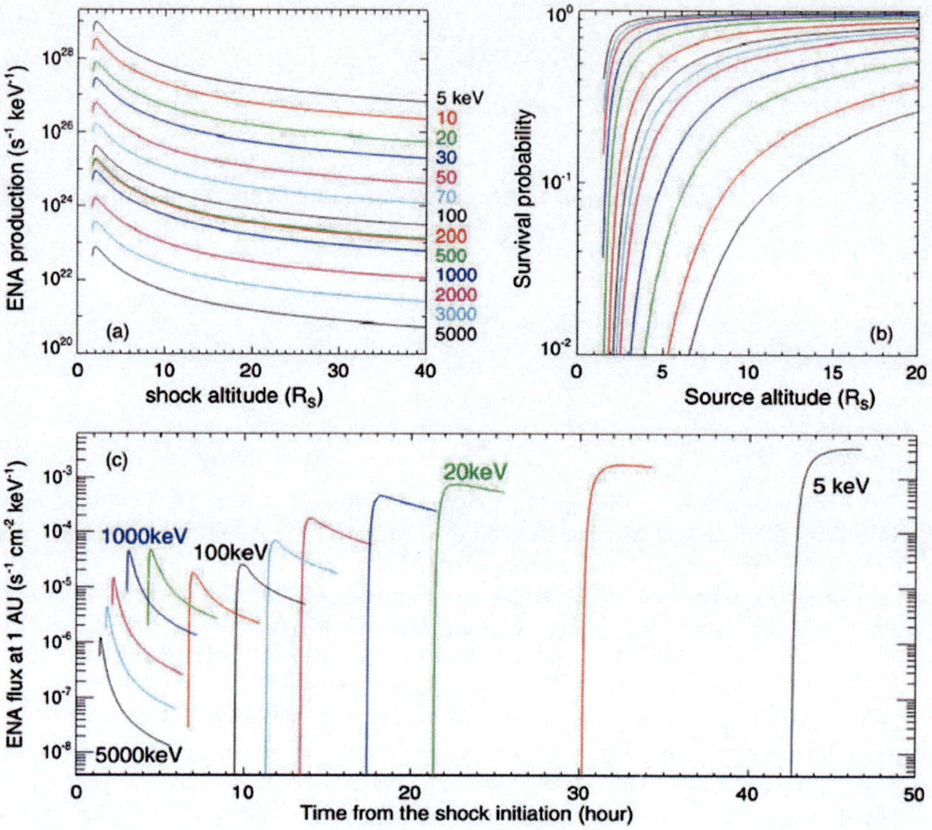

Fig. 1 **a** Total hydrogen ENA production rate vs. shock altitude r_{sh}, from 5 keV (topmost black curve) to 5 meV (bottommost black curve). **b** Hydrogen ENA survival probability from the source to 1 AU, from 5 keV (bottommost black curve) to 5 meV (topmost black curve). **c** Simulated hydrogen ENA flux vs. time at 1 AU from 5 keV (rightmost black curve) to 5 meV (leftmost black curve). The right end of each curve corresponds to r_{sh} = 40 RS. (A color version of this figure is available in the online journal.)

imaging area of the detector is selected as >400 cm^2, more than 1000 ENA counts can be accumulated within half an hour, so as to meet the statistical requirements of image inversion.

3 Payload Scheme

In 2002, *Integral* launched with four gamma ray coded modulation telescopes (Goldwurm et al. 1999; Bassani et al. 2005). The ENA coding aperture imager is shown in Fig. 2 left, and it works as shown in Fig. 2 right. The source is recorded by a 2D detector array through an aperture with an encoding modulation image. After then, the image is decoding and reconstructing. The process is shown from left to right in Fig. 2 right.

3.1 Spatial Resolution

The spatial resolution illustrates in Fig. 3, where tan (β) = d/h, Resolution ratio:

$$\beta = \arctan(d/h) \tag{1}$$

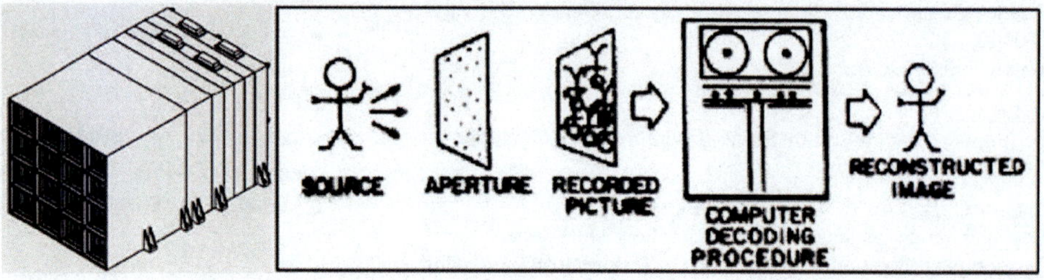

Fig. 2 Schematic diagram of ENA coding aperture imager (left) and the coding modulation process (right)

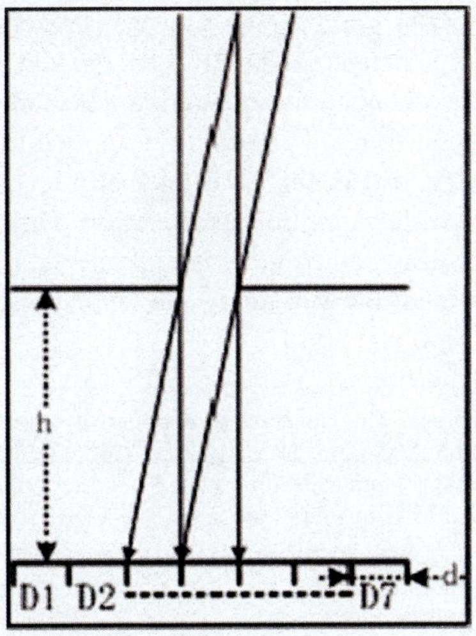

Fig. 3 Detector structure diagram

3.2 Coding Modulation Point Source Imaging and Image Inversion

Due to statistical requirements, at least 1000 ENA particle events should be accumulated for the inversion. Three examples are shown in Fig. 4.

3.3 Neutral Atom Coding Aperture Synthesis Imaging and Image Inversion

Synthetic imaging of ENA event is records versus time, position, and energy:

$$Pi(t, x, y, E) = Oi \times A \tag{2}$$

Far-field source, the probe bit difference can be ignored:

$$P = \sum Pi \tag{3}$$

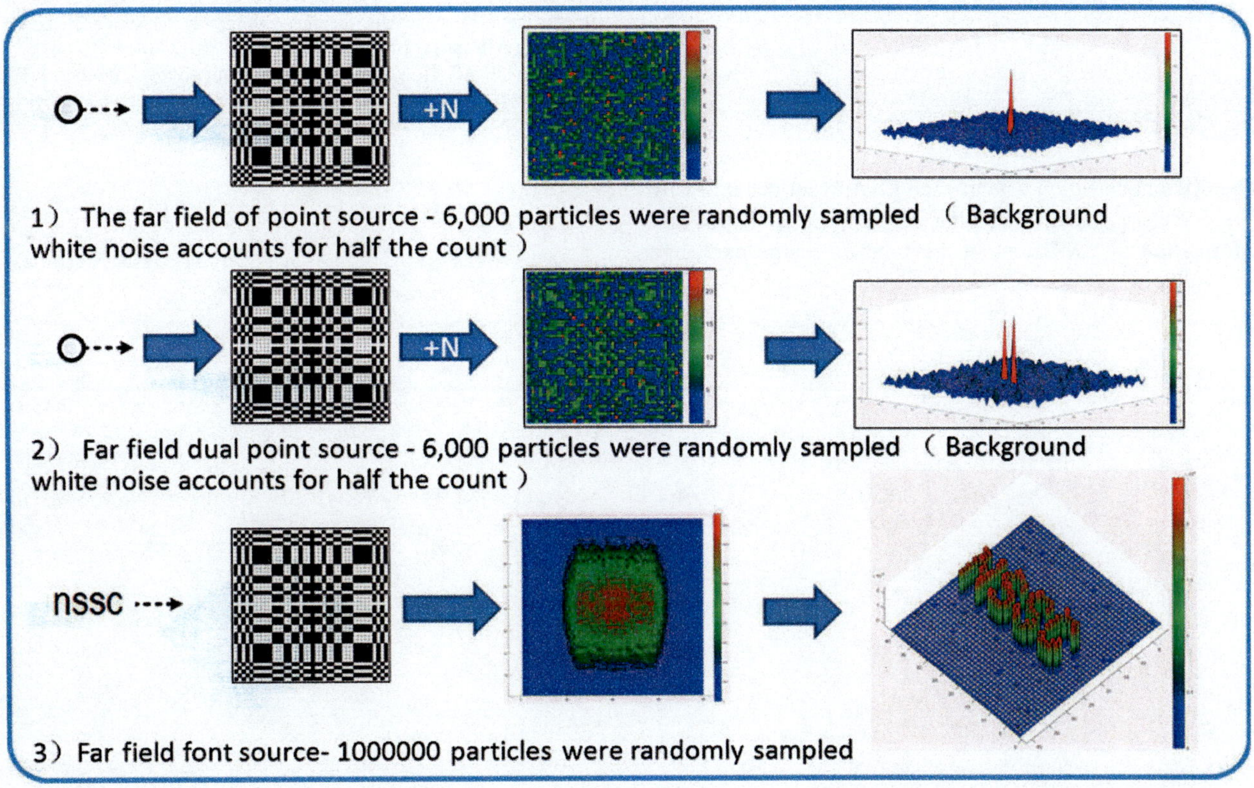

Fig. 4 Schematic diagrams of three observation cases

The point transfer function of coded aperture synthetic imaging is the same as that of single probe, and the same inversion operator is used for image decoding. The temporal resolution is improved by increasing the imaging area.

4 Key Indicators

1. **Technical index**
   (a) Imaging area: >400 cm^2
   (b) Temporal resolution: 1 ms for a single ENA particle event, 30 min for an inversion map.
   (c) Field of view: 20° × 20°;
   (d) Spatial resolution: 1°;
   (e) Energy range: 5–300 keV;
   (f) Energy resolution: <20% @ 100 keV;
2. **Engineering indicators**
   (a) Size: 410 × 400 × 300 mm^3;
   (b) Weight: 20 kg;
   (c) Power dissipation: 15 W;

5 Conclusions

The original detection data of each ENA event generated by interplanetary CME, including the energy and position information of CME ions at the time of charge exchange carried by ENA, $P_i(t, x, y, E)$, were recorded with a two-dimensional detector. The ENA energy spectrum image, $P(x, y, E)$, was obtained by integrating the modulation data. The image reconstruction algorithm is applied to reconstruct the target image, that is, the ENA telemetry image generated by CME-driven shock particle acceleration. The total field of view is 20° × 20°, and the center of the field of view is 10.27° away from the sun along the hemispheric line. The spatial resolution is 1° × 1°.

Acknowledgements This study was supported by the National Natural Science Foundation of China (Grant No. 41574152) and the Strategic Priority Program (SPP) on Space Science Advanced Research of Space Science Missions and Payloads (No. XDA15017100).

References

Bassani, L., De Rosa, A., Bazzano, A., et al.: Is the INTEGRAL/IBIS source IGR J17204–3554 a gamma-ray emitting galaxy hidden behind the molecular cloud NGC 6334? Astrophys. J. Lett. **634**(1), L21–L24 (2005)

Goldwurm, A., Goldoni, P., Laurent, P., et al.: Imaging simulations of the galactic nucleus with the ibis gamma-ray telescope on board integral. Astrophys. Lett. Commun. **38**(1), 1015–1016 (1999)

Wang, L., Li, G., Shih, A.Y., Lin, R.P., Wimmer-Schweingruber, R.F.: Simulation of energetic neutral atoms from solar energetic particles. Astrophys. J. Lett. **793**(2), L37 (2014)

OLS Modeling Using GIS for Aviation Safety Within a Challenging Topographic and Urban Neighborhood: The Case of Beirut Rafic Hariri International Airport

Amal Iaaly, Nael Alhassanieh, Michelle Nassar, Oussama Jadayel, and Angele Aouad

Abstract

Beirut Rafic Hariri International Airport, the only international airport in Lebanon, is situated within the heart of a high-density metropolitan area surrounded by hills in both east and south-east directions. Over the years, chaotic and unauthorized expansion of built-up structures surrounding the airport created a hazard to aviation safety. Aviation safety around an airport is maintained through well-defined virtual Obstacle Limitation Surfaces (OLS) based on Annex 14 of International Civil Aviation Organization. In this paper, a study of OLS for Beirut Rafic Hariri International Airport was presented. Geographical information systems (GIS) technologies were used to visualize these surfaces in two and three dimensions and assess any deviations through them. Results showed a large number of obstacles infringing the OLS surrounding the airport. Their locations and elevations were determined. Corresponding recommendations for mitigating those violations were presented.

Keywords

GIS • ICAO • OLS • Hazard • Aviation safety

1 Introduction

Beirut Rafic Hariri International Airport (BRHIA) is located on the Mediterranean coast, 9 km south of Beirut, the capital of Lebanon. The airport is situated in a highly populated area surrounded by buildings with the Ouzai region located between two of the airport runways and surrounded by the Aramoun and Naemeh hills. BRHIA has three intersecting runways. Based on the Lebanese Aeronautical Information Publication, this airport type is identified as precision approach, Cat I, Code 4E.

Aviation safety around an airport is maintained through well-defined virtual Obstacle Limitation Surfaces (OLS). The OLS is a series of virtual complex 3D surfaces that determines the volume of airspace around an aerodrome that shall be free from obstacles either natural or man-made in order to ensure the safety of aerial operations. Any object projecting through these OLS is considered as an obstacle and consequently becomes a hazard to aviation. The standards upon which the OLSs are established as specified by the International Civil Aviation Organization (ICAO) in its Annex 14 of the Convention on International Civil Aviation (ICAO 2004, 2008, 2016). The physical extent and shape of the OLS depend on many factors such as aerodrome dimensions, its intended use, available navigation facilities, and runway configurations. The OLS comprises the inner horizontal, conical, approach, transitional, takeoff, and climb surfaces.

The OLS visualization is by no means an easy task and becomes even harder in multirunway aerodromes particularly when the surfaces emanating from different runways intersect. The problem becomes more complicated when the OLSs are to be used to identify protruding obstacles or terrains (either human-made or natural). This is why it becomes absolutely imperative to harness the powers of geographical information systems (GIS) to map these surfaces and assess any protrusions through them (Chang 2013; Jadayel 2007). This paper illustrated how the GIS was used to model OLSs for Beirut Rafic Hariri International Airport (BRHIA).

2 The BRHIA Obstacle Limitation Surfaces Development

In order to develop the OLS, a seven-step methodology was adopted.

A. Iaaly · N. Alhassanieh (✉) · M. Nassar · O. Jadayel · A. Aouad
University of Balamand, El Koura, Lebanon
e-mail: amal.iaaly@balamand.edu.lb

H. El-Askary et al. (eds.), *Research Developments in Geotechnics, Geo-Informatics and Remote Sensing*, Advances in Science, Technology & Innovation,
https://doi.org/10.1007/978-3-030-72896-0_52

Step 1—BRHIA GIS Base Model: In order to develop an accurate OLS, the instigation of an accurate GIS model for the airport must be achieved. This was accomplished through the complete digitization of BRHIA from the most recent satellite imagery. Satellite imagery from a high-resolution commercial satellites (GeoEye-41 cm accuracy) covering a total area of 299 km^2 was acquired for that purpose (Fig. 1a). The geometric correction was carried out through the use of twenty-five ground control points (GCP) to achieve high spatial accuracy. The coordinate system and the projection used for the GIS work were WGS84 UTM Zone 36 N as per ICAO requirements. The ortho-rectified satellite imagery of the BRHIA airport was then used to digitize all airport runways and facilities. Also, the aerodrome had to be modeled conforming to Annex 14. To this end, it was essential to go deeply into the details of the aerodrome design to shape correctly some of the runway features (threshold markings, touchdown, centerline, etc.). These markings were drawn in compliance with ICAO standards irrespective whether they are actually present on the ground or not. Figure 1b depicts the GIS base map of BRHIA.

Step 2—Two-Dimensional OLS for BRHIA Runways: The dimensions for drawing the approach and takeoff two-dimensional surfaces for the runways as well as other airport relevant surfaces are specified in Annex 14. Figure 2a–c shows the required 2D surfaces of runway 21/03, while Fig. 2d shows the overall 2D configuration for all three intersecting runways.

Step 3—Digital Terrain Model (DTM) development: Contour lines with five meters' nominal resolution covering the area surrounding BRHIA were acquired. These contour lines were converted into a triangulated irregular network (TIN) using the 3D Analyst extension in ArcGIS in order to visualize the surface morphology around BRHIA. TIN is typically used for high-precision modeling of smaller areas; this is why they were chosen in this work.

Step 4—Three-Dimensional OLS for BRHIA Runways: To model 3D surfaces, calculations relating to elevations were generated for each section relevant to the OLS. These values were used to convert the 2D OLS into 3D TIN layers using the 3D Analyst toolbox. TIN layers were then converted into 3D raster surfaces using the TIN to Raster tool. Figure 3 shows the 3D model for the approach and takeoff for all runways.

Step 5—Setting the Floating Surface Base Height with respect to DTM: The 3D OLS layers were all displayed floating on the TIN surface of the digital terrain model.

Step 6—Built-up Structures Excursion: Spatial data extended over a radius of 30 km from the airport and pertaining to the locations and elevations of the built-up structures were acquired. Spatial distributions of these structures were of 1 m positional and 0.5 m accuracies and were modeled on the 3D layer and displayed on the floating TIN. They were extruded to their heights based on the Z value.

Step 7—Intersection of 3D OLS with Built-up Structures: The 3D OLS and the built-up units and elevated structures were overlaid together in order to get a visual picture about the obstacles that cause a threat to aviation safety.

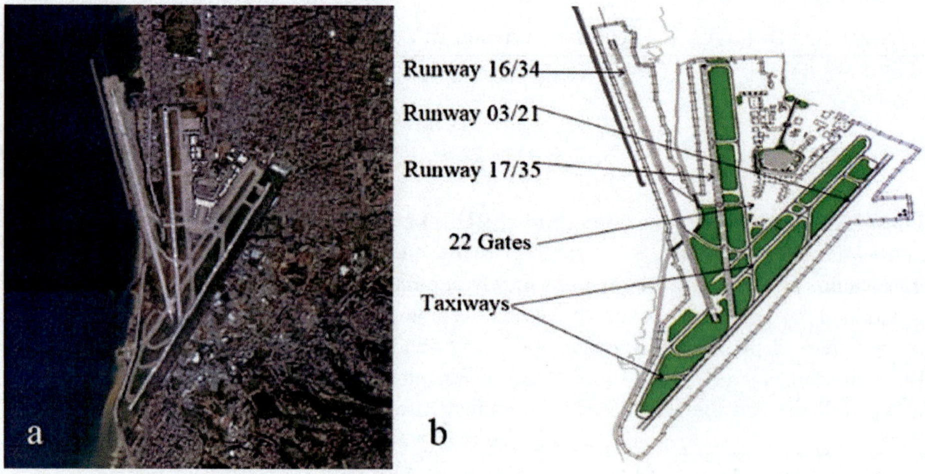

Fig. 1 **a** Satellite imagery (GeoEye-41 cm Accuracy) and **b** GIS base model for BRHIA

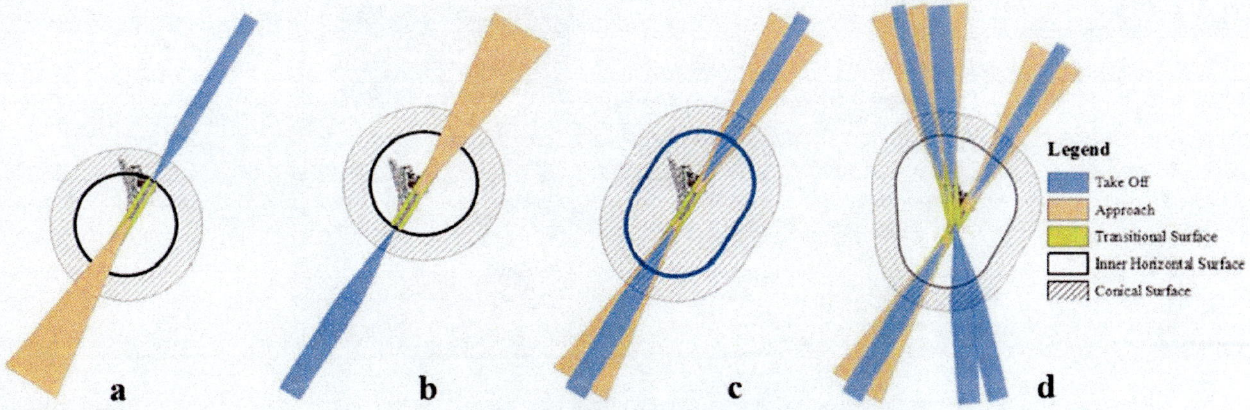

Fig. 2 **a–c** 2D OLS for runway 21/03 and **d** 2D OLS for all runways

Fig. 3 3D OLS for all BRHIA runways and built-up structures floating on DTM

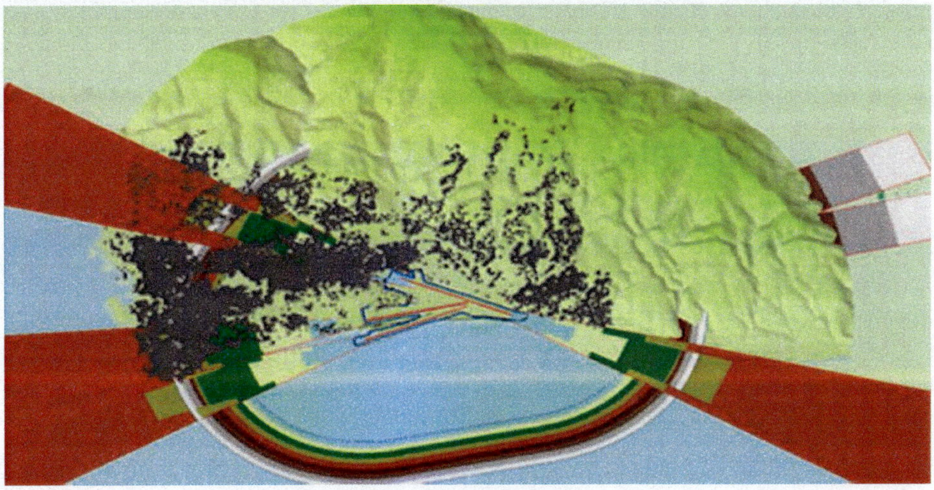

3 Results

The integration of the data used with the adopted methodology has surely been successful not only in the definition of the OLS but also in the visualization of the surrounding terrain and built-up structures.

Further GIS analysis enabled the OLS-obstacle interference to be assessed as shown in Figs. 3 and 4. Although the results shown herein are only concerned with the approach and departure zones, yet similar results are available for the other surfaces (conical, transitional, and inner horizontal). As such an overall picture of the OLS—obstacle interaction could be drawn.

A list with the exact locations of the violating buildings was extracted along with the height of penetration of the OLS which was calculated using GIS. The approach of runway 21 shows that 1823 built-up structures are penetrating the approach of runway 21. This approach is not currently operational since it is considered hazard to aviation safety. Additionally, takeoff of runway 03 is not used for the same reason.

4 Discussion

The obtained results enable the visualization of the airfield, the OLS, as well as the relevant obstacles, in addition to their correlations. These results, revealed for the first time, permitted an enhanced understanding of the potentials and constraints that may affect aerial operations within and around the airfield. Figure 1b, for example, provides an accurate base map for BRHIA with all its pertinent details which constitutes a starting point for any development activity for the airport, be it aeronautical, safety, security, or others. Figure 2, on the other hand, illustrates how different OLS can intersect or be overlaid be it due to displaced threshold or reciprocal runways or because of different

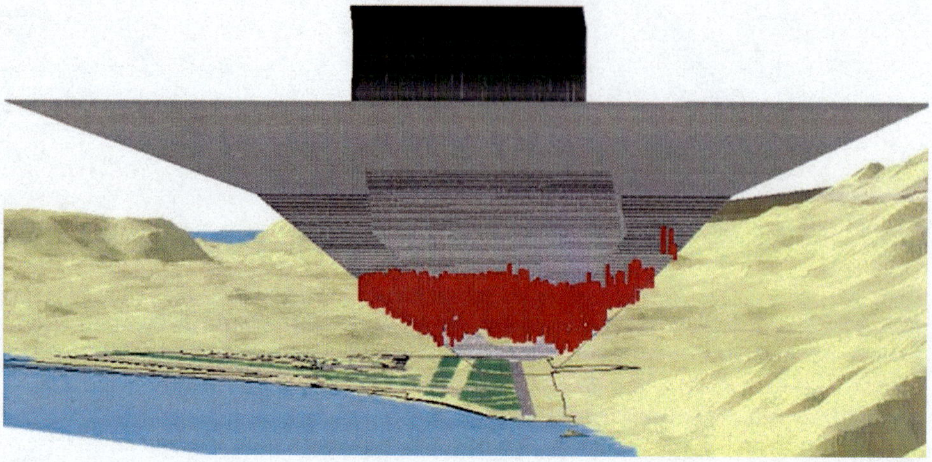

Fig. 4 Problematic buildings penetrating the app surface of runway 21 to 03

runways proximities. Figure 3 depicts an instant graphical representation of the OLS interference with surrounding terrain or man-made obstacles. With GIS, all these results may be dynamically visualized through 3D rotations, panning, and zooming in order to investigate finer details of any infringements.

The availability of these results and the ability to manipulate, integrate, and correlate them enable new areas to be explored. Perhaps the first could be the mitigation of these obstacles between the aviation authority and the respective municipalities where the infringements are encountered. Additionally, these results could always be incorporated in the airport development plans whether in terms of infrastructure or aerial operational improvements. They can also be used by policymakers in planning airfield expansions and enhancing safety and security.

5 Conclusions

The work presented in this paper enables the following conclusions to be drawn:

- GIS has proved instrumental not only for the definition and visualization of OLS but also for the analysis and mitigation of existing infringements.
- The identification of aviation safety-threatening obstacles necessitates the implementation of new regulations and physical provisions in their vicinity.
- The results found can further be used to set out new safety provisions, flight procedures, and recommend reviews for operational runway lengths.
- Highlights the need of continuous data updates in order to ensure that an accurate picture of the airfield and its surrounding is always available.

References

Chang, S.W.: A GIS model for analyzing airspace obstructions and safety near airports. J. Civ. Eng. Archit. **10**(1), 553–562 (2013). https://doi.org/10.17265/1934-7359/2016.05.004

ICAO: Doc 9848, Assembly resolutions in force, http://www.icao.int/publications/Documents/9848_en.pdf (2004)

ICAO: Annex 14, Aerodromes, aerodrome design and operations. 7th edition, Vol.1, http://www.ssd.dhmi.gov.tr/getBinaryFile.aspx?Type=3&dosyaID=920 (2016)

ICAO: Doc 7670. Standing rules of procedure of the assembly of the international civil aviation, http://www.icao.int/publications/Documents/7600_6ed.pdf (2008)

Jadayel, O.C.: Aerial zoning using GIS: Application for Kleia'at airport Lebanon. In: Feature article, GIM International, vol. 21, n° 4, pp. 39–41, Geomares, Netherland (2007)

Determination of a Local Geoid Model in the Region of "Grand Tunis" (Northeast of Tunisia)

Mohamed Hafedh Hamza, Mohamed Ajmi, Mohamed Ali Yahmadi, Sobhi Jellouli, and Mongi Belarem

Abstract

The main goal of this study was to test several methods to determine a local geoid model in the greatest metropolitan area in Tunisia, the region of "Grand Tunis." To this end, 56 leveled GPS points were used as well as the global geoid models EGM96 and EGM2008. The geoid was compiled by calculating a correction applied to the two global geoid models by least squares and kriging methods. The computation of the local model of geoid has been carried out on a regular 0.05′ by 0.05′ grid in a 40 km by 40 km quadratic study area. It finally turned out that the use of the EGM2008 global geoid model combined with a correction using the kriging method is the best combination. The RMS fit obtained is around 12 cm. Similar studies were performed in other countries (Gwaleba in J Geomat 12(2), 2018; Oluyori et al. in Int J Sci Res Publ 8(10), 2018).

Keywords

Geoid • Leveling • GPS • EGM2008 • EGM96 • Least squares • Kriging

M. H. Hamza (✉) · M. Ajmi · M. Belarem
Department of Geography and Geographic Information Systems, Faculty of Arts, King Abdulaziz University, Jeddah, Kingdom of Saudi Arabia
e-mail: mhhamza@kau.edu.sa

M. H. Hamza · M. Ajmi · M. A. Yahmadi · S. Jellouli
Geomatics, Earth, Environment and Planning Section, Department of Geology, Faculty of Science of Tunis, University of Tunis El Manar, Tunis, Tunisia

M. Belarem
Department of Geography, Faculty of Arts of Sfax, University of Sfax, Sfax, Tunisia

1 Introduction

Nowadays the estimation of a high-precision geoid is an important goal in the geodesy. Geoid is an equipotential surface which closely represents the mean sea level and can be used in combination with GPS observations to estimate orthometric heights. This is of particular relevance, since it can be achieved in a faster and cheaper way than using spirit leveling, although at a lower precision (which is, however, sufficient in many practical applications).

To calculate a local geoid model in the "Grand Tunis" area, 56 GPS/leveling points (as shown in Fig. 1) and the global geoid models EGM96 (Earth Gravitational Model 1996) and EGM2008 (Earth Gravitational Model 2008) were used. Many tests were carried out in order to find the optimal method to determine a local geoid model and find the adequate global equipotential model to use between EGM96 and EGM2008. In this case, the geometric approach was adopted. Two interpolation methods were used and tested: least squares and kriging.

2 Materials and Methods

2.1 Evaluation of the Global Geoid Models EGM 2008 and EGM96 in the Study Area

The EGM96 and EGM2008 used in this work are geoid models created in the form of grids by the US National Geospatial-Intelligence Agency (NGA). They have the height above the ellipsoid in meters and should range up to 100 m in absolute value (Roman et al. 2010).

To analyze the accuracy of these global geoid models in the study area, a comparison with 56 GPS/leveling points was made as shown in Fig. 2 and Table 1.

The EGM96 and EGM2008 global models are biased and do not have a high-quality precision compared to the GPS/leveling points. The EGM2008 is slightly more

H. El-Askary et al. (eds.), *Research Developments in Geotechnics, Geo-Informatics and Remote Sensing*, Advances in Science, Technology & Innovation,
https://doi.org/10.1007/978-3-030-72896-0_53

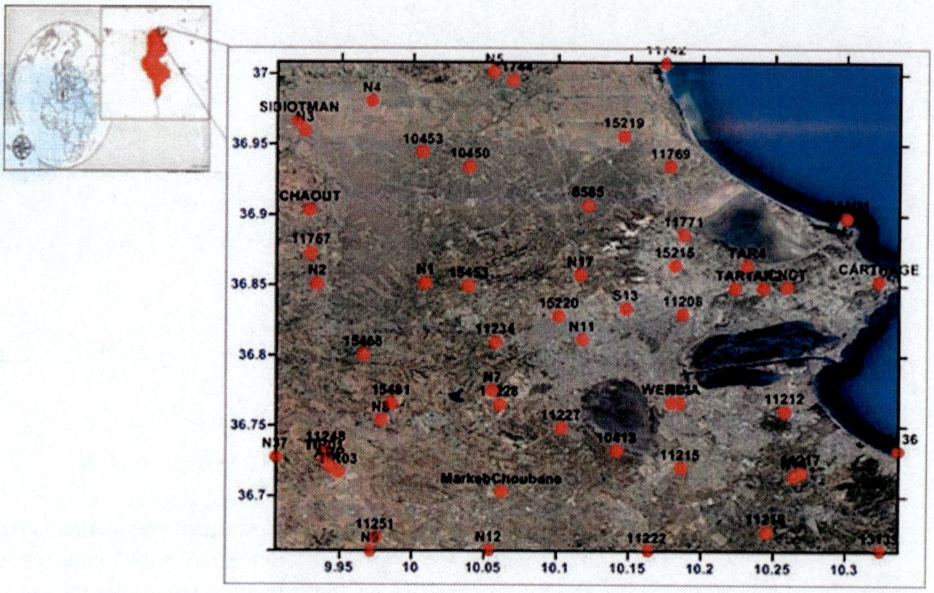

Fig. 1 Study area and distribution of GPS/leveling points

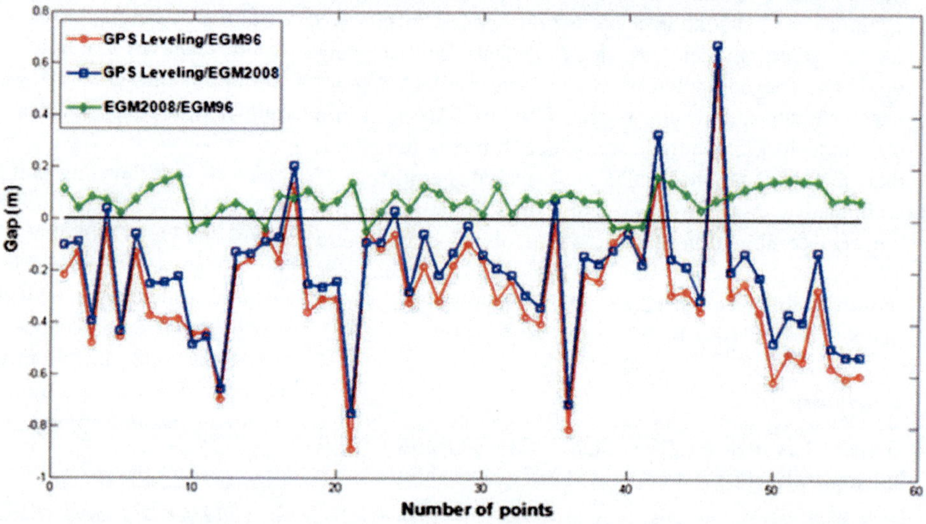

Fig. 2 Difference between EGM96 and EGM2008 and GPS/leveling points

Table 1 Comparison between EGM96, EGM2008, and GPS/leveling points

	GPS/leveling EGM96	GPS/leveling EGM2008	EGM96/EGM2008
Mean (m)	−0.28	−0.21	0.07
Std (m)	0.25	0.23	0.06
Min (m)	−0.89	−0.75	−0.04
Max (m)	0.60	0.68	0.17

accurate than the EGM96 (Sadasiva Rao et al. 2008). The use of such models to convert ellipsoidal GPS height is not accurate enough in topography applications, which made us elaborate a correction grid. To fit a global geoid model on GPS/leveling point, and therefore to improve the accuracy of the model and eliminate the bias, the following steps were followed: Step 1: Calculate the undulation in each GPS/leveling points. The used formula is the following: $N = h - H$, where h is the ellipsoidal height, H is the orthometric height, and N is the geoid undulation. Step 2: Interpolate the undulation from the global geoid models EGM96 and EGM2008 for the same GPS/leveling points.

The global geoid models used are presented in a regular grid. The bilinear interpolation was used to interpolate the undulation. It is an extension of linear interpolation for interpolating functions of two variables (e.g., x and y) on a rectilinear 2D grid (Chang 2009). Step 3: Calculate in each point the difference between undulation from GPS/leveling point and the undulation from global geoid model; Step 4: Estimate a correction model to fit global geoid model on GPS/leveling points (Soycan 2014). Two approaches were used: least squares and kriging. Step 5: Estimate a correction in each node of the grid of the global geoid model and calculate the new value of undulation by the following formula: $N\text{corrected} = N\text{gravitational} + C(\lambda, \varphi, P)$, where *N*corrected is the new correction of undulation, *N*gravitational is obtained by interpolation of the gravimetric geoid model, and C is the correction in the grid depending on the coordinates of the points (λ, φ, P) and the calculation parameters, and which is calculated in each point via the used adaptation model. The final result will be a grid of altimetric conversion which is unbiased and more accurate than the global model. A comparison between the least squares method (three parameters, four parameters, and five parameters) and the kriging method (with estimation of an optimal variogram model) was performed to find the optimal model to calculate the grid conversion.

The computation of the local model of geoid has been carried out on a regular $0.05' \times 0.05'$ grid in the area (between 9.9° and 10.35° in longitude and between 36.65° and 37.05° in latitude as already shown in Fig. 1). To carry out the calculation of a local geoid model, 56 GPS/leveling points have been collected for fitting and checking: 35 points used for the calculation of the model of the local geoid, 15 check points, and 6 other points have been eliminated by the criteria of three sigma (point having a residue exceeding three sigma in absolute value).

3 Results

Undulation value is obtained in every GPS/leveling point. In order to generate the altimetry grid transformer, many tests were done to select the optimal method and the global geoid model to use: Test 1: Correction model calculated by least square method with three parameters using EGM96 and GPS/leveling points, Test 2: Correction model calculated by least square method with four parameters by using EGM96 and GPS/leveling points, Test 3: Correction model calculated by least square method with five parameters by using EGM96 and GPS/leveling points, Test 4: Correction model calculated by kriging method by using EGM96 and GPS/leveling points, Test 5: Calculate a local geoid model using kriging method with only GPS/leveling points, Test 6: Correction model calculated by least square method with three parameters using EGM2008 and GPS/leveling points, Test 7: Correction model calculated by least square method with four parameters using EGM2008 and GPS/leveling points, Test 8: Correction model calculated by least square method with five parameters using EGM2008 and GPS/leveling points, and Test 9: Correction model calculated by kriging method using EGM2008 and GPS/leveling points.

4 Discussion

Test 1, Test 2, Test 3, Test 6, Test 7, and Test 8 show the difference in accuracy between the global geoid models EGM96 and EGM2008 and the fitting model by the least squares method using three, four, and five parameters. Hence, we can conclude that the accuracies of the fitting model with four and that with five parameters are the same. However, the three-parameter fitting model is less precise than the two other models. The residue graphical analysis also shows that the use of a fitting model by the least squares method using four and five parameters with EGM2008 and gives more precise results with the used data on the study area.

Table 2 shows the difference between each solution compared to the 15 check GPS/leveling points distributed on the whole region.

From Table 2, we can deduce that Test 9 gives the most accurate and unbiased solution. The use of the global geoid model EGM2008 that fits this surface on GPS/leveling point by calculating a correction using the kriging method is the optimal solution for the data used in the tested region. Figure 9 represents the new grid of altimetric conversion. It is the result of the conversion of the ellipsoidal GPS height into altimetric height, with an accuracy of 12 cm.

Table 2 Summary of statistical information related to all the tests done compared to the check GPS/leveling points

	Test 1	Test 2	Test 3	Test 4	Test 5	Test 6	Test 7	Test 8	Test 9
Std dev (m)	0.18	0.15	0.14	0.18	0.38	0.17	0.15	0.14	0.12
Bias (m)	0.05	0.05	0.05	0.18	0.04	0.05	0.05	0.05	0.00
Min (m)	−0.17	−0.23	−0.21	−0.05	−0.60	−0.20	−0.23	−0.22	−0.16
Max (m)	0.36	0.28	0.27	0.62	0.68	0.34	0.29	0.26	0.24

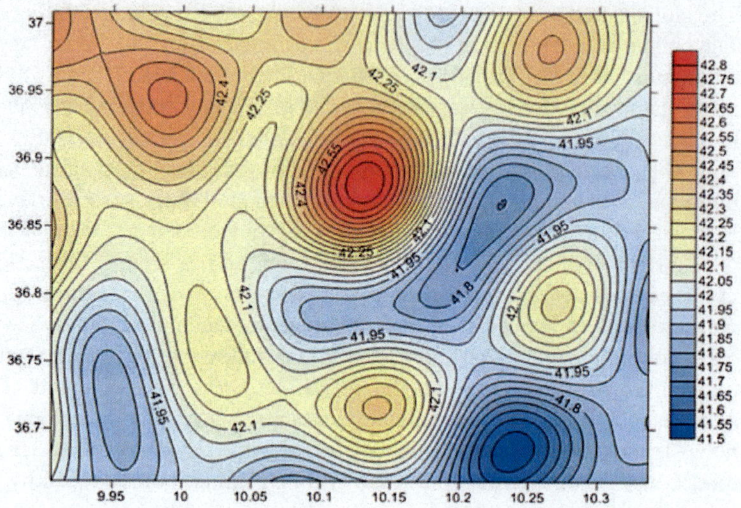

Fig. 3 New local geoid model obtained using Test 9 (correction model calculated by kriging method by using EGM2008 and GPS/leveling points)

5 Conclusions

A local geoid model was calculated in the region of "Grand Tunis" in a square study area 40 km by 40 km on a regular grid 0.05' by 0.05'. The local geoid model was tested using 15 GPS/leveling points that has indicated an RMS fit of 12 cm. Figure 3 displays the new surface of the local geoid model. It allows converting the ellipsoidal GPS height into orthometric height with an accuracy of 12 cm.

The optimal method is the one in which the calculation of the correction is made using the kriging method and the global model EGM2008. The approach used in this paper is the geometric method based on the calculation of the undulation which is the difference between the orthometric height and the GPS observed ellipsoidal height. The accuracy of the geoid established with the geometric approach depends on several factors such as the distribution and the number of GPS/leveling points, the characteristics of the geoid in the region, and the method of interpolation. This approach gives good results in a relatively small and flat area as the "Grand Tunis" zone. To improve the accuracy of this local geoid, the gravimetric approach which incorporates gravity measurements and a great density of local terrain information (GPS/leveling points) can be applied in future works in the study area.

References

Chang, K.T.: Computation for Bilinear Interpolation. Introduction to Geographic Information Systems, 5th edn. McGrawHill, New York (2009)

Gwaleba, M.J.: Comparison of global geoid models against the GPS/levelling-derived geoid heights in Tanzania. J. Geomat. **12** (2) (2018)

Oluyori, P.D., Ono, M.N., Eteje, S.O.: Computations of geoid undulation from comparison of GNSS/levelling with EGM 2008 for geodetic applications. Int. J. Sci. Res. Publ. **8**(10) (2018)

Roman, D.R., Wang, Y.M., Saleh, J., Li, X.: Geodesy, geoids, & vertical datums: a perspective from the U.S. National Geodetic Survey. FIG Congress 2010 Facing the Challenges—Building the Capacity, Sydney, Australia (2010)

Sadasiva Rao, B., Anil Kumar, G., Gopala krishna, P.V.S.S.N.P., Srinivasulu,V., Raghu, V.: Evaluation of EGM 2008 with EGM96 and its utilization in topographical mapping projects. J. Indian Soc. Remote Sens. **40**, 335 (2012)

Soycan, M.: Improving EGM2008 by GPS and leveling data at local scale. Bol. Ciênc. Geod. **20**(1) (2014)

Population Estimation Using Remote Sensing Data and GIS Techniques. Application to Jeddah City, KSA

Atef Belhaj Ali

Abstract

This paper presented a method allowing the estimation of the population at the local scale from the data resulting from the interpretation of the satellite images. The aim was to estimate the number of inhabitants in each habitable area for the year 2017. This estimate is based on the data census of 2010 and the interpretation of images gathered by the satellite SPOT in 2010 covering Jeddah city. A relationship was established between housing areas and the number of inhabitants using correlation techniques based on ordinary least square method. The obtained equation was applied to the data obtained by visual interpretation of the Quickbird images gathered in 2017 covering the study area, thus making it possible to estimate the population of the city on this date.

Keywords

Population estimation • Remote sensing • Areal interpolation • GIS • Correlation • OLS

1 Introduction

The estimation of a population and its distribution over a geographical space is important for decision-making support, especially for urban projects, socioeconomic studies, regional development, etc., population census is usually achieved every ten years and afflicted with the problem of the so-called 10-year gap (Kubanek et al. 2010). In contrast, cities and urban areas, such as Jeddah, the object of our present study, are evolving in a sustained way. This study was conducted on the city of Jeddah which consists of 113 urban districts grouped into 13 sub-municipalities.

The study presented in this paper shows the possibility of establishing a link between the surfaces of habitable spaces and the number of inhabitants and to put in place an equation that could be generalized to estimate the population. In fact, we used statistical modeling methods based on correlation techniques to establish a functional relationship between the housing built-up areas and the population size; these techniques are based on the ordinary least square method.

2 Data and Methodology

Jeddah is located in the Western part of Saudi Arabia and land bordered from three sides (north, south and east) and Red Sea to the west side ($N = 21° 34'$ E = 39° 10′). Jeddah city covers **1267 km^2** and has around **3,425,999** inhabitants in 2010. Its territory is divided into **13** sub-municipalities that comprise **113** districts. According to the United Nations Department of Economic and Social Affairs (https://population.un.org/wup/Country-Profiles/), during the last decade, the population growth rate in Saudi Arabia is around **2.3%** after having recorded a maximum peak of **6.52%** in 1982.

Several data sources were used to fulfill this study. We used the 2010 census data collected from the general authority of statistics of KSA; we relied on these data since they reflect the exact number of inhabitants in 2010; all the other data provided by the same administration after 2010 are just projections and remain uncertain. Satellite images were also used, from which the footprint of housing areas in 2010 and 2017 was extracted. The master plan of Jeddah city was also used as an ancillary data to help better refine the visual interpretation of images and exclude all buildings that are not for use as dwellings such as schools, mosques and industrial buildings.

A. B. Ali (✉)
Department of Geography and GIS, Faculty of Arts and Humanities, King Abdul-Aziz University, Jeddah, Kingdom of Saudi Arabia
e-mail: amalhabeb@kau.edu.sa

H. El-Askary et al. (eds.), *Research Developments in Geotechnics, Geo-Informatics and Remote Sensing*, Advances in Science, Technology & Innovation,
https://doi.org/10.1007/978-3-030-72896-0_54

Fig. 1 Methodology flowchart

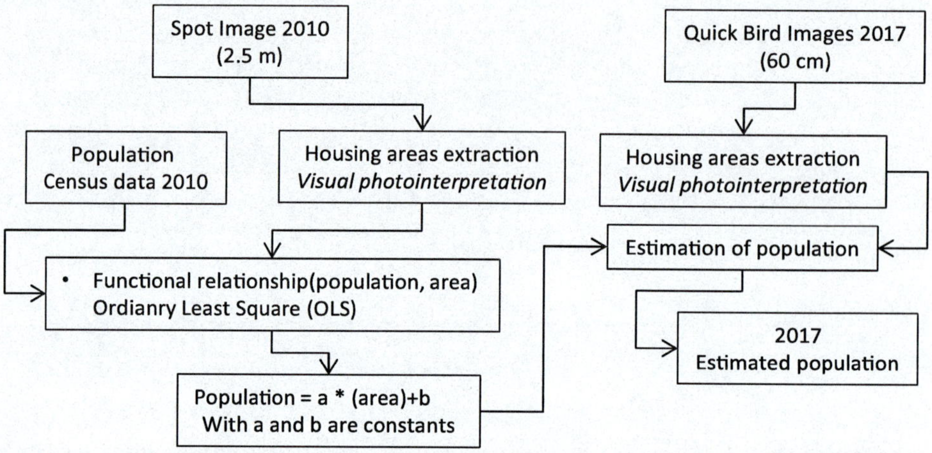

The adopted methodology in this study is illustrated by Fig. 1

3 Results and Discussion

The total and annual evolution rates were calculated based on measurements of housing footprints surfaces of both dates. The results show that the housing footprint represents **10.97%** of the total area of the city in 2010 against **13.44%** in 2017. Table 1 and Fig. 2 show the results of the visual interpretation of the 2010 and 2017 satellite images.

The results also show sustained this urban expansion during the period 2010–2017 in the north of the Jeddah city (sub-municipalities of Dhahban, Obhur and Teebah) and in the south (sub-municipalities of Al Janub and Al Moulyssa) and to a smaller degree in the east (sub-municipalities of Breeman and Om Salam). In terms of the entire city, the total evolution rate of housing areas between 2010 and 2017 was around **19%** with an annual evolution rate of **2.5%**.

Inspired by the biological law of allometric growth, many studies tried to establish a relationship between urban areas and population size (Huxley 1932; Nordbeck 1965). According to his work on US towns, (Tobler 1969) concluded that the built-up area (A) of a settlement is proportional to its population (P) raised to some power. The tests conducted on our study area showed that this type of relationship did not give complete satisfaction. In addition, as part of this study, regression techniques were used to establish a relationship between the population and the inhabitable units. There exists many methods to estimate the parameters of the population regression equation. The method of ordinary least squares (OLS) was used. The results of the interpretation as well as the data of the

Table 1 Housing footprints areas in 2010 and 2017

Sub-municipality	Housing footprint 2010 (m^2)	Housing footprint 2017 (m^2)	Total evolution rate (2010–2017) (%)	Annual evolution rate (%)
Obhur	19,082,778	22,874,859	19.87	2.62
Al Balad	4,910,383	5,020,831	2.25	0.32
Al Jameeaa	14,320,963	15,056,755	5.14	0.72
Al Janub	8,623,551	14,945,000	73.30	8.17
Al Sharafyia	8,186,436	8,812,761	7.65	1.06
Al Azizyia	12,431,422	12,934,340	4.05	0.57
Al Matar	22,860,734	23,723,655	3.77	0.53
Al Moulyssa	6,163,489	7,508,346	21.82	2.86
Om Salam	11,192,907	14,028,612	25.33	3.28
Breeman	8,994,246	12,283,790	36.57	4.55
Jeddah Al Jadeeda	14,257,166	15,081,969	5.79	0.81
Dhahban	4,658,459	9,252,288	98.61	10.30
Teebah	3,255,080	8,705,164	167.43	15.09

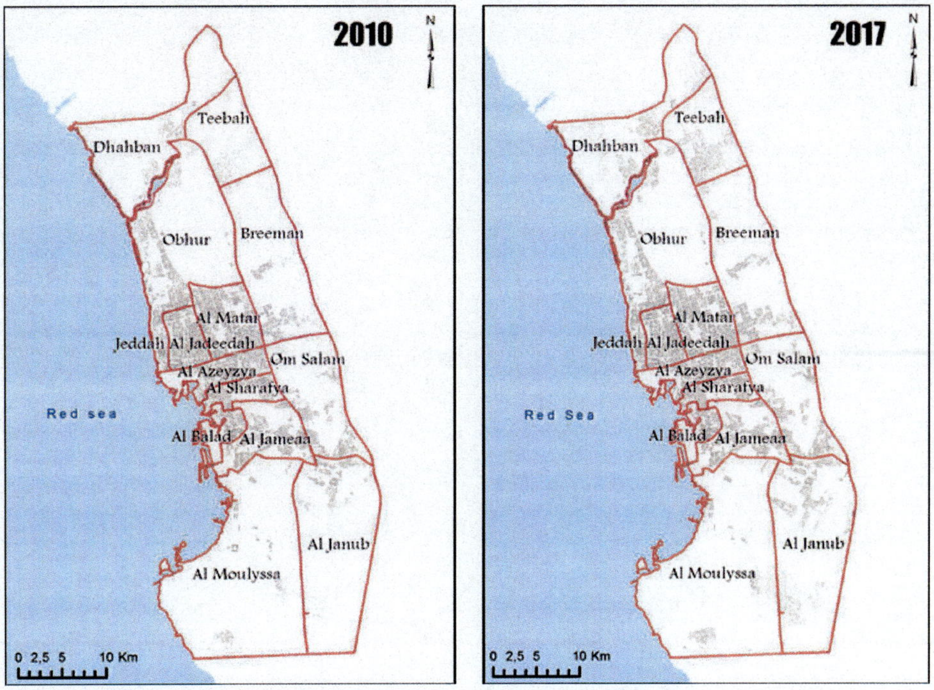

Fig. 2 Jeddah city housing footprints in 2010 and 2017

population of the districts show a good correlation (***R*^{2} = 0.5683**) Fig. 3. The relationship established relation between the housing footprint of 2010 and the 2010 population displayed as follows:

$$Population = 0.0222(HousingFootprintArea) + 2933.8 \quad (1)$$

4 Conclusions

The obtained results, in particular, the relationship between housing areas in 2010 and the number of inhabitants, were applied to the results of the visual interpretation from the satellite imagery of the year 2017. According to the established model, the population in 2017 has reached **4,099,326** inhabitants, which could be expressed by an annual growth rate of **2.6%** which is close to the average rate provided by United Nations Department of Economic and Social Affairs. In this study, information about the height of dwellings was not considered, and only the footprint of the building was used. For this, we recommend to introduce information about the nature of the buildings by the use of exogenous documents and ancillary data to classify the cities into homogeneous groups (sub-communities) according to their characteristics (land use, socioeconomic characteristics, etc.) and apply the method described in this work to each homogeneous sub-community.

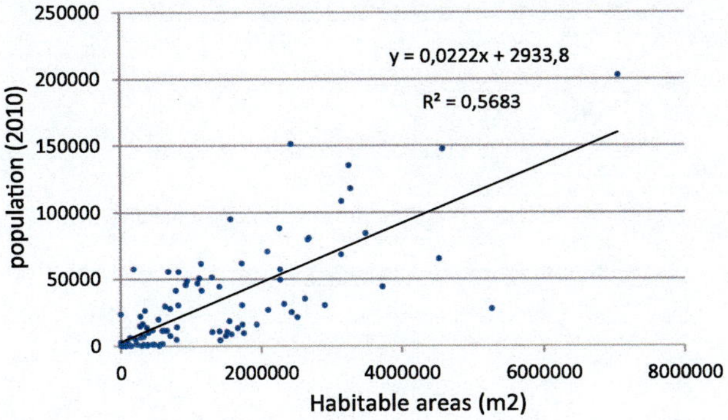

Fig. 3 Correlation between housing area and population

References

Huxley, J.: Problems of Relative Growth, p. 356. Methuen, London (1932)

Kubanek, J., Nolte, E.-M., Taubenböck, H., Kappas, M., Wenzel, F.: Modelling of Population Dynamics: GIS Versus Remote Sensing—A Case Study for Istanbul. Institute of Transport Research Publications (2010)

Nordbeck, S.: The Law of Allometric Growth. Michigan Inter-University Community of Mathematical Geographers, Discussion Paper No. 7 (1965)

Tobler, W.R.: Satellite confirmation of settlement size coefficients. Area **1**(3), 30–34 (1969)

Building Pattern Identification for Map Generalization

Nardjes Hamini

Abstract

This study investigates a new method used to detect identical patterns in the domain of cartographic generalization. As opposed to the traditional method which consists of only identifying linear, grid and previously saved patterns, the proposed technique allows for any repeated pattern to be identified regardless of the shape and without being previously saved in the software's memory. Through grouping the patterns at first and then using a comparative function based on three geometric operations: Translation, mirrorization and rotation, this technique is able to detect any pattern. The results found through testing on samples in Dubai and through the validation of the test in Algeria are promising as to the liability of these tools.

Keywords

Algorithm • Building • Cartographic • Clustering • Detection • Generalization • GIS • Pattern

1 Introduction

The process of producing a map of lesser scale from a larger scale is called generalization. When we reduce the scale of a map, the symbolized objects that are there can be super-imposed, they can shrink to the point of being imperceptible, and roads can get tangled and the map becomes congested. The generalization corrects these flaws with an amalgamation of algorithms that perform tasks such as coverage, selection or displacement (Mackaness 2007). The algorithms can be applied to an individual object or a group of objects. The patterns represent a typical cartographic property of an area that must be kept after generalization.

Pattern detection is useful for cartographic generalization, because instead of treating buildings individually, they could treat as much as a set. The generalization algorithms are used to see a group of objects as one. Example: For a typification, instead of removing buildings in an asymmetrical way, we can delete entire pattern occurrences. For a displacement, we move the entire occurrences of the pattern.

In literature, there are many algorithms that aim to group buildings. We have clustering algorithms based on the proximity of the buildings to each other (Cetinkaya et al. 2014). But the risk with these algorithms is to take in the group of buildings belonging to another pattern, or downright buildings that do not belong to the patterns because it is difficult to isolate the patterns from the rest of the buildings. There are also methods that identify specific patterns such as alignment pattern or grid pattern (Deng et al. 2017). Finally, we find methods (Yang 2008) that identify patterns by comparing groups of buildings with models existing in memory, and thus a pattern that is not in the database cannot be identified.

There is a need for an algorithm that automatically identifies any pattern. We have defined a new method to solve this problem based on the search of repetition in the similarity of groups of buildings. This similarity is found in the shape, size, orientation and relative position of the buildings vis-a-vis each other, while the grouping method is based on the search of proximity by an object Voronoi diagram.

2 The Methods

Before proposing the method, it is important to specify some properties and rules for our patterns. These latters can be modified and adapted thereafter: A building cannot be included in more than one pattern or pattern occurrence.

N. Hamini (✉)
Laboratory of Computer Science and Mathematics (LIM), Amar Telidji University of Laghouat (UATL), Bp 37 C, Ghardaia, 03000 Laghouat, Algeria

H. El-Askary et al. (eds.), *Research Developments in Geotechnics, Geo-Informatics and Remote Sensing*, Advances in Science, Technology & Innovation,
https://doi.org/10.1007/978-3-030-72896-0_55

Two patterns or pattern occurrences are 100% distinct. Each building of a pattern occurrence has corresponding matches in the occurrences of that same pattern. It is its equivalent in shape, size, position, distance and orientation (relatively speaking). A pattern occurrence is equivalent to another occurrence of the same pattern oriented from 0° to 360° or its mirror oriented from 0° to 360°. In other words, the occurrences of a pattern must be stackable like layer sheets. However, these rules do not concern homotheties of patterns. For pattern homotheties, distances differ by a fixed factor k.

In regards to the different terms used in this method, model is the design, an occurrence is a group reproducing the model, and a pattern is a set of occurrences reproducing the same model. Our method consists of creating groups of neighboring buildings of different sizes and look for similarities between these groups. The steps are as follows (Fig. 1):

1. Division of the zone in blocks: It consists of dividing the geographical zone according to the network of roads and streets. These sub-zones are called blocks.
2. Neighborhood Search: A Voronoi diagram for objects was applied to each block. This allowed us to group the buildings into pairs of buildings, where each couple (*bi*, *bj*) means that the building bi is adjacent to the building *bj* and vice versa. We call the set of groups of size two S2.
3. Elimination of duplication: (*bi*, *bj*) is similar to (*bj*, *bi*).
4. Classification of the groups: We look for similar groups using the function of similarity. We group these groups together. These represent potential patterns. We call the set of potential patterns of sizes 2, PP2.
5. Elimination of unrepeated groups (whose number of occurrence is equal to 1) from sets S2 and PP2.
6. Formation of groups of larger size, i.e.: size 3 from the set S2. We look for two groups of size two having a building in common and we merge them. Then, we obtain the set S3.
7. Steps 3–6 are then repeated with Si and PPi the set of groups and potential patterns of sizes i. For step 6, if we have groups of size i and we want to form groups of size $i + 1$, we look in groups of size I having $n - 1$ common buildings.
8. We stop when the obtained PPi is equal to the empty set that is to say that there are no more repeated groups.
9. We obtain the sets Pi which represent the patterns of sizes i. This is by eliminating groups from PPi which do not respect the rules on patterns mentioned above.
10. We choose the patterns that have the largest number of buildings and the smallest number of occurrences. When there is a multitude of patterns on the same buildings.

The similarity function consists in three geometrical methods: the translation, the rotation and the mirrorization.

Fig. 1 Our pattern extraction process

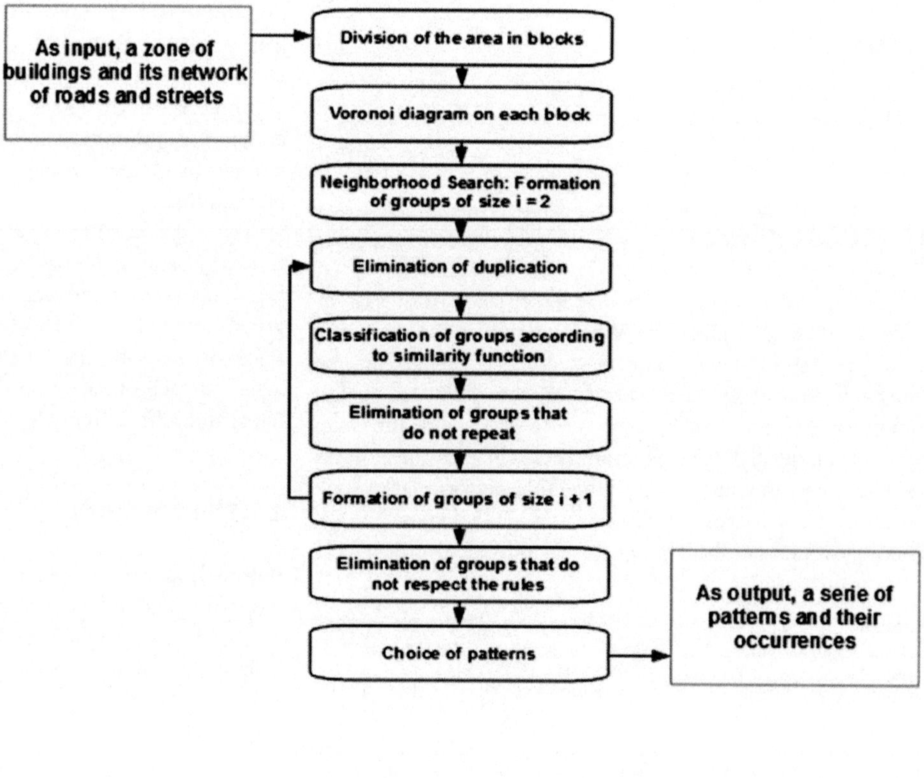

The process of comparison translates to simply calculate the proximity between points of the two groups of buildings.

2.1 Comparative Function

This function compares two groups of buildings ($G1$–$G2$) to verify if they are occurrences of the same pattern. It is based on three geometric operations:
Translation: The translation of $G1$–$G2$ is given by:

$x' = x + x'_0 - x_0, y' = y + y'_0 - y_0 / c_1(x_0, y_0), c_2(x'_0, y'_0)$ are the centers of $G1$ and $G2$.
Mirrorization: The mirrorization of $G1$ by the line $x = x_0$ (Passing $c1$) is given by:

$$x' = 2.x_0 - x, y' = y$$

Rotation: we use the polar coordinates regarding $c1$, a rotation of G1 by α is given by:

$x' = r\cos(\beta + \alpha), y' = r\sin(\beta + \alpha)/r$: the distance between a point and $c1$.

β is given by the arctangent function in 2π.

3 Results

For the implementation of the project, we chose the QGIS software. It has the usual GIS tools. Moreover, it offers the programming language PyQGIS which is python language enriched with QGIS spatial libraries. The data comes from OpenStreetMap and Google Maps. Samples of data are from the map of the city of Dubai.

In Fig. 2a, we see the geographical area on which the tests were made, so we have an urban area covered with buildings and streets. In Fig. 2b, we have the division of the zone into blocks. Figure 2c shows the application of the Voronoi diagram for objects on one block of the zone, and Fig. 2d shows the patterns detected in the block. The buildings of the patterns were colored in salmon. The numbers represent the different buildings belonging to the same pattern occurrence.

We note that the algorithm has well detected the different patterns and their occurrences, and the latters are well isolated from each other. The validation test was conducted on the Saharan region of Algeria (Laghouat) and confirmed the results previously obtained on the sample of Dubai (Fig. 3).

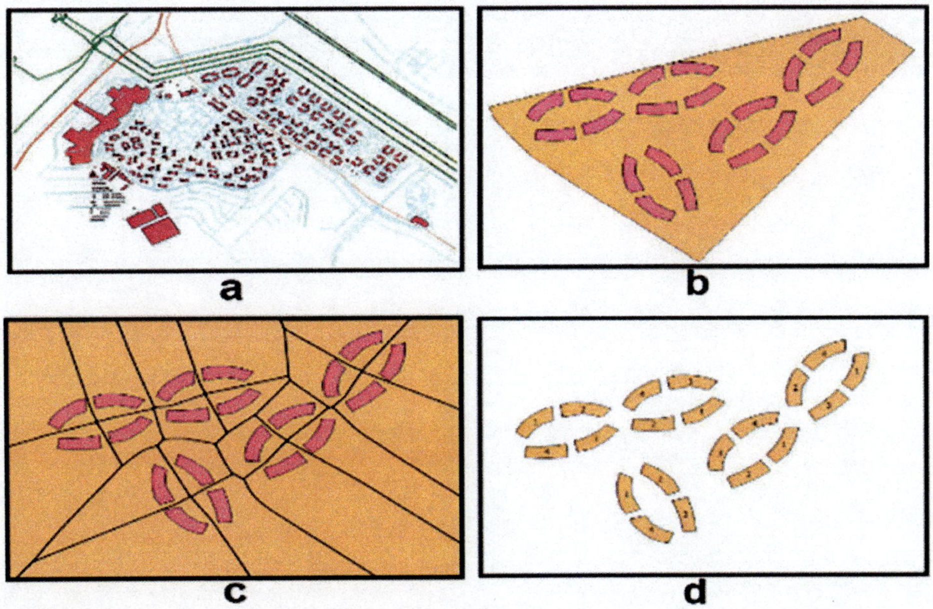

Fig. 2 Results of our pattern detection method

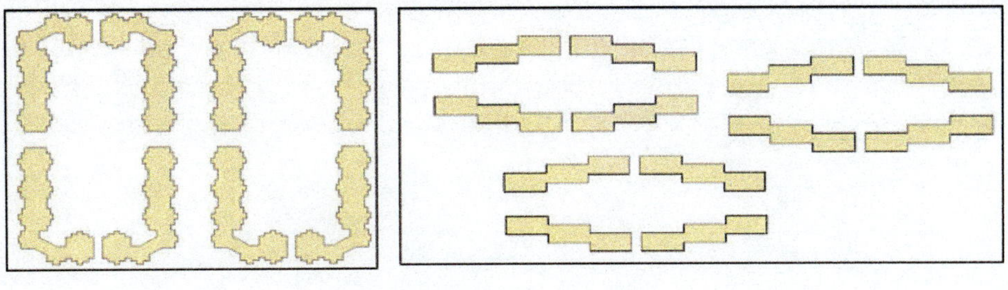

Fig. 3 Validation of the method

The accuracy of the results depends on the precision, while providing the software with the input. It has a function regarding the margin of error, if too large it allows an over detection and if too small the detection will be limited. It is a scaling factor. This algorithm is reactive to size. Sometimes there are patterns that may look similar to the naked eye but their sizes are not the same.

4 Conclusions

In this paper, we defined a new algorithm that aids to solve the problem of the detection of patterns of any form. The algorithm was applied to the area of the city of Dubai and was able to detect patterns and their occurrences. The algorithm provides very good results and allowed for the detection of occurrences of patterns even if they were oriented in different directions, a mirror or homothety of the model of the pattern.

The algorithm can work with other pattern identification algorithms, such as linear identification and grids identification for more complete pattern detection.

We hope that this modest work can be improved later by the research on hierarchies of patterns (a pattern included in a higher pattern).

References

Cetinkaya, S., Basaraner, M., Burghardt, D.: Proximity-based grouping of buildings in urban blocks: a comparison of four algorithms. Geocarto Int. **30**(6), 618–632 (2014)

Deng, M., Tang, J., Liu, Q., Wu, F.: Recognizing building groups for generalization: a comparative study. Cartogr. Geogr. Inf. Sci. **45**(3), 187–204 (2017)

Mackaness, W., Ruas, A., Sarjakoski, L.: Generalisation of Geographic Information: Cartographic Modelling and Applications (2007). https://doi.org/10.1016/B978-0-08-045374-3.X5000-5

Yang, W.: Identify building patterns. In: The International Archives of the Photogrammetry, Remote Sensing and Spatial Information Sciences, vol. XXXVII. Part B2. Beijing (2008)

Automatic Drawing Using Topological Codes

Dalibor Bartoněk

Abstract

The paper dealt with the possibility of automated drawing from the list of points with coordinates, obtained by geodetic measurement in the field. The text contains a design of topological codes, which makes this automatic drawing possible. Part of the solution is also creation of Python application for generating drawings in geographic information systems (ArcGIS, QGIS). The result of the automated process is the conversion of spatial data into a vector form, maximizing the amount of information contained in the input data with a minimum of manual work. The proposed method is original and actively leads to automate the creation of a 3D model.

Keywords

List of points with coordinates • Topological codes • 3D model

1 Introduction

When creating a 3D model of buildings or other structures, we usually follow the flow chart—see Fig. 1.

The bottle neck in the processing in Fig. 1 is the interconnection of the imported points, which is manually achieved in many cases. There are currently several applications that are trying to solve this problem. It is mainly about adding codes to the list of points that are used for an automated drawing directly from the list of points (CAD 2017). The application works in 2D and is primarily focused on use in the cadaster. In terms of 3D modeling, these codes are inadequate because they reflect the characteristics of particular geographic objects (trees, lamps ...) but lack of information in topological terms. The aim of the project was to add some measured points to the list of coordinates as attributes that would make it possible to create a drawing both in 2D and 3D in suitable software. Part of the solution would be the application design for acquisition of a topologically pure drawing. The result of the automated process would then be to transform the spatial data into a vector form, maximizing the amount of information contained in the input data and minimizing the amount of manual work.

2 Materials and Methods

High-quality technical drawing of the object should meet the highest standards, i.e., 3rd level of topology according to the DIGEST standard (https://www.dgiwg.org/digest/). In addition to topological entities, attributes, rules and relationships, it is necessary that the drawing meets these 3 topological concepts: (1) connectivity, (2) area definition and (3) neighborhood.

The goal was to design a code system for each point in the list that contains all of the three of these topological concepts. In Fig. 2, the quadrilateral is described by nodes (P), edges (L) and areas (A).

Each topological entity in Fig. 2 can be described by a unique code having 2 parts: 1 geometry and 2 number. Geometry is indicated by the letter P—point, L—line, A—area (polygon). A similar methodology is used in publication (Zalik 1999) (in CAD systems) and (Ladner et al. 1999) (general use). In the list of measured points, two types of points need to be distinguished from the topological point of view: (1) a topological point or intersection that indicates the beginning or end of a line and (2) a geometric point that determines the geometry of the line or its shape. Based on the above considerations, we can design topological codes to complement each point in the list. In the point identifier, we

D. Bartoněk (✉)
Brno University of Technology, Veveří 95, 602 00 Brno, Czech Republic
e-mail: Bartonek.D@Fce.Vutbr.Cz

H. El-Askary et al. (eds.), *Research Developments in Geotechnics, Geo-Informatics and Remote Sensing*, Advances in Science, Technology & Innovation,
https://doi.org/10.1007/978-3-030-72896-0_56

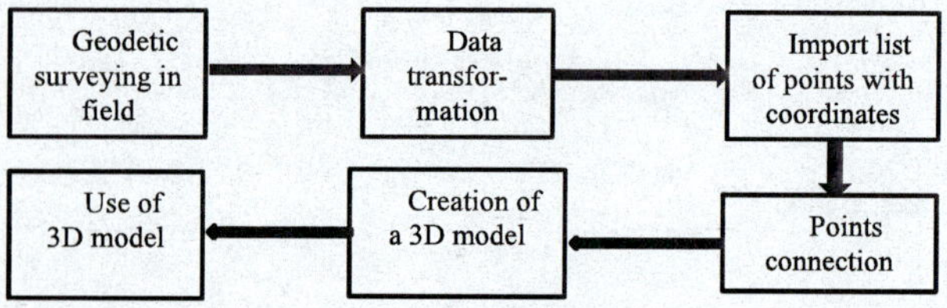

Fig. 1 Process flow chart of 3D model creation

Fig. 2 A quadrilateral topological model

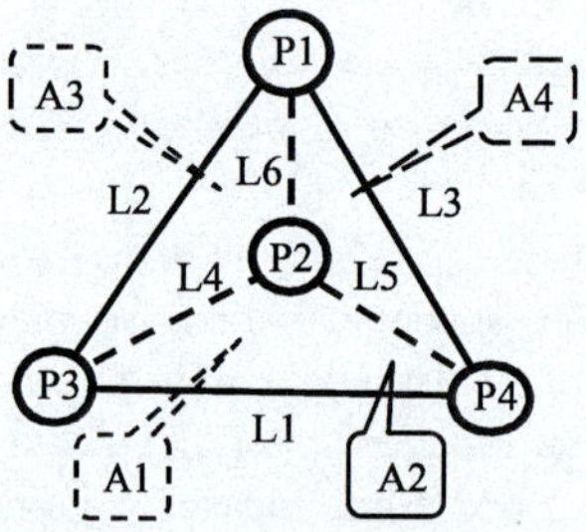

can omit the letter "P" for simplicity, because the point is implicitly determined by the line in the list and uniquely identified by the point number—see Fig. 3.

For each line in the list (Fig. 3), we enter the line and the area code as follows: Line code: LX/Y, where "L" indicates a line, X is a unique line number, Y indicates a point number on that line. The lowest sequence number indicates the start point of the line, and the highest sequence number indicates the endpoint of the line.

The area code is: AX, where "A" denotes the area (polygon) and X is the unique area number. An example of a list of coordinates with embedded codes (C) is shown in Fig. 3b.

3 Results

The system of topological codes proposed in the previous section allows for automatic drawing directly from the list of coordinates of the measured points. The method was tested in the geodetic surveying of the church of St. Peter of Alcantara in Karvina (Czech Republic).

Figure 4 is a sketch illustration and a portion of a point coordinate list with topological codes. This list was imported into QGIS, and an automated drawing was created using a

Fig. 3 List of points with coordinates

```
ID,   Y,          X,            Z
6005,600013.5779,1157910.6975,243.3187
6004,600055.6559,1157930.3558,240.6313
4003,600010.6320,1157946.7980,241.6320
4002,600050.8680,1157942.9820,240.0280
4001,600048.1800,1157901.5720,241.9250
4001,600048.1800,1157901.5720,241.9250
6003,600392.6220,1156284.1557,279.0464
5009,600374.9220,1156347.0340,278.0070
5008,600397.7750,1156290.3010,279.0570
6002,600300.9493,1156423.4698,278.6154
6001,600261.8212,1156589.2134,279.0689
5002,600258.3660,1156582.7810,278.8520
5001,600305.7540,1156468.5970,278.4140
```

a) List without topological codes

```
ID,   Y,          X,            Z         C
6005,600013.5779,1157910.6975,243.3187,L1/1L2/2A1
6004,600055.6559,1157930.3558,240.6313,L3/2A2
4003,600010.6320,1157946.7980,241.6320,L4/2A2
4002,600050.8680,1157942.9820,240.0280,L5/2A3
4001,600048.1800,1157901.5720,241.9250,L5/1A3
4001,600048.1800,1157901.5720,241.9250
6003,600392.6220,1156284.1557,279.046416,L6/1A1
5009,600374.9220,1156347.0340,278.0070,L6/2A3
5008,600397.7750,1156290.3010,279.0570,L6/3A3
6002,600300.9493,1156423.4698,278.6154,L7/1L8/2A4
6001,600261.8212,1156589.2134,279.0689,L7/2A4
6001,600261.8212,1156589.2134,279.0689
5002,600258.3660,1156582.7810,278.8520,L7/3A4
5001,600305.7540,1156468.5970,278.4140L8/1
```

b) List with topological codes

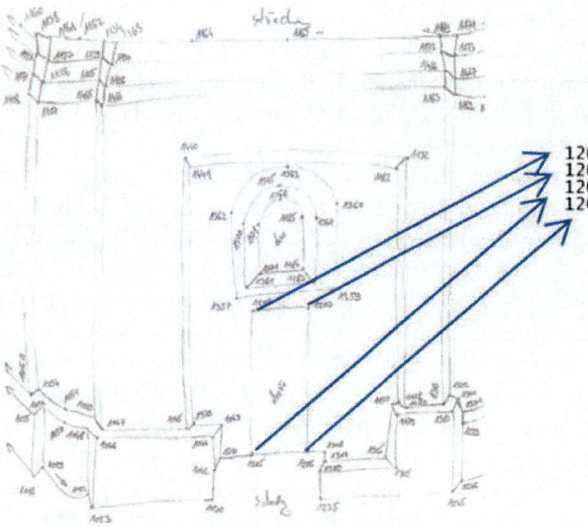

1208	455781.4936	1102921.9187	233.2225	L47/2L48/1A12
1207	455782.3089	1102921.8886	233.2146	L46/2L47/1A12
1205	455781.4755	1102921.7999	231.0791	L45/1L48/2A12
1206	455782.2589	1102921.7383	231.0825	L45/2L46/1A12

Fig. 4 Survey sketch and list of points with coordinates and topological codes

Fig. 5 Wire model in MicroStation software (zoom)

special Python script. This drawing was converted into DXF exchange format and imported in the MicroStation graphics program, where a wireframe model of the church was created after interactive editing—see Fig. 5.

4 Discussion

The proposed method of creating automatic cracks significantly reduces the manual work to create a 3D model. The disadvantage of this solution was the need to add topological codes during field measurements. To this end, topological entities (lines and polygons) have to be marked in the survey sketch. Because topological codes are very simple, this disadvantage is insignificant. The editing of the model (e.g., the insert of one more node) involves rebuilding the topology as well. This problem can be solved by an application to write back the topology into the source list.

5 Conclusions

The proposed method is completely original, significantly reduces the manual work in the CAD system and changes it into field measurement. The quality of the results depends on the quality of geometry and topology during the field survey.

References

CAD, GIS systems for facility management and projection. GISoft, 1995–2017 [cit. 2017-05-21]. Available on: http://www.gisoft.cz/ (in Czech)

Digital Geographic Information Exchange Standard (DIGEST), available on: https://www.dgiwg.org/digest/

Ladner, R, Abdelguerfi, M., Shaw, K.: A framework for 3d synthetic environment construction and rendering. In: Proceedings of the Society of Photo-Optical Instrumentation Engineers (SPIE), vol. 3694, pp.: 4–11 (1999)

Zalik, B.: A topology construction from line drawings using a uniform plane subdivision technique. Comput.-Aided Des. **31**(5), 335–348 (1999). https://doi.org/10.1016/S0010-4485(99)00034-2

GNU Radio-Based FMCW Ground Penetrating Radar for Range Detection

Nabiha Ben Abid, Rim Ghozzi, Samer Lahouar, and Chokri Souani

Abstract

Development of software-defined radio (SDR) that implements signals processing component in the software is rapidly developing now. The SDR is typically designed for communication applications in embedded systems used to emit and receive signals. In this paper, frequency-modulated-continuous-wave (FMCW) ground penetrating radar (GPR) was implemented using the SDR-based methodology relying on the open-source software GNU Radio. An FMCW GPR was developed and used to measure the depth of buried object. The source signal at frequency 259 kHz was transmitted using VCO with sensitivity of 420.7 MHz and a system sample rate of 500 MHz. The simulation results show that this technique can estimate the underground object depth with a relative error (RE) of 4.5%.

Keywords

Frequency-modulated continuous wave (FMCW) • Ground penetrating radar (GPR) • Software-defined radio (SDR) • GNU Radio

N. B. Abid (✉) · R. Ghozzi · S. Lahouar · C. Souani
Laboratoire de Microélectronique et Instrumentation, Faculté des Sciences de Monastir, Université de Monastir, 5000 Monastir, Tunisie

R. Ghozzi
Université de Sousse, Ecole Nationale dIngénieurs de Sousse, 4023 Sousse, Tunisia

S. Lahouar
Technopole of Sousse, Center for Research in Microelectronics and Nanotechnology, CRMN, 4054 Sousse, Tunisia

C. Souani
Université de Sousse, Institut Supérieur des Sciences Appliquées et de Technologie de Sousse, 4003 Sousse, Tunisia

1 Introduction

Ground penetrating radar (GPR) technology has been successfully used to detect, visualize and identify underground structures. It is a non-destructive prospecting technique. GPR has been used for a diverse range of applications including Earth sciences and engineering domains (Liu et al. 2019; Yuan et al. 2018; Ghozzi et al. 2019). In general, the SDR technology provides new opportunities to boost future radar research and development to provide enhanced GPR system capabilities (Ralston and Chad 2012).

Two types of GPR systems can be identified depending on the way data is acquired: either in time domain or in frequency domain (Annan 2009). There are several kinds of GPR systems: impulse (Daniels 1996), stepped frequency continuous wave (SF-CW) or frequency hopping continuous wave (FH-CW) (Carey et al. 2017), M-sequence (Pecovsky et al. 2017) and frequency-modulated continuous wave (FMCW) (Yarleque et al. 2017a, 2017b). The latter uses a linear augmented emitter frequency mixed with reflections; a continuous carrier signal is modulated and transmitted. It is easier to implement and is able to compensate for the imperfections of the radio-frequency (RF) electronics of the antenna and the power supply system.

This paper describes a GPR using a frequency-modulated continuous wave (FMCW) technique. The design and simulation of FMCW GPR for range detection were proposed using GNU Radio. The FMCW GPR is based on sending a periodic frequency-modulated continuous wave in the environment to be studied, using a transmitting antenna (Tx). This wave is scattered off the target on the interface of different dielectric properties and collected by the receiver antenna (Rx).

2 Overview of FMCW GPR

During a sweep period of time 'T_s', the transmitted signal sweeps linearly as a ramp waveform (Yarleque et al. 2017b).

H. El-Askary et al. (eds.), *Research Developments in Geotechnics, Geo-Informatics and Remote Sensing*, Advances in Science, Technology & Innovation,
https://doi.org/10.1007/978-3-030-72896-0_57

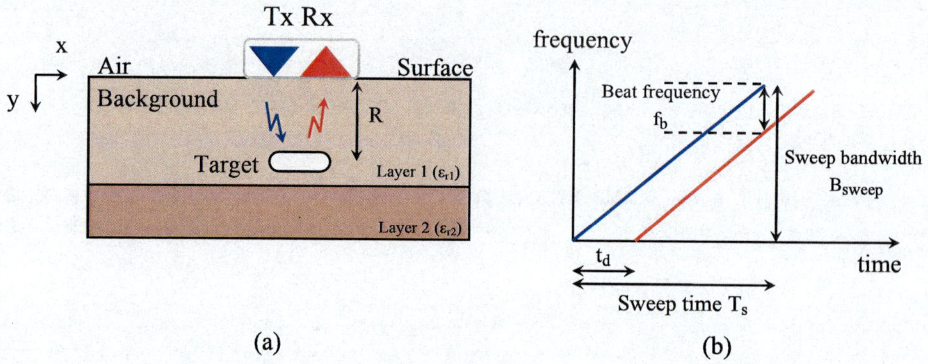

Fig. 1 **a** FMCW GPR test system. **b** An up-chirp linear frequency-modulated signal

The received signal was used to determine the depth '*d*' of the underground target by measuring the delay time 't_d' between the emission of the signal and the recording of its echo. The following equation is obtained by applying the side splitter theorem (see Fig. 1):

$$\frac{t_d}{T_s} = \frac{f_b}{B_{\text{sweep}}} \tag{1}$$

The delay time 't_d' would be determined using the equation of the EM wave, which is given by the following equation (Ghozzi et al. 2018):

$$t_d = \frac{2\sqrt{\varepsilon_{r,1}}R}{c} \tag{2}$$

where *c* is the speed of light in vacuum ($c = 0.3$ m/ns), and $\varepsilon_{r,1}$ is the dielectric constant of layer 1.

By developing the previous Eqs. (1 and 2), the distance to the GPR target '*R*' is given in Eq. 3. We can calculate the depth of the buried object '*R*' by detecting the beat frequency 'f_b'. Figure 2 shows the design of FMCW GPR system. The approach for echo processing in a FMCW GPR system is to mix the received signal with the transmitted signal (Zych 2011).

$$R = \frac{f_b c T_s}{2\sqrt{\varepsilon_{r,1}} B_{\text{sweep}}} \tag{3}$$

3 Simulation, Results and Discussion

To successfully simulate our FMCW GPR system; we need the beat frequency value; to calculate fb we fix the distance $R = 2$ m and $\varepsilon_{r,1} = 4$) ; in this case we apply Eq. (3) (Fig. 3). The specification of FMCW GPR using GNU Radio is reported in Table 1. The input signal for FMCW GPR is obtained from voltage-controlled oscillator (VCO). After applying the IFFT, the simulation result is depicted in

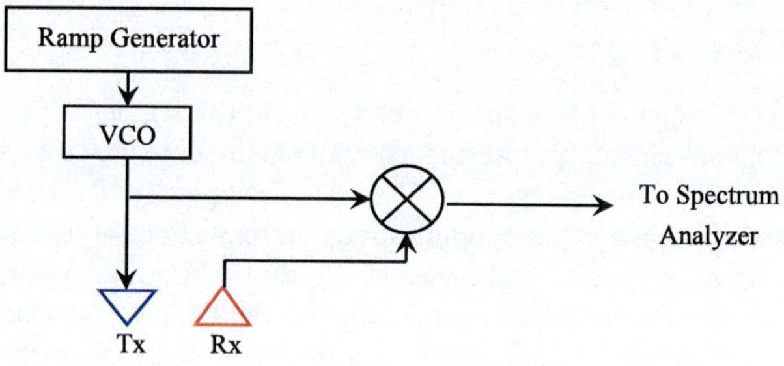

Fig. 2 FMCW GPR system

Fig. 3b. The beat frequency 'f_b' is equal to 441.69 kHz. So, the estimated depth is equal to 1.91 m.

In this paper, FMCW GPR was implemented using an SDR-based methodology and a GNU Radio. It is used to measure the depth of a buried object. The technique has a 4.5% relative error (RE) in estimating depth. The simulation of the FMCW GPR has not generated the correct result which probably affected by the setting delay.

4 Conclusions

GNU Radio was used in this article to design and simulate an FMCW GPR that detects the distance between two targets and achieve a higher and reasonable radar resolution because the simulation can be performed without any extra cost and with little complexity. The FMCW GPR prototype with a 500-MHz sampling frequency, a 420.76-MHz sensitivity and a 259-kHz signal source frequency. Based on the characterization results, it has been demonstrated that the software-defined FMCW GPR system prototype can perform distance detections for a buried object. The simulation results show that this technique can estimate the depth of the underground object with a relative error (RE) of 4.5%. The improvement of the simulation results using a GNU Radio is still in progress and will be presented in a future work. From the research, this GPR system could be very useful for

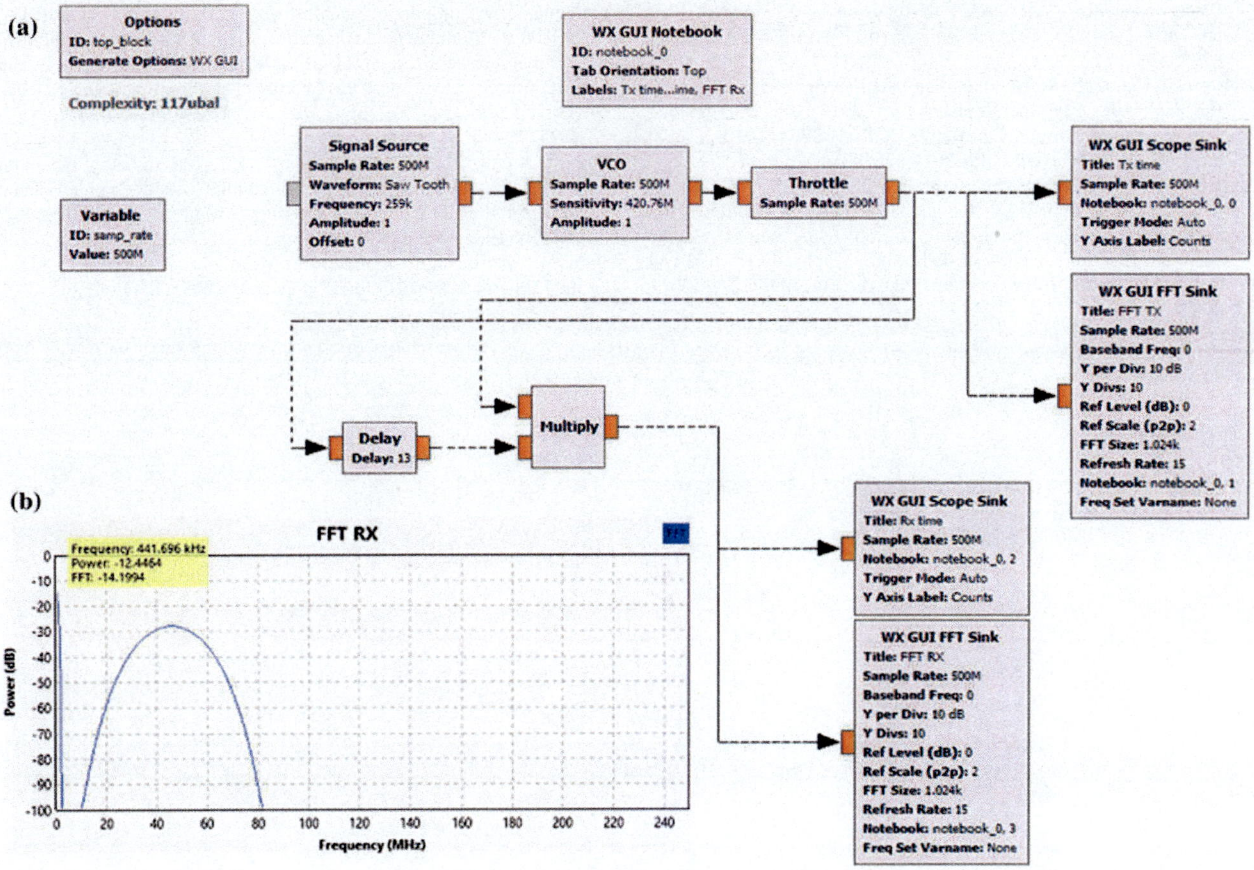

Fig. 3 **a** Block diagram of FMCW GPR using GNU Radio, **b** IFFT result

Table 1 Specification of FMCW GPR

Specification	Sample rate	Sensitivity	Frequency of signal source	Delay	B_{sweep}
Value	500 MHz	420.76 MHz	259 kHz	13	67 MHz

various detection applications. Future work could improve the performance of the GPR system with real-time FMCW GPR using GNU Radio and USRP N210 connected to an antenna to detect more targets and increase the GPR range and resolution.

References

Annan, A.P.: Electromagnetic principles of ground penetrating radar. In: Ground Penetrating Radar: Theory and Applications, 1st edn. Elsevier, Amsterdam. https://doi.org/10.1016/B978-0-444-53348-7.00001-6

Carey, S.C., Scott, W.R.: IEEE: Software defined radio for stepped-frequency, ground-penetrating radar. In: 2017 IEEE International Geoscience and Remote Sensing Symposium (Igarss), pp. 4825–4828 (2017). https://doi.org/10.1109/IGARSS.2017.8128082

Daniels, D.J.: Surface-penetrating radar. Electron. Commun. Eng. J. **8**, 165–182 (1996). https://doi.org/10.1049/ecej:19960402

Ghozzi, R., Lahouar, S., Souani, C.: The estimation of buried empty cylindrical tubes characteristics using GPR. In: Advanced Technologies for Signal & Image Processing (ATSIP'2018). IEEE, Sousse, Tunisia (2018). https://doi.org/10.1109/ATSIP.2018.8364486

Ghozzi, R., Lahouar, S., Souani, C.: An innovative technique for estimating the radius of buried cylindrical targets using GPR. In: Advances in Remote Sensing and Geo Informatics Applications. Springer, Cham (2019). https://doi.org/10.1007/978-3-030-01440-7_35

Liu, X.B., Cui, X.H., Guo, L., Chen, J., Li, W.T., Yang, D.D., Cao, X., Chen, X.H., Liu, Q.X., Lin, H.: Non-invasive estimation of root zone soil moisture from coarse root reflections in ground-penetrating radar images. Plant Soil **436**, 623–639 (2019). https://doi.org/10.1007/s11104-018-03919-5

Pecovsky, M., Sachs, J., Galajda, P., Slovak, S., Kocur, D.: IEEE: Electrically short antenna with ground-coupling measurement for M-sequence ground penetrating radar. In: 2017 Progress in

Electromagnetics Research Symposium—Fall (Piers-Fall), pp. 1404–1408 (2017). https://doi.org/10.1109/PIERS-FALL.2017.8293349

Ralston, J., Chad, H.: Software defined radar: an open source platform for prototype GPR development. In: 2012 14th International Conference on Ground Penetrating Radar (GPR). IEEE (2012). https://doi.org/10.1109/ICGPR.2012.6254855

Yarleque, M.A., Alvare, S., Martinez, H.J., IEEE: FMCW GPR radar mounted in a mini-UAV for archaeological applications: first analytical and measurement results. In; 2017 International Conference on Electromagnetics in Advanced Applications (ICEAA), pp. 1646–1648 (2017). https://doi.org/10.1109/ICEAA.2017.8065606

Yarleque, M.A., Alvarez, S., Martinez, H.J., Canelo, A.C.: IEEE: FMCW GPR radar for archaeological applications: first analytical and measurement results. In: 2017 Xxxiind General Assembly and Scientific Symposium of the International Union of Radio Science (Ursi Gass) (2017). https://doi.org/10.23919/URSIGASS.2017.8105225

Yuan, C.X., Li, S., Cai, H.B., Kamat, V.R.: GPR signature detection and decomposition for mapping buried utilities with complex spatial configuration. J. Comput. Civ. Eng. **32**. https://doi.org/10.1061/(asce)cp.1943-5487.0000764

Zych, M.: Measuring experiment of FMCW ground penetrating radar. In: 12th International Radar Symposium (IRS). IEEE (2011)

Geoprocessing of the Hydro-morphometric Index and Erosion Risk Modeling in the Oued Tessa Catchment

Nassira Zouaoui, Radhia Mansour, and Abdessallem El Ghali

Abstract

Hydric erosion is a natural phenomenon accelerated by humans, which causes upstream and downstream damages. The main causes of this erosive phenomenon are the topography (length and degree of slope), the types of rock, the soil surface's shape, the rain's aggressiveness and the anthropogenic action (Mounia et al. in quantification of water erosion and sedimentation using empirical models in the Tahaddart watershed (Northwestern Rif, Morocco), pp 87–101, 2017). This natural process requires a multidisciplinary mapping highlighting the spatiotemporal evolution of this erosive risk. The geoprocessing of hydro-morphometric indexes, the analysis of satellite image with high spatial resolution and the combination of multi-source data made it possible to establish multifactorial maps (linear erosion, gully erosion, water erosion, soil losses and rain erosion) that show the spatio-temporal evolution of the erosive phenomenon of Oued Tessa catchment in northwestern Tunisia.

Keywords

Hydro-morphometric index • Risk • GIS • Hydric erosion • PAP/CAR • DEM multi-date

1 Introduction

Erosion is a natural process, active through geological time, forming the earth's surface (Rozos et al. 2013). The Oued Tessa catchment is located in northwestern Tunisia. As in the humid Mediterranean regions, this watershed is characterized by torrential and irregular erosive rains combined with deforestation and population growth which causes intense erosion. Erosion rates are strongly related not only to geology but also to climatic conditions. These climates characterized by marked seasonal rainfall or thermal contrasts are the most aggressive (Paskoff 1985). This pluviometric-topographic contrast shows erosion rates that continue to increase from one wet season to another and affect the entire watershed. Water erosion is a risk for property and people, an economic issue, an issue for drinking water and also an agricultural issue (Ouvry 2012). For this reason, vulnerable areas', such as areas with steep slopes and/or degraded soils, natural hazards that often occur can cause significant damage and lead to loss of life and property (Bathrellos et al. 2012).

In this context, the aim of this study is to produce multi-factor maps of erosive risk by exploiting the data offered by digital cartography and remote sensing, which requires the use of several parameters that vary in time and space at different scales. To achieve the main objective, an appropriate methodological organization chart was followed to highlight the areas at potential risk of water erosion.

2 Materials and Methods

The present study consists of mapping erosive risk zones in the Oued Tessa watershed in northwestern Tunisia. To achieve the main scientific goal, it is useful to use monthly climate data, multi-date digital terrain models and satellite imagery using an appropriate methodological diagram (Fig. 1).

Vector and matrix geoprocessing in a GIS such as morphometric clues (erosive power index, sediment transport index, flow debris index, topographic moisture index and slope length and slope factors) and the surface indices (gloss index) make it possible to highlight the local erosive risk. To validate the results and gather the maximum amount of information, it is necessary to go through: (1) the PAP/CAR model for zoning according to the degree of susceptibility to

N. Zouaoui (✉) · R. Mansour · A. E. Ghali
Faculty of Sciences of Bizerte, Carthage University, 7021 Bizerte, Tunisia

H. El-Askary et al. (eds.), *Research Developments in Geotechnics, Geo-Informatics and Remote Sensing*, Advances in Science, Technology & Innovation,
https://doi.org/10.1007/978-3-030-72896-0_58

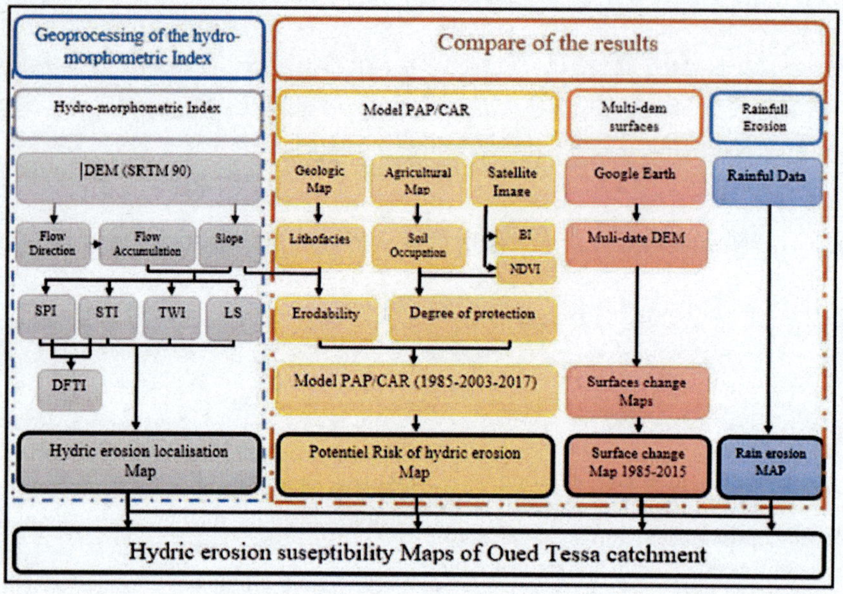

Fig. 1 Adopted methodologic flowchart

erosion, etc. (2) the muli-date surface maps highlighting the soil losses, etc. (3) the rainfall erosion map (Arnoldus index and the rain aggression index) to determine the zones sensitive to water erosion.

3 Results

Understanding such a phenomenon requires geometric and kinematic study at different scales. The irregularity of precipitation in relation to the nature of lithofacies and the relief accentuates the erosive risk. The results obtained by changing the surface area of the Oued Tessa catchment by stretching the multi-date DEM between 1985 and 2015 show a significant loss of soils which reaches 49,057 ha between the years 2000 and 2005, and this period is marked by the floods of 2003, and 67,627 ha between 1985 and 2015 a spatio-temporal evolution of erosion phenomenon and therefore surfaces of the study area (Table 1).

Generally, the most threatened areas by erosion are the steeply sloping areas of the Oued Tessa watershed, which show them the resulting maps of the hydro-morphometric indexes and validate by the PAP/CAR model (Fig. 2).

The PAP/CAR method allows the study of erosion sensitivity where the main factors controlling the erosive functioning are: vegetation cover, slope, and lithology. The results make it possible to spatialize vulnerable areas, classify areas according to their degree of sensitivity to erosion, and predict future behavior in the study area (Ben Rhouma et al. 2018).

4 Discussion

The multifactorial derived maps of erosive risk are the result of the combination of elementary maps (geological, lithological, morphometric, hydrological, climatic, etc) by spatial algebra: ARC GIS calculator highlighting the possibility of evaluation of lands susceptible to erosive risk. The SPI, STI, DTII, TWI and LS hydro-morphometric indexes have made it possible to highlight the areas susceptible to water erosion. These results are validated by the PAP/CAR model for the years 1985, 2203 and 2017, soil losses between 1985 and 2015 and the rainfall erosivity map.

The Oued Tessa watershed has 31 rainfall stations. The Arnoldus index calculation takes into account the monthly

Table 1 Table of soil loss in the Oued Tessa catchment between 1985 and 2015

Year	1985–1190	1990–2000	1995–2000	2000–2005	2005–2010	2010–2015	1985–2015
Accumulation area surface (ha)	59,256	72,051	38,775	16,521	44,421	60,740	29,874
Erosion area surface (ha)	27,764	27,488	26,725	49,057	36,810	19,856	67,627

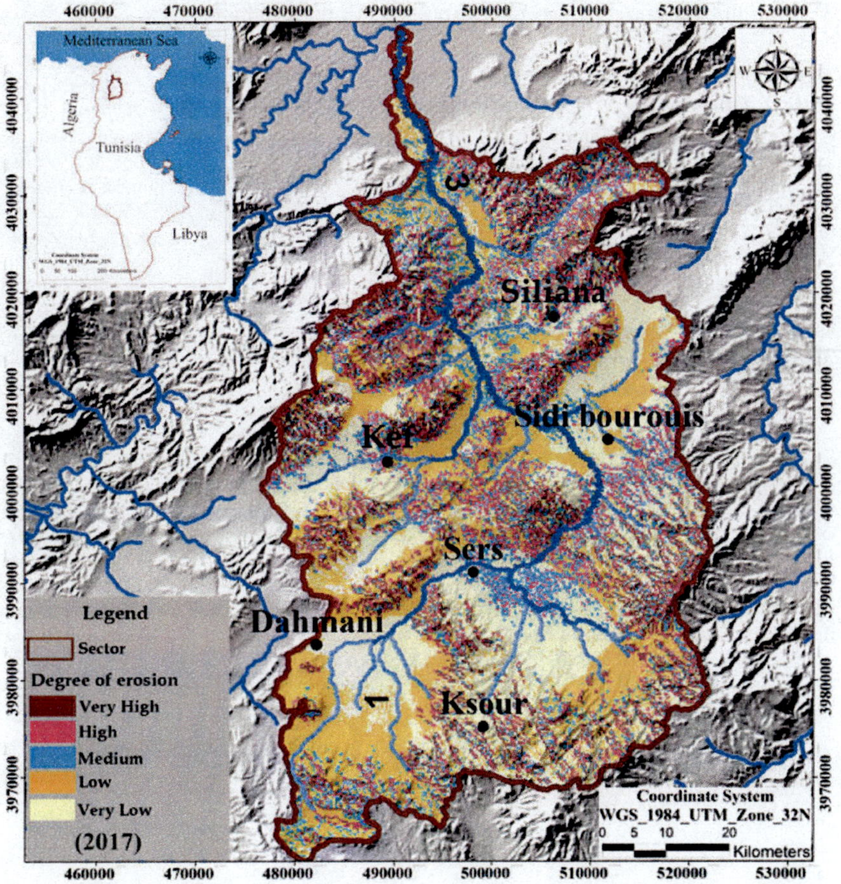

Fig. 2 Potential risk of hydric erosion Map (PAP/CAR) of Oued Tessa catchment (2017, SRTM 90)

rainfall of all the months of the year which makes the values closer to reality (Maamar-Kouadri et al. 2016), which is why this study was based on a series of monthly rainfall between 1987 and 2015. The analysis of precipitation data makes it possible to understand current climate trends and to seek an explanation for the erosive phenomenon affecting the Tessa wadi watershed and evaluating morpho-film system responses to climate change (Giaccone et al. 2015). Most precipitation indexes show statistically significant trends in the region with an upward trend, especially for extremely heavy rainfall, very rainy days, and maximum rainfall of 1 or 5 consecutive days (Feki 2016). The forms of slope erosion (diffuse and linear) generally occur under conditions of intense rain on bare ground or little covered. Gully erosion forms occur in all areas where concentrated flow velocity is greater than the soil resistance to pull-out (Ouvry 2012).

The high SPI values indicate the erosive zones, while the low values indicate the deposition zones.

The STI index shows that the higher the speed of the river, the higher the transport capacity.

The DFTI index is used to initiate the debris flow.

The index "LS" influences the flow and accumulation zones of the flows. Its effect is remarkable thanks to the most powerful "STI" and "SPI" indexes (Dehni et al. 2015).

5 Conclusions

The combination of basic multi-source and multi-date data organized according to the adopted methodological chart allows the development of multi-area maps highlighting the area most at risk of erosion risk in the Oued Tessa catchment.

This study is qualitative–quantitative showing the effectiveness of the PAP/CAR method in making a general diagnosis of the potential risk of water erosion and the effectiveness of the multi-DEM method in determining the erosion rate highlighting that steeply sloped areas are the most threatened by water erosion which shows them hydromorphic indexes.

This study constitutes a document of reference in the adjustments to envisage for the watershed of Oued Tessa.

To limit the nuisances related to erosion and to preserve the soil resources, several means of erosion control are available. Collective developments help to protect downstream areas by preventing erosion in upstream areas, including collective reasoning of the location of crops on a watershed scale and the development of retention ponds at the watershed downstream (Antoni et al. 2015).

References

Antoni, V., Darboux F.: Soil erosion. https://hal.archives-ouvertes.fr/hal-01189811 (2015)

Bathrellos, G.D., Gaki-Papanastassiou, K., Skilodimou, H.D., Papanastassiou, D., Chousianitis, K.G.: Potential suitability for urban planning and industry development by using natural hazard maps and geological—geomorphological parameters. Environ. Earth Sci. **66**(2), 537–548 (2012)

Belaid, H.: Study of the heights and erosivity of the rains at two stations of LaMedjerda, Bousalem and Medjez El Bab with reference to the torrential rains of 1973 and 2003. Review of Arid Regions n°41 (1/2017) (2016)

Ben Rhouma, A., Hermassi, T., Bouajila, K.: Water erosion modeling using the PAP/CAR qualitative method: case of the Sbaihia catchment, Zaghouan. J. New Sci. Agric. Biotechnol. **51**(11), 3225–3236 (2018)

Dehni, A., Lounis, M., Hassani, M.I.: Geoprocessing of hydro-morphometric indices for the automation of sedimentary and erosive models (Application on the BV of Tafna—North-West Algeria) Conference (2015)

Feki, H., Dridi, S.: Variability of extremes precipitation indices over Medjerda basin. Review of Arid Regions n°41 (1/2017) (2016)

Giaccone, E., Vergari, F., Del Monte, M., Fratianni, S.: The impact of climate on morphological dynamics in Tuscany (Central Italy). In: XXVIIIth Symposium of the International Association of Climatology, Liège (2015)

Maamar-Kouadri, K., Kouri, L., Chebouti, Y.: Using the Arnoldus Indice to map the risk of rainfall Erosivity in the Tell Oran (Algeria). Geo-Eco-Trop. **40**(4), 287–296 (2016)

Mounia, T., Hassan, T., Fatima, E., Mhamed, A., Abdelfatah, T., Hassan, E.: Quantification of water erosion and sedimentation using empirical models in the Tahaddart watershed (Northwestern Rif, Morocco), pp. 87–101 (2017)

Ouvry, J.-F.: Soil Degradation by Water Erosion: What Remedies in the Field of Field Crops. https://www.cairn.info/revue-pour-2012-1-page-163.htm (2012)

Paskoff, R.: Géographie de l'Environnement. Publications de l'université de Tunis, 227p (1985)

Rozos, D., Skilodimou, H.D., Loupasakis, C., Bathrellos, G.D.: Application of the revised universal soil loss equation model on landslide prevention. An example from N. Euboea (Evia) Island, Greece. Environ. Earth Sci. **70**(7), 3255–3266 (2013)

Assessment of Water Soil Erosion by RUSLE Model Using Remote Sensing and GIS in Wadi Cheliff Basin (Algeria)

Samir Toumi, Mohamed Meddi, and Gil Mahé

Abstract

Soil degradation is a serious and extensive problem in many areas in Algeria. Information on soil loss is essential to support agricultural productivity and natural resource management. In this study, remote sensing and GIS are used in order to map and assess water erosion in the Wadi Cheliff basin. The study area is characterized by a great irregularity in rainfall which is mainly generated during storm events. The region is also known for its steep slopes, a marly lithology and a very weak vegetation cover, which makes it more exposed and vulnerable to erosion. The RUSLE model can be described as the combination of six factors that feature the specific characteristics of the watershed. Data from remote sensing analysis and GIS are used to individually assess and map each of these factors. The integration of thematic maps for the RUSLE factors in the GIS is used to identify the impact of each factor on soil loss, as well as to rank the relative importance of the erosion zones, and further quantify the soil loss for the investigated region. The range of soil loss in the study area varies from 0 to 120 t/ha/year, and the area at high risk of erosion is located in the north and center parts of the region.

Keywords

Soil degradation • GIS • Remote sensing • Wadi Cheliff basin • Algeria

S. Toumi (✉) · M. Meddi
Ecole Nationale Supérieure dHydraulique, 29 route de Soumaa, BP 31 09000 Blida, Algeria
e-mail: s.toumi@ensh.dz

G. Mahé
IRD and Université Montpellier, Laboratoire HydroSciences Montpellier, Montpellier, France

1 Introduction

Knowing seriousness of soil erosion and its impacts, an appropriate management measure must be taken. In order to dress a global management plan, a good knowledge of the spatial soil erosion variations is necessary when planning conservation efforts (Tamene et al. 2006). The Universal Soil Loss Equation (USLE) developed by Wischmeier and Smith in 1978 (Wischmeier and Smith 1978) is an empirical soil erosion model that is used most frequently by technicians to predict soil loss by water erosion (Vezina et al. 2006; Trinh 2015; Nguyen 2011; Mc Cool et al. 1987). Revised Universal Soil Loss Equation (RUSLE) (Renard et al. 1996) simulated by GIS and remote sensing techniques is used to estimate and map the spatial pattern of annual soil loss rate. The approach used was initially intended to detect the factors triggering erosion. The collected data are used to evaluate and map each factor individually. Erosion data for some indicators are collected also, calibrated, and added to a GIS database, after which it will be spatially modeled to represent the risk of soil erosion in any element of the landscape. In a raster image embedded in a GIS, the RUSLE model can predict the erosion potential on a one-pixel basis. It presents distinct advantages when attempting to identify spatial patterns of soil loss present in a large area. GIS can then be used to isolate and query the databases created to obtain crucial information about the role of individual factors.

2 Materials and Methods

In this study, RUSLE model was integrated with GIS and remote sensing techniques to conduct cell by cell calculation the mean annual soil loss rate estimation (t ha^{-1} year^{-1}) and to identify and map soil erosion risk areas over wadi Chéliff. Raster map of each RUSLE parameters derived from different data sources was produced and discussed as follows. RUSLE model is a worldwide used model where soil erosion

H. El-Askary et al. (eds.), *Research Developments in Geotechnics, Geo-Informatics and Remote Sensing*, Advances in Science, Technology & Innovation,
https://doi.org/10.1007/978-3-030-72896-0_59

by water has to be estimated (Laflen and Moldenhauer 2003). It is described by the following equation:

$$A = R.K.LS.C.P$$

where

A is the average annual soil loss (tons ha^{-1} year^{-1}),

R is the rainfall erosivity (MJmm ha^{-1} h year),

K is the soil erodibility factor (tons ha^{-1} MJmm ha^{-1} h year),

LS is the topographic factor (dimensionless),

C is the cropping management factors (dimensionless), and

P is the practice support factor (dimensionless).

The map of R factor was extracted from the map of the spatial rainfall erosivity distribution over northern Algeria developed by Meddi et al. (2016). This approach has been used widely to map climate effect on soil erosion (Meusburger et al. 2012; Parajka et al. 2005) in areas where the number of gauging locations is deemed too low (Zhang and Chiew 2009).

To estimate the K factor, FAO HWSD database was used to determine the percentage of each component (Renard et al. 1996). For each texture class, clay, silt, and sand were estimated on the basis of assigned data. The profile permeability and structure can be estimated based on soil texture and organic matter content with the following table based on the Roose (1996) methodology. The following equation can be used to calculate K (Wischmeier and Smith 1978):

$$K = 27.66m^{1.14}.10^{-8}.(12 - a) + (0.0043.(b - 2)) + (0.0033.(c - 3))$$

In which K = soil erodibility factor (tha/MJmm) m = (silt (%) + fine sand (%))(100-clay (%)) a = organic matter (%) b = structure code: (1) very structured or particulate, (2) fairly structured, (3) slightly structured and (4) solid c = profile permeability code: (1) rapid, (2) moderate to rapid, (3) moderate, (4) moderate to slow, (5) slow, and (6) very slow.

DEMs derived from SRTM data (Jarvis et al. 2008), with 30 m spatial resolution, were employed to obtain the topographic factor.

The factor (C) is estimated from the normalized difference vegetation index (NDVI) using the equation developed by as follows:

$$C = (-\text{NDVI} + 1)/2$$

NDVI is determined using a Landsat-TM 8 satellite image 30 m resolution (date: May 8, 2015), such as

$$\text{NDVI} = \text{NIR} - \text{RED}/\text{NIR} + \text{RED}$$

The layers necessary for the calculation of NDVI are: the visible band of the red spectrum (RED) and the near infrared band (PIR) of the Landsat-TM 8. Both tapes are extracted from the same scene (path and row, 197/35-176/36) and taken on the same date (May 8, 2015).

Due to the lack of information on the erosion practices and surface characteristics of the study area, we have set the value of the P-factor to 1 over the entire watershed. We opted for this approach because support practices cover small areas as well and this work could not be observed on Landsat-TM satellite imagery.

3 Results

The erosive risk map is determined by multiplying the 30 m resolution layers representing the RUSLE parameters through a matrix format. Figure 1 not only shows the loss of soil for each pixel, but also provides information on critical soil loss risk areas.

4 Discussion

The low erosive risk values cover the northern and southwestern part of the basin. The land in these areas is almost flat (slope less than 10%). While strong values are located in the eastern and the western region of the territory. The erosive risk is important in these regions because of the steep slope of the land, but also the high values of the factor R despite the existence of vegetation in certain regions (protective factor). According to the superimposition of the layers of the different interacting factors on water erosion in Wadi Cheliff basin, the latter was divided into three regions according to the degree of erosive risk. The southern region is characterized by a consolidated geological formation, of weak slope and with more or less low vegetal cover. This area is considered as low risk erosion zone. The area at high risk of erosion is located in the north and center of the region, its geological formation is essentially marl ($K > 0.1$ tons ha^{-1} MJmm ha^{-1} h year), the soil is globally bare ($C < 0.5$), steep (LS > 10) and more exposed to agricultural exercises. Other areas have flat terrain of less erosive risk.

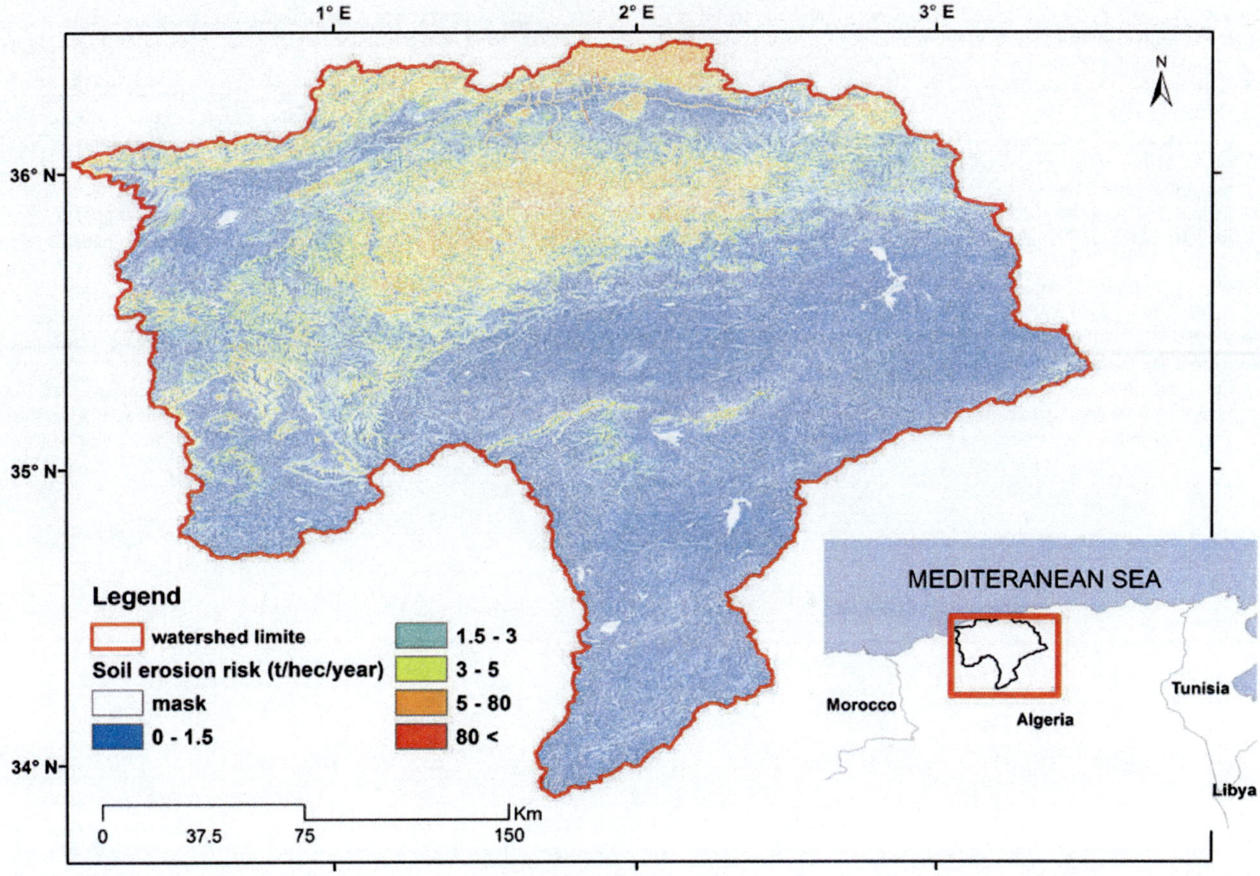

Fig. 1 Final map of soil loss risk

5 Conclusions

The watershed of Wadi Cheliff is facing a serious problem of soil degradation, like most parts of the Western region of Algerian tell. Water erosion is stronger in this region, especially because rainfall is mainly thunderstorms, which are very intense and have a high erosive power at the beginning of the rainy season. This is also explained by the presence of a sparse vegetation cover. This problem causes rapid siltation of many dams located in the area. This also affects local agriculture, causing huge losses of arable land, leading farmers to rural exodus. Natural factors (precipitation, topography, geology, lithology and vegetation) have played an important role in the erosion processes in the study area, but the fact remains that the acceleration of these phenomena largely depends on management and use of space.

The current soil erosion risk map at the Wadi Cheliff watershed scale is surely the best that can be obtained from the available data. The results can be improved by using a more detailed digital terrain model, satellite data that have better spectral and geometric characteristics, and more detailed soil information. The analysis of land-use dynamics can also provide us with additional information on the impact of climate and anthropogenic change on erosion.

Acknowledgements Thanks to ANRH for the hydrological and rainfall data. This work was done in the framework of Young Teams Associated with the IRD of ENSH Blida (Jeunes Équipes Associées à l'IRD).

References

Jarvis, A., Reuter, H.I., Nelson, A., Guevara, E.: Hole-filled SRTM for the globe Version 4, available from the CGIAR-CSI SRTM 90m Database (http://srtm.csi.cgiar.org) (2008)

Laflen, J.M., Moldenhauer, W.C.: Pioneering soil erosion prediction: the USLE story. World Association of Soil and Water Conservation, Beijing, China, p. 54 (2003)

Mc Cool, D.K., Brown, L.C., Foster, G.R., Mutchler, C.K., Meyer, L. D.: Revised slope steepness factor for universal soil loss equation. Trans. Am. Soc. Agric. Eng. **30**(5), 1387–1396 (1987)

Meddi, M., Toumi, S., Assani, A.: Spatial and temporal variability of the rainfall erosivity factor in Northern Algeria. Arab. J. Geosci. **9**, 282 (2016). https://doi.org/10.1007/s12517-015-2303-8

Meusburger, K., Stee, L.A., Panagos, P., Montanarella, L., Alewell, C.: Spatial and temporal variability of rainfall erosivity factor for Switzerland. Hydrol. Earth Syst. Sci. **16**, 167–177 (2012)

Nguyen, M.H.: Application USLE and GIS tool to predict soil erosion potential and proposal land cover solutions to reduce soil loss in Tay Nguyen. In: Proceedings of FIG Conference: Bridging the Gap between Cultures. Marrakech, Morocco, 18–22 May 2011 (2011)

Parajka, J., Merz, R., Blschl, G.: A comparison of regionalization methods for catchment model parameters. Hydrol. Earth Syst. Sci. **9**, 157–171 (2005)

Renard, K.G., Foster, G.R., Weesies, G.A., McCool, D.K., Yoder, D. C.: Predicting soil erosion by water: a guide to conservation planning with the revised universal soil loss equation (RUSLE). Hand book. no. 703. Washington, DC: USDA, Agricultural (1996)

Roose, E.: Land Husbandry -Components and strategy. 70 FAO Soils Bulletin, Food & Agriculture Organization of the UN, Rome, Italy (1996)

Tamene, L., Park, S., Dikau, R., Vlek, P.L.G.: Analysis of factors determining sediment yield variability in the highlands of Ethiopia. Geomorphology **76**, 76–91. Author, F.: Article title. Journal **2**(5), 99–110 (2016)

Trinh, C.T.: Soil erosion in Vietnam (The case of Buon Yong catchment). Scholars Press. ISBN 978-3-639-51692-0 (2015)

Vezina, K., Bonn, F., Pham, V.C.: Agricultural land-use patterns and soil erosion vulnerability of watershed units in Vietnam's northern highlands. J. Landsc. Ecol. **21**(8), 1311–1325 (2006)

Wischmeier, W.H., Smith, D.D.: Predicting Rainfall Erosion Losses: A Guide to Conservation Planning. USDA Agriculture Handbook No. 537, USDA, Washington DC (1978)

Zhang, Y.Q., Chiew, F.H.S.: Evaluation of regionalization methods for predicting runoff in ungauged catchments in southeast Australia. 18th world IMACS/MODISM Congress, Cairs, Australia (2009)

Tectonic Settings and Neo-tectonics of the Arabian Peninsula as Deduced from Satellite Altimeter and Gravity Data

Khaled Zahran

Abstract

The Arabian Peninsula is an interesting region from both tectonic and seismic perspectives. It shows an active geologic structure attributed to the tectonic movement of the rifting of the Red Sea from the west and the collision of the Arabian Plate with the Persian Plate to the northern and northeastern boundaries. To the south, similar rifling running in a more East–West direction through the Gulf of Aden has separated the Arabian Peninsula from Africa. These complicated tectonics attributed with seismological activities indicate seismological activities along the neo-tectonic zones. Radar altimetry data have been used to derive gravity and its variations over the oceans. It is considered as an excellent tool for mapping sea floor structures, including tectonics, sea mounts and rifts. Besides, the Gravity Recovery and Climate Experiment (GRACE), satellite mission has widely demonstrated its sensitivity to ongoing mass redistribution within the various sub-systems of the earth. Finally, Gravity field and steady-state Ocean Circulation Explorer (GOCE) satellite is the first satellite mission that observes gradient of the Earth gravity field from space. Integrated satellite gravity data have been used to delineate the tectonic settings and the neo-tectonics of the Arabian Peninsula. Results show important zones of mass discontinuity in this region correlated with the seismological activities and temporal gravity variation. The current study indicates that satellite gravity data is a valuable source of data in understanding the geodynamical behavior of the studied region.

Keywords

Arabian Peninsula tectonics • Satellite altimeter • Satellite gravity

K. Zahran (✉)
National Research Institute of Astronomy and Geophysics, Helwan, Cairo, Egypt

1 Introduction

The Arabian plate separated from Africa approximately 25 million years ago, resulting in the closure of the subducting Tethys Sea in the northeast (Stern and Johnson 2010). The rifting of Africa and Arabia opened the Red Sea as well as the Gulf of Aden, bordering the plate on the southwest and southern sides, respectively. The Red Sea has a spreading half rate of roughly 10 mm/yr, and the Gulf of Aden is believed to be a propagating rift opening westward at a rate of approximately 3 mm/yr (Vita-Finzi 2001). As the Red Sea continued to open, the Arabian plate eventually collided with Eurasia, and this collision zone is delineated by the Zagros Mountains. This mountain chain has a shortening rate of 9 ± 3 mm/yr in the SE and 5 ± 3 mm/yr in the NW (Hessami et al. 2006). The Dead Sea transform fault borders the Arabian plate in the west and has experienced approximately 107 km of left-lateral displacement since the mid-Miocene, which translates into roughly 6–10 mm/yr of slip on the fault (Vita-Finzi 2001). In addition, the continuous seismicity attributed to its tectonic settings, which affect almost all countries surrounding this region, indicates the activities of these tectonics. Tectonic boundaries of the Arabian Plate are given at Fig. 1. Thus, tectonics and geodynamics of this region have always been the attention of those interested in earth sciences.

2 Methodology

The main objective of this study is to figure out the tectonic settings and the geodynamics behavior of the Arabian Peninsula using integrated recent satellite data. To this end, radar altimetry data were used to derive gravity and its variations over the world's oceans and an excellent tool for mapping sea floor structures, including tectonics, sea mounts and rifts. Satellite altimeter data were used to determine spatial gravity map of the marine region of the studied area and to delineate the attributed to the stress–strain

H. El-Askary et al. (eds.), *Research Developments in Geotechnics, Geo-Informatics and Remote Sensing*, Advances in Science, Technology & Innovation,
https://doi.org/10.1007/978-3-030-72896-0_60

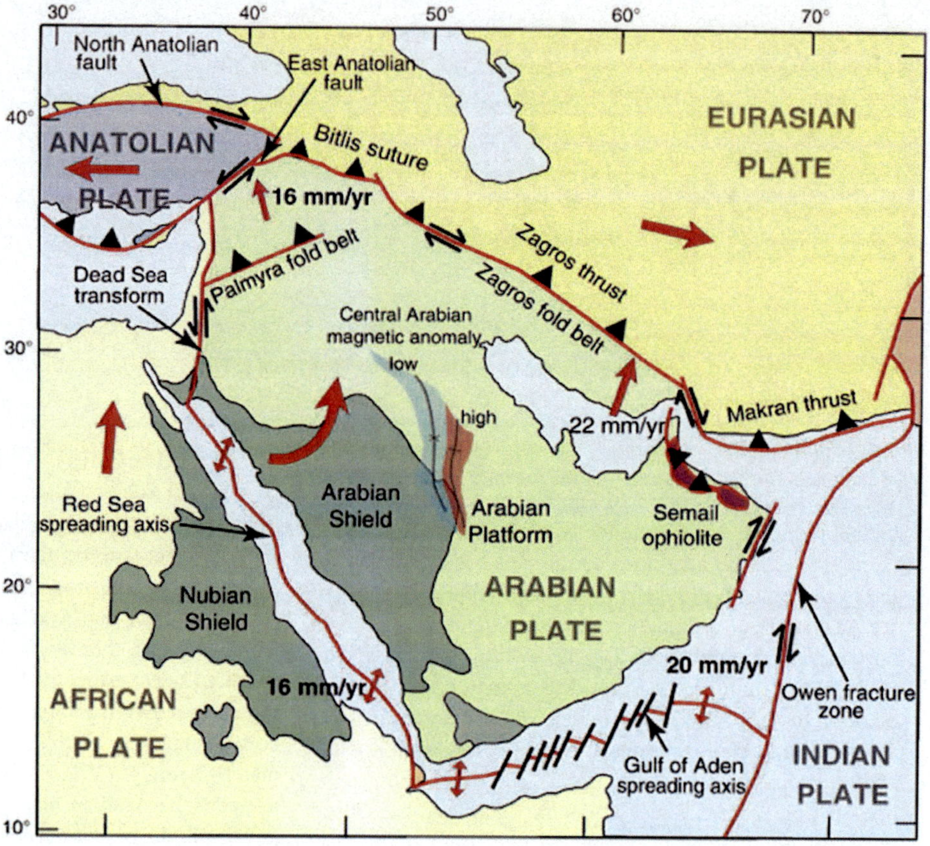

Fig. 1 Arabian plate boundaries adapted from Stern and Johnson (2010)

accumulation and its relation to the earthquake occurrence. The Gravity Recovery and Climates Experiment (GRACE) were monitoring time-varying changes of the earth's gravitational field on a near global scale since 2003. Tapley et al. (2004) showed the potentiality of GRACE measurements of mass variability in the Earth system including mass redistribution of tectonically seismic active region. 179 monthly gravity field solutions (RL06 unconstrained solutions) were used for the period 2003 to 2016 from the GRACE database provided by the Center of Space Research of the University of Texas. The gravity field solutions were processed as follows: (1) The temporal mean was removed; (2) correlated errors were reduced by applying de-striping methods developed by Swenson and Wahr (2006); (3) spherical harmonic coefficients were converted to grids (0.5° × 0.5°) of equivalent temporal gravity using a Gaussian smoothing function with a radius of 400 km. Furthermore, the use of GOCE data provides better information for the determination of regional gravity field. It can be used to predict surface gravity anomalies in unsurveyed areas of the Earth. Data from GOCE gravity mission will be used to determine the tectonic boundaries of the selected region. Finally, integrated satellite data will be utilized to figure out the geodynamics behavior of the studied region and its neo-tectonic activities.

3 Results

The satellite altimeter free-air gravity anomaly of the Arabian Plate Sea region is given in Fig. 2. The gravity map of the studied region delineates the tectonic settings clearly. Interpretation of the spatial gravity field variation shows important zones of mass discontinuities and delineate sub- and minor plates responsible of seismological activities.

Annual temporal gravity variation of the Arabian Peninsula from 2003 to 2016 computed from GRACE satellite data is given at Fig. 3. The figure includes earthquakes with magnitude more than 3. The significant temporal gravity variation indicates that the seismic activities attributed to a significant mass redistribution around the seismo-active zones. The wide range of the gravity anomalies which appear on the figure suggests deep mass discontinuities at the lower crust and the upper mantle. The plate conjunction is seen as negative gravity anomaly at the African side met by a positive gravity anomaly at both sides. Finally, Bouguer anomaly map of the studied region as computed using GOCE gravity data is given at Fig. 4. The figure shows important zones of mass discontinuities correlated well with the tectonic settings of Arabian Peninsula.

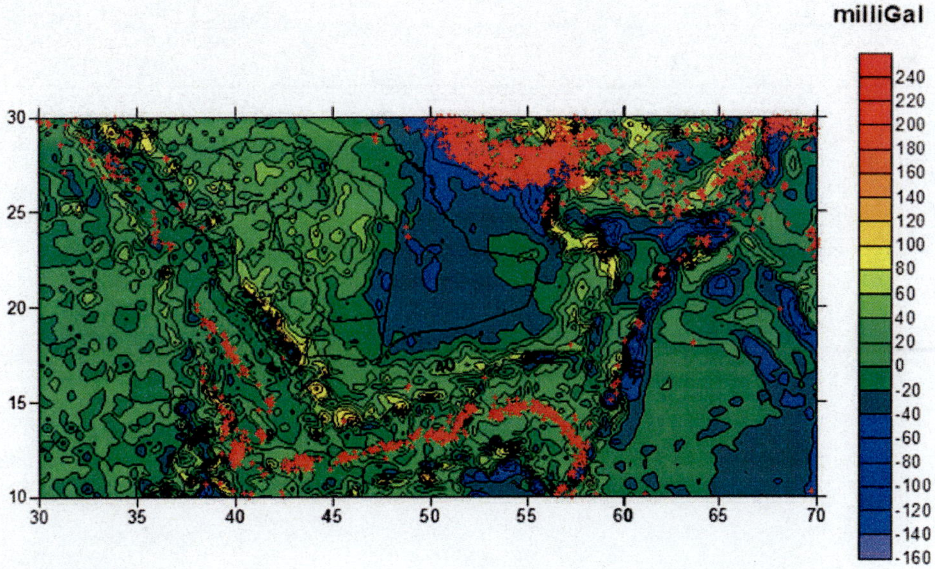

Fig. 2 Free-air gravity anomaly map of the Arabian Peninsula derived from satellite altimeter

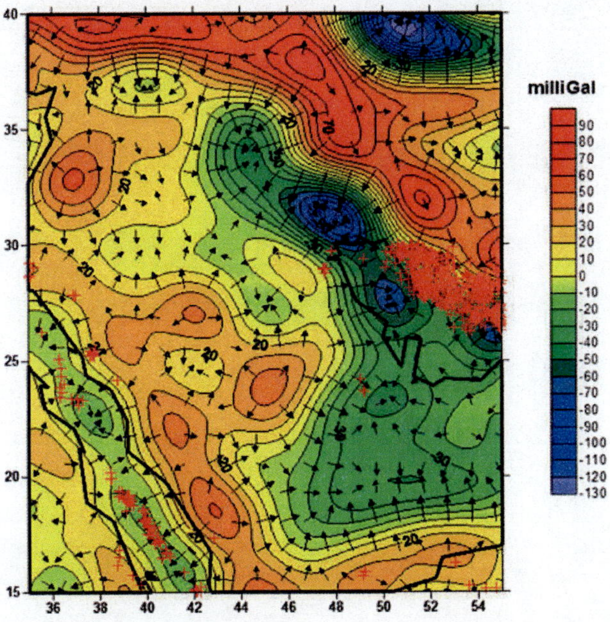

Fig. 3 Annual temporal gravity variation from 2003 to 2016 of the Arabian Peninsula from GRACE

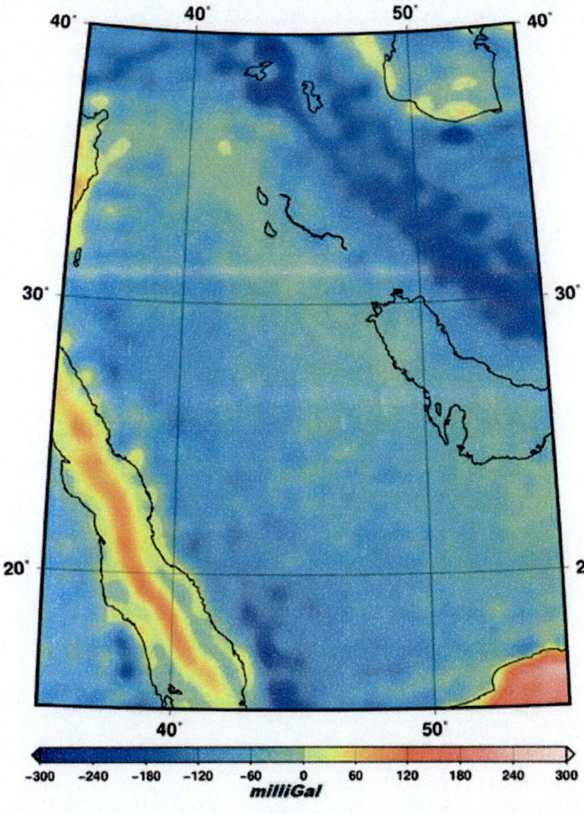

Fig. 4 Ann spatial gravity variation of the Arabian Peninsula from GOCE

4 Conclusions

The study highlights the importance of considering spatial and temporal gravity to the deformation and tectonic studies for completeness of the geodynamic studies. Generally, it can be stated that the satellite data offers additional and complementary data sets to help the geoscientists to determine the Earth's internal structure and tectonics. Determination of spatial and temporal earth's gravity field from satellite opens new perspectives on earth sciences and provides valuable information of the geodynamic studies.

References

Hessami, K., Nilforoushan, F., Talbot, C.J.: Active deformation within the zagros mountains deduced from GPS measurements. J. Geol. Soc. **163**, 143–148 (2006)

Stern, R.J., Johnson, P.: Continental lithosphere of the arabian plate; a geologic, petrologic, and geophysical synthesis. Earth-Sci. Rev. **101** (1–2), 29–67 (2010)

Swenson, S., Wahr, J.: Post-processing removal of correlated errors in GRACE data. Geophys. Res. Lett. **33**, L08402 (2006). https://doi.org/10.1029/2005GL025285

Tapley, B.D., Bettadpur, S., Ries, J.C., Thompson, P.F., Watkins, M. M.: GRACE measurements of mass variability in the Earth system. Science **305**, 503–505 (2004). https://doi.org/10.1126/science.1099192

Vita-Finzi, C.: Neotectonics at the arabian plate margins. J. Struct. Geol. **23**(2–3), 521–530 (2001)

Reconstructing the Palaeohydrology of the Eastern Sahara Using Spaceborne Remote Sensing

Robert Corrie, Cordula Robinson, and Thomas Farr

Abstract

Multimodal data are applied towards palaeohydrological analysis of the eastern Sahara, including Egypt, Libya, Sudan and Chad. Relevant applications include inventorying groundwater resources for long- and short-term planning through archaeological investigations and mapping ancient caravan routes. Results suggest that five major outflows occurred during the past pluvial periods, including the Sahabi, Gilf Kebir and the Kufra, in addition to two that are newly identified. All northern channels appear to discharge into a previously unknown large inland drainage area with possible outflow to the Mediterranean Sea. Visible near infrared false color composites and thermal infrared spectral analysis further identify seven potential areas with lake deposits throughout the region, four of which are newly mapped.

Keywords

Remote sensing • SAR • GIS • Palaeohydrology • Sahara

1 Background

Observations from NASA's first Shuttle Imaging Radar (SIR-A) mission on-board the Space Shuttle Columbia, in 1981, revealed a network of previously unknown palaeorivers buried beneath the desert sand sheet of Egypt and Sudan (McCauley et al. 1982). These ancient drainage channels, also known as "radar rivers", indicate major climate shifts occurred across north Africa, transforming a once vegetated landscape into the hyper-arid desert it is today (Kuper and Kröpelin 2006). Causes for the climate shifts are not well understood, but variations in solar insolation due to slow shifts in the Earth's orbital parameters—the so-called Milankovitch variations—have been proposed (Berger 1978). The work described here extends the early analysis by McCauley et al. (1982) to produce a comprehensive water resource management inventory throughout the eastern Sahara and North Africa. New semi-automated approaches are applied that include the use of PALSAR data, SRTM DEM mosaics to artificially flood the palaeolandscape and the application of flow models to visualize the extent and distribution of the water.

2 Data

The data set comprises (1) near- and sub-surface palaeorivers (observed in C-band SRTM and L-band PALSAR images); (2) slope and aspect (derived from the SRTM DEM); (3) oases (interpreted from ASTER and Landsat ETM + NDVI); (4) palaeolake carbonate deposits (derived from ASTER TIR band ratios); (5) palaeolake topographic depressions and associated shoreline features (from SRTM DEM); (6) caravan routes and other surface lineaments (observed in ASTER VNIR and TIR) and (7) archaeological remains, including habitation sites (from VNIR imagery).

3 Methodology

The assembly of previously described co-registered data provides inputs for hydrological modelling and analysis. SRTM DEM mosaics are used to flood the extracted palaeolandscape with "artificial water", and a GIS flow

R. Corrie (✉)
The Queen's College, Oxford University, Oxford, UK
e-mail: robert.corrie@queens.oxon.org

C. Robinson
Northeastern University, Boston, MA 02115, USA

T. Farr
Jet Propulsion Laboratory, California Institute of Technology, Pasadena, CA 91109, USA

H. El-Askary et al. (eds.), *Research Developments in Geotechnics, Geo-Informatics and Remote Sensing*, Advances in Science, Technology & Innovation,
https://doi.org/10.1007/978-3-030-72896-0_61

algorithm is applied to model fluvial features and visualize the extent of the water flow across the surface. This is achieved by computing an eight-direction surface run-off algorithm (D8) for flow direction, accumulation and volume. This involves necessary QA/QC and smoothing of flow direction considering locations of streams and to eliminate sinks and to recondition elevation points. Palaeolakes are mapped based on the spectral characteristics of their carbonate deposits using Yoshiki Ninomiya's (Geological Survey of Japan) ASTER SWIR and TIR band ratios (Corrie et al. 2010). Additionally, SRTM 1 arcsecond (30 m) DEM data are used to search for lake basins, where water may have collected, in addition to their shoreline features, which are often observed surrounding many palaeolake systems (Pachur and Rottinger 1997).

4 Results

Image processing confirms that SRTM image and PALSAR data penetrate beneath the surface of the sand sheet and reveal the remains of river systems that once flowed across the Sahara. Producing a contour map from topographic data emphasizes a general decrease in elevation with increasing latitude. Excluding any major tectonic event, it is therefore likely that the general direction of water flow was northwards towards the Mediterranean Sea, giving a basic estimation of palaeoflow direction. Results suggest that there were at least five major outflows (palaeorivers) during the key humid phases, including the Sahabi, Gilf Kebir and the Kufrah (Ghoneim et al. 2012) (Fig. 1). All northward channels appear to discharge into a large drainage region with a possible connection to the Mediterranean Sea.

The results also suggest a strong correlation between the location of archaeological sites and their proximity to mapped sources of water (Fig. 2), indicating that early people had a good knowledge of where to find surface and sub-surface water. Caravan routes are observed to run in proximity to water sources, suggesting that people planned their routes through the desert based on the available sources of water.

False color composites using Landsat ETM+ 7, 4, 2 band combination and ASTER TIR spectral analysis identify seven potential areas for lake deposits. Some are already known, e.g. the Northern Darfur Megalake/West Nubian Lake, Lake Megachad and Lake Bodélé (Pachur and Rottinger 1997; Bristow et al. 2009) (Fig. 3). However, the remaining four are presumed new. Both SRTM DEM and ASTER TIR are complementary in searching for past endorheic lake basins.

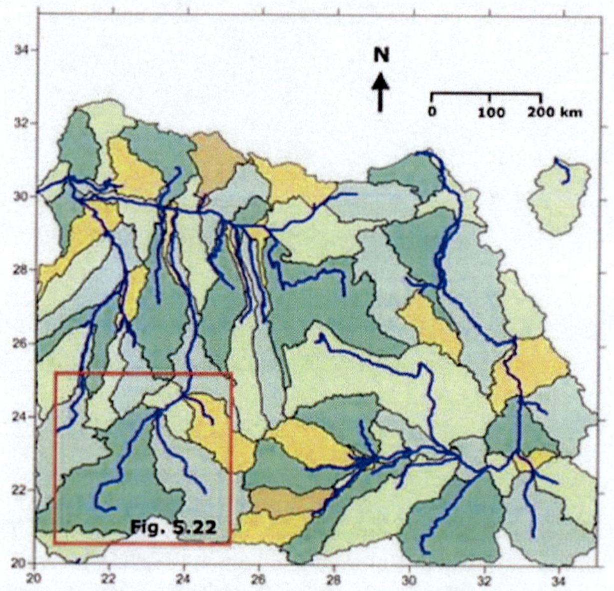

Fig. 1 Radar rivers of Egypt and Libya from SRTM with colors representing catchment basins. Kufrah system indicated in red

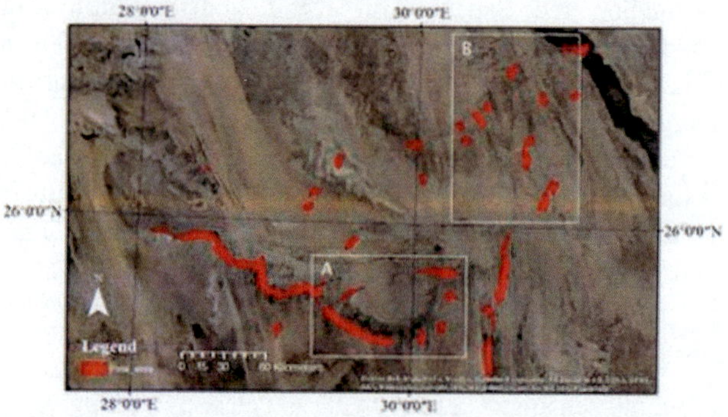

Fig. 2 GIS predicted locations for archaeological sites based on several criteria, including the location of water and caravan tracks

5 Conclusions

Results emphasize the complexity of the palaeohydrological desert landscape and a heterogeneous Nubia aquifer system punctuated with confined local aquifers and groundwater accumulations. Water is a vital natural resource in the eastern Sahara, and the social pitfalls of water scarcity can sometimes be glaringly evident, as it was in the Darfur Province of Sudan, where infighting among the people arose in part because of the lack of available freshwater. The present study contributes to the comprehensive evaluation of

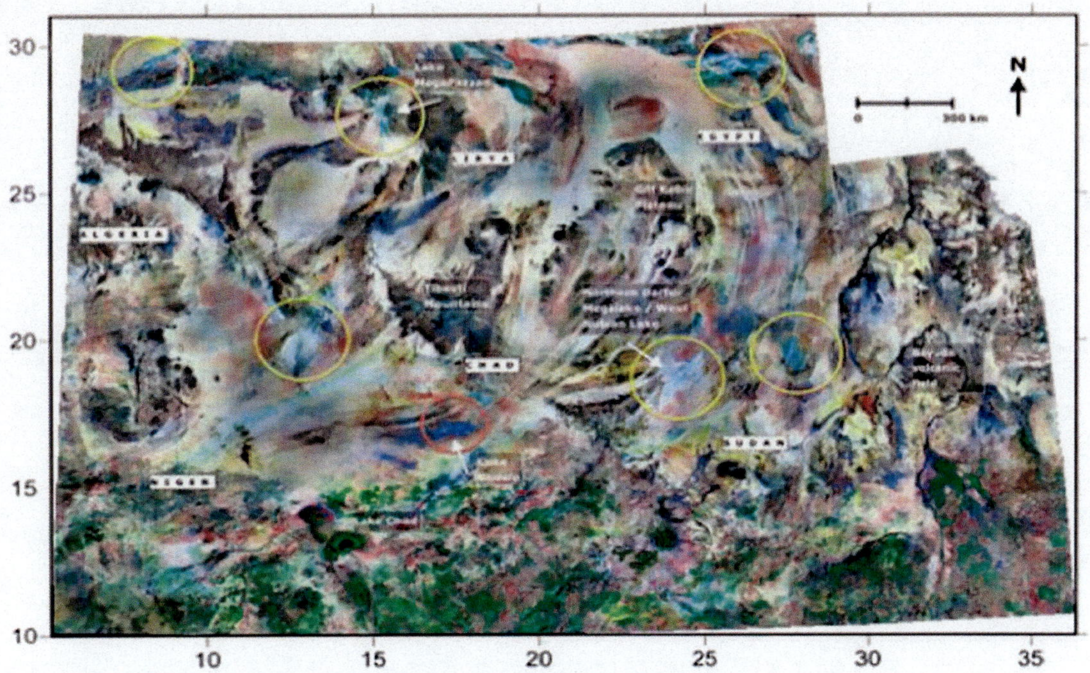

Fig. 3 Landsat ETM+ FCC (7, 4, 2) indicating possible lake deposits (turquoise color)

the water resources that incrementally builds on work of Robinson et al. (2007) to improve the understanding of aquifer dynamics and local groundwater situations, in addition to the propensity for archaeological discovery. Methods are cost effective, non-destructive and multispectral (MS) data that facilitate surface feature identification and appraisal of (close to) present-day conditions, including accessibility of areas of interest.

References

Berger, A.: Long-term variations of daily insolation and Quaternary climatic changes. J. Atmos. Sci. **35**, 2362–2367 (1978)

Bristow, C.S., Drake, N., Armitage, S.: Deflation in the dustiest place on Earth: the Bodélé Depression, Chad. Geomorphology **105**, 50–58 (2009)

Corrie, R., Ninomiya, Y., Aitchison, J.: Applying advanced spaceborne thermal emission and reflection radiometer (ASTER) spectral indices for geological mapping on the Tibetan Plateau. Int. Arch. Photogramm., Remote Sens. Spat. Inf. Sci. **38**, 464–469 (2010)

Ghoneim, E., Benedetti, M., El-Baz, F.: An integrated remote sensing and GIS analysis of the Kufrah Paleoriver, Eastern Sahara. Geomorphology **139–140**, 242–257 (2012)

Kuper, R., Kröpelin, S.: Climate-controlled Holocene occupation in the Sahara: motor of Africa's evolution. Science **313**(5788), 803–807 (2006)

McCauley, J.F., Schaber, G.G., Breed, C.S.: Subsurface valleys and geoarchaeology of the eastern Sahara revealed by Shuttle radar. Science **218**(4576), 1004–1020 (1982)

Pachur, H.J., Rottinger, F.: Evidence for a large extended palaeolake in the eastern Sahara as revealed by spaceborne radar lab images. Remote Sens. Environ. **61**, 437–440 (1997)

Robinson, C.A., Werwer, A., El-Baz, F., El-Shazly, M., Fritch, T., Kusky, T.: The Nubian aquifer in southwest Egypt. Hydrogeol. J. **15**, 33–45 (2007)

Development of Web Application for Rehabilitation of Limestone Quarry

Soumaya Ben Fredj, Fetheddine Melki, Kamel Jridi, and Haithem Alaya

Abstract

A guideline on quarry rehabilitation (wbcsd 2011) supports the argument that the rehabilitation of quarries can contribute to sustainable development. Yet, little work has been devoted to the quarry rehabilitation in Tunisia. In this respect, this study aims to develop a web application for proposing an efficient rehabilitation for limestone quarry in Tunisia to improve the field of re-development of aggregates quarries. This web application allows client users to find their own quarry, which is based on a web map server for geographic location, to the creation of a technical sheet. This paper is part of our research, proposing rehabilitation scenarios for limestone quarries with 3D modeling and 3D print: Case of Jbel El Oust (Tunisian Atlas).

Keywords

Rehabilitation • Quarry • Limestone quarry • Technology • Web application

1 Introduction

Over the last years, many quarries in Tunisia have been constantly opening due to the growing demand for construction materials. According to the Ministry of Equipment, Housing and Territorial Development, in 2014, Tunisia had 309 quarries including 187 quarries of massive rocks. These quarries have become a real barometer for economic and industrial development for the country. However, the exploitation of these quarries, especially the limestone quarry, can have a strong environmental impact on the landscape, causing vegetation losses, soil losses (Abdelaid et al. 2004) and changes in landscape topography. The Tunisian legislation imposed an impact study (law no. 89–20 of February 22, 1989) carried out by the operating company to leave the quarry in a good state (Ben Fredj et al. 2017). Nevertheless, many quarries are abandoned without being redeveloped and integrated into their natural environment, although the quarry rehabilitation in Tunisia can be a real composition and practical example of preserving and returning the natural resource in a state compatible with the surrounding environment (Halil et al. 2017). Furthermore, the technology, like the web application, can contribute to the advancement of this sector to fight against the abundance of quarries after their exploitation and to protect the nature reserve. Indeed, the web applications in geology are the subject of numerous works (Fang Qui et al. 2012; Defu et al. 2015; Newman et al. 2017 etc.). Recently, there are applications available on the web such as Geologisches 3D Modell Wien which is a viewer to visualize geological 3D models. From this perspective, a web application for quarry rehabilitation (WebQR) is created, scalable, highly available and easy to use for users of rehabilitation of limestones quarries. The purpose of this web application is to propose a rehabilitation scenario of limestones quarry in the area and integrate it in its natural environment. Also, users can interactively perform a calculation for the slope stability for safety of quarry that will always be considered as paramount for the rehabilitation plan (wbcsd 2011). Then, WebQR provides the ability to add a 3D model, designed by the user, to the map. In addition, WebQR can allow the user, such as design offices or departments, to monitor and track the creation of new sections and organize information about them and therefore to know the exact location of the quarry where

S. B. Fredj (✉) · F. Melki
Laboratory of Geodynamics, Geonumeric and Geomaterials (LR18ES37), Faculty of Sciences of Tunis, Universite de Tunis El Manar, 1060 Tunis, Tunisia
e-mail: soumaya.benfredj@fst.utm.tn

K. Jridi
SOROUBAT Group 34, 36 Avenue de la Gare Mégrine Riadh, Ben Arous, Tunisie

H. Alaya
Software Engineer, Paris, France

H. El-Askary et al. (eds.), *Research Developments in Geotechnics, Geo-Informatics and Remote Sensing*, Advances in Science, Technology & Innovation,
https://doi.org/10.1007/978-3-030-72896-0_62

they all appear on the map (Fig. 1). The solution described in this paper is part of a research project between SOROUBAT quarry and the Faculty of Science of Tunis that aims to develop the sector of quarry rehabilitation.

2 Materials and Methods

Web application development is the creation of application programs that reside on remote servers and are delivered to the user's device over the Internet. A web application does not need to be downloaded and is instead accessed through a network (Halil et al. 2017). In the present work, the technologies chosen to develop WebQR are as follows: "Java EE (JEE)" which is Java framework based on oriented language programming. "Spring framework" provides a comprehensive programming and configuration model for modern Java-based enterprise applications, and it is friendly with any kind of deployment platform. "Google maps API", "JSF" (Java server faces) as a client-side framework, "Apache Tomcat Server" as a deployment platform and "Postgresql" is a free and open-source relational database provide advanced functions for geo-spatial data. After the application is designed, it is installed on a server and the administrator interacts with the platform using web pages designed for specific functionalities (Fig. 1). WebQR enjoys the benefits of cloud computing like flexibility, availability or security, it can be accessed at any time, in any place by only using an Internet connection, and the number of users is undetermined (Madalin et al. 2018).

3 Results

3.1 Web Application Description

WebQR provides a web-based platform which can be accessed through either a local area network (LAN) or a wide area network (WAN) environment. The system can be configured to run inside a corporate network on a dedicated

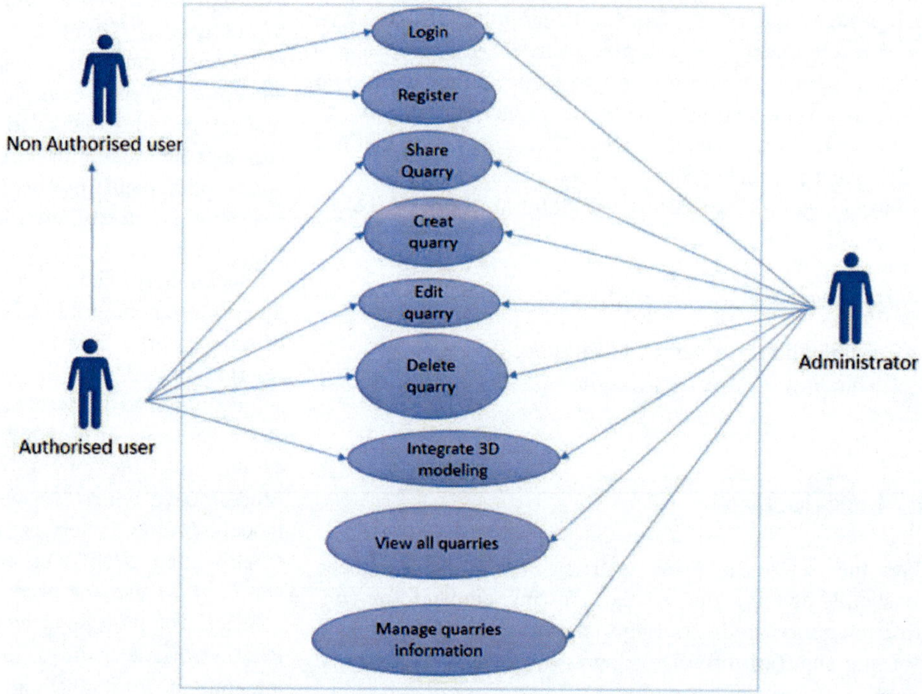

Fig. 1 Use case diagram

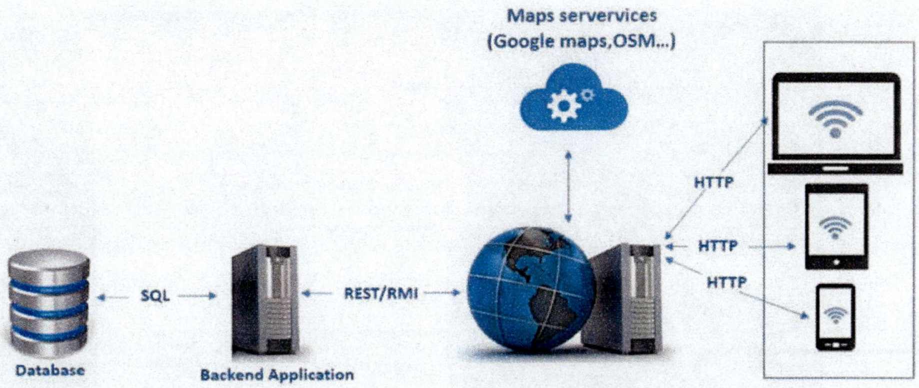

Fig. 2 Multiple tier architecture as implemented by the WebQR

server or through a server located in the cloud (Newman et al. 2017). It has been designed so that it can take advantage of a public or corporate network infrastructure and can provide access to a user based on authentication credentials. Through the adoption of a multiple tier architecture for this web application, the user can access and login regardless of the device used (laptop, PC or phone) (Fig. 2).

4 Discussion

4.1 The Application Workflow

The WebQR platform has been broken down into three main components: create quarry, rehabilitation of quarry and integrate 3D model design (Fig. 2). The front-end application exposes to the user performance and intuitive interface through which he can access all the services exposed by the backend application (Fig. 3). Upon logging, the "Create" button allows to access to interface for creating a new quarry. The frame allows the user to create their own quarry defined with GPS coordinates. To select a previously created quarry, the user accesses the page by selecting "view" button. The user is able to create, edit or delete a given quarry. Within the create quarry page, users are able to introduce the general information about the quarry, the geotechnical characteristics of rocks and calculate the slope stability (Fig. 3). Furthermore, the "Rehabilitation" button allows the user to access a web form filled with data sets such as existing animal species and plant. With a data selected, users are directed to a rehabilitation scenarios page where they are able to select a successful rehabilitation scenario suitable for its location information. Moreover, if the user has a 3D modelization for his quarry, he can able to integrate its modelization with "visualization 3D" button.

5 Conclusions

This study is part of a research project between SOROUBAT and the Faculty of Science of Tunis. At this stage, a web application for quarry rehabilitation is proposed. In the future works, multiple features will be developed within the web to further assist users in more accurate scenarios for successful rehabilitation. Furthermore, developing an application to print the 3D models of rehabilitations is a future objective of this research.

(a)

(b)

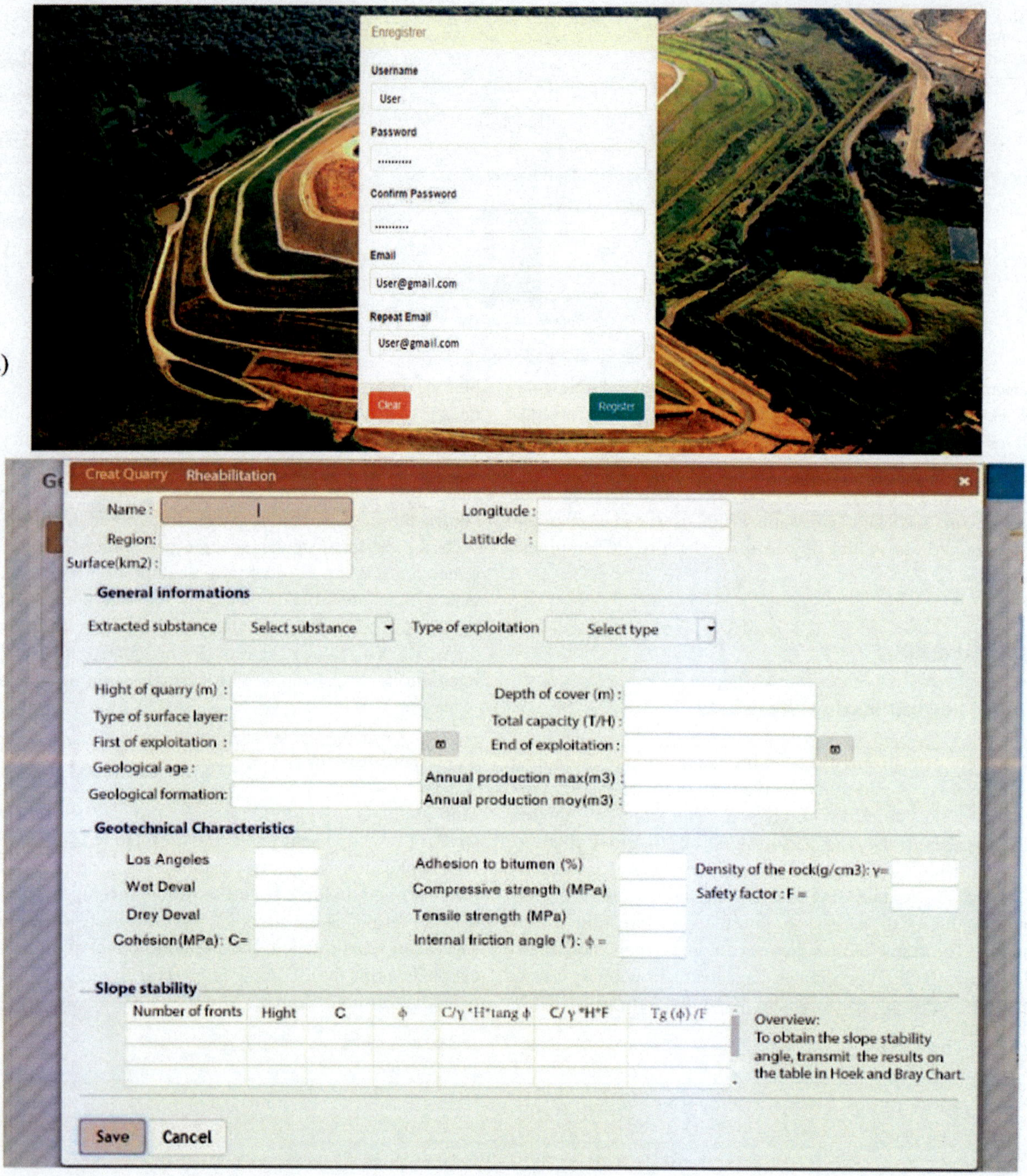

Fig. 3 WebQR interface, **a** edit user information; **b** edit information for quarry

References

Abdelaid, C., Christiane, W., Cristina, M., Maria Salomé, C., Maria Amélia, M.L., Otilia, C.: Restoration of a Limestone Quarry: Effect of Soil Amendments on the Establishment of Native Mediterranean Sclerophyllous Shrubs. Restoration Ecology, vol. 12, pp. 20–28. Society for Ecological Restoration, Washington (2004)

Ben Fredj, S., Melki, F., Jridi, K.: Proposing rehabilitation scenarios for limestone quarries with 3D modeling and 3D print: case of Jbel El Oust (Tunisian Atlas). In: Recent Advances in Geo-environmental Engineering, Geomechanics and Geotechnics, and Geohazards, pp 85–88. Springer, Heidelber (2017)

Defu, W., Chiam, K., Kin Seng, Y.: Building an interactive web-based GIS system forplanning of geological survey works. Acad. Sci. Eng. Technol.Int. J. Civ. Architect. Eng.**9**(3) (2015)

Halil, B., Jenuz, B., Shkurtë, G., Kostas, K., John, M., Kostis, D.: Case study on quarry rehabilitation and land resettlement in Dimce Quarry. In: Li, Z.X., Agioutantis, Z., Zou, D.H. (eds.) Proceedings of the 8th International Conference on Sustainable Development in the Minerals Industry, vol. 2, pp. 271–274 (2017)

Madalin, C., George, M., Florin, P., Catalin, N., Alexandru, D.: CLUeFARM: integrated web-service platform for smart farms. In: Computers and Electronics in Agriculture, vol. 154, pp. 134–154. Elsevier (2018)

Newman, C., Agioutantis, Z., Schaefer, N.: Development of a web-platform for mining applications. Int. J. Min. Sci. Technol. **28**, 95–99 (2017)

Qiu, F., Ni, F., Chastain, B., Huang, H., Zhao, P., Han, W., Di, L.: GWASS: GRASS web application software system based on the GeoBrain web service. Comput. Geosci., **47**, 143–150 (2012)

wbcsd: Guidelines on Quarry Rehabilitation, Biodiversity and land stewardship, December, https://docs.wbcsd.org/2011/12/GuidelinesOnQuarryRehabilitation.pdf.2019/09/12

The Minimum-Break Algorithm Applied to a Series of Road Turns

Nardjes Hamini

Abstract

Cartographic generalization is the process of small-scale map production from large-scale maps. When the size of the map is reduced, the objects present in the latter can entre in conflict with each other. There may be superimpositions of objects, coalescenced roads or imperceptibility problems. Generalization solves these problems through algorithms. However, the generalization is not yet automated due to the lack of algorithms. In this article, a new algorithm for processing turn series on a road to correct the coalescence is investigated. Tests as well as comparison with existing algorithms is conducted. This research has proven to be beneficial to the field of cartography as there is a lack of algorithms treating complex series of bends when producing a map. The sample used for the study is a road in china.

Keywords

Cartographic generalization • Coalescence • Road • Turn series • Algorithms

1 Introduction

Producing a small-scale map today still costs time and money. The operation of producing a small-scale map from a larger scale map is called generalization. Reducing the scale of the map involves reducing the space to represent the objects and reducing the objects themselves. The map becomes cluttered, the information becomes imperceptible, and the roads become coalescenced.

The generalization possesses a battery of tools to measure and correct all these problems like the selection of information, the amplification of the forms and the displacement of object to name a few. The latter is performed by cartographers in a manual or interactive way. During the generalization, the appearance of the map and its readability are favored instead of the geographic accuracy. To correct a superposition, the generalization proceeds to a displacement of objects, thus changing the real position of the objects.

There is a lot of work on automation of generalization (Mackaness et al. 2007; Zhilin 2007), but as the process is laborious and subjective, there is no fully automated method. This is due to non-availability of advanced algorithms and generalization measures. In this article, a new approach to dealing with series of road bends is proposed to solve the coalescence. Its principle is based on a tried-and-tested algorithm called minimum break (Lecordix et al. 1997). Originally, this algorithm was created to handle an isolated bend but not a series of turns; this study aims to make it suitable for the latter. The choice of this algorithm is based on its ability to correct the coalescence while minimizing the field gaps (added space) of the result. A mere application of this method on a series of bends is ineffective as it will produce unusable results. For a valid map, a number of alterations are needed such as lengthening the skeleton and truncating the shape.

2 The Method

The algorithm developed is aimed at correcting the coalescence in a series of road turns. It is based on the algorithm of minimum break (Lecordix et al. 1997) which was created for a single road turn.

The process of choosing a site to apply this algorithm was conducted as the following:

Since a single bend in a road does not present any issues while producing a map, the only logical and organic following

N. Hamini (✉)
Laboratory of Computer Science and Mathematics (LIM), Amar Telidji University of Laghouat (UATL), PB 37 Ghardaia road, Laghouat, Algeria

H. El-Askary et al. (eds.), *Research Developments in Geotechnics, Geo-Informatics and Remote Sensing*, Advances in Science, Technology & Innovation,
https://doi.org/10.1007/978-3-030-72896-0_63

step would be to select a geographical site that presents problems to cartographers. It is generally acknowledged that roads on mountain or that enjoy any type of height are quite laborious to map. A sampling of different mountainous roads was conducted such Stelvio Pass in Italy and Col de Turini in France. However, Tianmen Mountain in China presented the most intricacies while mapping the region.

2.1 Minimum-Break Algorithm

This algorithm uses a cartographer trick which consists of folding the edge of the turn on itself to form a line as explained in Mustière (2001). To mimic this trick in computer processing, the algorithm prefers to extract the skeleton of the bend and then expand this skeleton line to get the new bend (Mustière 2001). The details of the principle are as follows (Mustière 2001):

The line constituting the turn is closed by a segment to make it into a surface.

Apply a Delaunay triangulation on this surface.

Eliminate the triangles that are outside the surface.

A skeleton tree of the surface is constructed in the following way:

- *If a triangle has 2 neighbors, the middle of the common sides are connected*
- *If a triangle has 3 neighbors, the center of gravity of the triangle is connected to the middle of its three sides.*

Choose the longest skeleton line in the skeleton tree and lightly smooth it.

In the end, the line on each side is deleted to obtain the final bend.

2.2 The New Algorithm

The algorithm receives as input the series of turns (in the form of a polyline) and an array of n inflection points for delimiting the bends. Two consecutive inflexion points represent the start and the end of a turn. The output consists of a generalized line. The detailed steps of the algorithm are as follows:

1. Split the line according to the points of inflection in n-1 separated turns. A turn is a subset of successive points in the line that begins with an inflection point and ends with the next inflection point.
2. Extraction of the skeleton line from each turn as in Min-break algorithm (Mustière 2001).
3. The skeleton line is possibly packed from the top. Because the flaw dilation of minimum break adds thickness all along the line, even at the top of the turn, the latter takes height and may exceed the top of the original turn. In this case, a settlement is made from the top of the skeleton line. Thus, the height of the turn will be reduced to the original one or to a reasonable size (Fig. 1).
4. The skeleton line is lengthened from below to facilitate the next step.
5. A buffer with a parameter specific to each turn is applied and will be detailed in the next section.
6. The shape obtained is truncated at the bottom using a line supporting the segment connecting the two points of inflection which surround the bend. This step is easier because the skeleton line has been lengthened from below.
7. Changing the surfaces obtained to lines to finally have the final turns.
8. Aligning turns one at the end of the other, by translation, in order and respecting distances.
9. Assembling the turns into a new line.
10. We apply smoothing and aesthetic treatments to the new line.
11. Calibrating of the new line around the endpoints of the original line.
12. Reintegrating of the line in the map. The line can overflow on the sides because of a small augmentation of the size. Propagation of the results can be necessary. This can be done through a simple displacement of the lines around.

The bends have the same dilation factor to preserve the harmony of the series. This factor is that of the turn that has undergone the widest gauge ratio. The ratio between the dilation parameter and the original turn width, given by the distance between the two inflection points surrounding the bend. The dilation of a turn must be at least equal to the width of the symbol plus the visibility threshold. For this method to be fruitful, it needs to be applied to mountainous roads and certain cost lines presenting height of some sort, where roads are built with a series of bends. As opposed to urban and rural roads which have a straight line feature.

3 Tests

The data is obtained from OpenStreetMap and Google Maps. In Fig. 2, it can be observed the application of the algorithm on the road of Mount Tianmen District of Yongding, Zhangjiajie China. (a) represents the road on Google Maps. (b) represents the road after symbolization. It is easy to see the apparition of the coalescence. (c) is the zone where we will apply the algorithm. (d) is the result after application of the algorithm and propagation of the results in parts outside.

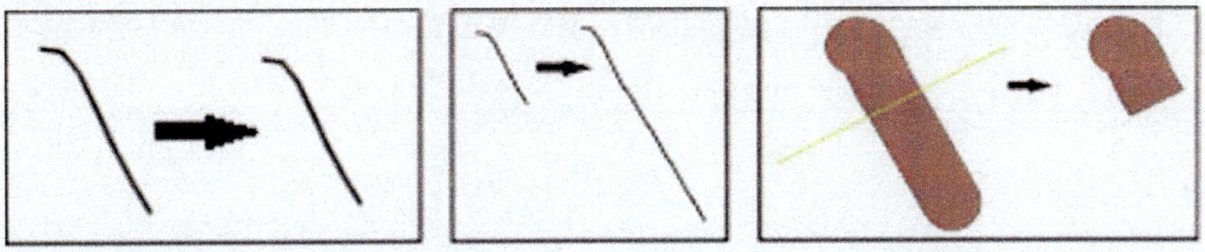

Fig. 1 Steps 3, 4 and 6 of the algorithm

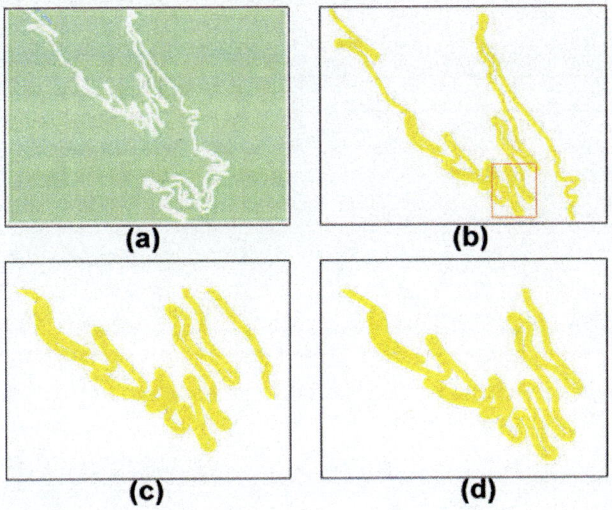

Fig. 2 Minimum-break results

At the bend level, the algorithm removes the coalescence on different forms of turns. It is particularly effective on long narrow bends. It helps to keep the general shape while limiting the grip. At the level of the series, the algorithm does not suffer from the deformation of the road axis. It also allows to keep the harmony of the sizes of the different turns.

The most efficient existing algorithms in road turn series are accordion (Lecordix et al. 1997) and schematization (Lecordix et al. 1997); thus, they are chosen for this comparison. The figure reveals the results of application of the three algorithms on a road turn series. In Fig. 3, (a) represents the application of the accordion algorithm, (b) is application of schematization algorithm and (c) is the application of our new algorithm.

Table 1 represents the values of evaluation of the results of the three algorithms. It is noticeable that the accordion algorithm conserves the initial number of turns; however, it provokes a 70% augmentation of the length of the initial axis as well as an augmentation of the envelop which may cause issues regarding map production with the surrounding objects. For schematization, no change is noticeable in relation to the length of the resulting axis as opposed to the envelope which decreases in size of 37%, but it reduces the number of turns. The new algorithm makes it so that the number of turns remains the same and augments moderately the axes length as well as the envelope.

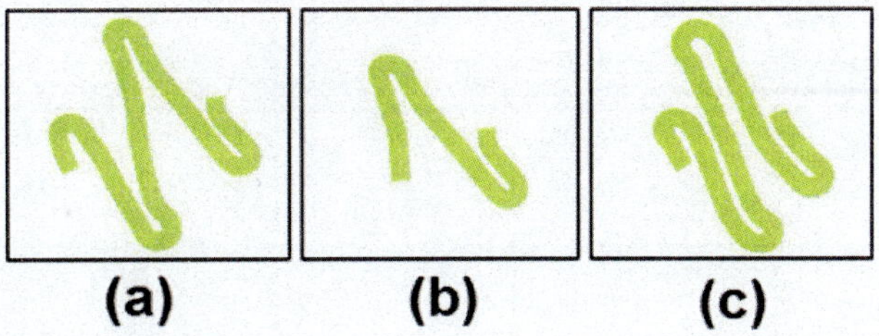

Fig. 3 Comparison

4 Conclusion

In this article, it was shown that the number of existing generalization algorithms on turn series could not solve all the problems of a coalescenced road. Accordion allows to widen a series of turns, but it does not minimize the gap of the result. Schematization is effective in minimizing the grip but removes a number of turns and does not apply to series of two or three turns. The proposed algorithm uses the principle of minimum-break algorithm and allows to correct the coalescence on series whose forms of turn are complex while minimizing the grip. However, the generalization is a complex process and needs more algorithms to handle it; this

Table 1 Comparison values

	Accordion	Schematization	The new algorithm
Axe length	+70%	+0%	+16%
Convex hull	+56%	−37%	+25%
Turn number	4	2	4

endeavor will be partially undertaken in the bulk of the research.

References

Lecordix, F., Plazanet, C., Lagrange, J.P.: A platform for research in generalization: application to caricature. GeoInformatica **1**(2), 161–182 (1997)

Mackaness, W., Ruas, A., Sarjakoski, L.: Generalisation of Geographic Information: Cartographic Modelling and Applications (2007). https://doi.org/10.1016/B978-0-08-045374-3.X5000-5

Mustière, S.: Apprentissage Supervisé pour la Généralisation Cartographique. Thèse de Doctorat de l'université Pierre et Marie Curie (Paris VI) (2001)

Zhilin, L.: Algorithmic Foundation of Multi-Scale Spatial Representation (2007)

Surface Lineament Density and Its Correlation with the Subsurface Permeable Zones at Patuha Geothermal Field, West Java, Indonesia

Mohamad Nur Heriawan, Awwab Hafizh, Suryantini, Arie Naftali Hawu Hede, and Chevy Iskandar

Abstract

In this study, the lineament density extraction and geostatistical modeling are integrated to analyze the correlation between surface lineament and subsurface permeable zone at Patuha geothermal field, West Java, Indonesia. The present work uses digital elevation model (DEM) (spatial resolution: 1 arcsecond or 30.7 m), given oblique irradiation with two directions and the lineament extracted using segment tracing algorithm (STA). Lineament density with grid size 1 × 1 km is generated by applying the ordinary kriging (OK) method. Total lost circulation (TLC) data from some wells are employed to represent the model of subsurface permeable zones where their probability is estimated using indicator kriging (IK) method. The high probability of TLC from IK estimation spatially is correlated to the surface lineament density. The high surface lineament density spatially has correlation with the main local faults, mapped in 2014 and the presence of geothermal manifestations such as crater lakes, fumaroles, and hot springs. The probability values (larger than 0.1) from IK method is vertically projected to the surface. The visual analysis suggests that high lineaments density has spatial correlation with the high probability TLC zones which are assumed as permeable zone and act as fluid path for geothermal reservoir.

Keywords

STA • Geostatistics • Lineament density • Permeable zones • Geothermal

M. N. Heriawan (✉) · A. Hafizh · Suryantini · A. N. H. Hede
Bandung Institute of Technology, Jl. Ganesha 10, Bandung, 40132, Indonesia
e-mail: heriawan@mining.itb.ac.id

C. Iskandar
PT Geo Dipa Energi (Persero), Jl. Warung Jati Barat No. 75, Jakarta Selatan, Indonesia

1 Introduction

The exploration and development of geothermal resources as a renewable and green energy have grown rapidly in Indonesia in the last 10 years. The utilizing of geothermal resources requires the suitable exploration techniques with high accuracy and low cost. Therefore, improving research and methodology for geothermal exploration is indispensable at this time. Geological structures such as faults and their associated fracture zones are an important aspect in the exploration of geothermal systems. Following previous studies which were conducted by Saepuloh et al. (2017) and Haeruddin et al. (2016), three indices—lineament intersection, lineament frequency, and lineament length—are adopted as lineament density for assessing the surface permeability zones. Surface lineaments on the satellite image can indicate the presence of geological structures such as fractures, faults, fissures, joints, and lithological contact. These structures are key elements in the evolution of geothermal systems, acting as the ascending and descending paths of geothermal fluids with high permeability, thus inducing the fluid circulation (Saepuloh et al. 2017). Sentinel-2A data are a satellite data previously used in making the lineament maps for presenting the geological information. The structural study uses directional filters to define the lineaments. The use of Sentinel-2A for geological discrimination gave good results for geological mapping in particular when it was combined with the field data (Tamani et al. 2009).

This study aims to enrich the geothermal exploration methodology using two approaches, i.e., surface and subsurface assessments. The segment tracing algorithm (STA) was introduced by Koike et al. (1995) to extract the surface lineament features from satellite images. The advantage of this method consists in the ability to observe the linear features of the low contrast and parallel to the sun's azimuth. Another advantage is that the extracted lineament is not biased; thus, it is representative in all areas.

H. El-Askary et al. (eds.), *Research Developments in Geotechnics, Geo-Informatics and Remote Sensing*, Advances in Science, Technology & Innovation,
https://doi.org/10.1007/978-3-030-72896-0_64

Some previous studies employing the STA were also undertaken to extract the lineament features and densities from remote sensing images conducted by Haeruddin et al. (2016) and Saepuloh et al. (2017). The present study conducts the probability estimation of permeable zones using indicator kriging (IK) method for the limited data of total lost circulation (TLC) in 14 wells. The results of this estimation are evaluated by correlating it with the lineament density map. The lineament density maps are validated using local geological mapping conducted by PT Geo Dipa Energi (Persero), a geothermal operator in Indonesia.

The present study analyzes the correlation between surface lineament density and subsurface permeable zone at Patuha geothermal field. It is located about 50 km southeast of Bandung City, West Java, Indonesia (see Fig. 1). Patuha geothermal is a vapor-dominated with steam zone underlying deep liquid reservoir. The reservoir temperature is about 215–230 °C covering area of 20 km^2 (Ashat et al. 2019). The Patuha field is surrounded by a volcanic center or vents which are distributed along a west to northwest trending structure (Layman and Soemarinda 2003). Based on the field mapping reports, the major structure trends that developed in this area are NE–SW and NW–SE.

2 Methodology

2.1 Lineament Features Extraction Using Segment Tracing Algorithm (STA)

Segment tracing algorithm (STA) is a method for (i) detecting a line of pixels as a vector element by examining the local variance of the gray levels in the digital image and (ii) connecting retained line elements along their expected directions. STA was used to extract the linear feature structures which were interpreted as faults or fractures based on digital elevation model (DEM) from the Shuttle Radar Topography Mission (SRTM) data with the resolution of 30.7 × 30.7 m. The extracted lineaments from STA produced three types of lineament densities, i.e., number density, total length density, and intersection density within grid size of 1000 × 1000 m. This grid size was the most effective to quantify the correlation between lineament density and geothermal features in the study site. The grid sizes with smaller or larger than 1000 × 1000 m for satellite imagery were found to produce much exaggeration and smoothing effect of density distribution, respectively (Saepuloh et al. 2017).

The STA is a non-filtering approach used to detect a line element (segment) which identifies a lineament by two principles (Haeruddin et al. 2016). Firstly, the orientation of a valley was determined by examining the local deviation of gray levels $Z(x)$ along a line:

$$\epsilon(x) = \int_{-a}^{a} w(x+h)\{Z(x) - Z(x+h)\}^2 \mathrm{d}h \qquad (1)$$

with $w(x)$ is a weight coefficient, h is a distance from x, and a is the maximum distance from the x. The second was to evaluate the variation of $Z(x)$ by the differential:

$$\pi = \left\{\frac{\mathrm{d}^2 Z(x)}{\mathrm{d}x^2}\right\}^2 / Z(x) \qquad (2)$$

Dividing the secondary differential by $Z(x)$ equalized the variations for areas of high and low contrasts in the image. The differential was carried out in the perpendicular direction to the valley.

After getting lineament extraction from STA, data 1 (ascending) and data 2 (descending) were grouped using the minimum and maximum searching radius of 800 m and 1500 m, respectively. The ordinary kriging (OK) interpolator was applied to determine the spatial variability of the indices of three lineament densities by interpolating and mapping it. The linear features derived from DEM in Patuha geothermal field showed three major directions, i.e., N65° E, N95° E, and N135° E.

2.2 Permeable Zones Modeling Using Indicator Kriging (IK)

Indicator kriging (IK) is an algorithm for non-parametric interpolator used to estimate the conditional cumulative distribution function at any location given neighboring conditioning data. Indicator kriging is yet another form of kriging, applied to variables that are binary indicators of the occurrence of an event:

$$I_k(\mathbf{u}) = \mathbf{1}, \text{if the even } k \text{ occurs at location } \mathbf{u}\text{; otherwise} = \mathbf{0} \qquad (3)$$

40 locations of total lost circulation (TLC) data were obtained from 14 wells and spread out in different depths which were arranged as indicator variables. The massive losses of drilling mud that were flowing from the well into the surrounding formations were good indicators of the presence of natural permeable fractures. It was often used to indicate the reservoir feed zone in the geothermal reservoir.

The IK estimation was run using SGeMS software with the following parameters: grid size 50 × 50 × 25 m, minimum and maximum conditioning data 2 and 10, respectively, searching radius 2000 m, and marginal probability

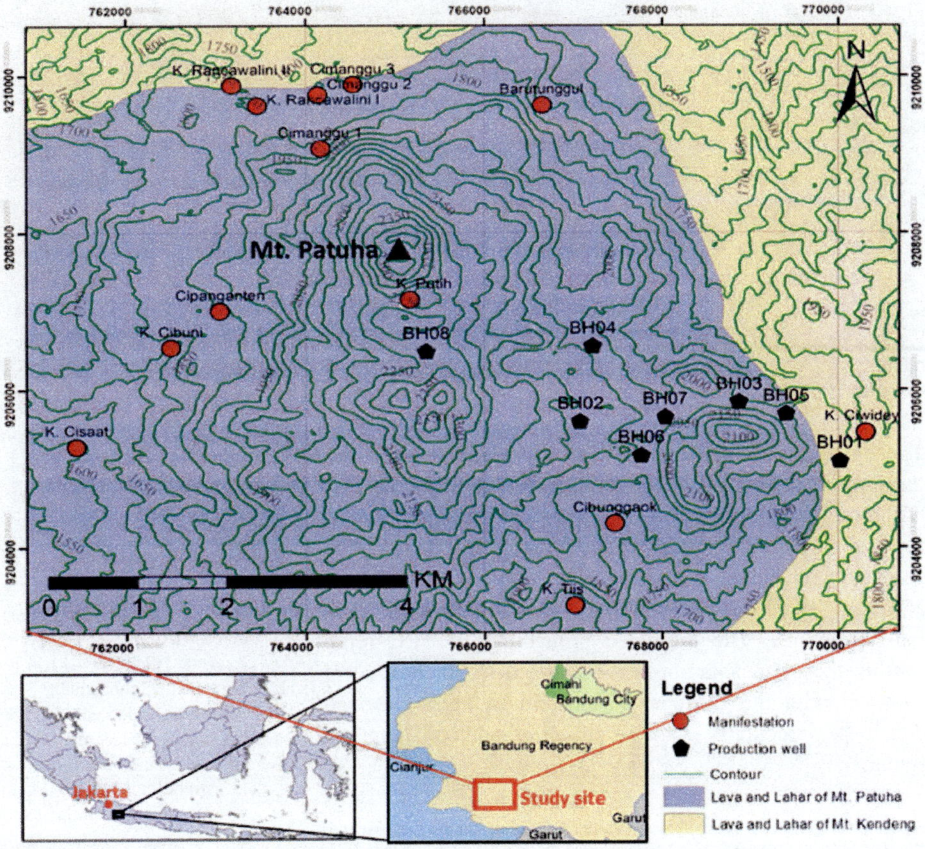

Fig. 1 Geological map of study area overlaid with geothermal manifestations and wells

0.09. To model the spatial structure, an experimental semivariogram of TLC with indicator data (0 and 1) was made using 3D omnidirectional with lag separation of 30 m. The semivariogram was well-fitted using the spherical model with range, nugget, and sill variances 100 m, 0.10, and 0.70, respectively.

3 Results and Discussion

3.1 Surface Lineament Density

High lineament density is indicated in red color overlaid with the local fault structures, production wells, and geothermal manifestations. The geothermal manifestations were located at medium to high density lineament. The geothermal manifestations probably formed on the most permeable fractures which extended between the reservoirs and the near surface fractures (Saepuloh et al. 2017). Therefore, the extracted lineament density seems to have good spatial correlation with the geothermal manifestations (see Fig. 2). The surface lineament density, mainly the number and total length also seem to coincide with the main local fault in Patuha. A similar study was addressed by Saepuloh et al. (2017) at Wayang Windu geothermal field (West Java, Indonesia) regarding the distribution of extracted lineaments from ALOS PALSAR images by modified STA which showed good spatial agreement between the concentration of long lineament and the surface geothermal manifestation and local faults.

3.2 Probability of Permeable Zones

The result of IK estimation represents the probability of the presence of TLC which, in turn, indicates the probability of subsurface permeable zones. The subsurface permeable zones could be represented by the fractures. When the probability values (>0.1) of the presence of subsurface permeability zones are projected vertically to the surface, it seems that they correspond with the zones where the lineament has medium to high densities. The high probability of permeable zones is exactly located in the left or SW side of the local faults (see Fig. 2). If the fault is dipping to the SW, then the subsurface fracture zones might be connected to the surface fault. Later, this could be confirmed by geophysical surveys such as the magneto-telluric (MT) method to identify the subsurface structures by resistivity values.

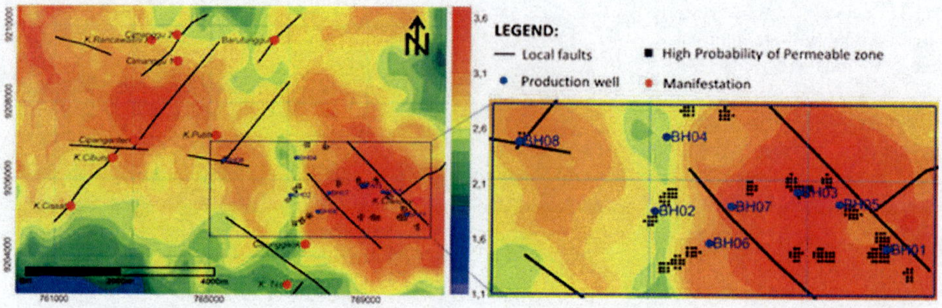

Fig. 2 Number of lineament density overlaid with local fault structures and probability of permeable zones above 0.1

4 Conclusions

The extraction of lineaments from the straight valleys and the large changes in surface roughness and slope dip from DEM images were successfully performed using STA. The linear features derived from DEM in Patuha geothermal field showed three major directions, i.e., N65° E, N95° E, and N135° E which coincided with the major direction of local fault structures. The estimated probability of TLC representing the permeable zones (>0.1) from IK method corresponded with the location of medium to high lineament densities, particularly with the number and total length of fractures.

Acknowledgements The authors sincerely thank Japan Science and Technology (JST) and Japan International Cooperation Agency (JICA) through the Beneficial and Advanced Geothermal Use System (BAGUS) Project in the framework of Science and Technology Research Partnership for Sustainable Development (SATREPS) with Grant No. JPMJSA1401. The acknowledgments should be extended to Faculty of Mining and Petroleum Engineering of ITB for the financial support under P3MI 2019 Research scheme and PT Geo Dipa Energi (Persero) for supporting this study.

References

Ashat, A., Itoi, R., Pratama, H.B.: Updating conceptual model of Ciwidey-Patuha geothermal using dynamic numerical model. OP Conf. Ser. Earth Environ. Sci. **254** (2019)

Haeruddin, Saepuloh, A., Heriawan, M.N., Kubo, T.: Identification of linear features at geothermal field based on Segment Tracing Algorithm (STA) of the ALOS PALSAR data. IOP Conf. Ser.: Earth Environ. Sci. **42**(1) (2016)

Koike, K., Nagano, S., Ohmi, M.: Lineament analysis of satellite images using a Segment Tracing Algorithm (STA). Comput. Geosci. **21**(9), 1091–1104 (1995)

Layman, E.B., Soemarinda, S.: The Patuha vapor-dominated resource West Java, Indonesia. In: 28th Workshop on Geothermal Reservoir Engineering. Stanford University, Stanford (2003)

Saepuloh, A., Haeruddin, Heriawan, M.N., Kubo, T., Koike, K., Malik, D.: Application of lineament density extracted from dual orbit of synthetic aperture radar (SAR) images to detecting fluids paths in the Wayang Windu geothermal field (West Java, Indonesia). Geothermics **72**, 145–155 (2017)

Tamani, F., Riheb, H., Amor, H., Younes, H.: Integrating remotely sensed and GIS data for the detailed geological mapping in semi-arid regions: case of Youks Ies Bains area, Tebessa province, NE Algeria. Geotechnical and Geological Engineering, Springer Nature, Switzerland AG (2009)

Landfill Site Selection Using GIS and Multi-criteria Decision-making AHP and SAW Methods: A Case Study in Sulaimaniyah Governorate, Iraq

Karwan Alkaradaghi, Salahalddin S. Ali, Nadhir Al-Ansari, and Jan Laue

Abstract

Lack of land for waste disposal is one of the main problems facing urban areas in developing countries. The Sulaimaniyah Governorate, located in Northern Iraq, is one of the main cities of the country in the Kurdistan Region, covering an area of 2400 km^2. Currently, there is no landfill site in the study region that meets the scientific and environmental requirements, and the inappropriate dumping of solid waste causes adverse effects to the environment, economic and urban aesthetic. To overcome this phenomenon, it is crucial to suggest a landfill site, even in countries that recycle or burn their waste to protect the environment. Landfill sites should be carefully selected taking into account all regulations and other restrictions. The integration of geographic information systems and the multi-criteria decision analysis were used in this study to select suitable landfill locations in the region. To this end, thirteen layers prepared according to their importance including slope, geology, land use, urban area, villages, rivers, groundwater, slope, elevation, soil, geology, road, oil and gas, land use, archaeology and power lines. Two different methods (simple additive weighting and analytic hierarchy process) were implemented in a geographical information system to obtain the suitability index map for candidate landfill sites, where all these conditions satisfied the scientific and environmental criteria adopted in this study. The comparison of the maps resulting from these two different methods demonstrates that both methods produced consistent results.

Keywords

Criteria weights • GIS software • Simple additive weighting (SAW) method • Analytic hierarchy process (AHP) • Landfill site

1 Introduction

Solid waste management is considered to be a significant issue in developing nations. Despite the use of many efficient processes such as reuse and recycling, appropriate landfill disposal is still the most prevalent method of minimizing adverse effects on the environment and waste management (Moeinaddini et al. 2010). The process of selecting a landfill site is still considered complicated task. Currently, there is no landfill site in the study area that fulfils the scientific and environmental requirements to resolve the waste dump site issue. Two techniques have been used in this multi-criteria decision-making (MCDM) research; these are the analytic hierarchy process (AHP) and the simple additive weighting (SAW) methods for selecting landfill sites (Teknomo 2006; Şener et al. 2011; Uyan 2014). The geographic information system (GIS) with AHP and SAW methods are the most common techniques in MCDM exhibiting a high capacity to manage complicated problems with large data during the decision-making process (Office of UN Resident Co-Ordinator 2002). This study aimed to compare both methods to find the best candidate for landfill sites that complies with the environmental and scientific criteria.

K. Alkaradaghi · N. Al-Ansari (✉) · J. Laue
Lulea University of Technology, 97187 Lulea, Sweden
e-mail: nadhir.alansari@ltu.se

K. Alkaradaghi · S. S. Ali
Department of Geology, College of Science, Sulaimani University, 460013 Sulaimaniyah, Iraq

K. Alkaradaghi
Kurdistan Institution for Strategic Studies and Scientific Research, 460013 Sulaimaniyah, Iraq

S. S. Ali
Komar University of Science and Technology, 460013 Sulaimaniyah, Iraq

H. El-Askary et al. (eds.), *Research Developments in Geotechnics, Geo-Informatics and Remote Sensing*, Advances in Science, Technology & Innovation,
https://doi.org/10.1007/978-3-030-72896-0_65

2 Materials and Methods

Sulaimaniyah is among the major cities in the region of Kurdistan and Iraq. The city is situated northwest of Iraq between latitude 35° 45′ 0″ N, 36° 0′ 0″ N and longitude 44° 45′ 0″ E, 45° 45′ 0″ E covering an area of around 2400 km^2. The administrative boundary unit of the Sulaimaniyah Governorate is shown in Fig. 1 (Office of UN Resident Co-Ordinator 2002). All types of waste are dumped without any treatment in an open area overlooking the Tanjaro River.

To proceed suitable landfill site, thirteen criteria as layer maps were prepared using GIS spatial analysis tools as an input for the AHP and SAW models. Each criterion was classified into sub-criteria and assigned a suitability rating value from zero to ten (Saaty 1980). Criteria rating and importance of their priority were specified on the basis of restrictions, literature and scientific experts. A pairwise comparison was implemented in the matrix for all criteria through the priority of the importance intensity of one activity over another using a numerical scale of nine points (Saaty 1980). The upper triangular matrix is filled with the comparative criteria values, and the lower triangular matrix is completed with the upper reciprocal values (Teknomo 2006; Şener et al. 2011). The eigenvalue was calculated by multiplying the value for each criterion in each column in the same row in the matrix of the pairwise comparison. The priority vector (Pr_i) was determined by normalizing the eigenvalue to one (Saaty 1980), as follows:

$$Pr_i = (Eg_i) / \left(\sum_{i=1}^{n} Eg_i \right) \tag{1}$$

where Eg_i = eigenvalue for the row (i) ($Eg_i = \left(a_{11} \times a_{12} \times a_{13} \times \cdots \times a_{1n} \right)^{1/n}$); n = number of elements in row (i).

The consistency index was calculated according to Saaty (1980).

$$CI = (\lambda \max - n)/(n - 1) \tag{2}$$

where CI is consistency index and n is size or order of the matrix and (λmax) is equivalent to the priority vector in the decision matrix (Saaty 1980).

The consistency ratio (CR) depends on the size of the matrix (n = 13); thus, the random index value (RI = 1.56) (Saaty 1980).

$$CR = (CI)/(RI) \tag{3}$$

The simple additive weighting (SAW) is a ranking method defined as a weighted linear combination or scoring method (Saaty 1980).

$$W_i = \frac{A_i}{\sum_{j=1}^{n} A_j} j = 1, 2, \ldots, n \tag{4}$$

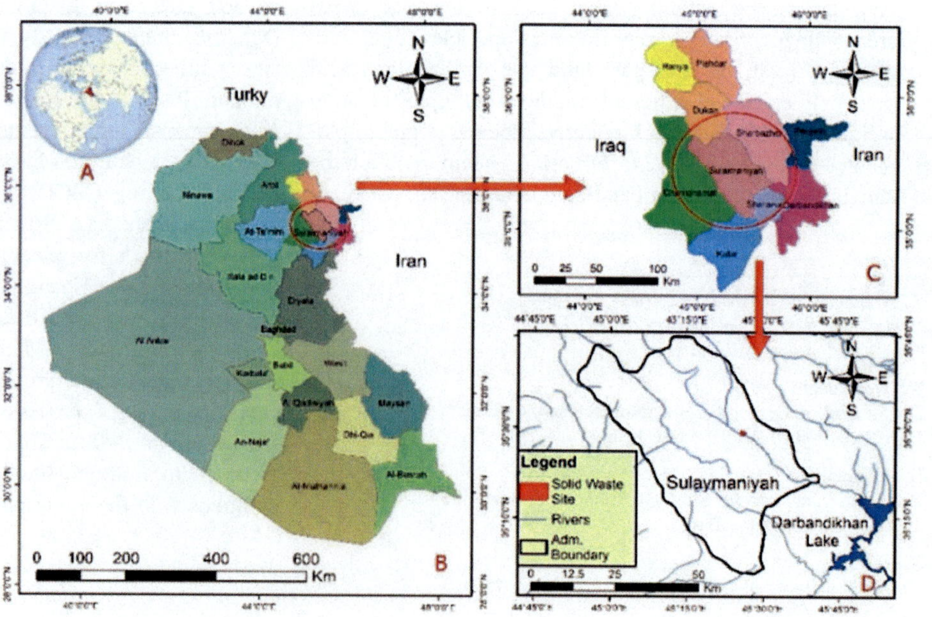

Fig. 1 Location map of the study area

Fig. 2 Pairwise comparison matrix (A): Urban area; (B): Villages; (C): Rivers; (D): Groundwater depth; (E): Slope; (F): Elevation; (G): Soil types; (H): Geological formations; (I): Roads; (J): Oil and gas field; (K): Land use classification (L): Archaeological site; (M): Power lines. NW = normalized weight

	A	B	C	D	E	F	G	H	I	J	K	L	M	NW	AHP	SAW
A	1	2	2	3	4	4	5	5	6	6	7	8	9	0.211	0.21	0.124
B	0.50	1	1	2	3	3	4	4	5	5	6	7	8	0.149	0.15	0.111
C	0.50	1.00	1	2	3	3	4	4	5	5	6	7	8	0.149	0.158	0.111
D	0.33	0.50	0.5	1	2	2	3	3	4	4	5	6	7	0.029	0.10	0.099
E	0.25	0.33	0.33	0.50	1	1	2	2	3	3	4	5	6	0.099	0.074	0.086
F	0.25	0.33	0.33	0.50	1	1	2	2	3	3	4	5	6	0.099	0.074	0.086
G	0.20	0.25	0.25	0.33	0.50	0.50	1	1	2	2	3	4	5	0.065	0.048	0.074
H	0.20	0.25	0.25	0.33	0.50	0.50	1.00	1	2	2	3	4	5	0.065	0.048	0.074
I	0.17	0.20	0.20	0.25	0.33	0.33	0.50	0.50	1	1	2	3	4	0.015	0.032	0.062
J	0.17	0.20	0.20	0.25	0.33	0.33	0.50	0.50	1	1	2	3	4	0.043	0.032	0.062
K	0.14	0.17	0.17	0.20	0.25	0.25	0.33	0.33	0.50	0.50	1	2	3	0.043	0.022	0.049
L	0.13	0.14	0.14	0.17	0.20	0.20	0.25	0.25	0.33	0.33	0.50	1	2	0.021	0.016	0.037
M	0.11	0.13	0.13	0.14	0.17	0.17	0.20	0.20	0.25	0.25	0.33	0.5	1	0.012	0.012	0.025

where W_i is the normalized weight of each criterion; A_i is the weight of each criterion of area (i) under criterion (j) n is the criteria number.

3 Results and Discussion

The matrix of pairwise comparisons with SAW and AHP weighs are presented in Fig. 2. The maximum lambda (λmax) = 13.51, CI = 0.04 and CR = 0.027, if CR is less than 0.1.The ratio indicates a reasonable consistency level in the pairwise comparison (Sólnes 2003). The final map shows the suitability index for landfill sites in Sulaimaniyah Governorate, which was divided into four categories of suitable areas, including: unsuitable, moderately suitable, suitable and the most suitable areas (Uyan 2014). The suitability index with areas for all categories of the SAW and AHP methods is shown in Fig. 3.

4 Conclusion

This research used the MCDM techniques with the GIS method to evaluate the suitable selection of landfill sites in the study region. The results show the index values that have been categorized into four areas calculated using the pixel calculation in GIS. The results indicate that the most suitable site covered the area of 16.37 and 24.35% or 392.92 and

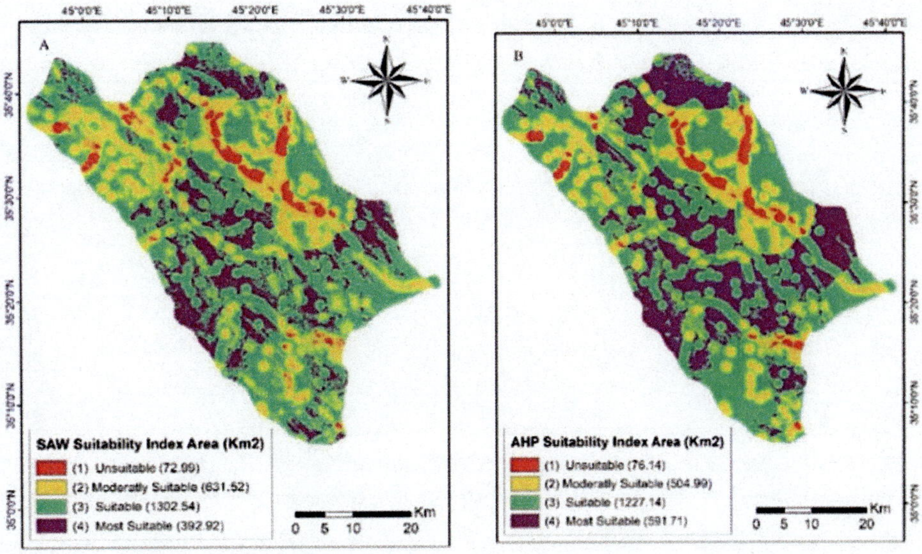

Fig. 3 Suitability index area for landfill site using SAW and AHP methods

591.71 km^2, respectively, in SAW and AHP methods. The compatibility of the most suitable area in both methods is 91.71%, while the compatibility of all zone areas in both methods is 99.8, 94.7 and 96.85%, respectively, for unsuitable, moderately suitable and suitable, respectively.

References

Moeinaddini, M., Khorasani, N., Danehkar, A., Darvishsefat, A.A., Zienalyan, M.: Siting MSW landfill using weighted linear combination and analytical hierarchy process (AHP) methodology in GIS environment (case study: Karaj). Waste Manag. (2010). https://doi.org/10.1016/j.wasman.2010.01.015

Office of UN Resident Co-Ordinator: Common Country Assessment. 2002, http://www.sprep.org/att/IRC/eCOPIES/Countries/Vanuatu/3

Saaty, T.L.: The analytic hierarchy process: planning, priority setting. Resour. Alloc. 287 (1980). https://doi.org/10.3414/ME10-01-0028

Şener, Ş., Sener, E., Karagüzel, R.: Solid waste disposal site selection with GIS and AHP methodology: a case study in Senirkent-Uluborlu (Isparta) Basin, Turkey. Environ. Monit. Assess. (2011). https://doi.org/10.1007/s10661-010-1403-x

Sólnes, J.: Environmental quality indexing of large industrial development alternatives using AHP. Environ. Impact Assess. Rev. **23**, 283–303 (2003)

Teknomo, K.: Analytic hierarchy process (AHP). Meas. Bus. Excell. (2006). https://doi.org/10.1108/13683040210451697

Uyan, M.: MSW landfill site selection by combining AHP with GIS for Konya, Turkey. Environ. Earth Sci. **71**, 1629–1639 (2014). https://doi.org/10.1007/s12665-013-2567-9

Geotechnical Monitoring of Pipeline Systems Operating Under Conditions of Permafrost (Yakutia)

Galina Struchkova, Tamara Kapitonova, and Pavel Efremov

Abstract

Life safety on the territory of the Republic of Sakha (Yakutia) is directly related to the uninterrupted supply of energy and heat, which is caused by low climatic temperatures. Even in Central Yakutia, the average period of temperatures below 0 °C is about 210 days, below −20 °C—138 days. The minimum temperature is distributed in the range from −47.8 to −60.3 °C. Under such natural conditions, the release of any object in gas and heat supply chain leads to disastrous consequences. Several dozen emergencies of varying severity are recorded at gas and heat supply facilities throughout the year, mainly in winter. The system of oil and gas pipelines laid on the territory of the Republic of Sakha (Yakutia), mostly located in the permafrost zone, is affected by dangerous geological processes, which significantly affects its reliability and safety. For stable operation of oil and gas complex (OGC) of the Republic, an assessment of pipeline state system, based on geotechnical monitoring, is necessary. The monitoring system provides heterogeneous data, the analysis of which helps to clarify the interaction between pipeline and permafrost in various geological conditions and allows to evaluate and predict the condition of objects and environment.

Keywords

Industrial safety • Main gas and oil pipeline • Permafrost • Geocriogenic processes • Geotechnical monitoring

1 Introduction

Huge reserves of gas and oil are concentrated on the territory of the Republic, which makes upon an equal basis with the Irkutsk region the majority of oil and gas reserves of Eastern Siberia and the Far East. Natural gas has been used by consumers of Central Yakut Energy District since 1967, and West Yakut—since 1983. Three fields are in development: Sredneviluisky and Mastakhsky Gas Condensate Fields, as well as Srednebotuobinsky Oil and Gas Condensate Fields located in Western Yakutia (Information Material 2004; Fuel and Energy Balance of the Republic of Sakha (Yakutia) 2005). On the territory of the region, the construction and operation of section of the main oil pipeline Eastern Siberia-Pacific Ocean (VSTO) are under way. TS VSTO includes the construction of oil pipeline with a length of 2757 km and a diameter of 1067/1220 mm (Information Material 2004; Fuel and Energy Balance of the Republic of Sakha (Yakutia) 2005).

It has been established that the construction of pipeline route leads to a sharp activation of exogenous processes on the territory of technological corridor. Logging during corridor construction, disruption of vegetation cover during trench digging, disruption of natural drainage of water bodies during construction of parable road and pipeline logging are the main potentially dangerous geotechnical processes, during which it is necessary to adhere to the principle of “prevention of impact” on frozen landscapes (Kapitonova et al. 2013, 2018). The main reasons mentioned above, which cause the possibility of creating emergency situations on the main pipeline (MT), create the following factors of influence on the technical condition of pipeline system: creation of stresses in the structure body related to mechanical movement of the pipe due to the change of trench profile (cleaning and washing of soils, bare sections of the route, freezing of water-saturated soils); cyclic seasonal freezing processes cause soil emptying during freezing and precipitation during thawing, which results in the

G. Struchkova · T. Kapitonova (✉) · P. Efremov
V.P. Larionov Institute of Physical-Technical Problems of the North Siberian Branch of the Russian Academy of Science, Yakutsk, Russia
e-mail: kapitonova@iptpn.ysn.ru

H. El-Askary et al. (eds.), *Research Developments in Geotechnics, Geo-Informatics and Remote Sensing*, Advances in Science, Technology & Innovation,
https://doi.org/10.1007/978-3-030-72896-0_66

creation of a complicated deformed state of the structure and causes its possible destruction (Kapitonova et al. 2013, 2018; Bolshakov 2010). Geotechnical monitoring makes it possible to detect initial forms of dangerous natural and man-made processes and to predict the occurrence of accidents (Li et al. 2019; Wang et al. 2019).

2 Materials and Methods

The organization of networks for the monitoring of various dangerous geological processes (thermokarst, landslides, frost) at different stages considerably differ by the technique, amount of works, used equipment, devices, and obtained information. For example, at the initial stage of observation of deposits, drones are used. At the following step, thermal imaging aerial collection is carried out with the use of "Geoskan-401" complex, which allows to obtain a thermal imaging plan and orthophoto plans of investigated area in one flight. Electrotomography of deposits is performed by "Scala-48" electric exploration station, Slumberger set. Georadiolocation studies are performed by a set of equipment "OKO-2" manufactured by NPP "Logis". In the structure of geotechnical monitoring network (GTM) of sites equipped with special equipment, allowing instrumental methods to carry out control, using such means as thermometric and hydrogeological wells; depth benchmarks, which serve as initial altitude base, relative to which measurements of planned and altitude position of pipeline are carried out at control points on sections with difficult geological conditions and determination of displacements relative to previous measurement cycle. Natural cryogenic processes and phenomena are characteristic of pipeline route section: cryogenic soil beam, which is common in almost the entire territory of allocated geocriological route sections; formation that develops mainly in the northern part of the studied area within the permafrost development area; Erosion and thermal erosion processes developed to varying degrees at a number of geocriological sites and causing processes of sedimentation, collapse, sedimentation, landslides, and others; Soliflution and landslide formation, a thermocark which is formed in connection with thawing of ice-saturated soils and extraction of underground ice, leading to seeding of ground surface, emergence of negative forms of relief and their waterlogging.

The following complex of materials served as basic data for this work: materials of archival space shooting of high and average resolution; field geological, hydrogeological, geodesic survey materials; air visual survey materials; engineering survey materials.

3 Results

A database (DB) has been formed in order to solve the problems of monitoring and ensuring safe operation of oil and gas complex (OGC) facilities related to exogenous processes using geographic information systems (GIS). The information is presented in the database in the form of an array: numerical data; coordinate data; tables; test records; audio-, photo-, aero-, space-photography; cartographic data. The list of map database objects (spatial data) of GIS contains the following sections: hydrographic network; terrain; human settlements; vegetation; roads. The monitoring results allowed us to identify several zones with time-varying boundaries and parameters in order to display the results of analysis of the state of extended linear natural and technical systems in the territory where there is mutual influence of pipeline system elements and geological environment. Zoning defines the boundaries of areas with homogeneous features. It is carried out according to the type of geological hazards and the degree of danger. These zones are formed on the basis of decryption and determination of indicators from materials of aero- and space surveys, clarification of parameters of exogenous geological processes during field survey; Measurements of coordinates of sedimentary marks by geodetic methods. To evaluate the parameters of hazardous processes that disrupt the equilibrium state of geological environment, the following digital models are created: relief; water currents; vegetation; thermal field forming around the pipeline. The change of pipeline spatial position over time is estimated for further ranking. All listed objects, including the map base, piping system parts, monitoring tools, zones, and digital models, are DB content elements. As an example of the integrated use of the database and mathematical modeling, the result of ranking the sections of the main pipeline according to the degree of danger of developing the thermowell is given (Fig. 1).

4 Conclusions

The assessment of the interaction between the environment and main pipeline includes the following stages: (i) The analysis of the state of the territory where the pipeline is located; (ii) the assessment of pipeline and environment interlinkages, perennial soils and their consequences, and (iii) the identification of measures to mitigate or prevent negative interactions. The use of a geotechnical monitoring database allowed to assess the condition of pipeline and to predict the possibility of emergency situations and consequences depending on the environmental conditions.

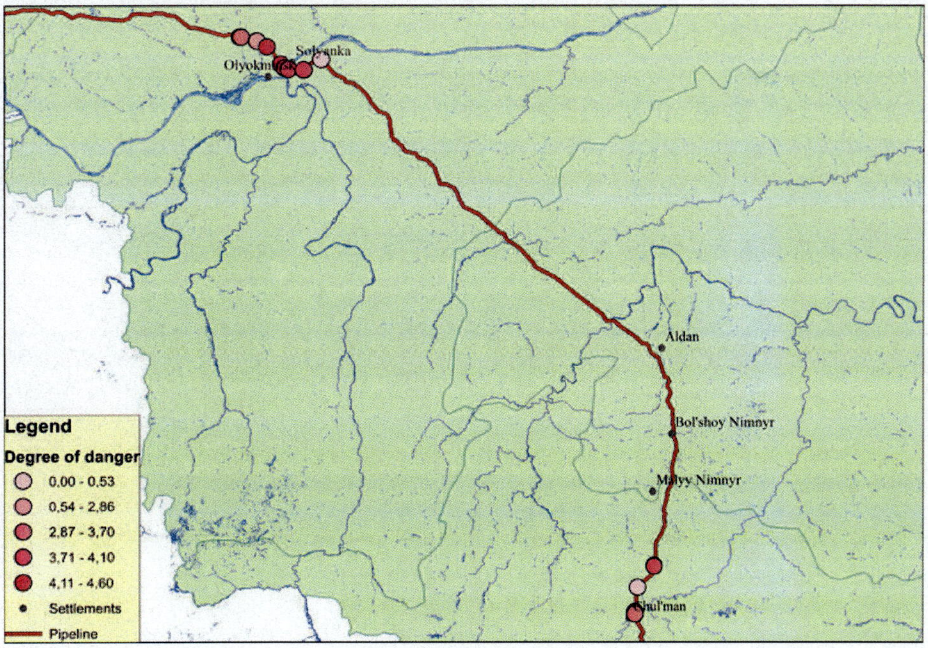

Fig. 1 Ranking of MT sections by degree of risk of thermal board development

References

Bolshakov, A.M.: Analysis of damage and defects in the main gas pipelines and tanks. North Gas Ind. **5**, 52–53 (2010)

Fuel and Energy Balance of the Republic of Sakha (Yakutia): Part I, 160 p. Sakhapoligrafizdat, Yakutsk (2005)

Information Material: Interregional Meeting in the Republic of Sakha (Yakutia) "Prospects of Development of Hydrocarbon Deposits in Eastern Siberia and the Far East", 30 p. Sakhapoligrafizdat, Yakutsk (2004)

Kapitonova, T.A., Struchkova, G.P., Tarski, E.L., Efremov, P.V.: Main geological risks for pipelines in cryolithozone conditions. In: VI Eurasian Symposium on Strength of Materials and Machines for Cold Climate Regions EURASTRENCOLD, vol. 3, pp. 126–141, Yakutsk (2013)

Kapitonova, T.A., Struchkova, G.P., Levin, A.I.: Risk assessment analysis of the Mastakh-Berge-Yakutsk main gas pipeline laid in the cryolithozone. Probl. Saf. Emerg. Situat. **6**, 34–43 (2018)

Li, H., Lai, Y., Wang, L., Yang, X., Jiang, N., Li, L., Wang, C., Yang, B: Overview of the current state: interactions between buried pipeline and frozen soil. Sci. Technol. Cold Reg. (2019)

Wang, F., Li, G., Ma, W., Wu, Q., Serban, M., Vera, S., Alexandr, F., Jiang, N., Wang, B.: Production–permafrost interaction monitoring system along the Sino-Russian oil pipeline. Eng. Geol. **254**, 113–125 (2019)

Development of Test and Calibration Platform for Space Electron Radiation Environment Exploration Payloads

Ping Zhou, Yizhong Ye, Wenjing Wang, Wentao Ji, Tao Jing, Aibing Zhang, Panlin Guo, Shugang Sheng, Yuhui Jing, Yueqiang Sun, Jingbao Liang, Guangwu Zhu, and Liyuan Ma

Abstract

In order to test and calibrate the satellite payloads-electron detectors, the medium- and high-energy electron accelerators with extremely weak beam current to simulate the space electron radiation environment have been developed, including the experimental terminal which has a vacuum chamber and a vacuum rotation plate. All above form a system which is so-called National Space Science Center-Space Electron Facility, for short "NSSC-SEF". The main specifications are electron beam energy range 10 ~ 2000 keV, beam current $10^3 \sim 10^9/(\text{cm}^2\ \text{s})$, beam spot Ø50 mm, beam un-uniformity < 10%, and vacuum < 5×10^{-5} Pa. This system has been used to test and calibrate more than 20 payloads for different missions since its building in October 2015.

Keywords

Space electron radiation environment • Electron accelerator • Extremely weak beam current

P. Zhou (✉) · Y. Ye · W. Wang · W. Ji · T. Jing · A. Zhang · Y. Sun · J. Liang · G. Zhu · L. Ma
National Space Science Center, CAS, Beijing, 100190, China
e-mail: pzhou@nssc.ac.cn

P. Zhou · Y. Ye · W. Wang · W. Ji · T. Jing · A. Zhang · Y. Sun · J. Liang · G. Zhu · L. Ma
Beijing Key Laboratory of Space Environment Exploration, Beijing, 100190, China

P. Zhou · Y. Ye · W. Wang · T. Jing · A. Zhang · Y. Sun
University of Chinese Academy of Sciences, Beijing, 100190, China

P. Guo · S. Sheng
Shanghai Hangye Vacuum Instrument Science and Technology Co., Ltd., Shanghai, 201806, China

Y. Jing
Institute of High Energy Physics, CAS, Beijing, 100049, China

1 Introduction

More than 50 years after Van Allen discovered the radiation belt (http://www.nasa.gov/vision/universe/features/james_van_allen.html; Hess 1968), both inside and outside the magnetosphere, space radiation still seriously restricts the safety of space flight in many aspects (Kamide and Chian 2010; Pisacane 2008). The influence of space radiation environment is more important especially for manned space flight (Tribble 1995). Therefore, it is necessary to monitor the radiation environment of the LEO space station and forecast the occurrence rate of solar events that may generate the radiation at the specific position of the orbit of the LEO space station with 57° inclination. The basic physical processes of solar cell box transistors encountering space electrons and protons were poorly understood in the early days, but this situation has been greatly improved. Nevertheless, the research on the effects of space radiation and space radiation on space electronic systems is still a critical field in space physics and space applications.

At present, there are more than 200 communication satellites in use in geosynchronous orbit, and many other satellites which are often closely related to national security are distributed in adjacent areas or elliptical orbits with different inclinations. In the process of satellite design and operation, it is essential to get accurate and timely detailed information of space radiation environment in the satellite operation area. Therefore, it is necessary to monitor and analyze the space radiation environment in real time.

2 Settings

The National Space Science Center of the Chinese Academy of Sciences (ab. "NSSC, CAS") has established a ground test and calibration test platform for space electronic radiation environment exploration payloads for the first time in China, according to the requirements of such kind of payloads

H. El-Askary et al. (eds.), *Research Developments in Geotechnics, Geo-Informatics and Remote Sensing*, Advances in Science, Technology & Innovation,
https://doi.org/10.1007/978-3-030-72896-0_67

development mission, which is so-called National Space Science Center-Space Electron Facility (NSSC-SEF) (Fig. 1). The platform consists of two accelerators, i.e., medium-energy extremely weak-current electron accelerator (10–200 keV) and high-energy extremely weak-current electron accelerator (100 keV–2.0 MeV), as well as test and calibration experimental terminal (Fig. 2) which consists of a vacuum chamber, vacuum rotating plate, beam diagnostic system, and central control interlocking system. Among them, the high-energy electron accelerator is a 2 MV singletron type extremely weak-current electron accelerator customized by NSSC and imported from High Voltage Engineering Europa B.V. (HVEE) of Netherlands. The medium-energy electron accelerator is designed jointly by the NSSC, IHEP (the Institute of High Energy Physics, Chinese Academy of Sciences) and Shanghai Hangye Accelerator Co., Ltd., and is manufactured by Shanghai Hangye Accelerator Co., Ltd.

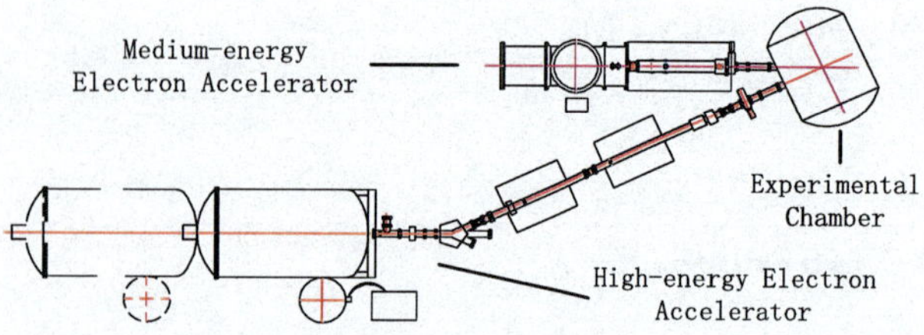

Fig. 1 Layout of test and calibration platform

Fig. 2 Experimental terminal

3 Results

3.1 Technical Specifications of the Test and Calibration Platform NSSC-SEF

The main technical indicators of the system are shown in Table 1.

3.2 Comparison the Main Technical Specifications of NSSC-SEF with USA GSFC-REF

The main technical specifications of the test and calibration system for intermediate and high-energy space electron exploration payloads, NSSC-SEF, are compared with those of GSFC-REF, a similar device of the Goddard Space Flight Center in the USA, as shown in Table 2.

Comparison of the main specifications of high test and calibration system in NSSC with the similar system GSFC-REF at Goddard Space Flight Center, USA, is shown in Table 2.

Table 1 Main technical specifications of the space electronic radiation exploration payloads test and calibration platform

Item	Results of NSSC-SEF
Beam energy	2 MeV electron accelerator: 0.1–2 MeV 200 keV electron accelerator: 10–200 keV
Beam flux	2 MeV electron accelerator: 10^3–10^7/(cm^2 s) 200 keV electron accelerator: 10^5–10^9/(cm^2 s)
Beam energy divergence	2 MeV electron accelerator: <0.1% (RMS) 200 keV electron accelerator: <2.0%
Beam spot	50 mm diameter
Beam stability	<5% in 30 min
Beam ununiformity Vacuum chamber size	<5% within 50 mm diameter beam spot Ø1500 mm × L1500 mm
Working vacuum	<5 × 10^{-5} Pa
Rotation accuracy	0.1°
Translational accuracy	0.1 mm

Table 2 Comparison of main specifications of NSSC-SEF with GSFC-REF of the same kind at Goddard Space Center, USA

Item	GSFC-REF	NSSC-SEF
Accelerator	2 MeV and 150 keV electron accelerator	2 MeV and 200 keV electron accelerator
Working vacuum	5×10^{-4} Pa	5×10^{-5} Pa
Beam spot size	≤ 51 mm diameter	≤ 50 mm diameter
Rotation accuracy	0.1°	0.1°
Translational accuracy	0.1 mm	0.1 mm
Beam measurement	Semiconductor detector	Semiconductor detector and Faraday cup (FC)

4 Discussion

By using special design to the weak electron beam, the two electron accelerators can provide a wide range electron beam flux. For 2 MeV electron, the beam flux range is 10^3–10^7/(cm^2 s), and for 200 keV it is 10^5–10^9/(cm^2 s). Another very important thing is that particularly designed very low resistance grounding system and very low ripple control of power supply system make the extremely weak beam measurement possible. That is why NSSC-SEF can work steadily and accurately as description above.

5 Conclusions

Extremely weak current electron accelerator is a special kind of accelerator. It belongs to a special type of accelerator with very low-energy accelerator and single particle accelerator. There is little information about it at home and abroad, and its development is very challenging. The development of test and calibration platform for space electron exploration payloads has been done in NSSC, CAS. The two electron accelerators in this system are high voltage electrostatic accelerators. They both have accurate beam energy and excellent beam stability. Moreover, through special design of weak beam and beam expansion, beam energy and flux are realized. The unique advantages of wide adjustment range and high uniformity of beam spatial distribution can meet the ground calibration test requirements of our space electron radiation research payloads.

At present, the platform has been accomplished, and more than 20 payloads test and calibration for different missions by using this system. Now it can provide relevant test and calibration experiments for domestic and foreign counterparts either.

References

Hess, W.: The Radiation Belt and Magnetosphere. Waltham Press (1968)

http://www.nasa.gov/vision/universe/features/james_van_allen.html [OL]

Kamide, Y., Chian, A.C.L.: Handbook of the Solar-Terrestrial Environment (2010)

Pisacane, V.L.: The Space Environment and Its Effects on Space Systems. The American Institute of Aeronautics and Astronautics, Inc. (2008)

Tribble, A.C.: The Space Environment: Implications for Spacecraft Design. Princeton University Press (1995)

Remote Sensing for Ground Investigations

Illustrating of a Landslide Site with Photogrammetric and LIDAR Methods

Aydın Alptekin, Mehmet Özgür Çelik, Yusuf Doğan, and Murat Yakar

Abstract

Landslides occur as a result of change in forces on soil and rock masses and cause loss of life and property. In this study, a landslide that was triggered by heavy rainfall in Mersin was investigated using terrestrial laser scanner (TLS) and unmanned aerial vehicle (UAV). A 3D model of a house in the landslide area was extracted from the measurements taken with TLS, and the amount of deformation caused by the landslide was determined. Pictures were taken overlay by UAV, orthophoto and actual digital elevation model (DEM) were generated and slope of terrain was determined using ground control points. As a result of the analysis, a 3D model and a base map of landslide site were prepared. An engineering project was modeled in a practical way. Achieving reliable data in a short time indicates that laser scanner and photogrammetric methods can be easily used for geological hazards.

Keywords

Landslide • Terrestrial laser scanner • Unmanned aerial vehicle • 3D model

1 Introduction

Landslides, which are frequently seen as natural geological hazards worldwide, may damage farm lands, transportation corridors, infrastructure, residential areas, and dams. Landslides occur as a result of change in forces on rock and soil masses due to excessive rainfall, earthquakes, and uncontrolled excavations. Mersin (Turkey) is continuously exposed to landslides because of the geomorphological setting.

A landslide occurred in Mersin due to excessive rainfall on February 17, 2019. There is no active fault in the vicinity of the study area. A typical mediterranean climate characterizes the landslide area, which is located in Adana Basin, Turkey. Geologically, the study area consists of Pliocene–Quaternary clay, silt, sand, and gravel. Skilodimou et al. (2018) have indicated that Pliocene aged sediments are more likely to cause landslides.

In this study, a landslide site was investigated with a terrestrial laser scanner (TLS) and unmanned aerial vehicle (UAV). A digital elevation model (DEM) of the study area and deformation amount was determined with high precision.

2 Methods

The TLS and UAV can be used in landslide detection, susceptibility mapping, modeling, and monitoring. Investigating a landslide requires a combination of different disciplines, such as geology, geophysics, geotechnics, geodesy, and photogrammetry. In recent years, modeling landslide sites has become easier with the developing technology. In this study, a landslide site was modeled using photogrammetric methods.

Nowadays, many engineering projects can be reliably modeled over a short time. Carvajal et al. (2011) used UAV to model a landslide site and showed that this vehicle can be used in limit equilibrium analyses. Ma et al. (2019) created DEM using an UAV in a landslide site. Gigli et al. (2014) used TLS to determine the volume of rocks in a rockfall analysis. Zeybek and Şanlıoğlu (2015) used TLS in two different times to determine landslide deformation and the amount of material. Santis et al. (2018) used the TLS to create inundation maps in a flood modeling. Wedmore et al. (2019) used TLS to determine fault displacement amount in an earthquake in Italy.

A. Alptekin (✉) · M. Ö. Çelik · Y. Doğan · M. Yakar
Mersin University, 33343 Mersin, Turkey
e-mail: aydinalptekin@mersin.edu.tr

H. El-Askary et al. (eds.), *Research Developments in Geotechnics, Geo-Informatics and Remote Sensing*, Advances in Science, Technology & Innovation,
https://doi.org/10.1007/978-3-030-72896-0_68

2.1 Terrestrial Laser Scanner

TLS sends signals to the objects and records the returning signals. A house that was affected by a landslide was scanned using Faro Focus^S 350 laser scanner, which can obtain data with ±1 mm accuracy. 6 Scans, each of which lasted 13 min, were performed with ¼ resolution, 4× quality without using targets. Point clouds taken with a laser scanner combined in SCENE 2018 software with cloud-to-cloud registration and 2× high dynamic range (HDR) pictures were imported into the point cloud.

2.2 Unmanned Aerial Vehicle

Using DJI Phantom 3 Standard, flights were performed at an average height of 80.8 m and 146 pictures were taken with an overlay amount 80% using the 4000 * 3000 resolution FC300C camera. The flight plan was prepared in Pix4D capture mobile phone application. Six ground control points were determined in the field and the coordinates were determined in the fixed position of the global navigation satellite system (GNSS) receiver with horizontal: ±3–5 cm and vertical: 5–7 cm accuracy. The error amount at ground control points is 2.2 cm. Pictures were combined in the Agisoft Metashape program, and the digital elevation model (DEM) was obtained with the slice middle meridian (DOM) 36° in the ITRF-96 coordinate system. The ground sampling distance of the DEM and orthophoto is 10.7 cm/pixel and 2.97 cm/pixel, respectively.

Fig. 1 Vertical and horizontal displacement

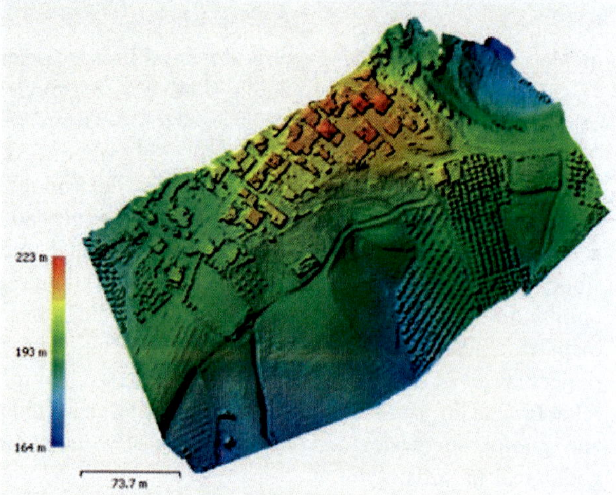

Fig. 2 Digital elevation map

3 Results

From the point cloud obtained from laser scans, a 3D model was built using SCENE software that has an image-processing capacity and 3D-modeling features. From the obtained model, horizontal and vertical motions were determined as 0.178 m and 0.5827 m, respectively (Fig. 1). The pictures taken from a low-cost UAV were processed in the Agisoft Metashape software. DEM (Fig. 2), slope map, and orthophoto of the study area were generated. According to DEM, the land is 164–223 m asl. In case a new landslide occurred at the site in future, the same survey procedure would be applied again. We subtracted two DEMs to determine the volume of the new landslide.

4 Discussion

Landslides near settlements may potentially cause loss of life and property. The reason of the landslide needs to be determined in order to manage the negative effects to society. There is no precaution against landslide in the study area. Engineers need to model and monitor the possible landslide sites. Therefore, the results of this study will help us to decide which method to be applied in order to decrease the negative effects of landslides. With the progressing technology, it has become easier to investigate landslides. TLS devices can get millions of points in a short time with a high resolution and accuracy without touching the object. Obtaining data is simple, whereas registering those data is troublesome. A computer with high standards is needed to process those data. In small areas, TLS can be used easily. However, in large areas, an airborne laser scanner is more useful. Actual digital elevation map (DEM) was generated from the data with high resolution obtained from the UAV. The TLS and UAV have some limitations, when the land has dense vegetation, and the weather conditions are bad; they cannot be used.

5 Conclusions

In this study, the TLS and UAV were used for the modeling of landslide area and a 3D model of the land was extracted. Actual DEM, slope map, orthophoto, and TLS data of the study area were produced. The conditions of the region were documented. As a result of the measurements to be made in the future, it will be determined how much material has shifted. Modeling of the land with TLS and UAV was achieved in a practical way and without damaging the land surface.

References

Carvajal, F., Agüera, F., Perez, M.: Surveying a landslide in a road embankment using unmanned aerial vehicle photogrammetry. Remote Sens. Spat. Inf. Sci. 201–206 (2011)

Gigli, G., Morelli, S., Fornera, S., Casagli, N.: Terrestrial laser scanner and geomechanical surveys for the rapid evaluation of rock fall susceptibility scenarios. Landslides **11**, 1–14 (2014)

Ma, S., Xu, C., Shao, X., Zhang, P., Liang, X., Tian, Y.: Geometric and kinematic features of a landslide in Mabian Sichuan, China, derived from UAV photography. Landslides **16**, 373–381 (2019)

Santis, R., Macchione, F., Costabile, P., Costanzo, C.: A comparative analysis of 3-D representations of urban flood map in virtual environments for hazard communication purposes. In: Ninth International Conference on Fluvial Hydraulics, vol. 40 (2018)

Skilodimou, H.D., Bathrellos, G.D., Koskeridou, E., Soukis, K., Rozos, D.: Physical and anthropogenic factors related to landslide activity in the Northern Peloponnese, Greece. Land **7**(85) (2018)

Wedmore, L.N.J., Gregory, L.C., McCaffrey, K.J.W., Goodall, H., Walters, R.J.: Partitioned off-fault deformation in the 2016 Norcia earthquake captured by differential terrestrial laser scanning. Geophys. Res. Lett. **46**, 1–7 (2019)

Zeybek, M., Şanlıoğlu, İ: Accurate determination of the Taskent (Konya, Turkey) landslide using a long-range terrestrial laser scanner. Bull. Eng. Geol. Environ. **74**, 61–76 (2015)

Assessment of the Risk of Landslides and Rockslides in Northern Morocco by Radar Interferometry Differential

Dahbia El Kenouss, Abdelhamid Rossi, Omar El Kharki, and Balkecem Bouhamdi

Abstract

The northern Rif is probably the region of Morocco the most affected by various types of natural hazards. Landslides and rock slides are widespread, mainly due to geological causes, such as the nature of facies, tectonics, climatic conditions, and urban planning activities in the region. These risks have been monitored using the DINSAR technique (Differential Radar Interferometry), which gives an accurate estimate of soil deformation rates typically from a few millimeters to a few centimeters over large areas. The DINSAR treatment method is performed using SNAP software, then the image-generated NDTs from the Sentinel 1A, 1B satellites are a family of Earth observation satellites launched respectively in April 2014, April 2016.

Keywords

DinSAR • Landslide • Rock slides • Interferometry • Risk reduction • Radar image • Sentinel-1 • Northern Rif

1 Introduction

Landslides consist of a mass descent of materials such as rocks, earth, mine waste, or debris on a slope. These phenomena are caused by heavy rains, earthquakes, volcanic eruptions, or various human activities, landslides occur without warning. They change the landscape and can destroy buildings and homes, roads and railways, cut off the power grid, water, gas, and sewer lines.

The Rif chain, located in northern Morocco, is an area highly exposed to landslides. This region of the kingdom is experiencing significant socio-economic development. Major infrastructure projects are either nearing completion or launch and development. These projects are expected to favor a tourist development of Mediterranean coasts and to create reception platforms. However, some areas are geologically unstable, characterized by landslides whose environment must be protected.

This study aims to identify landslide risk zones in northern Morocco using the DINSAR technique and their classification to define zonation maps according to the degree of risk using SAR images (Synthetic Aperture Radar). Synthetic Aperture Radar (SAR) is a radar imaging technology capable of producing high-resolution images of targets and fixed-surface terrain.

Radar interferometry is a recent satellite observation technique with applications in geophysical fields as varied as volcanology, seismology, or glaciology. In urban environment, it allows to characterize the vertical deformations of low amplitude soil (typically a few millimeters to a few centimeters) on large expanses.

Since the first description of the technique, many DInSAR applications have been developed. The most important results were obtained in different branches of geophysics: ice and glacier dynamics; volcanoes; and landslides, dam construction and management. Oil and gas extraction, mining activities, gas capture and storage. The DINSAR technique provides an accurate estimation of the soil deformation rate and also generates the land displacement map by means of the phase difference (interferogram) between two SAR image mode SLC (only complex look).

The study area is located in the northern rift between the Jebha area and the city of Al-Hoceima. It administratively belongs to the Tangier region Tetouan Al-Hoceima. In the study area, three sections are generally distinguished: the European plate, the field of Alboran, and the African Plate.

D. El Kenouss (✉) · A. Rossi · O. El Kharki · B. Bouhamdi
Tangier Faculty of Science and Technology, Abdelmalek Essaadi University, Tétouan, Morocco

H. El-Askary et al. (eds.), *Research Developments in Geotechnics, Geo-Informatics and Remote Sensing*, Advances in Science, Technology & Innovation,
https://doi.org/10.1007/978-3-030-72896-0_69

This is a key area for understanding the vertical movements associated with orogenic processes (Fig. 1).

From a geological point of view, the area of interest is represented exclusively by the internal domain also called Alboran domain, and the flysch domain. The inner domain located to the west of the Jebha major accident covers more than half of the area.

The landslide in our study area depends on the conjunction of several parameters, namely: lithology, fracturing (density), seismicity (isoprobe depth), slope gradient, hypsometry, slope exposure, hydrographic network, precipitation, and land use.

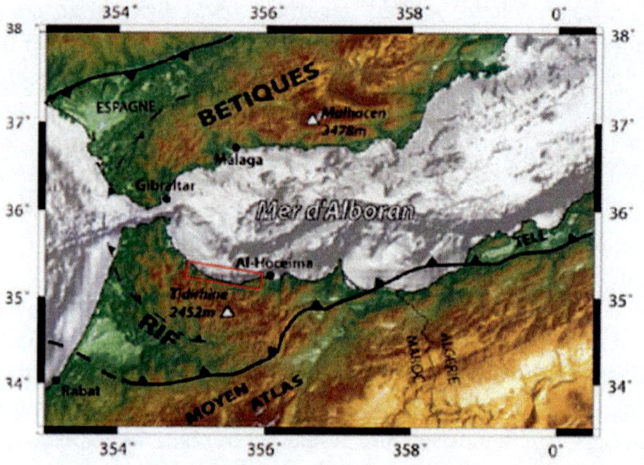

Fig. 1 Location and representation on a relief map of the study area

2 Materials and Methods

The study aims to determine the displacement map of instabilities of terrain between the two dates of the images. To this end, several types of treatments are followed. The steps are summarized in Fig. 2.

The generation of differential interferograms must go through several stages ranging from the selection of radar data, the generation of DTM digital elevation model, and the map of displacement of the ground. The different stages of the image processing are carried out using software snap (Sentinel Application Platform). The unwrapping is not yet implemented in SNAP. The unwrapping modules for SNAP are under development. For this reason, SNAPHU is used as a phase flow tool to follow the DInSAR step.

The first step consists in downloading radar images; the acquisition dates are adequate for the identification of landslides. Indeed, the seasonal difference between the two images acquired in 2018-08-08 and 2018-12-18 allows to study the displacement of the ground. It is seen that in summer periods, the grounds are relatively stable due to the absence of the triggering factors, while in winter, the area experiences heavy rainfall leading to soil instabilities.

The second step is the DinSAR (Differential Interference) processing. To this end, we will design a processing chain in which the final output could be differential interference with only deformation. This requires primarily remove the topographic induction phase of the original interferogram as the originals. Our images contain Speckle noise, thus, a "Multilook" treatment is applied to reduce the noisiness and

Fig. 2 The methodology of the work

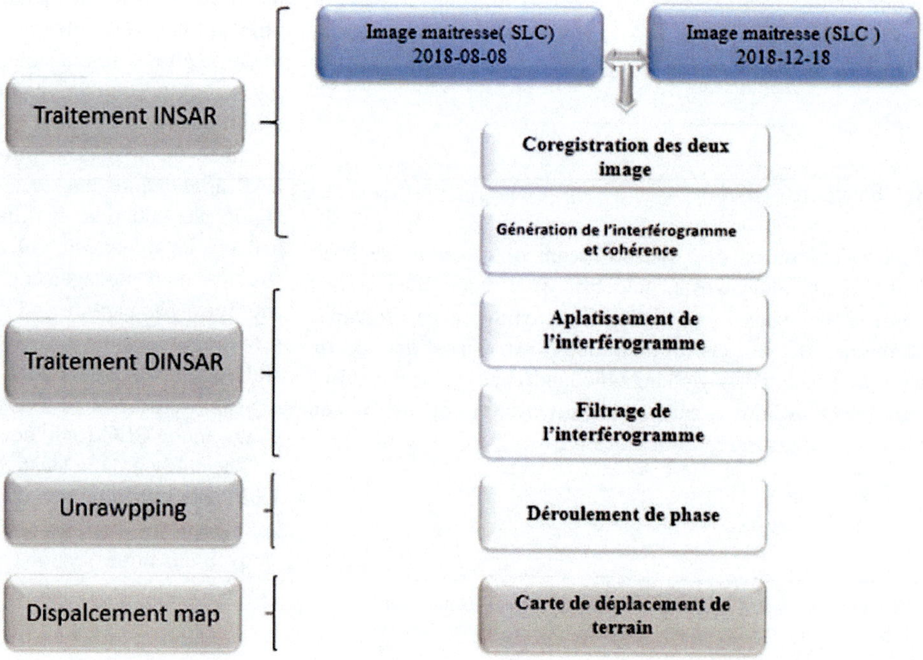

improve the interpretation capacity of the images. Then, phase filtering of the interferogram is performed to reduce phase noise for visualization purposes. This part of the processing ends by exporting the data in SNAPHU format (more details in the next step).

The unwrapping is not yet implemented in SNAP. The unwrapping modules for SNAP are under development. For this reason, SNAPHU is used as a phase flow tool to follow the DInSAR step.

3 Results and Discussion

The displacement of the land is calculated by the following equation:

$$\text{vert\_displ} = \frac{\phi_{\text{unw}} \cdot \lambda}{-4\pi \cdot \cos \theta_{\text{inc}}}$$

ϕ_{unw}: Phase en rad déroulée en rad

λ: la longueur d'onde

θ_{inc}: Angle d'incidence

Using C-band Sentinel-1 radar images, covering the entire period from August 2018 to December 2018, it was possible to obtain a preliminary characterization of landslides based on backscattering coefficient.

The map of displacement reveals subsidence and landslide estimated by −0.092 m and also elevation of the soil about 0.038 m.

4 Conclusions

It can be concluded that the different parameters and the direct and indirect causes responsible for the triggering of landslide hazards at this site are common between natural

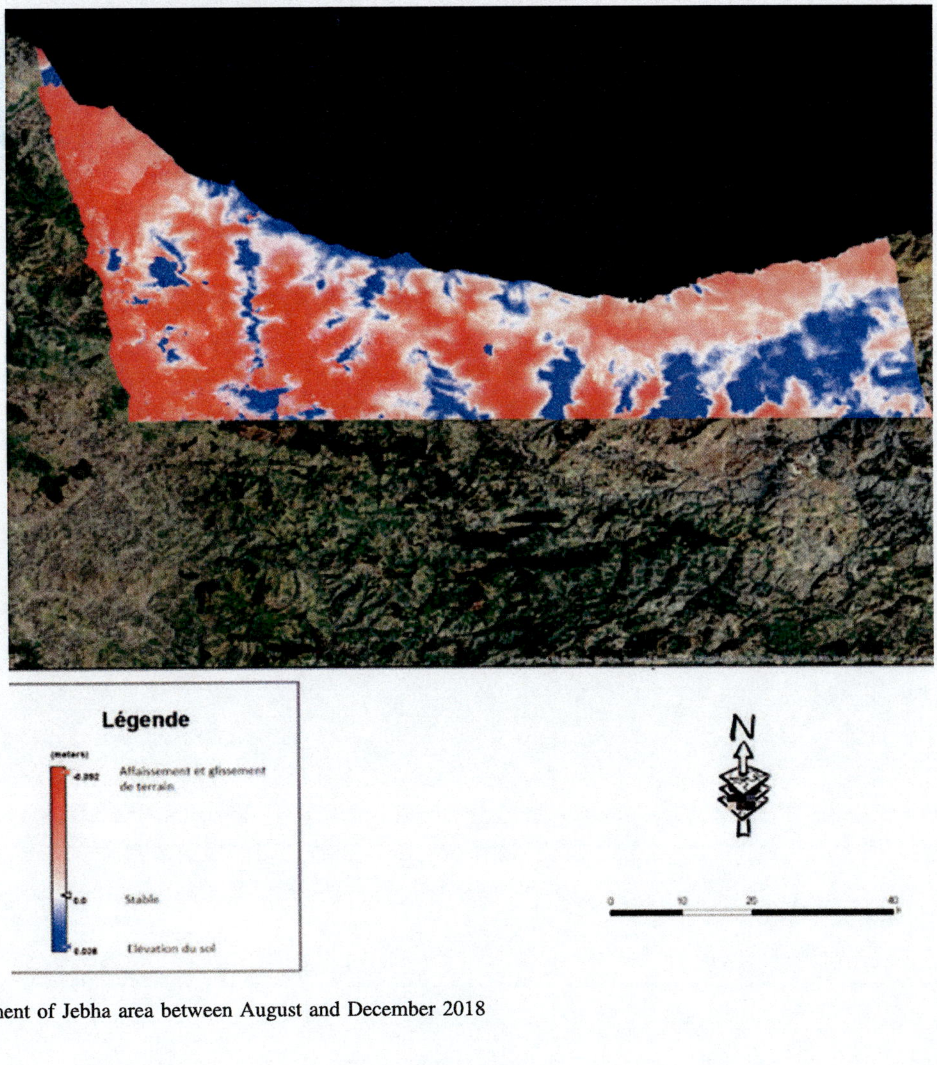

Map of displacement of Jebha area between August and December 2018

factors predisposed to this kind of instability associated with factors.

Radar interferometry, applied in an urban environment, provided a spatial vision of the deformations of the ground with millimeter vertical precision.

Bibliography

Crosetto, M.: Deformation monitoring using remotely sensed radar interferometric data. In: Proceedings, 11th FIG Symposium on Deformation Measurements, Santorini, Greece (2003)

Frédéric, A.: A ground uplift in the city of Paris (France) detected by satellite radar interferometry. Geophys. Res. Lett. **29**(17), 1853 (2002). https://doi.org/10.1029/2002GL015630

Harmouzi, H.: Geomorphological and geological analysis of Akchour landslide (Rif, Morocco). Geo-Eco-Trop. **42**(1), 19–32 (2018)

Mastere, M.: Cartographie de l'occupation des sols en relation avec les mouvements gravitaires et le ravinement dans le Rif nord-occidental (Maroc) 335–352 (2013)

Pateau, M.: From hazard to natural risk: Tangier-Tetouan region example (Rif, Morocco). Geo-Eco-Trop. **38**(1), 23–32 (2014)

SIRNAT-JPRN Orléans: Le satellite: un outil de mesure des faibles déformations du sol (2002)

Tahiri, M.: Quantification of water erosion and sedimentation using empirical models in the Tahaddart watershed (Northwestern Rif, Morocco) 2458-7184 (2017)

The *Mw* 7.1, Hector Mine, California earthquake surface rupture, surface displacement field, and fault slip solution from ERS SAR data. Earth Planet. Sci. **333**, 545–555 (2001)

Development of a Comprehensive Remote Sensing Technique for Drought Monitoring in Morocco

Noureddine Bijaber and Atmane Rochdi

Abstract

Because of its recurrence and the severity of its impacts, drought has become a structural component of the Moroccan climate. It is a real threat to agricultural activities, which represent nearly 20% of the Moroccan economy. The objective of this study was to develop a comprehensive method for drought monitoring based on remote sensing techniques. The main input parameters are derived monthly from satellite data at the national scale and are then combined to generate a composite drought index representing the different severity classes of drought. The input parameters are: Standardized Precipitation Index (SPI) calculated from satellite-based precipitation data (Climate Hazards Group InfraRed Precipitation with Stations, CHIRPS), anomalies in the day-night difference of Land Surface Temperature (LSTa) as a proxy for soil moisture, Normalized Difference Vegetation Index anomalies (NDVIa) from Moderate Resolution Imaging Spectroradiometer (MODIS) data and Evapotranspiration anomalies (ETa) obtained from surface energy balance modeling. The weighted combination of these input parameters into one composite indicator takes into account the importance of the rainfall-based parameter. The composite drought index maps were generated during the growing seasons going back to 2003. The results presented in this paper dealt with the 2018–2019 growing season which was a dry year.

Keywords

Drought monitoring • Remote sensing • Composite index

N. Bijaber (✉)
Royal Centre for Remote Sensing, 10000 Rabat, Morocco

A. Rochdi
Faculty of Sciences, University Ibn Tofail, 14000 Kenitra, Morocco

1 Introduction

Agricultural drought is a recurrent phenomenon in many parts of Morocco, primarily in the non-irrigated areas. It occurs when rainfall and soil moisture conditions are inadequate to support healthy crop growth to maturity during the growing season. This situation causes extreme crop stress and physiognomic changes within vegetation.

Morocco is characterized by a highly contrasting climate with a rainfall regime dominated by high irregularity in the spatial distribution and timing of precipitation across the country. The amount of precipitation varies extremely from the north-western plains to the southern regions of Morocco. In addition to this spatial variability, rainfall is seasonally irregular with considerable inter-annual variations in the onset and end of the rainy season, as well as the timing of the most humid month for a given region.

Historically, Morocco has experienced several years of drought with greater frequency and with high intensity. The impacts of these droughts on the national economy are of great importance.

The objective of this study was to develop a comprehensive remote sensing technique for drought monitoring in Morocco. The study is based on the use of several satellite-derived data sets that characterize relevant components of the hydrologic cycle related to drought into a new, gridded Composite Drought Indicator (CDI) that will be used to support various drought-related decision-making activities within the different concerned departments (Ministry of Agriculture, Department of Water, insurance companies).

2 Materials and Methods

The CDI developed in this study is based on a combination of four different parameters derived from various satellite-based earth observation products that are generated

H. El-Askary et al. (eds.), *Research Developments in Geotechnics, Geo-Informatics and Remote Sensing*, Advances in Science, Technology & Innovation,
https://doi.org/10.1007/978-3-030-72896-0_70

on a monthly basis to produce a monthly time series of CDI over Morocco for agricultural drought monitoring. This set of remote sensing-derived data inputs was chosen based on the state of the art and best practices of many agricultural drought monitoring systems throughout the world (World Meteorological Organization (WMO), Global Water Partnership (GWP) 2016). The used criteria to determine the most suitable inputs for analyses were based on: data access, data reliability, and long-term availability, the best operational resolution possible, and commonly accepted and operationally used variables by other agricultural focused drought early warning systems around the world. Several sources of information on the different indices and indicators being used nowadays around the world for drought monitoring and early warning exist and is supported by institutions of good reputation in the field, like the National Drought Mitigation Centre, the World Meteorological Organization, and the European Drought Observatory (World Meteorological Organization (WMO), Global Water Partnership (GWP) 2016). The used input indices in this study are described below:

- SPI calculated for a 2-month period from CHIRPS data set at 5 km spatial resolution data (Funk et al. 2014);
- NDVI anomalies derived from MODIS data (250 m) for a 10-day period using the maximum-value-composite method (Senay et al. 2015);
- ET anomalies calculated using the Operational Simplified Surface Energy Balance model (Senay et al. 2013) and acquired from Famine Early Warning System Network;
- Long-term MODIS day-night LST anomalies at 1 km spatial resolution data and used as a proxy of soil moisture proxy conditions (Hain et al. 2011).

The first step in the methodology was to have a uniform resolution for the drought indicator by sampling all these indices to a 5-km spatial resolution consisting of 25.589 grid cells over Morocco. The CDI is then calculated as given by Eq. (1):

$$\mathrm{CDI} = 0.4 \times \mathrm{SPI} + 0.2 \times \mathrm{NDVIa} + 0.2 \times \mathrm{LSTday\_night} + 0.2 \times \mathrm{Eta} \quad (1)$$

The weights assigned to each parameter in Eq. (1) were defined according to the degree of approximation to reality observed in a period of real drought (Bijaber et al. 2018).

3 Results

The drought indicator is calculated for each 5 km grid cell, and the CDI maps present a mosaic of pixels corresponding to different drought categories. A statistical zoning method was applied to simplify these maps for end users by assigning a drought severity class to each district or province. Figure 1 shows an example of CDI time series maps generated during the most sensitive period of the 2018–2019 growing season.

From Fig. 1, it can be observed that the 2018–2019 growing season was strongly affected by drought. All the severity of drought classes were detected during the first quarter of the current year.

4 Discussion

The persistence of the different drought classes during the most sensitive period of the agricultural cycle has negative impacts on non-irrigated cereal zones. These zones (about 5 million ha) are located in the center, the eastern and the northern parts of Morocco, and are very sensitive to drought. The 2018–2019 growing season recorded a rainfall deficit since the beginning of December compared to a normal year.

The evolution of the CDI during this specific period shows that more than 16% of the national territory of Morocco (northern part) experienced drought during December 2018. The proportion of the national territory concerned with drought increased to 91% during January 2019, and about 57% during February 2019. This seasonal drought causes losses to the cereal yields and affects the domestic water supply.

The generated monthly CDI maps are not a forecast. These products are an assessment of drought conditions based on a combination of separate parameters that can help decision makers to mitigate drought impacts. Also, the CDI classes may change depending on the structural vulnerability to drought of any region (irrigated, non-irrigated areas, forests, and other zones).

5 Conclusions

The risk of drought in Morocco has become a structural factor of the climate due to its recurrence and the sustainability of its impacts at different levels. The apprehension and monitoring of this phenomenon require new technologies and more effective instruments to produce and disseminate information about drought: earlier, reliable, gridded, and cost-effective.

The objective of this study was to develop an operational methodology for drought monitoring based on the use of remote sensing techniques combined with meteorological data and modeling. The combined drought index highlighted in this paper constitutes a new methodology of drought monitoring in Morocco. The added value is that we can

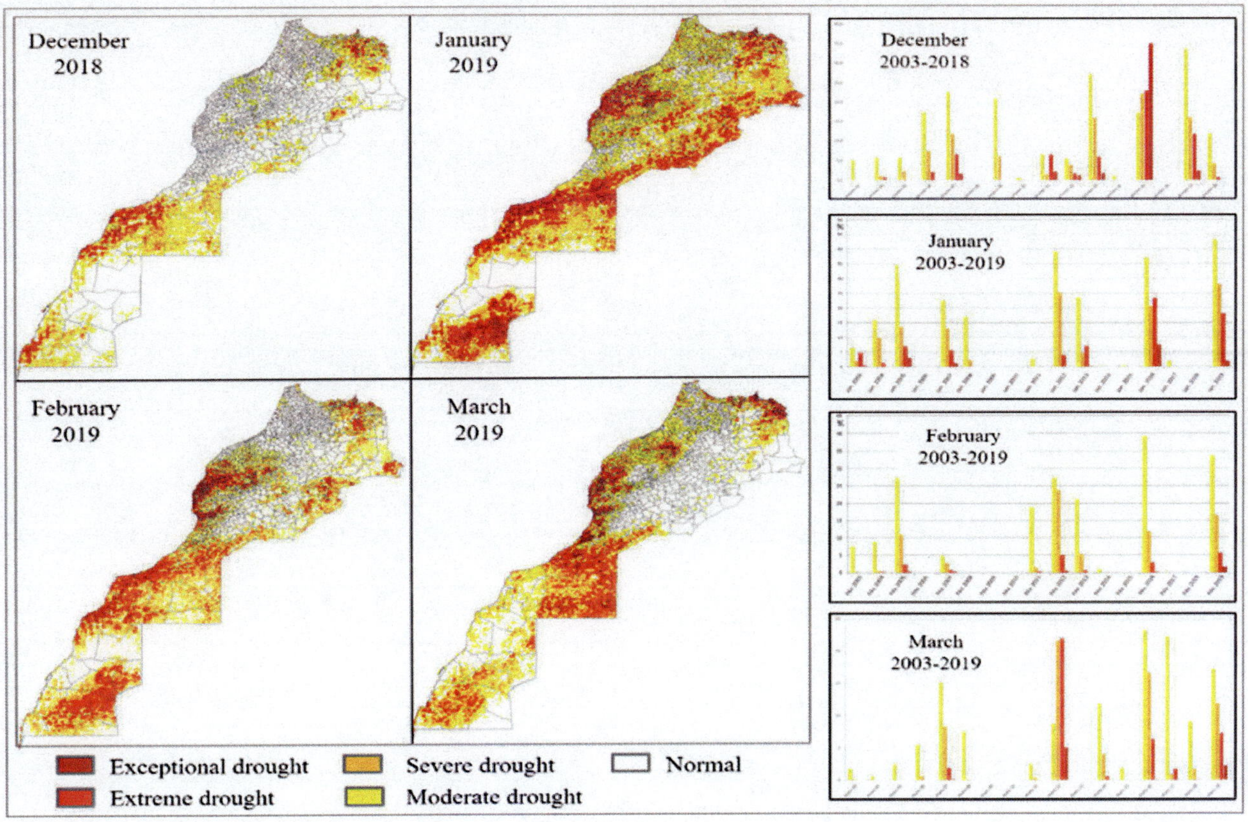

Fig. 1 Composite drought indicator maps (December 2018 to March 2019)

monitor the vegetation state at each 5 km grid cell every month by combining different input parameters.

The drought index maps are disseminated to the main users in Morocco (Ministry of Agriculture, insurance companies, and Department of Water). These maps can also be used by other stakeholders such as research institutes and private farmers.

References

Bijaber, N., El Hadani, D., Saidi, M., Svoboda, M.D., Wardlow, B.D., Hain, C.R., Poulsen, C.C., Yessef, M., Rochdi, A.: Developing a remotely sensed drought monitoring indicator for Morocco. Geosciences **8**, 55 (2018)

Funk, C.C., Peterson, P.J., Landsfeld, M.F., Pedreros, D.H., Verdin, J. P., Rowland, J.D., Romero, B.E., Husak, G.J., Michaelsen, J.C., Verdin, A.P.: A Quasi-Global Precipitation Time Series for Drought Monitoring. USGS Data Series 832. United States Geological Survey (USGS), Reston, VA (2014). ISSN 2327-638X

Hain, C.R., Crow, W.T., Mecikalski, J.R., Anderson, M.C., Holmes, T.: An intercomparison of available soil moisture estimates from thermal infrared and passive microwave remote sensing and land surface modelling. J. Geophys. Res. **116** (2011). https://doi.org/10.1029/2011JD015633

Senay, G.B., Bohms, S., Singh, R.K., Gowda, P.H., Velpuri, N.M., Alemu, H., Verdin, J.P.: Operational evapotranspiration mapping using remote sensing datasets and weather: a new parameterization for the SSEB approach. J. Am. Water Resour. Assoc. **49**, 577–591 (2013). https://doi.org/10.1111/jawr.12057

Senay, G.B., Velpuri, N.M., Bohms, S., Budde, M., Young, C., Rowland, J., Verdin, J.P.: Drought Monitoring and Assessment: Remote Sensing and Modelling Approaches for the Famine Early Warning Systems Network. USGS Staff—Published Research, Paper 858. Elsevier, Amsterdam (2015)

World Meteorological Organization (WMO), Global Water Partnership (GWP): Handbook of Drought Indicators and Indices; Integrated Drought Management Tools and Guidelines Series 2. In: Svoboda, M., Fuchs, B.A. (eds.) Integrated Drought Management Program (IDMP). Geneva (2016). ISBN 978-92-63-11173-9

On Building of a Deep Learning-Based Drought Forecasting System for the Sarab Region [Iran]

Raja Inoubli, Ali Ben Abbes, and Imed Riadh Farah

Abstract

The aim of this paper was to propose a long short-term memory model (LSTM) for the drought forecasting in Iran based on the standardized precipitation evapotranspiration index (SPEI). The model results were compared with three different machine learning methods. The same input data were used, including rainfall, temperature (mean, minimum, and maximum), humidity, pressure, evapotranspiration, and the normalized difference vegetation index (NDVI). The performance of all the models was compared using two statistical indices root mean square error (RMSE) and coefficient of determination (R^2). The comparative study results clearly indicate the good performance of the proposed method in forecasting SPEI values over multiple lead times.

Keywords

Drought forecasting • SPEI • NDVI • Deep learning • LSTM

1 Introduction

Drought is a natural hazard that exerts serious hydrological, agricultural meteorological, and socioeconomic damages (Tadesse et al. 2005). Thereby, over the last decade, many machine learning-based methods have been proposed for drought forecasting (Vicente-Serrano et al. 2010). Recently, a large and open set of big data sets has been made available to monitor drought over time. However, the existing methods are complex for analyzing a huge data set due to the lack of computing infrastructure. This paper proposed the use of LSTM networks (Hochreiter and Schmidhuber 1997) as a class of recurrent neural networks (RNNs) one of the most commonly used deep learning networks. In fact, this model has the advantage of handling multivariate time series data.

2 Materials and Methods

The study area is located in the eastern cities of East Azarbaijan province in Iran. The input data included the SPEI, rainfall, temperature (mean, minimum, and maximum), humidity, pressure, evaporation, and the NDVI in the period from 1988 to 2016 (Fig. 1).

The proposed methodology, illustrated in Fig. 2, consists of five major steps: (1) generating the input features and target data from the used data set; (2) passing data through a pretreatment process such as normalization; (3) subdividing the input data into training and testing data sets then fitting those splits into the model; (4) setting the proposed network is with the appropriate hyper-parameter choice, more details about the hyper-parameters are illustrated in Sect. 3.2; (5) evaluating the model results using two metrics: RMSE and R^2.

3 Results

3.1 The LSTM-Based Forecasting Framework

In order to forecast drought, we proposed the LSTM which was introduced for the first time by Hochreiter and Schmidhuber in 1997 for classifying, processing, and predicting tasks. The LSTM objective is learning long-term dependencies to overcome the RNNs problems such as vanishing

R. Inoubli (✉) · A. B. Abbes · I. R. Farah
Laboratoire RIADI, École Nationale des Sciences de l'Informatique, Manouba, Tunisia

A. B. Abbes
Centre d'applications et de Recherches en Télédétection (CARTEL), Université de Sherbrooke, Sherbrooke, QC J1K 2R1, Canada

I. R. Farah
Laboratoire ITI Department, IMT Atlantique, Brest-Iroise, France

H. El-Askary et al. (eds.), *Research Developments in Geotechnics, Geo-Informatics and Remote Sensing*, Advances in Science, Technology & Innovation,
https://doi.org/10.1007/978-3-030-72896-0_71

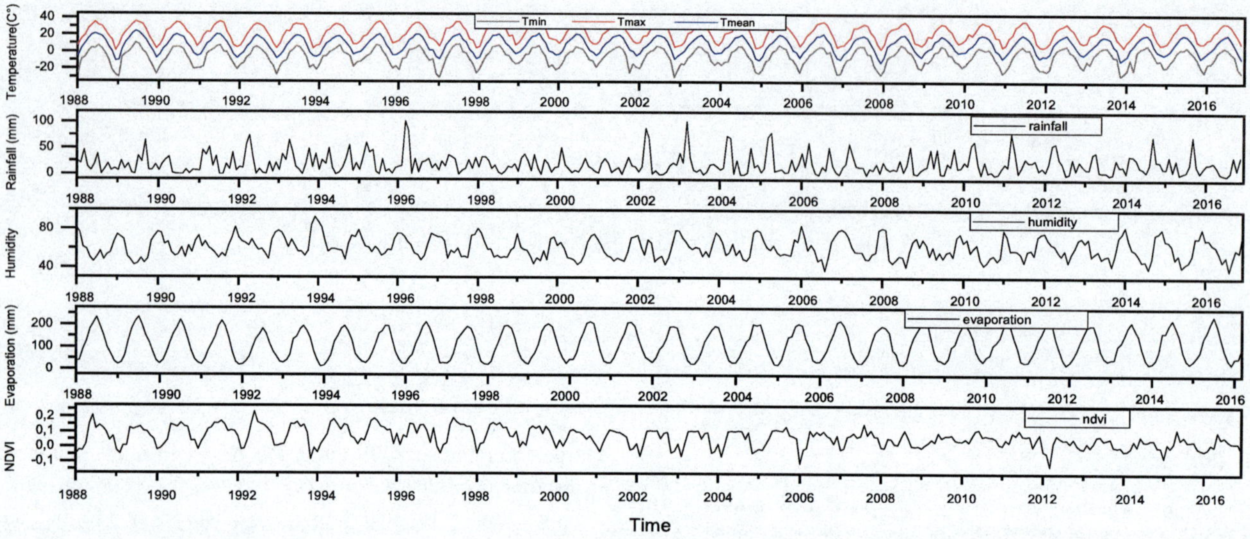

Fig. 1 Input data, including NDVI, evaporation, humidity, rainfall, temperature *(mean, minimum, and maximum)*, and pressure

Fig. 2 Flowchart of the proposed methodology

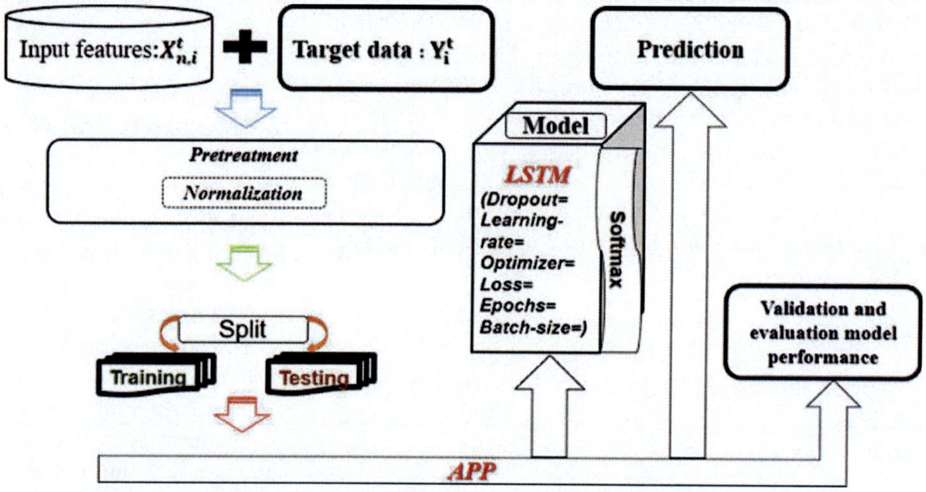

and exploding gradients. As a matter of fact, this advanced deep learning architecture has been used in several fields (Kumar et al. 2018; Elsheikh et al. 2019). The framework is built on a desktop PC with a 3.4 GHz Intel i3 processor and 6 GB of memory using the Keras, a package of Python software with TensorFlow and Pandas.

3.2 The LSTM Hyper-parameters

The proposed model was implemented with the use of the hyper-parameters that fit with our model objectives and ensure the TensorFlow's convergence. Therefore, we have used as the loss functions the mean square error. However, the weight initialization was a part of the Keras. According to the literature, Adam optimizer proved to work the best for our study. The learning rate is 0.001, while the batch size is 24 and the epochs are set to 200. To prevent over-fitting problems, the dropout is set on the LSTM layer.

3.3 Model Evaluation

In order to evaluate the model performance, we compared the proposed model to the neural network (NN), support vector regression (SVR), and the random forest regression (RFR) models. In Fig. 3, we displayed the SPEI prediction performance over time.

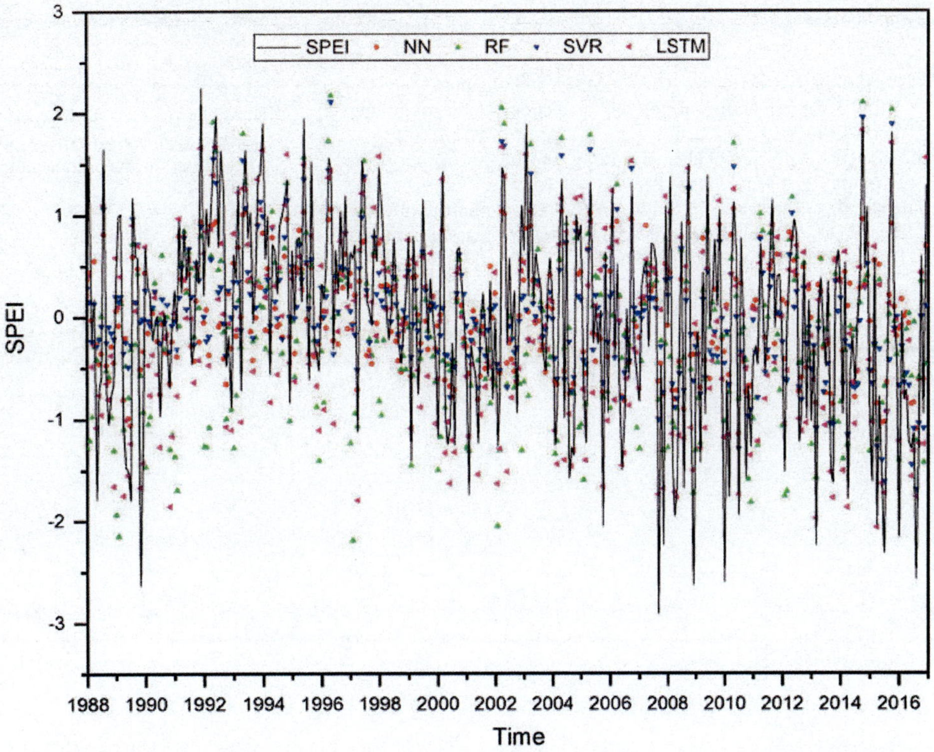

Fig. 3 LSTM, SVR, NN, and RFR forecasting results

Table 1 Comparison of performance of the proposed model with SVR, RFR, and NN model through the different time lags

Leading time	LSTMNN		SVR		NN		RFR	
	R^2	RMSE	R^2	RMSE	R^2	RMSE	R^2	RMSE
1-month	0,92	0,02	0,70	0,05	0,90	0,03	0,87	0,04
3-month	0,91	0,03	0,68	0,05	0,89	0,03	0,87	0,05
6-month	0,91	0,03	0,68	0,06	0,89	0,04	0,86	0,05
12-month	0,91	0,03	0,63	0,07	0,87	0,05	0,85	0,06

Table 1 provides the performance measurements.

4 Discussion

The proposed model proved its effectiveness in predicting SPEI at the different time lags compared to the other models. In fact, this result was validated by comparing the two-metric values. We have recorded minimal RMSE values for the LSTM (≤ 0.03) compared to the RMSE results (≥ 0.03) for the other machine learning methods. Additionally, in 1-month with R^2 equal 0,92, the LSTM outperforms the highest R^2 value recorded by NN which is equal to 0,90. Furthermore, the LSTM with R^2 is equal 0,91 overcoming all the other methods in 3-, 6-, and 12-months. Clearly, this gives a high improvement of the best results of the LSTM over the machine learning models. The above similar results are displayed in Table 1.

5 Conclusions

In this paper, the LSTM was applied in order to predict the SPEI at a different leading time (1-, 3-, 6-, and 12-) month in the Sarab region, Iran. The first step was to achieve a pre-processing step to fit the data to the model. Then we carried out a set of tests to find the best performing hyper-parameters. As a final step, the model was compared with three machine learning methods. The results revealed the effectiveness of our architecture for monthly prediction and the detected improvement through by long-term dependency over the input sequence.

References

Elsheikh, A., Yacout, S., Ouali, M.S.: Bidirectional handshaking LSTM for remaining useful life prediction. Neurocomputing **23**, 148–156 (2019)

Hochreiter, S., Schmidhuber, J.: Long short-term memory. Neural Comput. **9**, 1735–1780 (1997)

Kumar, J., Goomer, R., Singh, A.K.: Long short-term memory recurrent neural network based workload forecasting model for cloud datacenters. Procedia Comput. Sci. **125**, 676–682 (2018)

Tadesse, T., Brown, J.F., Hayes, M.J.: A new approach for predicting drought-related vegetation stress: integrating satellite, climate, and biophysical data over the US central plains. ISPRS J. Photogramm. Remote Sens. **59**(4), 244–253 (2005)

Vicente-Serrano, S.M., Beguería, S., López-Moreno, J.I.: A multiscale drought index sensitive to global warming: the standardized precipitation evapotranspiration index. J. Clim. **23**(7), 1696–1718 (2010)

Results of the Analysis of Water Stress Regimes for Irrigated Crops Based on Remote Sensing Data

Anatoly Zeyliger, Olga Ermolaeva, Danil Antonov,
Fedor Serebrennukov, Alexey Kravchuk, and Sergey Zatinatsky

Abstract

Results of the temporal analysis of the water stress regimes of the irrigated agricultural crops (alfalfa, corn, and soybean) are presented at the level of a typical agricultural agrocenosis with the use of remote sensing data of MODIS. To achieve this aim, remote sensing and ground data observed in 2012 in the fields of several different agricultural farms of the Saratov Zavolzhie (Marks Region, Saratov Oblast, Russia) have been used. Results of this study revealed good correlations between cumulated values of crop water stress with their bio-productivity.

Keywords

Agricultural crop • Irrigation • Evapotranspiration • Crop water stress • Crop bio-productivity • Remote sensing • MODIS

1 Introduction

New technologies of water consumption management of the irrigated agrocenoses (IA) are aimed at efficient use of nature resources (Mannini et al. 2013; Wenchao et al. 2017; Yanfei et al. 2018; Hongjun et al. 2018). These technologies are based on the combination of the computer modeling of agricultural crops growth and development (Feddes 1987; Zeiliger and Ermolaeva 2016) with the technologies of geo-information systems (GIS), remote sensing (RS), and global navigation satellite systems (GNSS) (Zeiliger and Ermolaeva 2016; SHul'ga et al. 2016).

Studies of the IA water consumption based on computer modeling were grounded on the use of physical and mathematical models of the vertical water and heat exchange of underlying cover with atmosphere (Feddes 1987; Bastiaanssen et al. 2005; Gowda et al. 2008; Muzylev et al. 2017, 2018; Zeyliger et al. 2019). Assimilation in the RS data was done for the ground area occupied by IA. Application of this data provides detailing of the results for estimation of parameters of water and heat regimes and leads to a significant increase of their accuracies (Bezerra et al. 2015). As a result, it allowed estimating vegetation parameters of IA growth and development, as well as its water use and biomass productivity (Overgaard et al. 2006). Finally, assimilation of the RS data of the IA, being subjected to modeling, significantly increases the quality, efficiency, and reliability of the modeling for diagnostic and prognostic purposes. It was realized by verification and adaptation of the water consumption (WC) modeling results, evapotranspiration (ET), and soil water storage in root zone (SWS-RZ) with the use of developed method, based on the estimation of the dynamics of the coefficient of cumulated water stress (CC-WS) K_w^c of IA.

The Saratov Zavolzhie (Marks Region, Saratov Oblast, Russia) is one of the largest agricultural producers in the southeastern zone of the Russian Federation Climate in this territory which is continental arid with cold winter and hot summer seasons. Duration of the frost-free period is 150–160 days per year. The sum of active temperatures is 2600–2800 that allows to cultivate a large range of agricultural crops. Average annual precipitation in these areas is 400–500 mm, and the major part of it occurs during the warm period of the year. Soil cover mostly consists of dark-chestnut soils, forming the typical background. Ground waters mostly lie at the depth below 10 m, although in some irrigated areas, the ground waters' level occurs at the depths of 2–3 m. Agriculture in this zone is carried out in arid

A. Zeyliger (✉) · O. Ermolaeva · D. Antonov
Russian State Agrarian University MTAA, Timiryazevskaya str., 49, 127550 Moscow, Russia

F. Serebrennukov · A. Kravchuk · S. Zatinatsky
Saratov State Agrarian University named after Vavilov, Teatralnaya Pl, 1, 410012 Saratov, Russia

H. El-Askary et al. (eds.), *Research Developments in Geotechnics, Geo-Informatics and Remote Sensing*, Advances in Science, Technology & Innovation,
https://doi.org/10.1007/978-3-030-72896-0_72

conditions, and this factor determines the volume and stability of the production of crop. According to the scientific institutions of the Volga region, from 1894 to 2018, that is, for 123 years, only 20 years (18%) were the most climatic, moderate—51 years (41%), arid, and extremely arid—52 years (42%). Consequently, agriculture is carried out in conditions where almost every third year is arid, and every fifth year is lean.

In this paper, we consider the coefficient of cumulated water stress variations of study plots for a 2012 year monitoring period using the moderate resolution imaging spectroradiometer (MODIS) ET product. Our major objective was to prove out the correlations between cumulated values of crop water stress and biomass productivity.

2 Materials and Methods

Studies of the assessment of CC-WS of IA were conducted for the irrigated crops of alfalfa, corn, and soya, which were cultivated in the territory of the Saratov Zavolzhie during 2012 vegetation period.

The method of evaluation of CC-WS of IA has been developed for conditions of the absence of ecological stress (soil salinization) and is based on the following equation for diurnal water stress coefficient K_w (Allen et al. 1998):

$$K_W = ET_a/ET_c, \quad (1)$$

where ET_a—actual daily evapotranspiration in the current conditions of WC and ET_c—daily evapotranspiration in conditions of the optimal WC.

The value of daily ET of IA in conditions of the optimal water use ET_c depends on meteorological parameters of the weather conditions, vegetation phase, as well as on a number of factors that influence decrease of the total ET (Allen et al. 1998). In case of the absence of factors, diminishing the total ET, the following relation of ET_c from ET_0 has been used (Allen et al. 1998).

$$ET_c = K_c \cdot ET_0, \quad (2)$$

where K_c—empirical coefficient of WC of IA that takes into account its type and vegetative features, and also its stage of development and ET_0—etalon daily ET from the surface of the grass cover, which has the volume of soil moisture in its root inhabited layer needed for the optimal WC.

In the developed method, calculation of the K_w values of IA is made for the daily timescale. To meet this aim, data of the standard ground-based hydrometeorological daily parameters (temperature and humidity of air, atmospheric pressure, and also wind speed) and data of measurements of the MODIS/EOS Terra and Aqua radiometer are used. Calculation of the data sets of etalon evapotranspiration $\{ET_0\}^t$, based on Penman method, is conducted using data of daily hydrometeorological series, gained at the meteorological stations. Calculation of the daily $\{ET_c\}^t$ values, which is based on the FAO-56 method (Allen et al. 1998), is carried out according to Eq. (2) with the use of empirical K_w equations of the corresponding IA.

Data of MODIS radiometer for the sites of IA are used for the calculation of $\{ET_a\}^t$ series according to SEBS model. Calculated $\{ET_a\}^t$ series, series of precipitation data, sowing dates, time of phases of vegetative development as well as irrigation water application data were used for the calculation of $\{K_w\}^t$ time series and K_w^c.

3 Results

Calculations of K_w^c were conducted for 17 alfalfa crops (most of them were of the second year of vegetation), for nine crops of corn, and ten crops of soya. As an example, Fig. 1 presents graphic image of time concomitant time series of the daily values of temperatures, dates and norms of rainfall, dates and norms of irrigation water applications, as well as of calculated values of background $\{ET_0\}^t$ and actual $\{ET_a\}^t$ evapotranspiration, and also of the water stress coefficient of alfalfa $\{K_W\}^t$.

Calculated time series of $\{K_W\}^t$ for alfalfa crops in the territory of OPX VolzhNIIGIM are shown by red lines in both parts of graphs, presented in Fig. 1, and form the basis for comparison. The aim of such presentation was to make visual comparison of these series with the calculated series of alfalfa crops that were situated in the territories of different farms. Three harvests were collected for each of these crops: The first harvest was made in the first decade of June, the second in the first decade of July, and the third in the second decade of August.

4 Discussion

Data analysis of the conducted calculations of $\{K_W\}^t$ indicates the water stress regimes, formed on alfalfa crops during 2012 irrigation period. In one hand, it is similar to each other in general course of curve, and on the other hand it is different from each other. Similarity of a number of water stress regimes is largely connected with the ownership of corresponding crops that apparently resulted in a uniform water irrigation application, almost in the same time and same irrigation norms. This is quite convincingly confirmed by the progress of crops' schedule in the farms ZAO Trudovoi, OPX VolzhNIIGiM, ZAO Meliorator, and others. At the same time, differences of the water stress regimes within the marked groups selected due to uniform management of water irrigation applications are apparently connected with

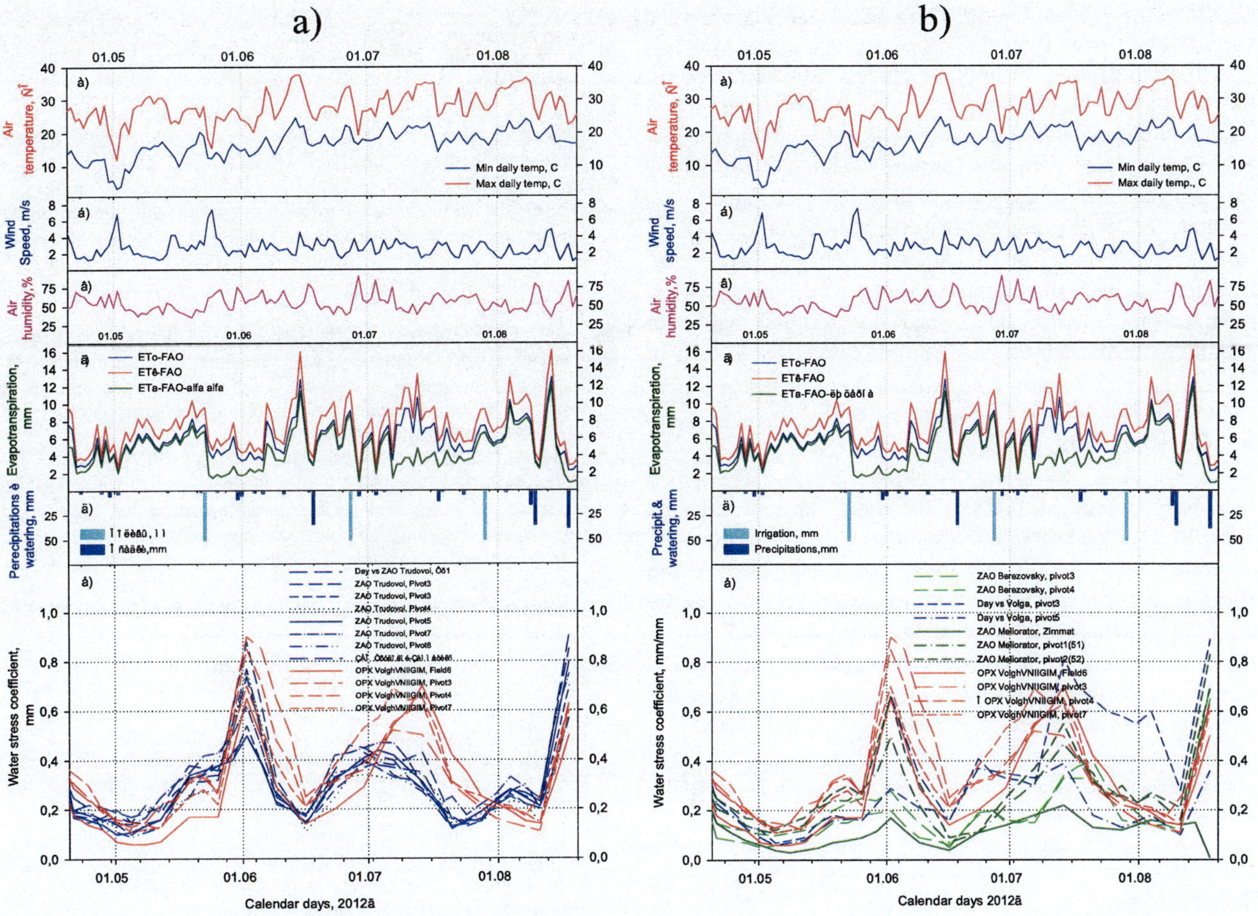

Fig. 1 Value of hydrometeorological parameters and CC-WS of IA at agricultural farms: **a** ZAO Trudovoy and OPX VolgNIIGiM; **b** ZAO Berezovsky, ZAO Volga, ZAO Meliorator, and OPX VolgNIIGiM

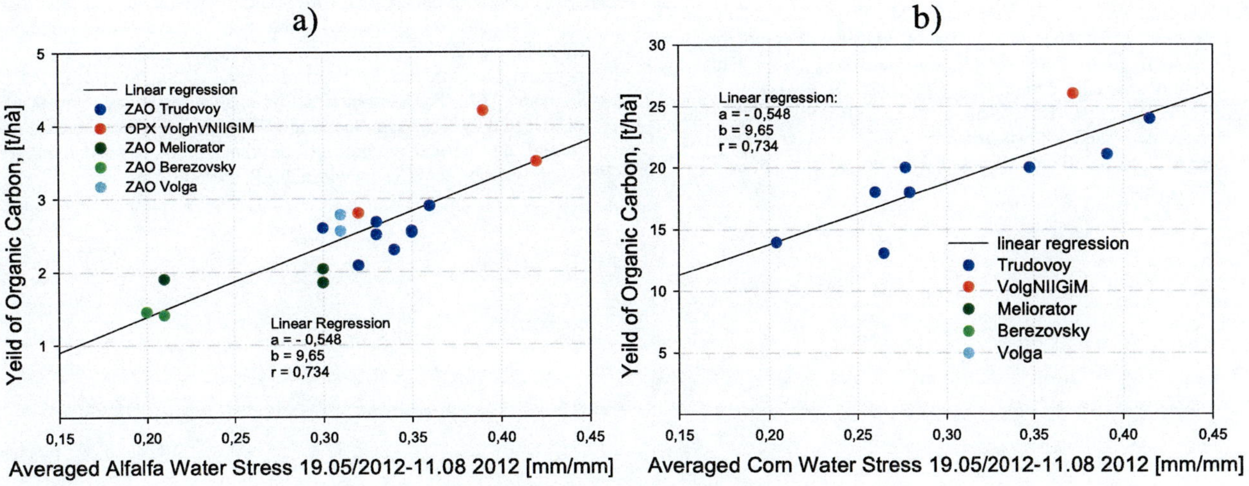

Fig. 2 Correlation between values of CC-WS and crop biomass: **a** alfalfa; **b** corn

the differences in soil and hydrogeological conditions of the corresponding land plots.

Collected data on the biomass productivity were used for the evaluation of the impact of water stress regimes of alfalfa, corn, and soya on their productivity of organic carbon. These values were used for the evaluation of correlation and dependencies of each crop with K_w of IA (Fig. 2).

Obtained linear correlations between productivity of irrigated agricultural crops and K_w^c value of IA for the irrigation period show quite close connections between biomass productivity and formed water stress regime in conditions of irrigation.

5 Conclusions

The study conducted to assess the water stress of irrigated agricultural crops using space monitoring data has shown the feasibility of using the developed method to analyze the effectiveness of irrigation. Further improvement of the developed method will be focused on using observed data by sensors on aerospace platforms and vegetation indicators in conjunction with forecast hydrometeorological data for the operational management of irrigated agricultural crops.

Acknowledgements The reported study was funded by RFBR according to the research project № 19-29-05261.

References

Allen, R.G., Pereira, L., Raes, D., Smith, M.: Crop Evapotranspiration: Guidelines for Computing Crop Water Requirements. FAO Irrigation and Drainage Paper 56. Food and Agriculture Organisation, Rome (1998)

Bastiaanssen, W.G.M., Noordman, E.J.M., Pelgrum, H., Davids, G., Thoreson, B.P., Allen, R.G.: SEBAL model with remotely sensed data to improve water resources management under actual field conditions. J. Irrig. Drain. Eng. **131**(1), 85–93 (2005)

Bezerra, B.G., Silva, B.B., Santos, C.A.C., Bezerra, J.R.C.: Actual evapotranspiration estimation using remote sensing: comparison of SEBAL and SSEB approaches. Adv. Remote Sens. **4**(3), 234–247 (2015)

Feddes, R.A.: Crop factors in relation to Makkink reference crop evapotranspiration. In: 'Evaporation and Weather', TNO Committee on Hydrological Research, Proceedings and Information, no. 39, pp. 33–46 (1987)

Gowda, P.H., Chavez, J.L., Colaizzi, P.D., Evette, S.R., Howell, T.A., Tolk, J.A.: ET mapping for agricultural water management: present status and challenges. Irrig. Sci. **26**, 223–237 (2008)

Hongjun, L., Jiazhen, L., Yanjun, S., Xiying, Z., Yuping, L.: Web-based irrigation decision support system with limited inputs for farmers. Agric. Water Manag. **210**(C), 279–285 (2018)

Mannini, P., Genovesi, R., Letterio, T.: IRRINET: large scale DSS application for on-farm irrigation scheduling. Procedia Environ. Sci. **19**, 823–829 (2013)

Muzylev, E.L., Startseva, Z.P., Uspensky, A.B., Volkova, E.V., Vasilenko, E.V., Kukharsky, A.V., Zeyliger, A.M., Ermolaeva, O. S.: Using remote sensing data to model water and heat regimes of rural territories. Sovrem. probl. distantsionnogo zondirovaniya Zemli iz kosmosa **14**(6), 19–47 (2017)

Muzylev, E.L., Zeyliger, A.M., Starceva, Z.P., Uspenskij, A.B., Volkova, E.V., Vasilenko, E.V., Ermolaeva, O.S.: Modeling processes of water and heat regime formation for agricultural region area (Russia) utilizing satellite data. In: Advances in Sustainable and Environmental Hydrology, Hydrogeology, Hydrochemistry and Water Resources. Proceedings of the 1st Springer Conference of the Arabian Journal of Geosciences (CAJG-1) Tunisia, 2018

Overgaard, J., Rosbjerg, J.D., Butts, M.B.: Land-surface modeling in hydrological perspective—a review. Biogeosciences **3**, 229–241 (2006)

SHul'ga, E.F., Kupriyanov, A.O., Hlyustov, V.K., Balabanov, V.I., Zejliger, A.M.: Management of Agricultural Enterprise with Use of Space Navigation Aids (GLONASS) and Remote Sensing of Earth, 282 p. RGAU-MSHA, Moscow (2016)

Wenchao, W., Yuanlai, C., Yufeng, L., Zenghuan, L., Junwei, T.: Web-based decision support system for canal irrigation management. Comput. Electron. Agric. Available online 22 Nov 2017. https://doi.org/10.1016/j.compag.2017.11.018

Yanfei, M., Shaomin, L., Lisheng, S., Ziwei, X., Yaling, L., Tongren, X., Zhongli, Z.: Estimation of daily evapotranspiration and irrigation water efficiency at a landsat-like scale for an arid irrigation area using multi-source remote sensing data. Remote Sens. Environ. **216**, 715–734 (2018)

Zeiliger, A.M., Ermolaeva, O.S.: Information technologies in monitoring of the rain fed and irrigated agrotsenoz. Sovrem. Nauk. Tekhnol. **10–1**, 62–66 (2016)

Zeyliger, A.M., Muzilev, E.L., Ermolaeva, O.S., Startseva, Z.P., Sukharev U.I.: The computer analysis of water stress regimes of an irrigated agrocoenosis using the SWAP model, and the data of ground and space monitoring. Sovrem. Probl. distantsionnogo zondirovaniya Zemli iz kosmosa **16**(3) (2019) (in press)

Detection and Modeling of Recent Tectonic Deformation by Remote Sensing and Geomorphometric Indexes: Example of the Watersheds of the North of Tunisia

Radhia Mansour, Nassira Zouaoui, and Abdessalem ElGhali

Abstract

The satellite image processing and morphometric analysis of SRTM-DEM data are two complementary approaches used to highlight active tectonics and the spatio-temporal landscape spatio-temporal, both in hilly areas and in flat morphologic zones. The purpose of this study is to determine the relationship between the current geomorphological landscape and the surface and subsurface tectonic activity. The tectono-geomorphological evolution is established in this work following: (1) the use of geomorphometric indexes such as the topographic index of humidity (TWI), the topographic position index (TPI), the relief amplitude (Ar), and slope gradient (S) derived from SRTM-DEM data (resolution 30 m); (2) the extraction of drainage anomalies resulting from the analysis of the hydrographic network; and (3) the detection of lineaments (Ld) extracted automatically by the directional filters applied to the principal component analysis of Landsat 8 satellite images, in particular the principal component 2 (PCA2). These operations allow us to extract the lineament maps of the large watershed of the middle valley of Oued Medjerda covering most of the north-west of Tunisia. The synthesis of these lineaments allowed the establishment of the morphostructural map and the understanding of the structural evolution of the region. The lineament produced map was then compared with field data (on the surface) and gravity maps reflecting the structural architecture of the same subsurface region. The data in this map are (1) in good agreement with all the results of the previous works, (2) provide geometrical details for some already known tectonic structures, and (3) allowed us to present a new diagram on the evolution of morphostructural structures in close relation with the recent tectonic activity of the region.

R. Mansour (✉) · N. Zouaoui · A. ElGhali
Faculty of Sciences of Bizerte, Carthage University, 7021 Bizerte, Tunisia

Keywords

Spatial modeling • IRAT • Lineaments • Anomalies • Geomorphometric indexes • GIS

1 Introduction

In the context of the recent geodynamics of the Mediterranean area and the converging movement of the African and Eurasian continents, tectonics have continuously shaped the North African landscapes and continue to have a significant effect on their geomorphology. Quantitative and qualitative analyses of the landscape, based on remote sensing and geomorphometric indexes, are useful to deduce the model and the spatial variation of the last deformations of the topographic envelope and the neotectonic activity. In this context, the development and interpretation of morphometric maps are important tools in studies related to the neotectonic–geomorphological interaction, where the responses of natural landscapes to the internal dynamics of the planet are often marked by the presence of anomalies in the drainage network and relief discontinuities. In northern Tunisia, the Quaternary neotectonic activity manifests itself not only by a deformation of the geomorphological surfaces, but also by changes in the hydrographic network (Rim 2014). In tectonically active regions, the drainage model is a potential parameter for tectono-morphological analysis. It is very sensitive and quick responsive to active processes. Tectonic "accidents" can be responsible for the accelerated incision of the river, the asymmetry of the basin, the geometry of drainage, and the complexity of the deviations of the river. Tectonic activity is also a driving force in the variation of morphology and landscape distribution. The latter results from a complex combination of the rejection of faults are related to the vertical and horizontal movements of the blocks and/or by surface erosion or deposition processes. Quantitative analysis and modeling of this recent tectonic

H. El-Askary et al. (eds.), *Research Developments in Geotechnics, Geo-Informatics and Remote Sensing*, Advances in Science, Technology & Innovation,
https://doi.org/10.1007/978-3-030-72896-0_73

activity have been established through the extraction of morphometric indexes. These represent the most capable indicators of decoding the response of the relief to deformation processes and have been widely used as recognition tools to differentiate deformed areas by active tectonics (Chen et al. 2003).

2 Materials and Methods

The approach adopted to achieve the purpose of this study is carried out using a methodological flowchart based essentially on the altimetry data (Fig. 1). This flowchart includes two successive steps, the first of which corresponds to a morphometric analysis and interpretation based on the extraction of morphometric indexes derived from the digital terrain model (DTM) such as integral hypsometry (HI), the index of variation slope (VSI), topographic humidity index (TWI), topographic position index (TPI), mountain front curvature (Smf), stream length gradient (SL), and the slope gradient (S) (Bikram et al. 2016; Reyaz et al. 2014; Shamrailtpam 2014). The second phase consists in determining the different hydrological indexes such as *T*, the drainage density (Dd), the bifurcation ratio (Rb), the circularity ratio (Rc), and the asymmetry factor (Af) (Esmaeil et al. 2017; Shamrailtpam 2014). This second phase is followed by a structural study in the field of the various anomalies. Microtectonic measurements of the principal lineaments and their kinematic identification are used to deduce the possible concordances between the anomalies and the nature of the movements of ground.

3 Results

The distribution and quantitative analysis of the relative tectonic activity (IRAT) (El Hamdouni 2008; Syed and Richard 2012) to the study area, formed by 24 sub-basins, results from the combination of a number of geomorphometric indexes and drainage. These indexes represent a quantitative approach to geomorphotectonic analysis based on the detection of anomalies affecting the river system, mountain fronts, and topographic slopes. The drainage anomalies were determined essentially following the calculation of the asymmetry index, which is related to the inclination of the main watershed network under the influence of either the tectonics or the lithology. The most important asymmetry characterizes the watersheds that are on the border of the major tectonic "accidents" of NE–SW of Cape Serate-Gardima (CSGF) and that of Gardima-Tunis (GTF) direction E–W. These drainage anomalies were also defined by the circularity index (Rc), with values between 0.003 and 0.475: note that low values characterize elongated basins. These anomalies follow two dominant directions: the

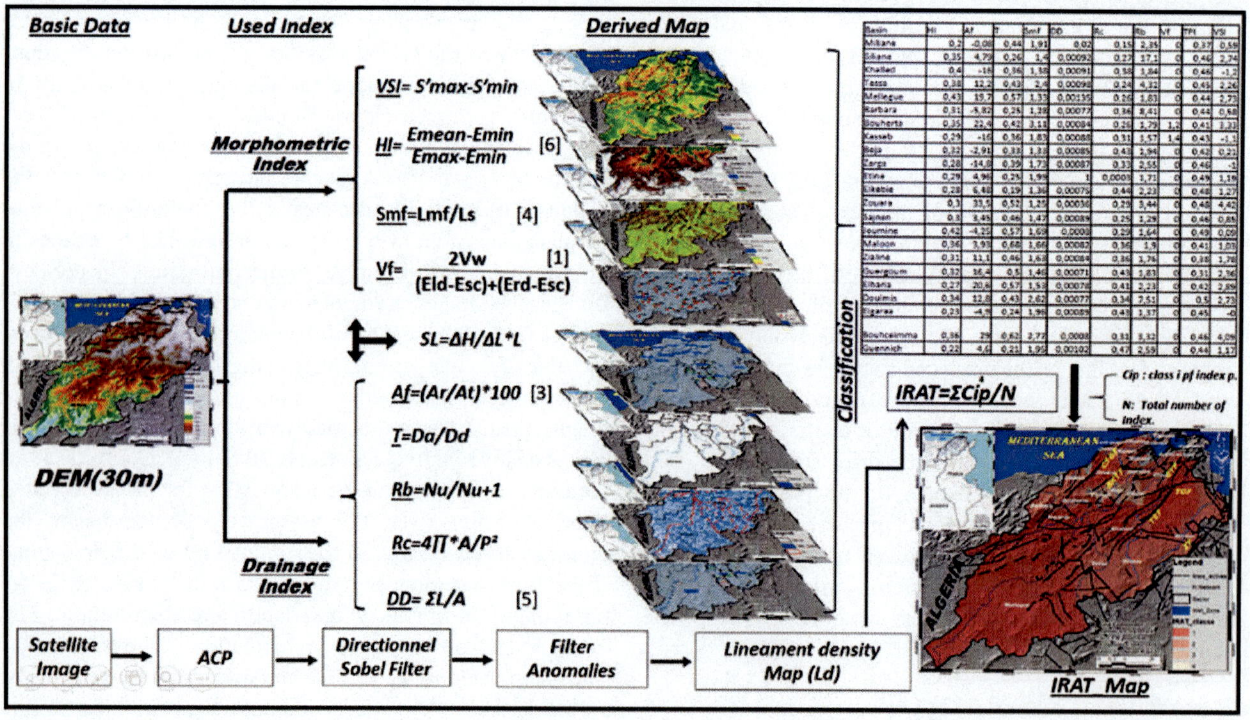

Fig. 1 Methodological flowchart

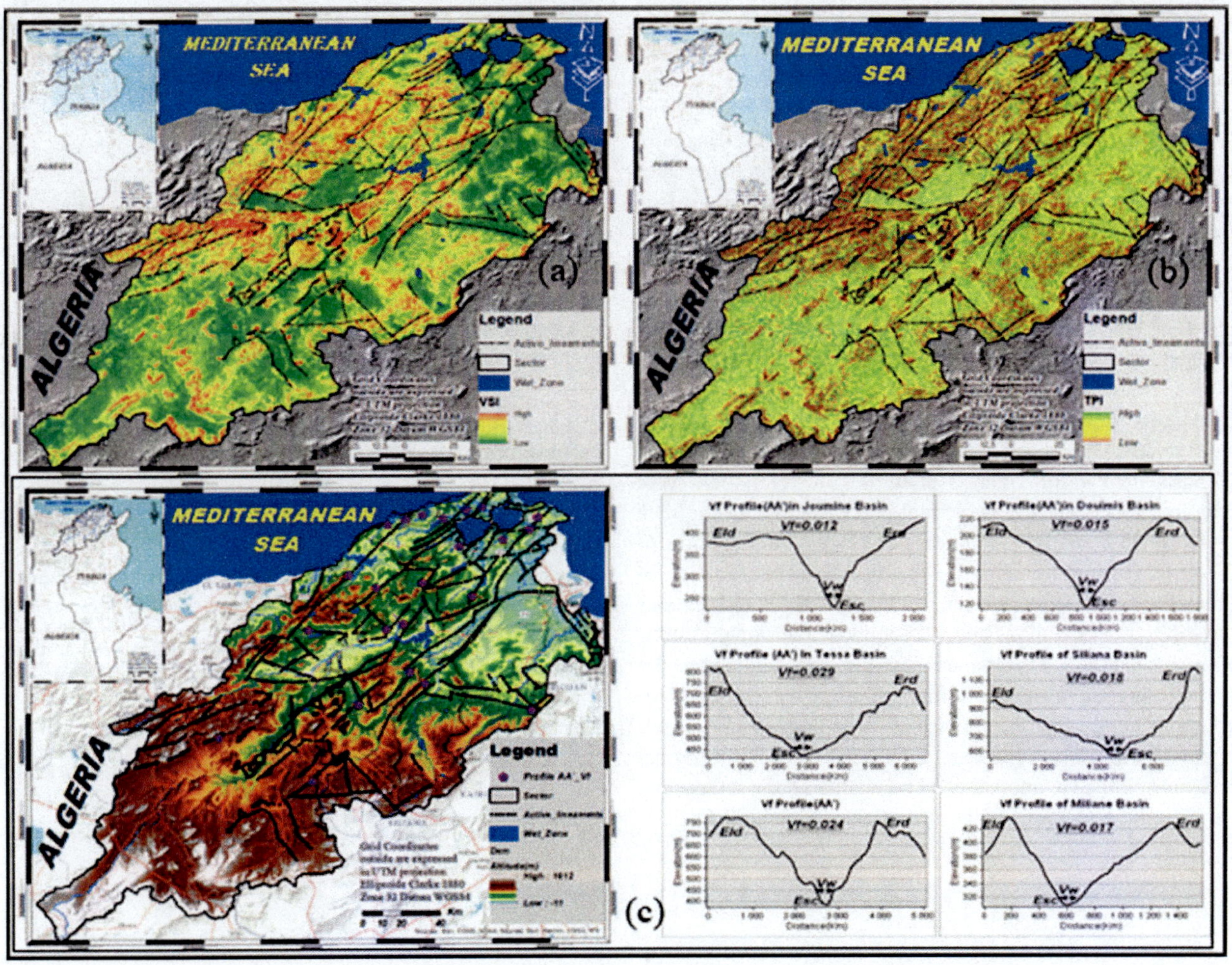

Fig. 2 Derived index maps **a** variation slope index (VSI), **b** topographic position index (TPI), and **c** valley width index

first is directed NE–SW and the second is oriented rather NW–SE. These anomalies coincide with most of the major faults in the study area, or even the whole of northern Tunisia, such as the Zaghouan (ZF), Alia-Teboursssek (ATF), and Cape Serate-Gardiaou (CSGF) scars. Morphometric abnormalities were determinated primarily from slope variation indexes (VSI), vallevel width index (Vf), and sinuosity index (Smf). The Vf values of our study area (Fig. 2c) vary from 0.01 to 1.4.

The high values of Vf are associated with tectonic stability due to the change in land cover and erosion. However, the low value of Vf indicates a rapid incision due to tectonic uplift (Dubey et al. 2017). Integral hypsometry (HI) also contributes to the determination of IRAT. This index describes the distribution of the elevation as a function of the corresponding surface, having values between 0.2 and 0.43. The combination of all these different indexes made it possible to quantitatively map the relative tectonic activity (IRAT) whose degree was subdivided into four classes: very high, high, medium, and low. These classes are characterized by a spatial extent of, respectively, 57%, 33%, 4%, and 6% of the total study area.

4 Discussion

The compilation and fusion of the resulting maps allow the development of a spatial distribution map of the relative tectonic activity index (IRAT). The latter shows that the zones characterized by the highest IRAT correspond to the faulty strikes, the incision of valleys, and the most rugged hydrographic network architecture. The confrontation of this synthesis map with the maps published in the previous works shows a great similarity, in addition to a good quality of precision, making it possible to detect the lineaments of small extent. The hiearchization of these detected anomalies is manifested by flush scars resulting from the reactivation and propagation of subsurface accidents and sealed fractures of subsoil.

After the compilation of the synthesis map compiled from all the analyses and the processing of the satellite image of the SRTM data, a geological and tectonic reconnaissance companion was carried out at certain points of control on field. The stereographic projection of the field measurements shows that all the fractures are organized in the form of quarter families of faults directed according to different quads: NE–SW, E–W, NW–SE, and N–S.

5 Conclusion

Geographical information system (GIS) techniques were adopted, following the numerical processing of all the data of reasonable spatial resolution, for (1) the combination of the different morphometric indexes and the determination of the degree of relative tectonic activity (IRAT) at the scale of the different watersheds and (2) the evaluation of the influence of tectonic disturbances on the landscape and the highlighting of the tectonic-hydrology and tectonic-geomorphology relationship sector. This has become possible thanks to the detection and interpretation of hydrological and morphometric anomalies. The results of this study can be an important tool for the assessment of natural hazards and risk management in areas with high urban concentrations.

References

Bikram, S., et al.: Morphotectonic analysis of the Madhumati watershed, northeast Kashmir Valley. Arab J. Geosci. (2016)

Chen, Y.C., et al.: Along-strike variations of morphotectonic features in the western foothills of Taiwan: tectonic implications based on stream gradient and hypsometric analysis. Geomorphology **56**, 109–137 (2003)

Dubey, R.K., et al.: Evaluation of relative tectonic perturbations of the Kashmir Basin, Northwest Himalaya, India: an integrated morphological approach. J. Asian Earth Sci. **148**, 153–172 (2017)

El Hamdouni, R.: Assessment of relative active tectonics, southwest border of the Sierra Nevada (Southern Spain). Geomorphol. J. **96**, 150–173 (2008)

Esmaeil, H., et al.: Assessent of relative active tectonics in the Bozgoush Basin (SW of Caspian Sea). J. Mar. Sci. **7**, 211–237 (2017)

Reyaz, A., et al.: Tectono-geoorphic study of the Karewa Basin of Kashmir Valley. J. Asian Earth Sci. **92**, 143–156 (2014)

Rim, G.: Apports de l'image sismique et des SIG à l'étude morphostructurale de la Jeffara (Sud Est Tunisien): implications géodynamiques et intérets pétroliers. Thesis, Université Paris-Est Marne La Vallée (2014)

Shamrailtpam, A.: Application of TecDEM in morphometric studies of Imphal River. Am. Int. J. Res. Sci. Technol. Eng. Math. (2014)

Syed, A., Richard, G.: Appraisal of active tectonics in Hindu Kush: insights fro DEM derived geomorphic indices and drainage analysis. Geosci. Front. J. **3**(4), 407–428 (2012)

Modelling Spatial Variability of Soil Particle Size Distribution in Mountainous Watershed of Bisha-Khamis, KSA

Roohul Abad Khan, Javed Mallick, and Rachida El Morabet

Abstract

Soil is a natural material with varying physical and chemical properties. The particle size distribution (PSD) plays an important role as it affects the interaction of soil with weathering agents as well as hydrological parameters. This study attempts to integrate information from various sources including geographic information system (GIS) and global positioning system (GPS) to accurately quantify and map the geotechnical design parameters in the watershed of Bisha-Khamis, KSA using geospatial and geostatistical analysis tools. The results obtained were validated against the experimental results from soil samples of the concerned study area. The suggested method here will help soil exploration agencies and practicing geotechnical engineers for immediate decision-making process about soil suitability for various application.

Keywords

PSD • Geospatial • Geostatistical • Experimental results • Decision making

1 Introduction

Particle size distribution (PSD) is a fundamental soil attribute. It plays a key role in defining the physical, chemical, and biological properties of soil and their interaction with natural and anthropogenic parameters. The soil PSD can quantitatively describe soil texture. Spatial and dynamic variations are necessary in order to establish planned development and land use decision-making process. Digital soil mapping uses quantitative relationships between soil properties and other spatially distributed environmental variables such as landslide (Mallick et al. 2018; Karim et al. 2019) and geological mapping (Tamani et al. 2019). At the watershed level (regional scale), the availability of pedological information and cartography is crucial for the development of projects related to soil sciences. In the Abha watershed of Aseer region, Saudi Arabia, the existence of geosoil properties information has not been conducted in the recent past. Besides, the challenge of understanding the large stores of soil data has led to the development of new tools in the field of statistics and spawned new areas such as data mining and machine learning (Hadji et al. 2014). In addition to this, in soil science, the increasing power of tools such as GIS, GPS, remote and proximal sensors and data sources are suggesting new ways forward. Fortuitously, this comes at a time when there is a global clamour for soil data and information for environmental monitoring and modelling. The geospatial properties of soil such as moisture content, wet density, dry density, void ratio, porosity, grain size distribution, organic content, pH and conductivity are essential for the determination of engineering properties of soil, which will help geotechnical engineering in the decision-making process of suitability of soil as foundation materials or construction materials (Mallick et al. 2014). If the properties of soil are properly studied and the results of soil exploration are correctly perceived and intelligently applied to the design and construction of geoearthworks and structural foundations, failures usually can be escaped. Previously developed digital models of soil were mostly based on empirical results and were restricted in use owing to complexity of terrains concerned (Ma et al. 2019). A review of past studies showed that no significant study of soil properties in ABHA has been conducted in the recent past. It

R. A. Khan (✉)
Faculty of Science and Technology, Himalayan University, Itanagar, Arunachal Pradesh, India

J. Mallick
Department of Civil Engineering, King Khalid University, Abha, Saudi Arabia

R. E. Morabet
Department of Geography, LADES, FLSH-M, Hassan II University of Casablanca, Mohammedia, Morocco

H. El-Askary et al. (eds.), *Research Developments in Geotechnics, Geo-Informatics and Remote Sensing*, Advances in Science, Technology & Innovation,
https://doi.org/10.1007/978-3-030-72896-0_74

is therefore decided to integrate information from various sources including GIS to accurately quantify and map the soil engineering properties in mountainous, ABHA using geospatial analysis tools. A case study area has been selected to demonstrate the concept of soil properties studies. The Bisha-Khamis watershed chosen for the proposed project is situated in ASIR region of kingdom of Saudi Arabia. It covers an area of 1208 km^2. The boundary of the study area lies between the latitude of 17.99 and 17.99 N and longitude of 42.56 and 42.56 E.

2 Data and Method Used

This study has modelled spatial variability of hydro-geological design parameters with a main focus on soil particle size distribution. The general approach used in this study is illustrated in Fig. 1 with the following data and tools used:

- Topographical data: Advanced Land Observing Satellite (ALOS): DEM 12.5 m resolutions https://vertex.daac.asf.alaska.edu/
- Software: ArcGIS 10.3 (incl. ArcHydro); ERDAS, ENVI and Microsoft office software and for statistics computation, SPSS software
- Navigator: GPS62sc Garmin
- Laboratory experiments: Sieve analysis, hydrometer analysis (soil texture), oven and muffle furnace (moisture content and organic content, respectively)
- Field survey conducted (four days): 16–19 November-2018; 15 January to 25 February 2019.

Soil samples were collected from various spatial locations in the watershed area. A total of sixty four (64) soil samples approximately 1 kg for aggregate stability (at the depth of 0–30 cm) was collected from the study area. The soil sampling was performed using stratified composite approach, where the region is subdivided into areas of similar topography, soil moisture, and land use/land cover. Thereafter, this area is then sampled separately and at each survey location, 2 replicates, two to three meters apart were collected. Individual samples were weighed and carefully sieved through a 2 mm screen and later analysed in laboratory for their soil grain properties, namely soil texture and particle size distribution. The topsoil samples were taken in order to assess

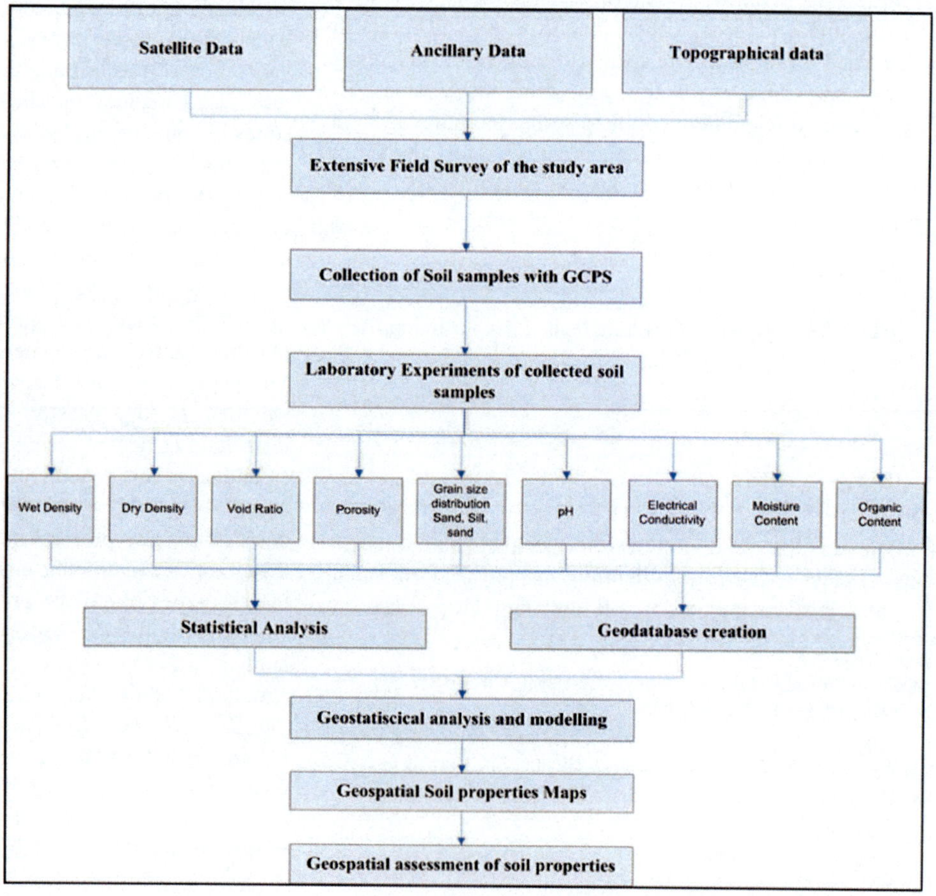

Fig. 1 General methodology

the spatial variability of soil moisture and organic matter content. Soil moisture was obtained after air-drying (102 °C for 24 h in oven) in the laboratory. Thereafter, the soil samples were crushed in order to pass a 2 mm-sieve and carefully homogenized. Organic matter content was obtained from high temperature of 350–600 °C for 2 h using muffle furnace. The precision of the measurements is specified to 1.5% of the detected amount, with a detection limit of 0.02% (Hill and Schütt 2000). The determination of the grain size distribution of soil particles is known as mechanical or particle size analysis. Soil texture is the composition of the soil particles expressed as the per cent of particles in the sand, silt and clay size separates after moisture and organic matter content are removed. Texture analysis of the soil samples was carried out by hydrometer method using Stokes' law (Sheldrick and Wang 1993). Stokes' law states that the rate which particles fall in viscous medium (water) is governed by the radius of the particles and the force due to gravity.

In order to assess the soil properties in the study area, various thematic maps, viz. sand, silt, clay, organic content and soil moisture content were generated using GIS and conventional data with the help of ArcGIS software.

Geostatistical mapping can be defined as the analytical production of maps by using field survey, auxiliary information and a computer program that generates predictions. In the geostatistical method, Kriging interpolation technique is applied in this study using geostatistical analyst tolls and ArcGIS 10.1 software. This method gives the most accurate results after validation and offers the possibility of flexibility in assumptions required from data handling. Kriging technique is widely used in geology, hydrology, environmental monitoring and other fields to interpolate spatial data.

3 Result and Discussion

Distributions of clay, silt and sand contents in the soil samples are shown in Fig. 2. Sand contents were generally between 33.17 and 72.33%. Most soils contained between 23.33 and 49.83% silt and from 1.00 to 19.50% clay. The analysis showed that organic contents varied from 0.567 to

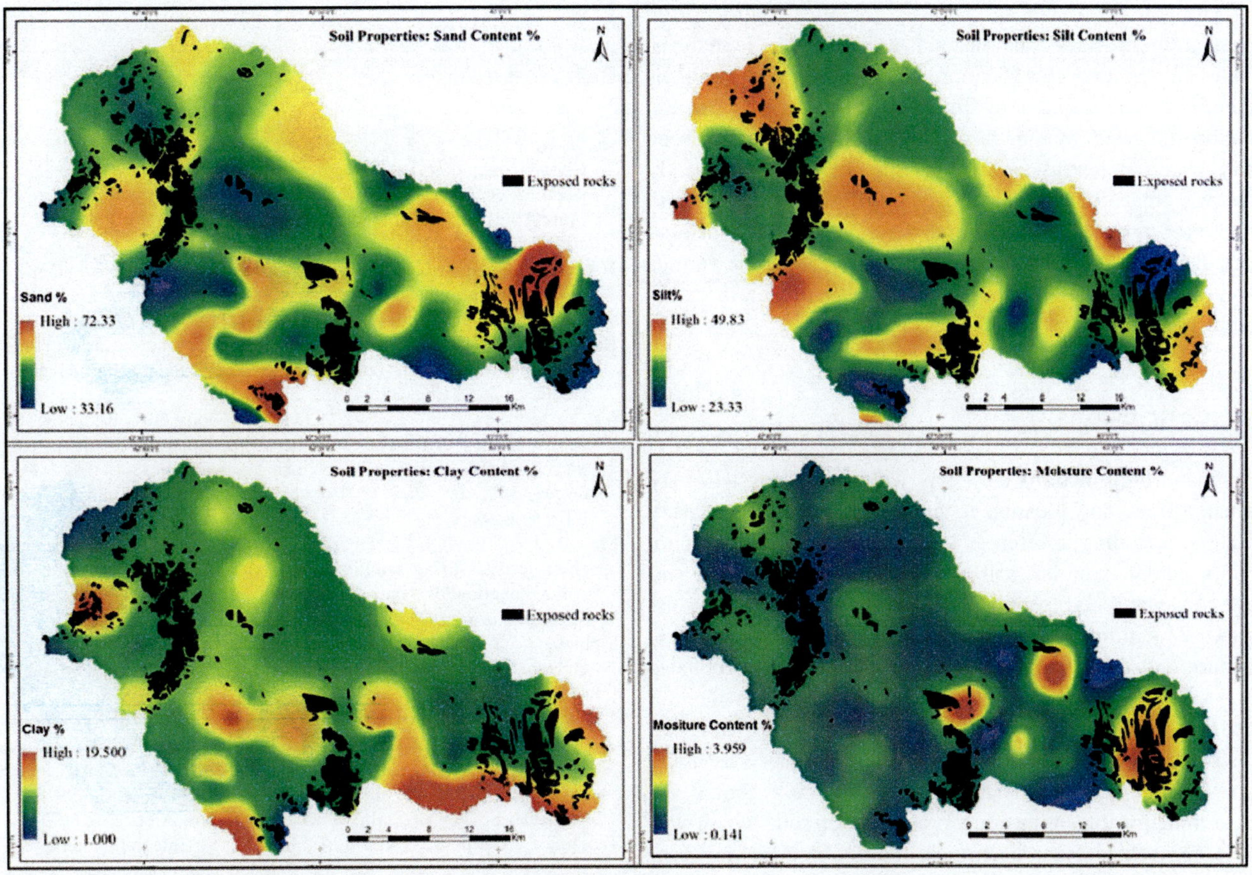

Fig. 2 Spatial distribution of soil properties (sand, silt, clay and moisture)

7.240%, with a mean of 2.703% and water content (moisture) 0.141–3.959% with a mean of 1.080%. These maps indicate that most of the study area had clay contents below 15% around some areas located near foothills and vegetated area. Unlike the clay, there was a wide spatial variability in silt contents across the study area, from 23 to 50%. Towards the north-west, western, central and east-north part, most soils had silt contents between 40 and 50%. A similar situation of such wide range was also observed for the sand contents. It is found towards the down hills and along the wadies (dry ephemeral riverbed). The organic matter content in soil is found less than 2.0% across half of the study area. Kriging interpolation technique, following trend removal, was carried out on the examined soil samples. The prediction (interpolated) map of each property was calculated, and trend was added back to the output surface. Many model parameters were studied to select the most appropriate model for each of examine soil sample or properties and the best semi-variogram that fits better experimental semi-variogram was chosen. The proportion of each particle size group (sand, silt or clay) in the soil is called the soil particle size distribution. Sand is the largest soil particle, silt is intermediate in size and clay is the smallest.

The soil texture is the gateway for all the information needed to evaluate any watershed hydrological analysis. Thus, the textural classification of the soil of the country is identified in this spatial hydrological soil database. The soil texture is the dominant factor in the relationship between soil–water potential–content. Cosby et al. also clearly showed that the soil texture could be associated with hydraulic properties. Therefore, the soil–water characteristics of the new soil database also help to compile the various soil–water parameters that play an important role in catchment hydrology modelling.

4 Conclusions

Soils and the functions they play within an ecosystem vary greatly from one location to another as a result of many factors, including differences in climate, existing ecosystem, soil's parent material, soil position on landscape and soil age. At a particular location, soils are distinguished from one another by differences in their structure, colour, consistence, texture and the amount of roots, rocks and free carbonates they contain. Laboratory analyses of particle size distribution and moisture content reveal differences among spatial soil distribution. The geospatial properties of soil such as moisture content and grain size distribution are essential for determination of engineering properties of soil, which will help geotechnical engineering in the decision-making process of suitability of soil as foundation materials or construction materials. The main objective of this study is to integrate information from various sources including GIS to accurately quantify and map the soil engineering properties in mountainous watershed of Bisha-Khamis, KSA using geospatial analysis tools. The findings showed that most of the study area had clay contents below 15% around some areas located near foothills and vegetated area. Unlike the clay, there was a wide spatial variability in silt contents across the study area, from 23 to 50%. Towards the north–west, western, central and east–north part, most soils had silt contents between 40 and 50%. A similar situation of wide range was observed for the sand contents. It is found towards the down hills and along the wadies (dry ephemeral riverbed). The organic matter content in soil is found less than 2.0% across the half of the study area. Fine-textured soils tended to have more organic matter than sandy soils. Much of the soil organic matter in soil was related with the occurrence of fine particles, i.e. clay and fine silt. These particles were most susceptible to erosion. Therefore, in this watershed study, it is possible that soils with high organic contents also contain large fractions of silt. Future challenges include the introduction of new data, primarily to improve the predictions of topsoil organic matter and refine the current dataset.

References

Hadji, R., Limani, Y., Boumazbeur, A., DemdoumA, Z.K., Zahri, F., Chouabi, A.: Climate change and their influence on shrinkage—swelling clays susceptibility in a semi-arid zone: a case study of Souk Ahras municipality, NE-Algeria. Desalin. Water Treat. **52**(10–12), 2057–2072 (2014)

Hill, J., Schütt, B.: Mapping complex patterns of erosion and stability in dry Mediterranean ecosystems. Remote Sens. Environ. **74**, 557–569 (2000). https://doi.org/10.1016/S0034-4257(00)00146-2

Karim, Z., Hadji, R., Hamed, Y.: GIS-based approaches for the landslide susceptibility prediction in Setif Region (NE Algeria). Geotech. Geol. Eng. **37**(1), 359–374 (2019)

Ma, Y., Minasny, B., Welivitiya, W.D.D.P., Malone, B.P., Willgoose, G. R., McBratney, A.B.: The feasibility of predicting the spatial pattern of soil particle-size distribution using a pedogenesis model. Geoderma **341**, 195–205 (2019). https://doi.org/10.1016/j.geoderma.2019.01.049

Mallick, J., Al-Wadi, H., Rahman, A., Ahmed, M.: Landscape dynamic characteristics using satellite data for a mountainous watershed of Abha, Kingdom of Saudi Arabia. Environ. Earth Sci. **72**(12) (2014). https://doi.org/10.1007/s12665-014-3408-1

Mallick, J., Singh, R.K., AlAwadh, M.A., Islam, S., Khan, R.A., Qureshi, M.N.: GIS-based landslide susceptibility evaluation using fuzzy-AHP multi-criteria decision-making techniques in the Abha Watershed, Saudi Arabia. Environ. Earth Sci. (2018). https://doi.org/10.1007/s12665-018-7451-1

Sheldrick, B.H., Wang, C.: Particle size distribution. In: Carter, M.R. (ed.) Soil Sampling and Methods of Analysis, pp. 499–511. Canadian Society of Soil Science. Lewis, Ann Arbor, MI (1993)

Tamani, F., Hadji, R., Hamad, A., Hamed, Y.: Integrating remotely sensed and GIS data for the detailed geological mapping in semi-arid regions: case of Youks les Bains Area, Tebessa Province, NE Algeria. Geotech. Geol. Eng. (2019)

Remote Sensing Characterization of Urban Expansion: A Case Study of Shenyang, Northeast China

Ziqi Yu, Longqian Chen, Long Li, Ting Zhang, Ruiyang Liu, and Zhiqiang Wang

Abstract

Urban expansion is a result of multiple factors and should be under control to guarantee sustainable socioeconomic development. This study examined the northeast Chinese city of Shenyang in Liaoning province using remote sensing data of 1992, 2000, 2008, and 2016. We characterized the urban expansion of the study area through spatial analysis performed in ArcGIS 10.3. Results show that from 1992 to 2016, the area of urban built-up land continued to increase and the expansion intensity gradually increased. The scale and rate of urban expansion were the largest during the period from 2008 to 2016. The findings of this study will provide useful insights into the urban planning of Shenyang.

Keywords

Urban expansion • Image classification • Spatial analysis • Shenyang

1 Introduction

Urban expansion refers to the expansion of a metropolitan or suburban area into the surrounding areas. With its acceleration in recent decades, particularly in developing countries, urban expansion has received growing attention from the research community (Fang et al. 2008). As the capital city of Liaoning province, Shenyang is a typical heavy industrial city in Northeast China, the old industrial basis. In recent years, Shenyang has seen an accelerated speed of urban expansion, which led to various socioeconomic and eco-environmental problems. Coordinating human-land conflicts and ensuring socioeconomic sustainability through proper guide are the goal of sustainable urban development. It is, therefore, important to understand the process of urban expansion in Shenyang in the past years and its underlying reasons.

The spatial expansion of urban built-up land is an important measure of urbanization (Zeng et al. 2004). To this end, we made use of remote sensing image data to characterize the growth of urban built-up land in Shenyang from 1992 to 2016 through spatial analysis. The objective of this study was to provide a better understanding of the change of urban growth over time and useful insights on the future development and urban planning of Shenyang.

2 Data and Method

Shenyang is the capital city of Liaoning province in Northeast China (Fig. 1), and its main urban area consists of the districts of Huanggu, Tiexi, Sujiatun, Hunnan, Yuhong, Heping, Shenhe, Dadong, and Shenbei New District (Fig. 2).

The used remote sensing data in this study were freely downloaded from the United States Geological Survey (USGS) website. They were 30-m-resolution Landsat 5 TM (1992, 2000, and 2008) and Landsat 8 OLI (2016) images. As they were acquired in July or August, seasonal variability was minimized for multi-temporal analysis.

We used the remote sensing software ENVI 5.3 for image processing and the geographical information system software ArcGIS 10.3 for spatial analysis. After image preprocessing, we performed a supervised classification (SVM, support vector machine) to produce a land use map with

Z. Yu · L. Chen (✉) · L. Li · T. Zhang · Z. Wang
School of Public Policy and Management, China University of Mining and Technology, Xuzhou, 221116, China
e-mail: chenlq@cumt.edu.cn

L. Li
Department of Geography and Earth System Science, Vrije Universiteit Brussel, 1050 Brussels, Belgium

R. Liu
School of Environmental Science and Spatial Informatics, China University of Mining and Technology, Xuzhou, 221116, China

H. El-Askary et al. (eds.), *Research Developments in Geotechnics, Geo-Informatics and Remote Sensing*, Advances in Science, Technology & Innovation,
https://doi.org/10.1007/978-3-030-72896-0_75

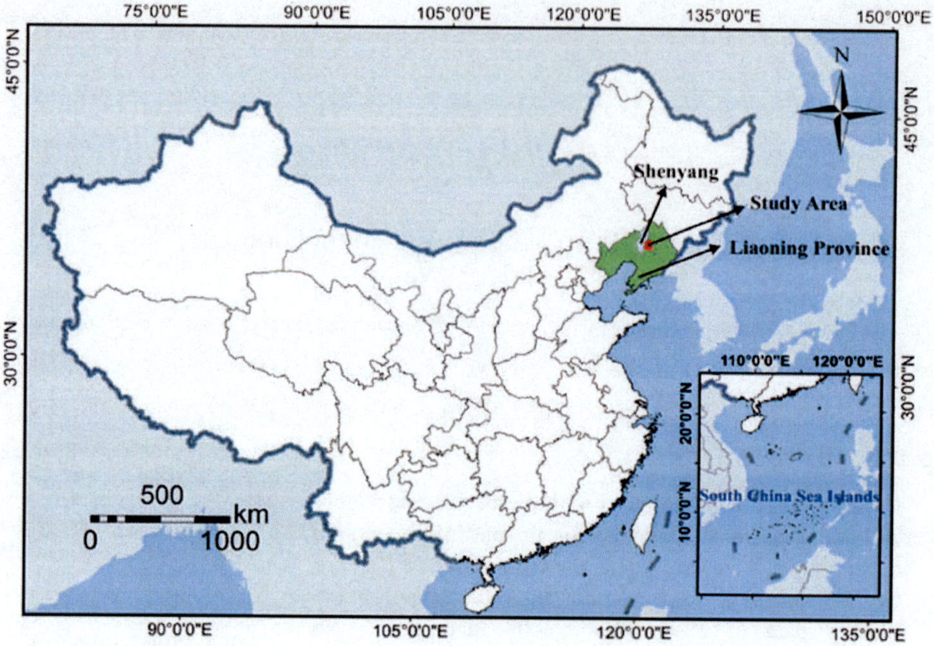

Fig. 1 Study area in China

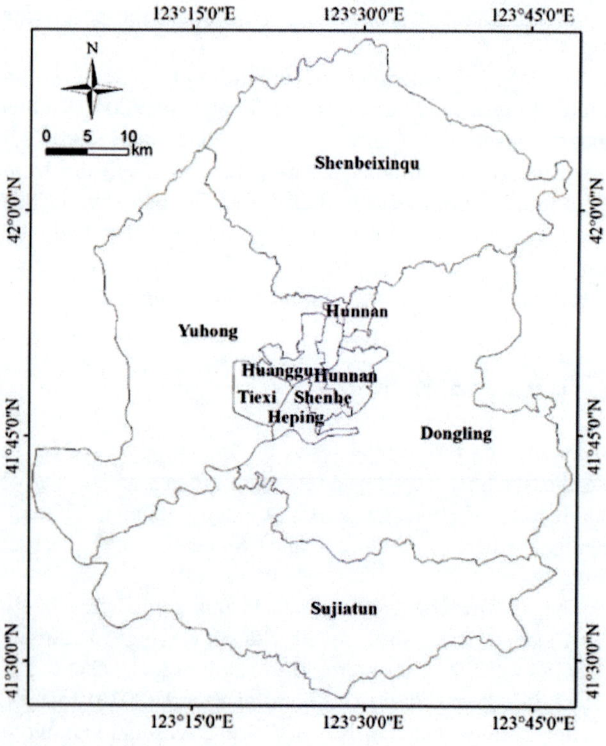

Fig. 2 Study area: Shenyang

good classification accuracy for each period. This method does not require a large number of samples to be selected which may result in a better classification accuracy in comparison with other classification methods.

In order to depict the directional characteristics of urban expansion in the study area, we defined the geometric center of Shenyang's urban boundary as the circular center and divided the study area into eight circular segments of equal angles. Then we calculated the area of built-up land in each segment for each period to characterize the urban expansion of the study area.

3 Results

There were four land use types that were considered in the classification, namely built-up land, water, forest land, and cultivated land. As the focus of the study is on urban expansion, we only needed to examine the built-up land in the study area. The overall accuracy and Kappa coefficient were 89.09% and 0.80, respectively, for the 1992 image, 87.36% and 0.75 for the 2000 image, 87.54% and 0.71 for the 2008 image, and 87.03% and 0.75 for the 2016 image. All the classification results were good and could be used for urban expansion analysis. The urban expansion maps were produced by overlaying the built-up land layer of each period in ArcGIS 10.3 (Fig. 3).

In order to fully reflect the spatiotemporal characteristics of urban expansion in Shenyang, three common indicators (expansion area, annual average expansion speed, and expansion intensity index) were used in this study (Wang et al. 2008).

Figure 4 shows the expanding directions in different periods. From 1992 to 2000, the peak of expansion appeared

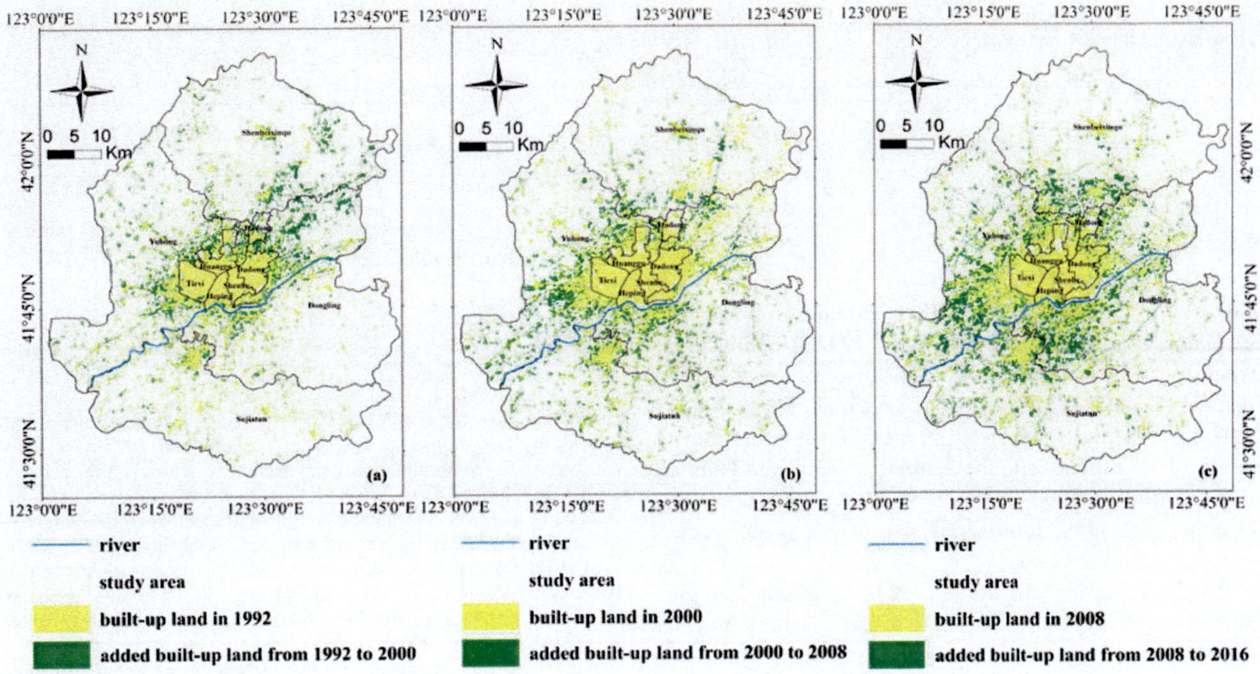

Fig. 3 Urban expansion maps **a** from 1992 to 2000; **b** from 2000 to 2008; and **c** from 2008 to 2016

Fig. 4 Expanding direction in different periods

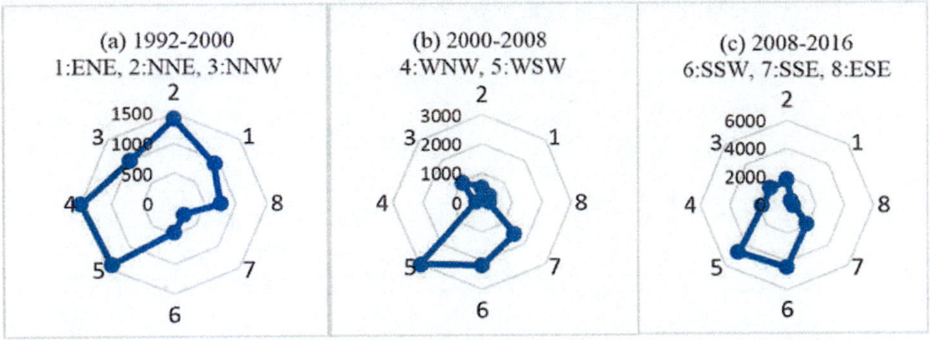

in the 4th segment. In the 2000–2008 period, the 5th segment saw the peak of expansion, followed by the 6th and 7th segments. From 2008 to 2016, the peak of expansion also appeared in the 5th segment.

4 Discussion

We understand that the 5th and 6th segments, where the districts of Tiexi and Hunnan are located, are the main directions for urban expansion in Shenyang.

When we examined the urban expansion characteristics of Shenyang, the approach we used for dividing the main urban area into equal segments has made it straightforward to understand which direction had the greater urban expansion in the study area.

From Table 1, the area of urban built-up land in Shenyang increased from 558.79 km^2 in 1992 to 664.46 km^2 in 2000 with an average annual expansion speed of 13.20 $km^2\ a^{-1}$. The intensity index was 2.36%. During the 1992–2000 period, the urban expansion speed of Shenyang gradually accelerated. From 2000 to 2008, Shenyang's urban expansion was relatively stable, with an expansion area of 72.03 km^2. During the period of 2008–2016, the speed of urban expansion increased rapidly, the expansion area reached 367.79 km^2, and the expansion speed reached 45.97 $km^2\ a^{-1}$. It has reached an unprecedented scale, and we predict that the urban expansion speed of Shenyang will be faster in the future.

Table 1 Urban expansion of Shenyang in different periods (1992–2016)

Period	1992–2000	2000–2008	2008–2016	1992–2016
Expansion area (km^2)	105.67	72.03	367.79	545.49
Expansion speed ($km^2\ a^{-1}$)	13.20	9.00	45.97	22.72
Expansion intensity index (%)	2.36	1.35	6.24	4.06

5 Conclusions

This study investigated the urban expansion process of Shenyang from 1992 to 2016. From 1992 to 2000, urban expansion was more uniform in all directions, but slightly focused on the 2nd, 4th, and 5th segments. From 2000 to 2008, the urban area expanded in the 5th and 6th segments, built-up land expanded to the districts of Tiexi and Hunnan. In the period of 2008–2016, as Shenyang's economy was developing steadily, the overall size of Shenyang's urban area gradually increased.

It is expected that the urban area of Shenyang will continue to grow in the future and urban planning policies should be in place to regulate this process. Reasonable urban expansion can reserve space for future urban and improve the land use efficiency. Further research should focus on analyzing the influencing factors affecting urban expansion, the relationship between urban planning and natural parameters and the potential applicability of natural parameters to urban development (Bathrellos et al. 2012, 2017).

References

Bathrellos, G.D., et al.: Potential suitability for urban planning and industry development by using natural hazard maps and geological-geomorphological parameters. Environ. Earth Sci. **66** (2), 537–548 (2012)

Bathrellos, G.D., et al.: Suitability estimation for urban development using multi-hazard assessment map. Sci. Total Environ. **575**, 119–134 (2017)

Fang, S., et al.: Regional ecosecurity pattern in urban area based on land use analysis: a case study in Lanzhou. Chin. J. Appl. Ecol. **16** (12), 2284–2290 (2008)

Wang, H., et al.: Analysis of urban spatial expansion process in Shenyang from 1979 to 2006. Chin. J. Appl. Ecol. **19**(12), 2673–2679 (2008)

Zeng, L., et al.: Spatial-temporal feature of urban land extension in Baoding City. Resour. Sci. **26**(4), 96–103 (2004)

Comparison of Pixel-Based and Object-Oriented Classification Methods for Extracting Built-Up Areas in a Coastal Zone

Chayma Kefi, Amina Mabrouk, and Haythem Ismail

Abstract

Monitoring the coastal environment using remote sensing and GIS techniques has been undertaken in this study, with a particular focus on the comparison between the classical and object-oriented image classifications of remote sensing imagery in coastal areas. In fact, the investigation was based on the testing of a coastal zone image classification, pixel-based image classifiers and an object-oriented image classifier. The method was later compared using Pleiades image. The use of reference data sets that were taken from high-resolution satellite images, aerial photographs and field investigation was considered as an effective way to assess the accuracy of this method. The result of classification and accuracy assessment shows that the object-oriented approach gave more accurate and satisfactory results.

Keywords

Satellite images • SVM classifier • Coastal zone • Object-oriented classification

1 Introduction

Urban extraction and classification of coastal zones is one of the most active research areas. Until the late 2000s (Hwang and Chiang 2011), this task has primarily been tackled using pixel-based approaches. As these methods consider pixels and pixel neighborhoods as the sole basic entities for analysis, they cannot account for the irregular boundaries of real-world objects. The object-based analysis emerging from the field of remote sensing has been proposed as an alternative approach and was successfully applied in case studies falling in the domains of urbanization (Wang et al. 2014). The current study will try to focus on the classification method for extraction of land-cover information from remotely sensed images in a coastal zone, with a particular interest on the built-up areas.

2 Methodology

The study region is located in the Northeast of Tunisia (A town called Kelibia covering 1100 Ha) and is an urban area opening on the sea. Pleiades image acquired in 2016 is provided as an available data for this area, along with the 2012 Urban Development Plan, the agricultural map, the coastal lines, the digitalized rivers map, government urbanization documents, etc. These data have been collected and generated with the help from urban experts. Obviously, the Pleiades image has been geometrically and optically corrected (Figs. 1 and 2).

The classification results as presented in the afore-mentioned details, the SVM algorithm is chosen owing to its flexibility with adaptive feature vector and kernel trick.

For each pixel, a feature vector consisting of different discriminating features is generated. The main classes are urban areas, vegetation, soil, routes, water and shadow. For textural signature, the mean, standard deviation, entropy, energy and Gabor wavelet decomposition selecting features were used to identify texture change. The Gabor functions are band-pass filters regarded as a Gaussian adjusted by complex sinusoid in the spatial domain. Also, we used radiometric values and specified indices, or spectral signature, such as: the Normalized Difference Vegetation Index (NDVI) which is used to discriminate vegetation regions, the Soil Brightness Index (SBI) which is used to characterize bare soil, the Difference Water Index (NDWI) and the Index

C. Kefi (✉) · A. Mabrouk
Laboratory of Sedimentary Basins and Petroleum Geology, Faculty of Sciences of Tunis, University of Tunis El Manar, 2092 Tunis, Tunisia

H. Ismail
Tunisian Center of Cartography and Remote Sensing, Tunis, Tunisia

H. El-Askary et al. (eds.), *Research Developments in Geotechnics, Geo-Informatics and Remote Sensing*, Advances in Science, Technology & Innovation,
https://doi.org/10.1007/978-3-030-72896-0_76

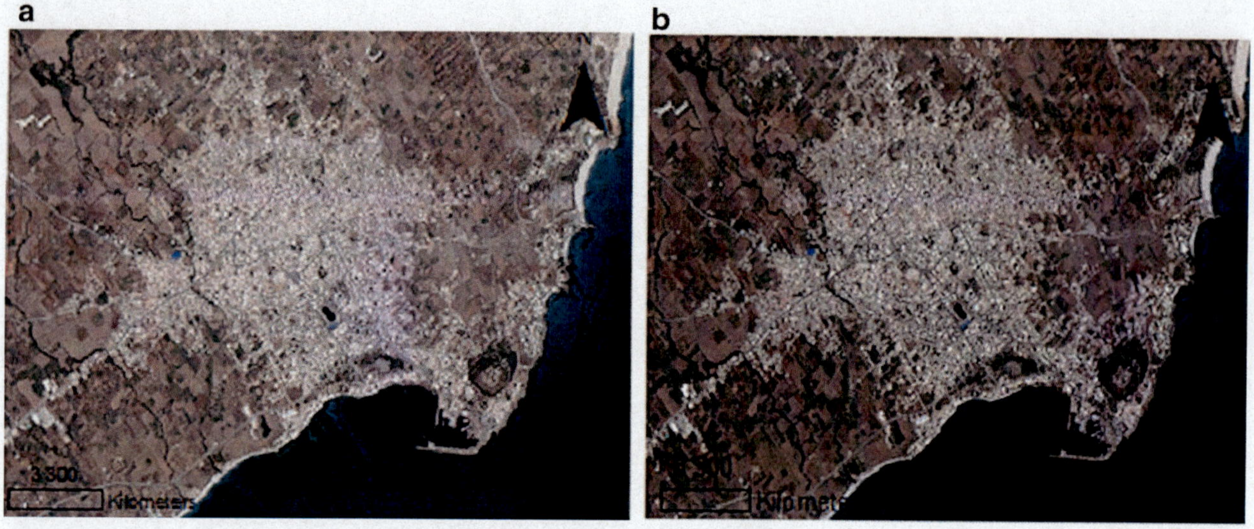

Fig. 1 **a** Pleiades image covering Kelibia region, **b** georeferencing and optical correction

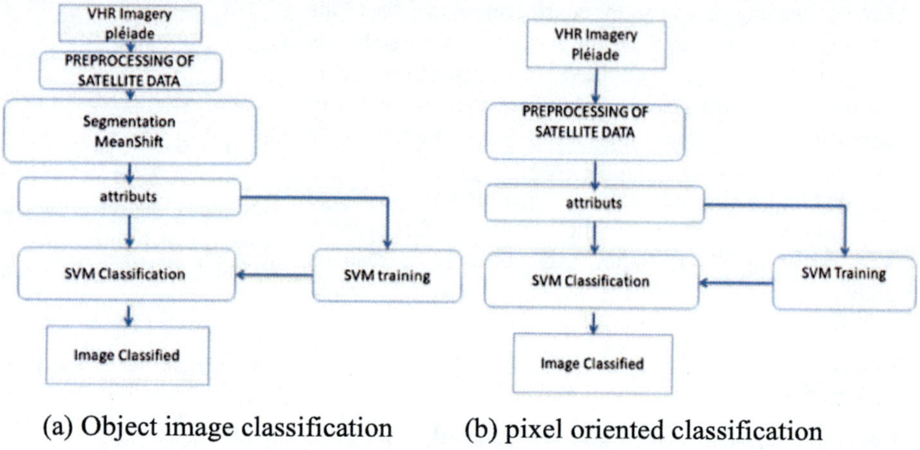

Fig. 2 **a** Object image classification, **b** pixel-oriented classification

Surfaces Built (ISU) which are intended, respectively, for water and urban areas. The variables R, G and NIR are, respectively, the red, green and near-infrared channels. Also, a and b are parameters (a = 100 and b = 25 for VHR).

$$\mathrm{NDVI} = \frac{\mathrm{NIR} - R}{\mathrm{NIR} + R} \tag{1}$$

$$\mathrm{SBI} = \frac{\sqrt{R^2 + G^2}}{2} \tag{2}$$

$$\mathrm{NDWI} = \frac{G - \mathrm{NIR}}{G + \mathrm{NI}} \tag{3}$$

$$\mathrm{ISU} = (a - b) \cdot \frac{R}{\mathrm{NIR}} \tag{4}$$

Given the fact that it is stained by some isolated and misclassified pixels, the spectral, spatial and textural features were then merged in a feature vector and utilized as input to the SVM algorithm, a post-classification majority filter was then used on the resulted image. Later, the classifier was supplied with all image feature vectors to derive the final classification map. Pleiade image was segmented into regions by the mean shift algorithm. Image segmentations of meaning objects were implemented with merged image. By test, the scale parameter was defined as 50 (Fig. 3).

The results of the pixel-based and the object-oriented classification of Pleaide selected randomly. The reference samples are ground truth data. Then error matrices have been generated, and the assessment indices are given in the above table.

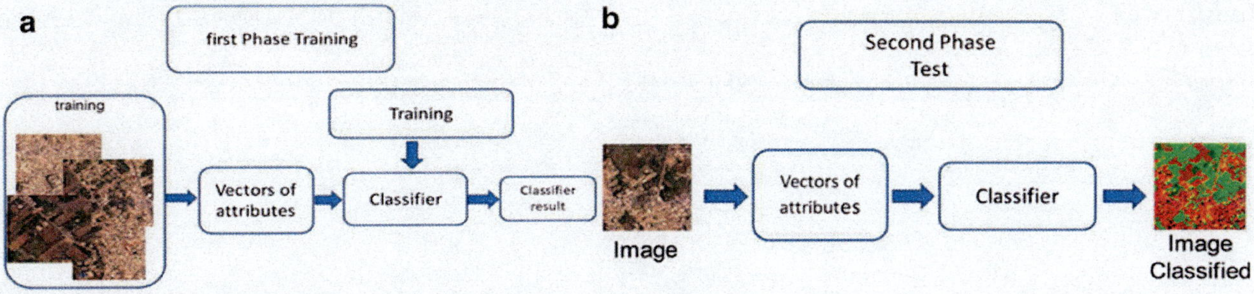

Fig. 3 **a** SVM first training phase, **b** SVM second-phase test

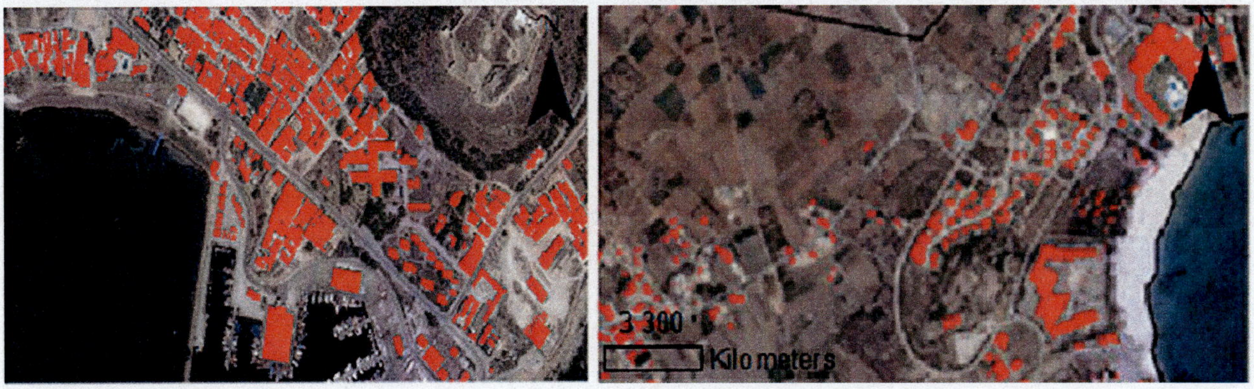

Fig. 4 Urban areas extraction

Table 1 Pleiades image classification results

Experiments	Accuracy	
	Over all accuracy (%)	Kappa (%)
Object-based image analysis	0.88	0.74
K-nearest neighbor	0.69	0.68
Random forest	0.81	0.58
SVM-based classification	0.79	0.71

3 Results

Figure 4 illustrates the obtained results, and Table 1 shows the SVM classification accuracy compared to other algorithms.

To compare the classification accuracy, Table 1 provided the good classification rate overall accuracy and kappa for the proposed approach based on the SVM object image classification proving that it is well classified in the three cases and demonstrating a kappa coefficient of 0.74, compared to 71% that was concluded from the conventional pixel-based method.

4 Conclusions

The obtained results through this study are satisfactory and underline the influence of size, density and coverage of buildings in the quality of their detection in a coastal area. The classification result could be improved with other segmentation, classification and feature extraction algorithms. We compared object-oriented with pixel-based classification approaches using Pleaides images in Kelibia city. The reason is that the image objects do not mean that a single pixel can offer important semantic information which is necessary to interpret an image. Object-oriented image analysis techniques are a perfect classification approach especially for high-resolution imagery.

Acknowledgements The Tunisian Center of Cartography and Remote Sensing are greatly acknowledged for their great help and support for the achievement of this study.

References

Hwang, J.T., Chiang, K.T.: Gabor texture feature SVM. In: I. CGeoInformatics, June 2011

Wang, L., Wang, S., Li, W.: Monitoring urban expansion 1985 to 2013 using Landsat images. In: IGARSS, pp. 4264–4267, Quebec, Canada, 13–18 July 2014

Prospects of Remote Sensing for Mineral Deposits Prediction in Kazakhstan

A. Baibatsha, E. Mamanov, and M. Kembayev

Abstract

Using remote sensing database (Landsat and Aster materials), promising areas for the detection of endogenous mineralization were identified. On documented cosmo-structural schemes, objects of linear morphology occur: faults, areas of increased fracturing, geological boundaries, bedding elements, dykes and others. Under conditions of weakly dissected relief, multidirectional gradients and various filtering methods are particularly effective for isolating linear structures. Arc and ring structures are recognized based on their boundaries between blocks with different textures of space materials as well as their morphology. Borders of landscape heterogeneity are also taken under consideration. Ring structures are conventionally divided into structures of the second order and smaller. Bodies of intrusive rocks are established using spectral libraries and textural features of satellite images.

Keywords

Remote sensing • Space image interpretation • Ore-controlling factors • Cosmo-geological schemes • Promising areas

1 Introduction

Remote sensing is a promising method for identifying potential mineral deposits areas. Here, we present our results on the Eastern Balkhash and Usharal districts.

A. Baibatsha (✉) · E. Mamanov · M. Kembayev
Kazakh National Research Technical University named K.I. Satpayev, Almaty, Kazakhstan

2 Methods

Archived data from Landsat ETM + and Aster satellite surveys, as well as digital elevation models—SRTM (Shuttle radar topographic mission, 2000) and AsterGDEM (Aster Global Digital Elevation Model, 2011) data were used for the present research.

Landsat ETM + archival materials were obtained from the satellite imagery library of the Maryland University (USA). The shooting was performed in August 1999. From the same library, the digital elevation model—SRTM was taken. The spatial resolution of the data was of 90 m. A digital elevation model (special resolution 25 m) was obtained from the same library according to AsterGDEM. Selected digital satellite images had the maximum spectral completeness that is, they covered the area of research in all possible spectral ranges. Space materials were selected taking into account climate peculiarities—during the period of oppressed vegetation, lack of snow cover, minimum humidity. These conditions met the requirements for such materials of remote surveys (Percov 2000; Baibatsha et al. 2016; Poceluev et al. 2012). Thus, the spatial resolution of the selected space materials, their spectral characteristics and the accuracy of spatial positioning fully corresponded to the target of our research.

To predict intrusive rocks occurrence, we collected data from the Johns Hopkins University Rock Library (http://speclib.jpl.nasa.gov/search-1) and the USGS Spectroscopy Laboratory (http://speclab.cr.usgs.gov/spectral-lib), as well as textural features of satellite images.

The overwhelming majority of the bodies of intrusive rocks were separated on the basis of Aster satellite imagery (Baibatsha et al. 2016; Poceluev et al. 2012). In terms of their composition, intrusive rocks were divided into ultrabasic, basic, and medium. Also, areas with traces of hydrothermal changes that probably control the position of ore mineralization were referred to as ore bodies.

H. El-Askary et al. (eds.), *Research Developments in Geotechnics, Geo-Informatics and Remote Sensing*, Advances in Science, Technology & Innovation,
https://doi.org/10.1007/978-3-030-72896-0_77

Data processing was performed using the licensed software Erdas Imagine 10.0 and ArcGIS 9.3. The purpose of this work is (i) to construct a space structural scheme through Remote sensing bases and (ii) to identify main ore-controlling factors, on the basis of which most promising areas for the detection of endogenous mineralization.

3 Results

The data analysis allowed the compiling the cosmo-structural schemes of the Eastern Balkhash (Arganaty) and Usharal districts, as shown in Fig. 1. The map shows the whole complex with selected elements.

In cosmic structural schemes, the objects of linear morphology are faults, areas of increased fracturing, various cleats, geological boundaries, bedding elements, dykes and others. In the course of deciphering and preparing cosmo-structural schemes, about 3 thousand lineaments of a geological nature were identified and interpreted.

Also rings and arc structures were found. Considering that there is a dependency between ring structures and type of mineral deposits (Baibatsha 2014; Rajendran et al. 2013), various mechanisms of their formation may be assumed: endogenous, exogenous, and cosmogenic. In the areas of the Eastern Balkhash and Usharal districts, more than 30 ring and arc structures with a radius from hundreds of meters up to 48 km were mapped. The radius ring structures were conventionally divided into structures of the second order and smaller. Structures with radii from 4 to 48 km are assigned to ring structures of the second order, and smaller structures with radii less than 4 km were assigned to small structures.

More than 2000 linear structures were identified within Usharal area alone, among which about 1400 got a geological interpretation. As for linear structures, it is suggested to make the distinction between faults, geological boundaries, elements of litified stratified complexes and dykes. 15 ring structures with a radius of 2–22 km were mapped in the Usharal district and in its surroundings. The intrusive rocks were divided into ultrabasic and basic rocks. The criterion for such a classification was the spectral characteristics of these bodies in the materials of the Aster spectral space imagery. The bodies of ultrabasic composition formed lenses stretching up to hundreds of meters.

The bodies of metasomatites were localized mainly in the central part of the site. The general direction of their strike was consistent with the strike of the main right-shear fault in

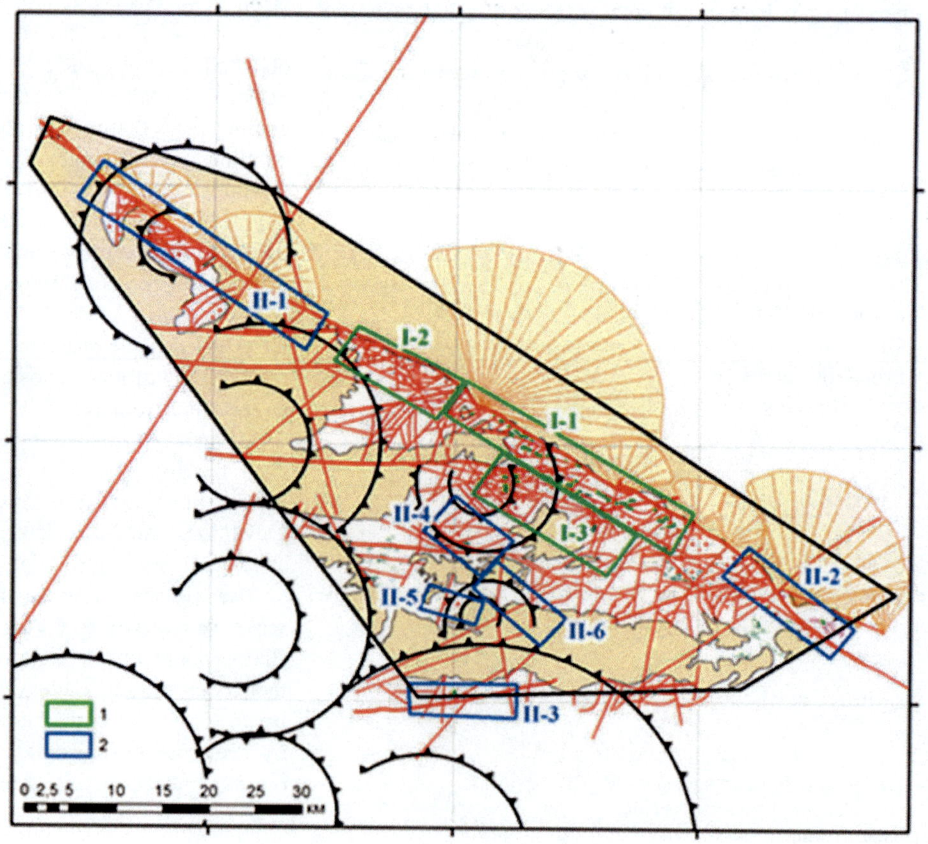

Fig. 1 Prospective sites for the detection of endogenous mineralization using remote sensing. Blocks for prospecting: 1—the first stage; 2—the second stage

the north-west. Meanwhile, the zones of metasomatic study were, as a rule, removed from the main shift at a distance of no more than 3 km. These facts suggested a paragenetic connection of mapped metasomatites with a shear violation.

4 Discussion

It is assumed that all structures are associated with endogenous processes and magmatism. Second-order ring structures are associated with intermediate intrusive bodies, both blind and exposed erosion. Higher-order structures can capture traces of the interaction of hydrothermal-fluid systems with host rocks.

Based on the factors of mineralization, within the limits of the Usharal site, as well as in its immediate vicinity, nine promising sites were identified for the detection of endogenous mineralization (see Fig. 1).

Spectral analysis of Aster plots indicated within metasomatic rocks the presence of newly formed hydromica minerals, chlorite, carbonate, and epidote. The areas of thermal impact on host rocks almost everywhere spatially coincided with the areas of development of intrusive rocks.

5 Conclusions

In this work, distant bases were prepared, and cosmo-structural schemes in a scale of 1: 100,000 for Arganata and Usharal regions were drawn up. The main cosmological and geological factors of mineralization were determined. Furthermore, recommendations on a complex of search works were given. Prospects for the identified areas were associated with the possibility of detecting endogenous mineralization (gold, copper, molybdenum, lead, zinc, tin, and tungsten), chrysotile asbestos and groundwater (Rajendran et al. 2013).

The proposed method of using remote sensing to identify areas that are promising for subsoil use is an innovative approach to solving the technology of searching for new deposits in areas covered by young sediments and deep areas.

Acknowledgements This work was supported by the scientific program "Comprehensive geological study of subsurface resources for the development of resource base and mining exploitation of new sources of ore raw materials in Kazakhstan."

References

Baibatsha, A.B.: Paleovalleys mapping using remote sensing. In: The International Archives of the Photogrammetry, Remote Sensing and Spatial Information Sciences, Vol. XL-5. ISPRS Technical Commission V Symposium, Riva del Garda, Italy, pp. 83–86 (2014). The Archives are listed in the ISI Conference Proceedings Citation Index (CPCI) of the Web of Science, SCOPUS, and the Directory of Open Access Journals (DOAJ). http://www.int-arch-photogramm-remote-sens-spatial-inf-sci.net/XL-5/83/2014/isprsarchives-XL-5-83-2014.pdf

Baibatsha, A.B., Mamanov, E.Zh., Bekbotayev, A.T.: Allocation of perspective ores on the areas Shu-Lle belt on the materials remote sensing. In: 16th international multidisciplinary scientific geoconference, SGEM2016, pp. 35–41, Albena (2016)

Percov, A.V.: Ajerokosmicheskie metody geologicheskih issledovanij, 316 p. Izd-vo SPb kartfabriki VSEGEI, St. Petersburg (2000)

Poceluev, A.A., Anan'ev, Ju.S., Zhitkov, V.G.: Distancionnye metody geologicheskih issledovanij, prognozirovanija i poiskov mestorozhdenij poleznyh iskopaemyh, 2nd edn, 304 p. STT, Tomsk (2012)

Rajendran, S., Nasir, S., Kusky, T.M., Ghulam, A., Gabr, S., El-Ghali, M.A.K.: Detection of hydrothermal mineralized zones associated with listwaenites in Central Oman using ASTER data. Ore Geol. Rev. **53**, 470–488 (2013)

Comparative Analysis of Lineaments Extracted from Landsat ETM+ and ASTER Images, and ASTER GDEM: A Study Based on Visual Interpretations in Tichka Massif (Western High Atlas, Morocco)

Hicham Si Mhamdi, Mohammed Raji, Soufiane Maimouni, and Mostafa Oukassou

Abstract

This work focuses on the application of optical remote sensing techniques, in particular, the comparison between different methodological approaches in order to highlight structural lineaments within the Tichka massif in the Western High Atlas of Morocco. Lineaments were extracted from Landsat ETM+ and ASTER Images, and ASTER DEM. The extracted lineaments were analyzed to understand which material is most suitable for the extraction of lineaments accurately. The results showed that the maximum number of lineaments could be extracted by using ASTER Images. However, this study suggests that the combination of all extracted lineaments from different entries provides the most detailed geological lineaments. The results denoted a good agreement with field data, confirming the potentiality of the applied techniques in geological studies. The combination of different findings allowed the establishment of structural maps supporting the geological interpretation of the Tichka massif characterized by different geodynamic settings.

Keywords

Remote sensing • Lineaments • Paleozoic • Tichka • Western High Atlas • Morocco

H. Si Mhamdi (✉)
Laboratory of Applied Geology, Department of Geosciences, Faculty of Sciences and Techniques Errachidia, Moulay Ismail University, Meknes, Morocco

M. Raji · S. Maimouni · M. Oukassou
Laboratory of Applied Geology Geoinformatic and Environment, Department of Geology, Faculty of Sciences Ben M'sik, Hassan II University of Casablanca, Casablanca, Morocco

1 Introduction

Fracture networks are central to various fluid flow processes, such as groundwater and hydrothermal circulation that may contain mineralization of high economic value. Accordingly, the study of fracture networks falls at a crossroads of several fields and is particularly significant in hydrogeological and mining explorations.

In lineament mapping, we distinguish between tectonic lineaments (faults, joints, cleavages) and geomorphological lineaments (ridgeline, rivers).

The techniques used for lineaments extraction included: (i) algorithm-based automatic extraction (Karnieli et al. 1996; Argialas and Mavrantza 2004; Thannoun 2013), (ii) extraction semi-automatic (Juhari and Ibrahim 1997; Lim et al. 2001; Jordan et al. 2005; Si Mhamdi et al. 2016), (iii) visual interpretations (photos-interpretation) of directional filters, color composite (False color), and bands Ratios (Morris 1991; Mah et al. 1995; Youan Ta et al. 2008; Si Mhamdi et al. 2017). Visual interpretation provided better cartography close to the reality of linear or curvilinear discontinuities.

In this study, the geological mapping through remote sensing was suitable to handle some outcrops in inaccessible areas. It consists of lineament mapping the granitoid and its host rocks in the Tichka massif. The computer methods applied in this study were accompanied by geological studies at the outcrop scale.

The Tichka Variscan plutonic complex—intruded 291 ± 5 Ma ago (Mrini 1985; Gasquet 1991; Gasquet et al. 1992)—crops out 120 km SW of Marrakech in the Western High Atlas (Morocco). It was emplaced within Lower Cambrian volcano-sedimentary sequences including andesites, rhyolites, tufts, pelites, and carbonates. Subject to metamorphose, these sequences turned into the green-schist, with partial melting in the vicinity of the pluton itself (Gasquet 1991; Gasquet et al. 1992; Termier and Termier 1971).

H. El-Askary et al. (eds.), *Research Developments in Geotechnics, Geo-Informatics and Remote Sensing*, Advances in Science, Technology & Innovation,
https://doi.org/10.1007/978-3-030-72896-0_78

Fig. 1 Methodology chart

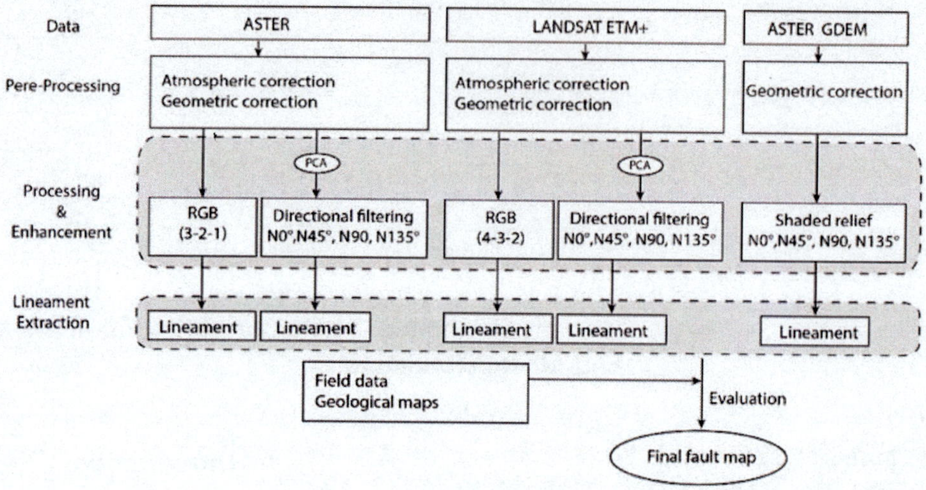

2 Materials and Methods

The data collection process compiled varied approaches of extraction methods to get maximum information: Various data required the lineaments segment mapping (potential fault zones mapping). Different image processing and enhancement techniques for different remote sensing data, including the Landsat ETM+ (EnhancedThematicMapper) satellite images and SRTM (The Shuttle Radar Topography Mission) were involved. The data contained topographic maps, geological maps, and field data as well. Figure 1 illustrates the main step of lineament extraction.

The results' validation and evaluation were carried out through the geological map and fieldwork, which enabled the identification and collection of a few faults, mainly as linear features (Strike/dip and striation).

3 Results

3.1 Qualitative Analysis of Structural Lineaments

Table 1 shows the number and the average length of the significant lineaments from each method. The results suggest that

Table 1 Statistics of lineaments extracted from different techniques

			Number of lineaments	Total length (km)
Landsat ETM+	Filter "PC 1"	N-S	140	279,64
		NE-SW	172	371,99
		E-W	124	276,78
		NW-SE	130	251,81
	RGB color composite		185	414,48
	Band Ratio		23	37,79
ASTER GDEM Shaded relief		N-S	148	339,07
		NE-SW	188	402,13
		E-W	181	495,88
		NW-SE	115	265,06
ASTER	Filter "PC 1"	N-S	92	191,14
		NE-SW	98	202,14
		E-W	109	244,55
		NW-SE	111	198,97
	RGB color composite		267	618,24

the extraction of more lineaments is possible by using the color composite of the Landsat ETM+ and ASTER. Etch techniques allow the recognition of three main directions of lineament with a dominance of the N–S and NE–SW systems.

The size, geometry, and density of the fractures varied according to several parameters including the lithology and the distance of the fractures compared to the major fault zone.

3.2 Synthetic Map

Following the compilation of different processing operations' results, a further examination step took place for the validation of the final map of the lineaments (Fig. 2). It resulted in a satisfactory agreement between the susceptibility fractures and field data.

Fractures map showed a preferential direction NE–SW with a dominance of fracturing in the external zones of the granite massif. Three fault systems could be recognized: E–W, NW–SE, and NE–SW fault systems (Fig. 2).

4 Discussion

The regional map of fractures in the study area was built from different data (Landsat ETM+ and ASTER Images, and ASTER DEM). The lineament analysis allowed a reasonable interpretation of the main structural geology and tectonics in the study areas. The findings showed at least three directional fracture systems of regional importance: N–S, NE–SW, E–W, and NW–SE. In the present study, based on the available observations of a fault plane with striations at the center of the granodiorite massif, the fractures revealed late

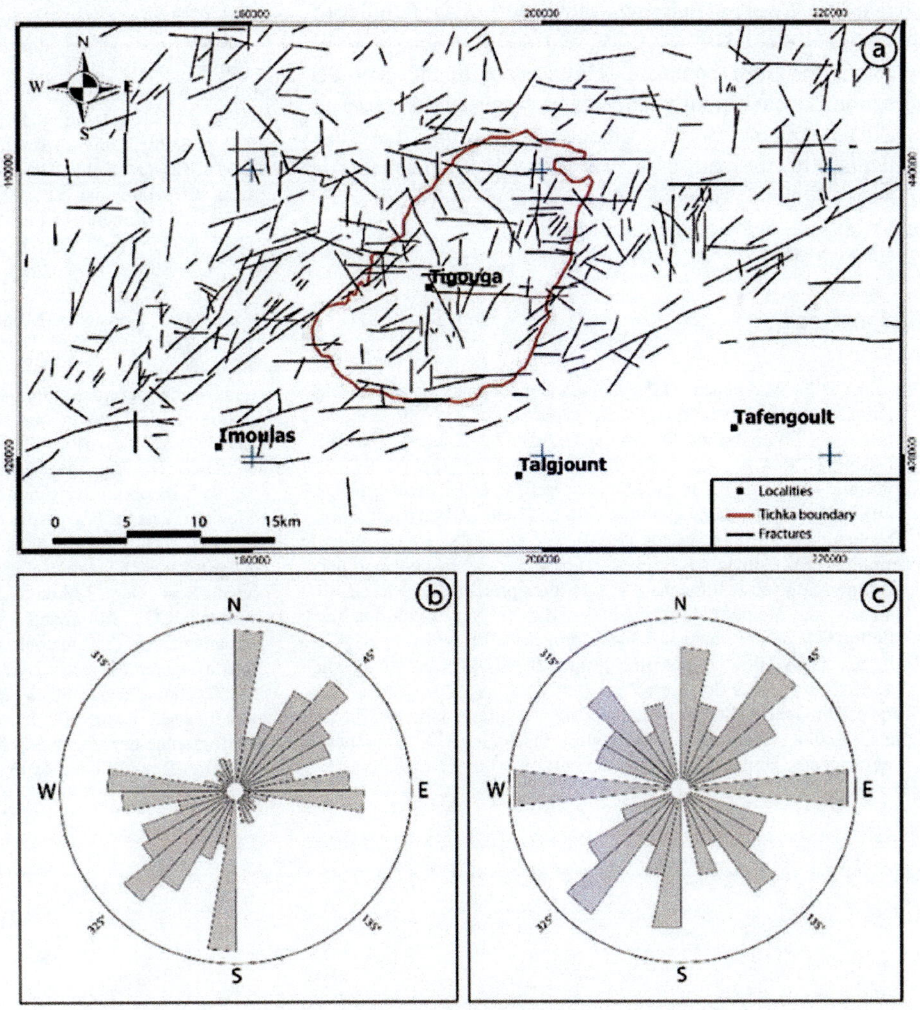

Fig. 2 Results after image processing. **a** Synthetic map of fractures. **b** Rose diagram of the trend of fractures after remote sensing. **c** Rose diagram of the fracture trends, after the fieldwork in the granitic massif

Variscan kinematics. It is generally accepted that the major faults were inherited from the Precambrian basement, and they reworked during the Variscan orogeny, the late Variscan event, and Atlas orogeny (Si Mhamdi et al. 2017). Similar tectonic events were described in the eastern part of the study area by Saber et al. (2007) and in other regions of Morocco, such as Imi-n-Tanoute region (Ferrandini et al. 1987), in the Rehamna massif (El Wartiti 1990; Saber et al. 2014), and in the Ida OuZal and OuedZat regions (Saber et al. 2014).

5 Conclusions

The applied methods proved to be efficient. This synthesis exposes a methodological approach combined with different types of data aiming to detect and analyze the network fractures.

The potential fractures map produced of the southern zone of the Western High Atlas shows three main families of average orientations N–S, E–W, and the direction NE–SW which is the most dominant. The absence of the NW–SE direction is noted in all treatments performed. This absence could be explained by the difficulty of detecting these lineaments with the data used. The density of lineaments is observed in the margins of the granitoid, as well as in some areas on both sides of the granitoid.

References

Argialas, D.P., Mavrantza, O.D.: Comparison of edge detection and Hough transform techniques for the extraction of geologic features. Int. Arch. Photogramm. Remote Sens. Spat. Inf. Sci. **34**, 790–795 (2004)

El Wartiti, M.: Le Permien du Maroc mesetien: étude géologique et implications paléogéographiques, 501 p. Thèse Es-Sciences, Univ. Mohammed VnRanat, Maroc (1990)

Ferrandini, J., Cornée, J., Saber, H.: Mise en évidence d'une compression sub-méridienne d'âge permien probable dans le massif ancien du Haut-Atlas occidental (Maroc). In: Évolution de l'ellipsoïde de déformation depuis l'orogenèse hercynienne jusqu'à la distension triasique, 304 II(20), pp. 1243–1248. Compte Rendu de l'Académie des Sciences, Paris (1987)

Gasquet, D.: Genèse d'un pluton composite tardi-hercynien, Le Massif du Tichka, Haut-Atlas occidental (Maroc), 413 p. Thèse Es-Sciences, Univ. Nancy, France (1991)

Gasquet, D., Leterrier, J., Mrini, Z., Vidal, P.: Petrogenesis of the Hercynian Tichka plutonic complex (Western High Atlas, Morocco): trace element and Rb-Sr and Sm-Nd isotopic constraints. Earth Planet. Sci. Lett. **108**, 29–44 (1992)

Jordan, G., Meijninger, B.M.L., Van Hinsbergen, D.J.J., Meulenkamp, J.E., Van Dijk, P.M.: Extraction of morphotectonic features from DEMs: development and applications for study areas in Hungary and NW Greece. Int. J. Appl. Earth Obs. Geoinform. **7**, 163–182 (2005)

Juhari, M., Ibrahim, A.: Geological applications of Landsat TM imagery: mapping and analysis of lineaments in NW Peninsula Malaysia. In: Proceedings of the 18th Asian Conference on Remote Sensing, pp. 132–140 (1997)

Karnieli, A., Meisels, A., Fisher, L., Arkin, Y.: Automatic extraction and evaluation of linear geological features from digital remote sensing data using a Hough transform. Photogramm. Eng. Remote Sens. **62**, 525–531 (1996)

Lim, C., Ibrahim, K., Tjia, H.: Radiometric and geometric information content of TiungSat-1 MSEIS data. In: TiungSAT-1: From Inception to Inauguration. Astronautic Technology, pp. 169–184 (2001). ISBN: 9789838671934 9838671932

Mah, A., Taylor, G., Lennox, P., Balia, L.: Lineament analysis of Landsat TM images, Northern Territory, Australia. Photogramm. Eng. Remote Sens. **61**, 761–773 (1995)

Morris, K.: Using knowledge-base rules to map the three-dimensional nature of geological features. Photogramm. Eng. Remote Sens. **57**, 1209–1216 (1991)

Mrini, Z.: Age et origine des Granitoides hercyniens du Maroc: apport de la géochronologie et de la géochimie isotopique (Sr, Nd, Pb), 156 p. Thèse, Univ. clermontferrand (1985)

Saber, H., El Wartiti, M., Hmich, D., Schneider, J.: Tectonic evolution from the Hercynian shortening to the Triassic extension in the Paleozoic sediments of the Western High Atlas (Morocco). J. Iber. Geol. **33**(1), 31–40 (2007)

Saber, H., Hminna, A., Jouhari, A., Rmich, A.: Lithostratigraphy and evidence of an extensive tectonic of Lower Permian age in the continental deposits of M'tal (Western Rehamna, Morocco). Am. Int. J. Res. Form. Appl. Nat. Sci. **7**(1), 81–87 (2014)

Si Mhamdi, H., Raji, M., Oukassou, M.: Utilisation de la télédétection dans la cartographie automatique des linéaments géologiques du granitoïde de Tichka (Haut Atlas occidental). Eur. J. Sci. Res. **142** (4), 321–333 (2016)

Si Mhamdi, H., Raji, M., Maimouni, S., Oukassou, M.: Fractures network mapping using remote sensing in the Paleozoic massif of Tichka (Western High Atlas, Morocco) (2017)

Termier, H., Termier, G.: Le massif granito-dioritique du Tichka (Haut Atlas occidental, Maroc). Volume I: Les régions et les terrains. Note Mém. Serv. Géol. Maroc, N216 (1971)

Thannoun, R.G.: Automatic extraction and geospatial analysis of lineaments and their tectonic significance in some areas of Northern Iraq using remote sensing techniques and GIS. Int. J. Enhanc. Res. Sci. Technol. Eng. Bull. **2** (2013)

Youan Ta, M., Lasm, T., Jourda, J.P., Kuamé, K.F., Razack, M.: Cartographie des réseaux de fractures du socle précambrien de Côte d'Ivoire. Rev. Télédét. **8**, 119–135 (2008)

Spatiotemporal Assessment of Land Cover Change and Vegetation Degradation Using Remote Sensing in Kebbi State, Nigeria

Muhammad Mansur Aliero, Mohd Hasmadi Ismail, Mohamad Azani Alias, Abubakr Sadik Ambursa, Atiku Muhammed, Iliyasu Umar, and Ridwan Mukhtar Bunza

Abstract

Spatiotemporal assessment of land cover (LC) trends and vegetation degradation (VD) using remote sensing (RS) and indicators such as anthropogenic activities and the socio-demographic information is essential for an effective and efficient planning. This paper evaluated land cover change and VD in Kebbi State, Nigeria, using Landsat data from 1986 to 2016. The RS data were classified into five thematic land cover classes as dense vegetation (DV), shrubs/built area (SB), farmland (FL), bare/grassland (BGL), and water body (WB) using maximum likelihood classifiers with acceptable accuracy. Pre-classification and post-classification change detection (CD) methodologies were performed using normalized difference vegetation index (NDVI) and image differencing, respectively. The study illustrated a decline in the DV and SB while FL and BGL increased; however, water WB remains unchanged. It is clear that major drivers of the change in the study area are the pressing need for farming land due to population growth and socioeconomic demands including fuelwood use and endemic poverty. Expansion of farming land, fuelwood consumption, and the need for construction materials were identified as key elements exerting pressure for the change. The state of the condition indicated a decline in the DV and SB and an increase in FL and BGL. The impacts included land degradation, the decline in the provision of ecosystem goods and services, and the loss of biodiversity.

Keywords

Land cover change • Vegetation degradation • Remote sensing • Kebbi state

1 Introduction

An ecosystem is a natural capital that not only ensures a steady and constant supply of goods and services that are useful to populaces (Gann and Lamb 2002) but also plays an essential role in the global carbon cycle (Kumar et al. 2010). However, it is in a state of permanent flux at various strategic spatial and temporal scales. The bases of these changes can be natural as well as anthropogenic or maybe a combination of both (Khaine and Woo 2014). They are believed to have caused an adverse change in the ecosystem, which subsequently leads to its degradation.

Therefore, monitoring the trends to which the ecosystem is changing by the use of indicators such as land cover and socio-demographic information was essential in order to make proper planning for a sustainable ecosystem management. Nowadays, the advancement of geospatial technology such as remote sensing (RS) and geographic information system (GIS) presents the best efficient tool for analyzing land cover change (LCC) (Aguilar et al. 2016; Lu et al. 2005; Pang et al. 2013; Zhao et al. 2013).

Geospatial technologies in addition to conventional inventories can facilitate the quantitative evaluation and provide a baseline for monitoring the extent, impacts, and trend of LCC. This paper attempted to evaluate LCC and VD in Kebbi State, Nigeria, using Landsat thematic mapper (TM), enhanced thematic mapper (ETM), and operational land imager (OLI) imagery from 1986 to 2016 by means of RS technique.

M. M. Aliero (✉) · M. H. Ismail · M. A. Alias
Universiti Putra Malaysia, 43400 Serdang, Selangor, Malaysia

M. M. Aliero · A. S. Ambursa · A. Muhammed · I. Umar
Kebbi State University of Science and Technology, Aliero, Nigeria

R. M. Bunza
Bayero University, Kano, Nigeria

H. El-Askary et al. (eds.), *Research Developments in Geotechnics, Geo-Informatics and Remote Sensing*, Advances in Science, Technology & Innovation,
https://doi.org/10.1007/978-3-030-72896-0_79

2 Methodology

2.1 The Study Area

See Table 1.

2.2 Data Collection Procedure

Satellite (Landsat) images for 1986, 1996, 2006, and 2016, Path and row 191/51 and 191/52 were downloaded from the United States Geological Survey (USGS) website. Anniversary dates were maintained in order to avoid biases that are due to the seasonality, this is essential for change detection (CD) analysis. Field and other secondary data were collected mainly to determine the human influences on LCC and VD. Simple random sampling survey technique was conducted by means of focus group discussion and interview with the respondent using a structured questionnaire. The secondary data (population density, growth rate and other socioeconomic variables) were collected from various existing literature and government agencies such as the National Population Commission of Nigeria and Bureau of National Statistics of Nigeria websites.

2.3 Data Analysis

The whole process of RS data analysis is shown in Fig. . The analysis of remotely sensed data involves three main steps; the first step is the preprocessing, then the images were classified into five thematic land cover classes as dense vegetation, shrubs/built area, farmland, bare/grassland, and water body in accordance with Anderson classification scheme level 1. Finally, the post-processing, which included the classification accuracy assessment (using 156 ground truth points) and CD, was achieved. The pre-classification CD using spectral vegetation indices (SVIs) (image differencing method) was computed because of its relationship with the net primary productivity (NPP) of the land and its accuracy as a classification method especially for phenology/vegetation studies using remotely sensed data (Jensen 2005; Singh 1989).

3 Results and Discussion

3.1 Land Cover Change Detection

The 1986 thematic land cover of the study area shows that the largest area was shrub/built area followed by farmland and bare/grassland. The lowest is dense vegetation and water body. In 2016, the change in all the classes was not very significant except for the class 'dense vegetation' that suffered a drastic decrease in the area.

The percentage of land area occupied by shrubs/built area in 1986 was about 43.1% and has substantially dropped to 32.0% in 2016. Likewise, farmland was 24.9% in 1986 when compared to 2016, it increased to 38.9%. Bare/grassland showed a slight increase from 20.8% in 1986 to 24.6% in 2016. Dense vegetation, which used to be 10.8% in 1986, has drastically relegated and went down to 4.2% in 2016. There was not a significant change in the water body as it occupied an area of 0.3% in both 1986 and 2016 themes.

Figure 3 shows the thematic map of classification analysis which indicates visually the changes that occurred in the area from 1986 to 2016 (30 years). In three (3) decades, dense vegetation has declined rapidly by about 61% with its larger part being converted into bare/grassland and farmland. It is clear that there is a downturn in the shrubs land by about 25.7% in the area. Within this timeframe (1986–2016), farmland has increased by 35.9%, which is not surprising since there is pressure imposed by the demand for food due to population increase.

There is no doubt that food supply increases but at the expense of the lost forest, biodiversity, and habitats for wildlife. Bare/grassland also increased by about 18.2% in the area, which is attributed to the factor, that most of the shrubs

Table 1 Descriptive summary of the study area

	Description
Extent	The area lies between Lat. 13° 54′ 58.925″ N–11° 7′ 27.002″ N and Long. 3° 32′ 57.995″ E–4° 53′ 19.708″ E, with the extent of about 20,657.5 km^2, in the extreme northwestern part of Nigeria
Climate	The mean annual temperature lies between 35 and 40 °C, average annual rainfall of about 850 in. and relative humidity ranges from 51–79% and 10–25% during rainy and dry seasons, respectively
Vegetation	The vegetation is Sudan Savannah type, and the soil is a semiarid type, characterized by frequent weathering and leaching due to poor soil structure and low organic matter content
Economic activity	The main economic activity is agriculture. Over (70%) of the people practice one form of agriculture or another

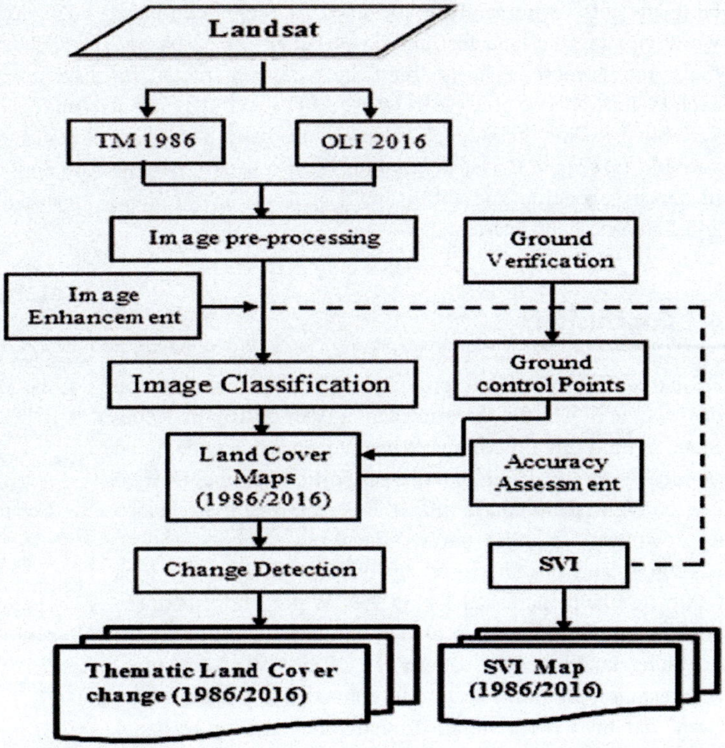

Fig. 1 Methodology flowchart for remote sensing data processing

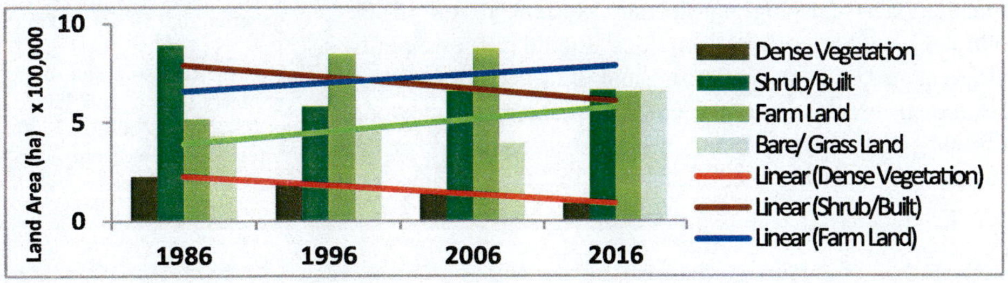

Fig. 2 Trends of land cover change in the study area (1986–2016)

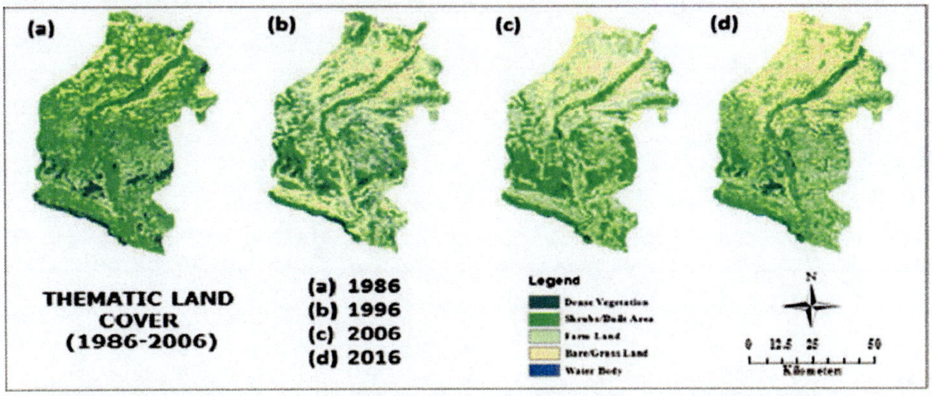

Fig. 3 Land cover maps of the study area (1986–2016)

land are cut-down mainly to be used as fuelwood, burnt while clearing the land for farming or being grazed by animals. Furthermore, a linear trend line was constructed for each thematic class; time (years) was used as an independent variable (x) while thematic classes were used as a dependent variable (y) (Fig. 2). The trends illustrated a steady decline in dense vegetation and shrubs/build areas while farmland and bare/grassland increased.

4 Conclusion

RS-based CD techniques provide the baseline information on LCC and VD for three decades (1986–2016) in Kebbi State. Within this time frame, dense vegetation has declined rapidly by about 61% with its larger portion being converted into bare/grassland and farmland. It is clear that there is also a downturn in the shrubs land by about 25.7%. Bare/grassland has increased by about 18.2% in the area while farmland increased by 35.9%. Water bodies remain unchanged. RS provides a unique explanatory platform for the understanding of the pattern of LCC and VD in the area. Afforestation programs need to be enhanced to a more robust form that involves planting drought- and grazing-resistant and economically viable tree species. There is a need to improve the livelihood of the rural people by providing other income sources since they depend heavily on land for sustenance. The farming agroforestry system needs to be adopted, and restoration activities should be essentially strengthened. It is also urgent to look for alternative energy sources in order to stop the colossal problem of fuelwood consumption. Further studies need to be conducted on the future land cover predictions so as to provide the decision-makers with synaptic information on the future LCC scenarios that are likely to occur. There is also a need to integrate the soil and climatic variables for a better understanding of the phenomena. The limitation of this study lies in the use of low-resolution data set, and thus, it is urgent to apply very high-resolution data for a better accuracy of the results.

References

Aguilar, F.J., et al.: A quantitative assessment of forest cover change in the Moulouya river watershed (Morocco) by the integration of a subpixel-based and object-based analysis of Landsat data. Forests **7**, 1–19 (2016)

Gann, G.D., Lamb, D.: Ecological restoration—a means of conserving biodiversity and sustaining livelihoods. International and the IUCN Commission on Ecosystem Management (2002)

Jensen, J.R.: Introduction to Digital Image Processing: A Remote Sensing Perspective. Pearson Prentice Hall, Hoboken (2005)

Khaine, I., Woo, S.Y.: An overview of interrelationship between climate change and forests. Forest Sci. Technol. **11**, 11–18 (2014)

Kumar, P., Rani, M., Pandey, P.C., Majumdar, A., Nathawat, M.S.: Monitoring of deforestation and forest degradation using remote sensing and GIS: a case study of Ranchi in Jharkhand (India). Rep. Opin. **2**, 55–67 (2010)

Lu, D., Mausel, P., Batistella, M., Moran, E.: Land cover binary change detection methods for use in the moist tropical region of the Amazon: a comparative study. Int. J. Remote Sens. **26**, 101–114 (2005)

Pang, C., Yu, H., He, J., Xu, J.: Deforestation and changes in landscape patterns from 1979 to 2006 in Suan County, DPR Korea. Forests **4**, 968–983 (2013)

Singh, A.: Review article digital change detection techniques using remotely-sensed data. Int. J. Remote Sens. **10**, 989–1003 (1989)

Zhao, X., Xu, P., Zhou, T., Li, Q., Wu, D.: Distribution and variation of forests in China from 2001 to 2011: a study based on remotely sensed data. Forests **4**, 632–649 (2013)

Relation Between Normalized Difference Built-up Index and Land Surface Temperature in Kolkata (India)

Uday Chatterjee and Sushobhan Majumdar

Abstract

Kolkata is one of the major cities in the eastern part of India which has experienced rapid changes in land use and land cover (LULC) as remarked when analyzing the satellite images and field verification. The land surface temperature (LST) of Kolkata City is increasing day by day, which explains the choice of this city as a case study. The objective of this article was to analyze the LST of Kolkata City with an increase in impervious surfaces. The other objective was to find out if there are any relationships between the impervious surfaces and LST in Kolkata City from 2005 to 2019. Landsat satellite images were used for the analysis of normalized difference built-up index (NDBI) and land surface temperature (LST) mapping. This study reveals that the percentage of built-up area in Kolkata City has been increasing over time and has very high positive relationship with the increase in LST which makes Kolkata City a heat island. The mean temperature of Kolkata City has increased by 12 °C. This study will be helpful to the concerned authorities and policymakers for the future planning of the city and to built-up various policies for its sustainable development.

Keywords

Land surface temperature • Impervious surfaces • NDBI • Heat island • Sustainable development

U. Chatterjee (✉)
Department of Geography, Bhatter College, Dantan (Vidyasagar University), West Bengal, Dantan, 721101, India

S. Majumdar
Department of Geography, Jadavpur University, Kolkata, West Bengal 700032, India

1 Introduction

Land surface temperature (LST) means the temperature of the ground or surface (Rajeshwari and Mani 2014; 'Mallick et al. (2008), Barsi et al. (2014), Jiménez-Muñoz et al. (2009)'. The increase in LST in Kolkata City has become one of the major concerns for environmentalists, researchers, and also policymakers (Ustin 2004). Kolkata City has been developing rapidly in the service sector providing huge employment opportunities. Because of the huge influx of population, vacant spaces in Kolkata City have decreased fast due to the rapid growth of built-up areas, which results into the increase of impervious surfaces.

From the various reports (Ghosh 2014; Sarkar 2013), it has been found that the temperature of Kolkata City (Fig. 1) has been on the increase. The objective of this paper was to analyze the change in surface temperature of Kolkata City from 2005 to 2019. Another objective was to find out if there are any relationships between the increase in impervious surfaces and the increased surface temperature.

2 Materials and Methods

For the analysis of Landsat satellite images, the Landsat 5 TM and 8 OLI of 2005, 2010, 2015, and 2019 have been used (same time period). For the surface temperature analysis at the time of layer stacking, the thermal bands were stacked. For measuring the LST and normalized difference built-up area index (NDBI), various statistics have been used. For the LST mapping, first radiance values have been calculated using this formula.

$$CV_{R1} = \frac{LMAX_\lambda - LMIN_\lambda}{(QCALMAX - QCALMIN) * (QCAL - QCALMIN) + LMIN_\lambda} \text{(For Landsat 5 TM)} \quad (1)$$

where CVR1 indicates cell value as radiance, QCAL is the digital number, LMIN indicates spectral radiance scales to

H. El-Askary et al. (eds.), *Research Developments in Geotechnics, Geo-Informatics and Remote Sensing*, Advances in Science, Technology & Innovation,
https://doi.org/10.1007/978-3-030-72896-0_80

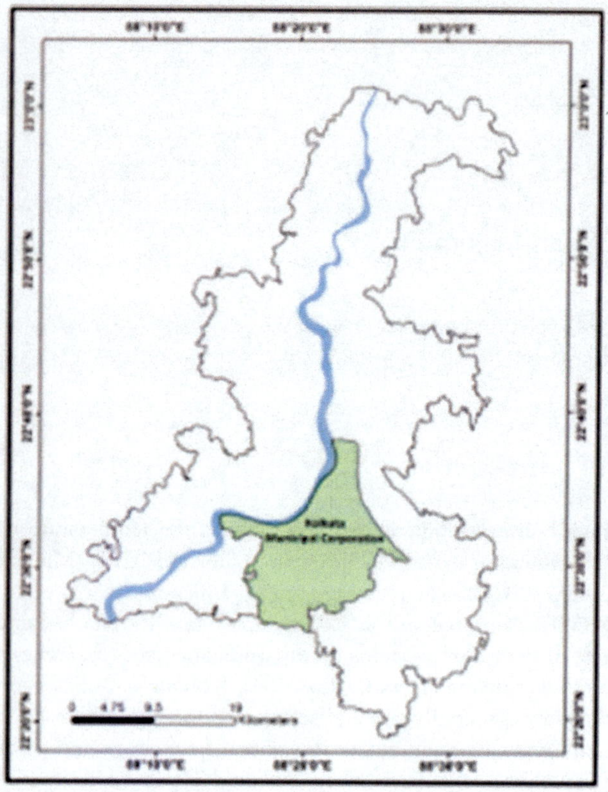

Fig. 1 Kolkata metropolitan area

QCALMIN, LMAX indicates spectral radiance scales to QCALMAX, QCALMIN = the minimum quantized calibrated pixel value of the images and QCALMAX is the maximum quantized calibrated pixel value of the images. From the radiance value of the images, the temperature of Kolkata City has been measured.

$$L\lambda = M_{\mathrm{L}} Q_{\mathrm{cal}} + A_{\mathrm{L}} - O_i \text{ (For Landsat OLI 8)} \tag{2}$$

where $L\lambda$ = TOA spectral radiance (W/(m^2 × srad × μm)), M_L = Band-specific multiplicative rescaling factor from the metadata (RADIANCE_MULT_BAND_x, where x is the band number), A_L = Band-specific additive rescaling factor from the metadata (RADIANCE_ADD_BAND_x, where x is the band number), Q_{cal} = Quantized and calibrated standard product pixel values (DN), and O_i is the correction for band 10. From the radiance value, LST has been collected by applying these statistics.

To find out the nature and trends about the impervious surface of Kolkata City normalized difference vegetation index has been calculated. For the Landsat thematic mapper only band 4 and band 3 have been taken, but in case of Landsat 8 only band 5 and band 6 have been considered.

$$\mathrm{NDBI} = \frac{\mathrm{Band\,4} - \mathrm{Band\,3}}{\mathrm{Band\,4} + \mathrm{Band\,3}} \text{(For Landsat TM)}; \frac{\mathrm{Band\,6} - \mathrm{Band\,5}}{\mathrm{Band\,6} + \mathrm{Band\,5}} \text{(For Landsat 8 OLI)} \tag{3}$$

3 Results

3.1 Mapping of Impervious Surfaces

After analyzing the impervious surfaces using NDBI index (Fig. 2) in Kolkata City, it has been found that the percentages of impervious surfaces (mainly artificial structures) have been increasing. By analyzing this index, it has also been found that the percentage of impervious surfaces have been increased in the peripheral areas of the city from 2005 to 2019.

For this reason, most of the wetland, agricultural land, barren land, and vegetation areas have been converted into artificial surfaces which impacted negatively the terrestrial ecosystem (Table 1).

3.2 Land Surface Temperature (LST)

By analyzing the land surface temperature of Kolkata, it has been found that the surface temperature of Kolkata has been increasing during the study period turning Kolkata into a "Heat Island." The mean temperature has also increased from 20.5 to 32 °C between 2005 and 2019 (mean temperature has been calculated from maximum and minimum temperature). From the field verification during this study, it has also been revealed that the impervious surfaces have increased (Fig. 3).

3.3 The Relation Between NDBI and LST

Because of the increase in an urban area or artificial area, the LST of Kolkata has increased recently. By analyzing NDBI and LST in Kolkata City, it has been found that with the time both of the surface temperature and percentage of built-up area increased during the study period. So, there is a high positive relationship between the NDBI and LST in Kolkata. The rate of correlation is high in the year 2010, followed by the years 2019, 2005, and 2015.

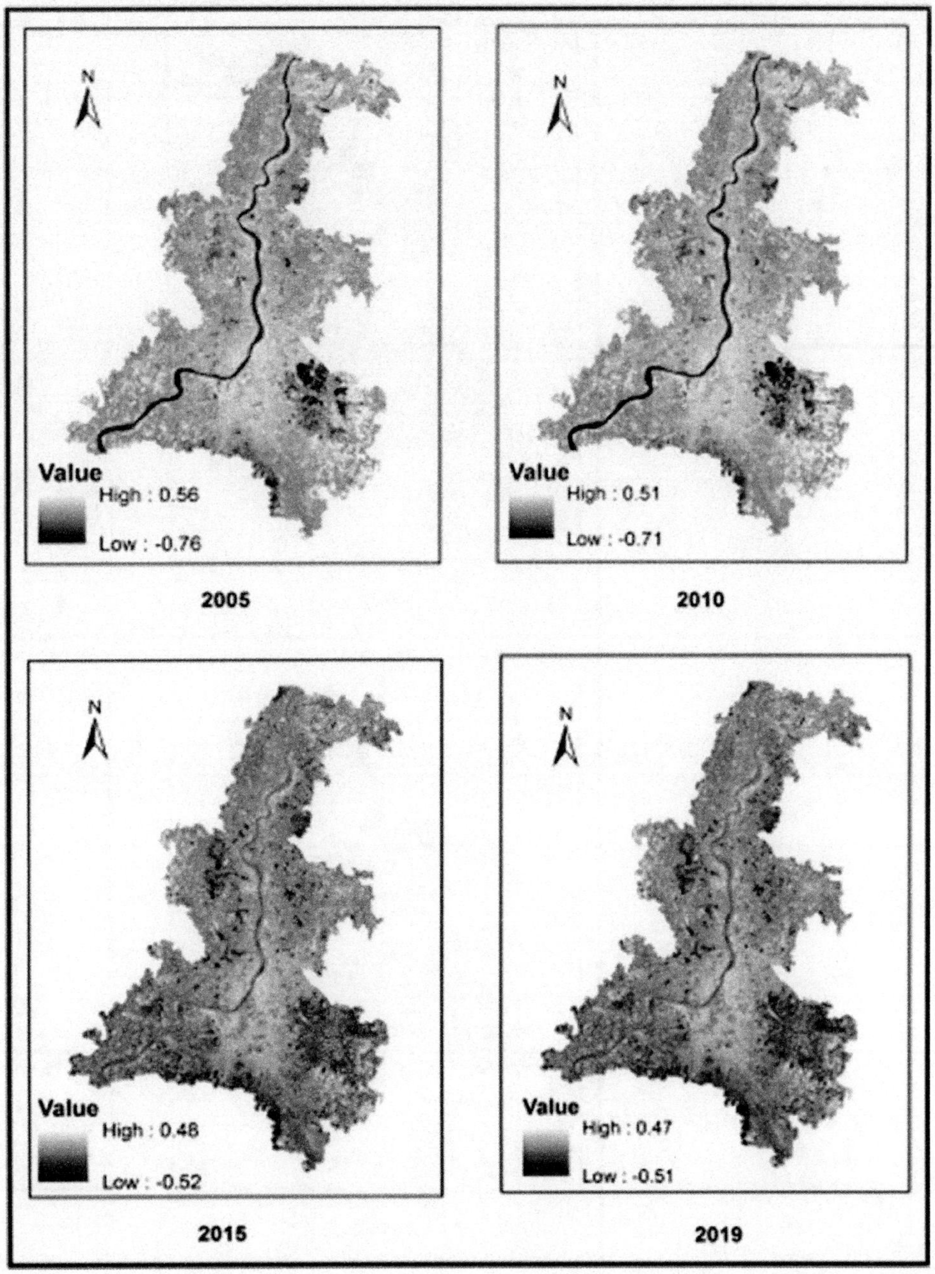

Fig. 2 Normalized difference built-up area index from 2005 to 2019

Table 1 Trends of LULC

Year	2005	2010	2015	2019
Built-up area	16.55	26.04	30.17	33.6
Vegetation	52.25	39.4	38.17	37.11
Cultivated land	7.89	5.58	3.89	3.26
Water bodies	9.32	5.26	4.78	3.8
Barren land	23.9	23.7	23.07	22.23

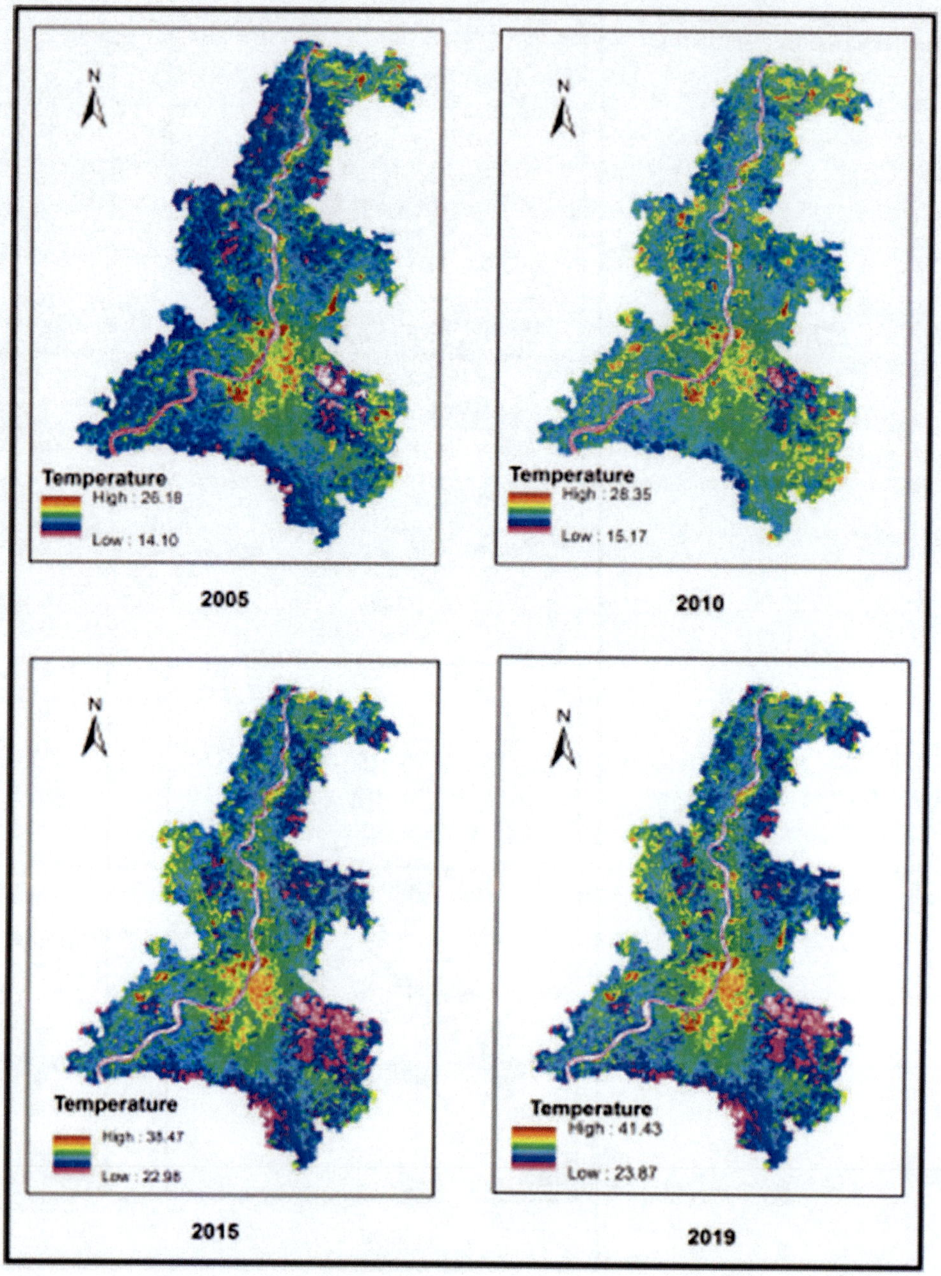

Fig. 3 LST from 2005 to 2019

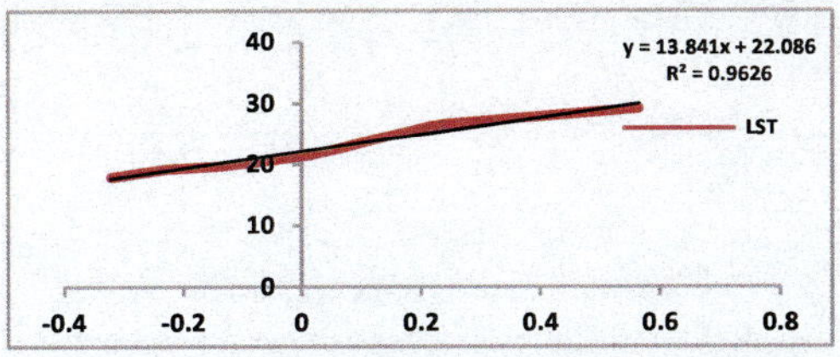

2005

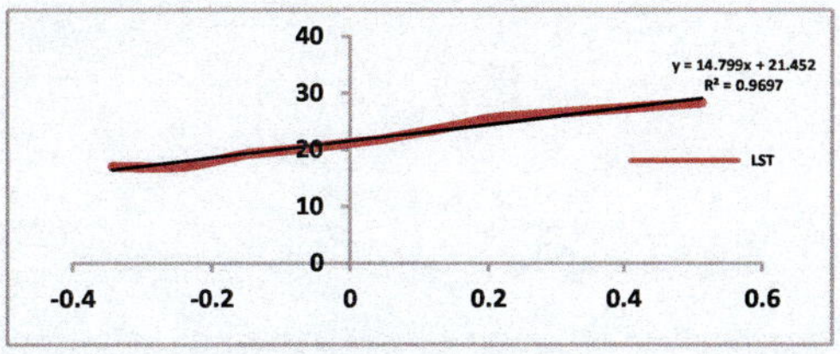

2010

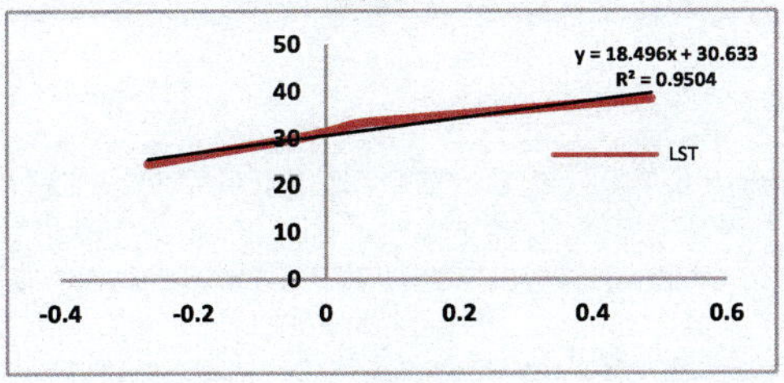

2015

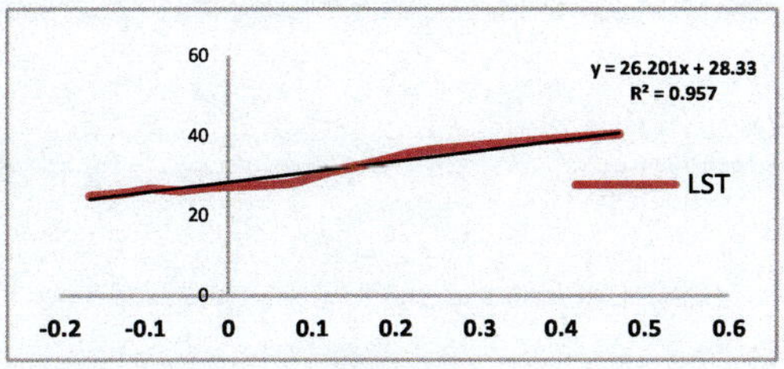

2019

4 Discussion

Most of the people in the surrounding areas of Kolkata want to live in the places nearer to city which accelerates the growth of built-up area or impervious surfaces. Because of the rapid changes in impervious surfaces in this area, the percentage of natural cover has been decreasing very rapidly. High temperature is mostly concentrated toward the main CBD areas and decreases outwards. The value of NDBI is very much less in the northern and southern portions of Kolkata City where the percentages of natural cover are relatively higher than the other areas. From the overview of land use and land cover changes in Kolkata City and its surroundings, it has been found that the vegetation area and wetland or water body areas are decreasing very fast which will create adverse impacts on the local ecosystems.

5 Conclusions

From the above analysis, it has been found that the LST of Kolkata has been increasing very fast (12 °C in 14 years). The increase in the impervious surface is playing a major role in this temperature increase. Kolkata is a very old city in India. The major problem of Kolkata City, as noticed during the data collection and field observation, is that there is no coordination among the developmental sectors. For this reason, unplanned and illegal construction has been common in Kolkata City. Most of the major constructions now do not respect any environmental rules and legislations of India. This will create adverse impacts on the land resources. For the future planning and management of Kolkata City, the authorities and policymakers should pass strict laws to prevent the green spaces of Kolkata from becoming impervious surfaces.

References

Barsi, A., Schott, J.R., Hook, S.J., Raqueno, N.G., Markham, B.L., Radocinski, R.G.: Landsat-8 thermal infrared sensor (TIRS) vicarious radiometric calibration. Remote Sens. **6**(11), 11607–11626 (2014)

Ghosh, A.: Climate Change in Kolkata City, KMDA, pp. 14–16 (2014)

Jiménez-Muñoz, J.C., Sobrino, J.A., Plaza, A., Guanter, L., Moreno, J., Martínez, P.: Comparison between fractional vegetation cover retrievals from vegetation indices and spectral mixture analysis: case study of PROBA/CHRIS data over an agricultural area. Sensors **9**(2), 768–793 (2009)

Mallick, J., Kant, Y., Bharath, B.D.: Estimation of land surface temperature over Delhi using Landsat-7 ETM+. J. Indian Geophys. Union **12**(3), 131–140 (2008)

Rajeshwari, A., Mani, N.: Estimation of land surface temperature of Dindigul district using Landsat 8 data. Int. J. Res. Eng. Technol. **3** (5), 122–126 (2014)

Sarkar, P.: LULC of Indian Mega Cities, KMDA, pp. 123–124 (2013)

Ustin, S.: Manual of Remote Sensing: Remote Sensing for Natural Resource Management and Environmental Monitoring. Wiley, NJ, USA (2004)

A Non-stationary NDVI Time Series with Big Data: A Deep Learning Approach

Manel Rhif, Ali Ben Abbes, and Imed Riadh Farah

Abstract

In this paper, a suitable method to forecast the normalized difference vegetation index (NDVI) time series (TS) is deep learning in the context of remote sensing big data. In fact, we proposed a non-stationary NDVI time series forecasting model by combining big data system, wavelet transform (WT) and long short-term memory (LSTM) neural network. In the first step, the MapReduce was investigated to extract NDVI TS. Then, WT and LSTM were applied for the analysis and forecasting of the NDVI. Our results show that our methodology gives a good result for forecasting NDVI time series in terms of root mean square error (RMSE) of 0.05.

Keywords

Non-stationary time series • Big data • Deep learning • Wavelet transform

1 Introduction

Recently, a large and open set of satellite imagery is available to track changes over time. However, the existing methods are complex for analyzing a huge dataset due to the lack of computing infrastructure. In fact, the analysis, interpretation, and forecasting of non-stationary big data in remote sensing is a challenging problem (de Assis et al. 2017). Taking these challenges into account, the main contribution of this paper was to present a new methodology for forecasting big data remote sensing data based on wavelet transform (WT) and long short-term memory (LSTM).

In this paper, we used a distributed file system (DFS) to stock moderate resolution imaging spectroradiometer (MODIS) images over Tunisia. Then, MapReduce, which is a programming model that can process big data in parallel on multiple nodes, presented the tool for the remote sensing of big data analytics. After that, the WT was applied in order to decompose NDVI TS into different components in order to analyze land cover land use change (Rhif et al. 2019). Finally, the deep learning LSTM model (Reddy and Prasad 2018) was investigated for forecasting non-stationary big data NDVI TS.

2 Material and Methods

2.1 Study Area and Data

The study area is located in the northwest of Tunisia. More precisely, four regions (Jendouba, EL Kef, Beja, and Bizerte) were considered. These regions cover 15,000 km^2 and have a typical Mediterranean climate. In this study, the MODIS product (MOD13Q1 collection 6) was investigated. It has 250 m^2 spatial resolution and a time period of 2001–2017, with 23 composite images per year.

2.2 Proposed Methodology

The proposed methodology is based on four steps for non-stationary big data NDVI TS. (1) The MODIS images were stocked using a DFS. (2) Data were processed by MapReduce model. (3) The WT was applied. (4) LSTM model was used for each obtained component for TS forecasting (Fig. 1).

- Distributed file system:
  As an input, MODIS images of the study area were used. Then, the MODIS images were stored by a sequence of

M. Rhif (✉) · A. B. Abbes · I. R. Farah
Laboratoire RIADI, École Nationale des Sciences de l'Informatique, Manouba, Tunisia
e-mail: manel.rhif@ensi-uma.tn

A. B. Abbes
Centre d'applications et de Recherches en Télédétection (CARTEL), Université de Sherbrooke, Sherbrooke, QC J1K 2R1, Canada

H. El-Askary et al. (eds.), *Research Developments in Geotechnics, Geo-Informatics and Remote Sensing*, Advances in Science, Technology & Innovation,
https://doi.org/10.1007/978-3-030-72896-0_81

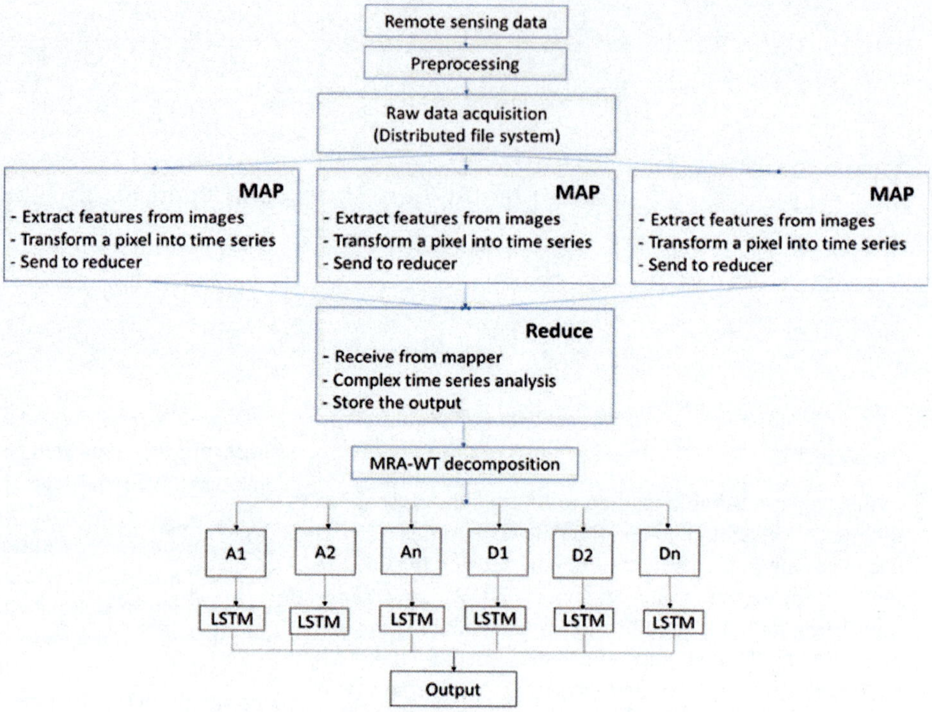

Fig. 1 Proposed methodology for forecasting non-stationary NDVI big data TS. $A_1, A_2, \ldots, A_n$: presents the approximation components and $D_1, D_2, \ldots, D_n$: are the detail components obtained by decomposing the original signal into n levels using MRA-WT

years using the DFS which made it easier to build the TS. In fact, the DFS has several advantages: it is able to store any type of data without any restriction. In our case, the schema-on-read approach offered a more adequate design.

- MapReduce Programming Model
  This model consists of two methods responsible for extracting the features from the images and processing the complex algorithms for remote sensing NDVI TS applications. The role of the map method is to filter and sort values into lines (de Assis et al. 2017). This step was done in order to obtain the value of each pixel with a specific key. In fact, the mapper gets the images and extracts features of images (⟨key, pixel value⟩). It builds the TS by getting the id and the values for each pixel. Finally, the NDVI TSs are sent to the reducer which performs and stores the result for complex analysis.
- Wavelet transforms
  The multiresolution analysis (MRA-WT) is one of the most used types for NDVI TS analysis. It decomposes the signal into a mutually orthogonal set of wavelets (Rhif et al. 2019). It is a discrete set of the wavelet scales and translations. The mother wavelet is scaled by power two, $a = 2^j$, and translated by integer $b = k2^j$, where k is a location index, which runs from 1 to $2^{-j}N$ (N is the number of observations) and j runs from 0 to J (J is the total number of scales). In fact, the discrete wavelet transform (DWT) is defined as: $\psi_{j,k}(t) = 2^{-j/2}\psi(2^{-j}t - k)$.
- Long short-term memory (LSTM)
  LSTM is a powerful deep learning method for TS forecasting (Liu et al. 2019). It is a feedforward network based on backpropagation algorithm. The LSTM contains memory blocks which contain memory cells in order to remember the information. The LSTM contains special multiplicative units called gates to control the information flow. To predict the future values of NDVI TS, the dataset was divided in training, test, and validation sets. In addition, different parameters were required for the LSTM (e.g., optimizer learning rate, etc.). More details are presented in Reddy and Prasad (2018). In this paper, the LSTM model was applied for each component obtained from WT (details and approximations) in order to forecast the NDVI TS.

3 Results

The organization of the MODIS data using the big data enabled us to build an infrastructure able to extract, transform, and load all the images. After that, the MRA-WT was applied to extract the different components and analyze the trend.

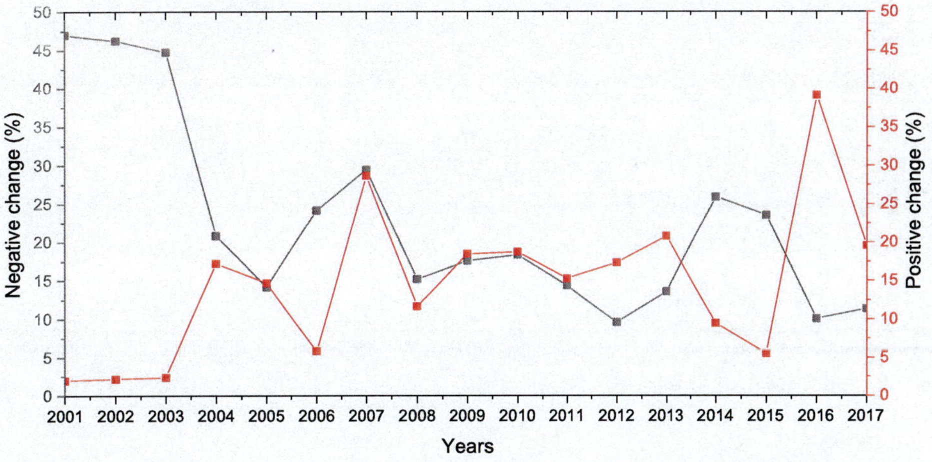

Fig. 2 Statistical results of area percentages of negative and positive changes for each year of the study period

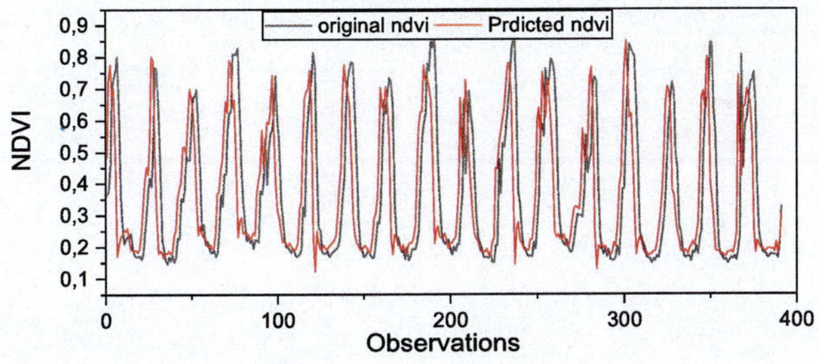

Fig. 3 Original and predicted NDVI time series using the LSTM

In fact, we analyzed the change for the whole area (different regions). As presented in Fig. 2, the browning area was larger than the greening one. In addition, the largest browning areas occurred in 2002 (46.22%), whereas the largest greening areas occurred in 2016 (39.13%).

To predict the future values of NDVI TS, the LSTM was used. Studies have shown that dividing the data in 7:2:1 ratio gives a better assessment of the network (Reddy and Prasad 2018). In this paper, the used parameters are Adam optimizer, learning rate 0.001, and batch size 50. In addition, the RMSE was calculated in order to measure the accuracy of the prediction by the LSTM. The original NDVI TS versus predicted TS in the training phase of the LSTM network is shown in Fig. 3. The obtained RMSE (0.05) shows that the LSTM network is satisfying at predicting the NDVI TS by learning the different components obtained after wavelet decomposition. In addition, there is a displacement between the original and predicted TS due to the fact that a decomposed data series provide additional information than that obtained directly from the original data series.

4 Conclusions

In this study, a non-stationary NDVI forecasting model in the north-western of Tunisia was studied for the period of 2001–2017. The map reducer was used in order to extract different TS. Then, the MRA-WT was investigated to analyze non-stationary TS. The obtained results indicate several changes faces in the study area. In addition, our model gives a low RMSE value. As a future works, we are planning to evaluate this approach using different data types (temperature, precipitation, NDVI, etc.) and multivariate LSTM and focusing more on data, memory, and CPU intensive tests.

References

de Assis, L.F.F.G., de Queiroz, G.R., Ferreira, K.R., Vinhas, L., Llapa, E., Sanchez, A.I., Victor, M., Câmara, G.: Big data streaming for remote sensing time series analytics using MapReduce. Braz. J. Cartography **69**(5), 991–1000 (2017)

Liu, Y., Guan, L., Hou, C., Han, H., Liu, Z., Sun, Y., Zheng, M.: Wind power short-term prediction based on LSTM and discrete wavelet transform. Appl. Sci. **9**(6), 1108 (2019)

Reddy, D.S., Prasad, P.R.C.: Prediction of vegetation dynamics using NDVI time series data and LSTM. Model. Earth Syst. Environ. **4** (1), 409–419 (2018)

Rhif, M., Ben Abbes, A., Farah, I.R., Martínez, B., Sang, Y.: Wavelet transform application for/in non-stationary time-series analysis: a review. Appl. Sci. **9**(7), 1345 (2019)

Assessment of Vegetation Changes in Saudi Arabia Using Harmonic Analysis and ENSO Indices

Wenzhao Li, Hesham El-Askary, Mohamed Qurban, K. P. Manikandan, Mohammad H. Makkawi Ashri, Thomas Piechota, and Daniele C. Struppa

Abstract

Arid environment with limited vegetation and scare water resources characterizes Saudi Arabia (SA). Therefore, studying the changes and trends of vegetation conditions is of high importance, especially considering the impact that climate change is likely to have on the living standards and food security in the near future. In this research, the enhanced vegetation index (EVI) is applied as an indicator of the drying trends over the SA region in the past 20 years. The soil moisture products, vegetation, and precipitation-based products are assessed to monitor and predict the variability of vegetation distribution. Overall, the vegetation in SA has a declining trend in most of its habitats, with the exception of regions with cropland reclamation. Most of the wild sparse vegetation areas (such as grassland and shrubland) suffer decreasing trends in the southwestern mountains. The findings suggest that soil moisture and EVI of five major vegetation categories react differently to the precipitation. Interestingly, a comparably positive relationship between the SA's NDVI and Trans-Niño Index (TNI) is shown over the other El Niño indices, with increasing lag correlation from 0 to 11 months. This may indicate the teleconnection between El Niño events, eastern Pacific El Niño in particular, and vegetation richness in SA.

Keywords

Vegetation • Saudi Arabia • Harmonic analysis • Soil moisture • ENSO

1 Introduction

Climate in arid/semi-arid regions has undergone significant changes, making it increasingly drier over the past 100 years (Narisma et al. 2007). By the end of this century, arid/semi-arid regions will increase in area by about 10% (Feng and Fu 2013). The region of Saudi Arabia (SA) is among the most arid/semi-arid areas globally due to the lack of precipitation and high temperature (Funk et al. 2015; Hulme 1996). SA is characterized by its dry climate, scarce rainfall, lack of water resources, and extremely fragile ecological environment, sensitive to human activities and global climate change (Reed et al. 2012). To obtain the spatial distribution of vegetation changes in the SA region, vegetation indices such as normalized difference vegetation index (NDVI) and enhanced vegetation index (EVI) can be used (Liu and Huete 1995). On the other hand, soil moisture is also considered to be a key variable to assess vegetation in drought conditions, with its wide range of applications including wildfire and drought monitoring, climate forecasting, and agriculture productivity (Sazib et al. 2018). Moreover, the Saudi vegetation conditions could also be impacted by certain climate events such as ENSO (El Niño-Southern Oscillation) for its global impact. Standardized sea temperature anomalies for various regions across Pacific Ocean could be used as indices to measure the severity of such ENSO events. These include Nino 1 + 2 (90° W–80° W, 0° S–10° S), Nino 3 (150° W–90° W, 5° S–5° N), Nino 4 (160° E–150° W, 5° S–5° N), Nino 3.4 and ONI (170° W–120° W, 5° S–5° N), and Trans-Niño Index (TNI) which measures the gradient in SST anomalies between the central and eastern equatorial Pacific (Nino 4

W. Li · H. El-Askary (✉) · T. Piechota · D. C. Struppa
Chapman University, Orange, CA 92866, USA
e-mail: elaskary@chapman.edu

H. El-Askary
Alexandria University, Moharam Bek, Alexandria, 21522, Egypt

M. Qurban
Ministry of Environment, Water and Agriculture, Riyadh, 11195, Saudi Arabia

K. P. Manikandan · M. H. M. Ashri
King Fahd University of Petroleum and Minerals (KFUPM), Dhahran, 31261, Saudi Arabia

H. El-Askary et al. (eds.), *Research Developments in Geotechnics, Geo-Informatics and Remote Sensing*, Advances in Science, Technology & Innovation,
https://doi.org/10.1007/978-3-030-72896-0_82

minus Nino 1 + 2). This research aims to investigate vegetation changes in SA region with datasets from multiple satellite observations in synergy with modeling approach to study its drought conditions. Finally, this work proposes a possible teleconnection between SA's vegetation variation and ENSO events, which is worth further investigation.

2 Data and Methods

The datasets used in this paper consisted of two satellite-derived global soil moisture datasets generated by integrating the European Space Agency (ESA)'s Soil Moisture Ocean Salinity (SMOS) (Kerr 1998) and the National Aeronautics Space Agency (NASA)'s Soil Moisture Active Passive (SMAP) (Entekhabi et al. 2010) in spatial/temporal resolution of 0.25° and 3 days, respectively. The vegetation health is closely connected with the soil properties such as soil moisture and salinity (Kim et al. 2014; Whitney et al. 2018). Both datasets were used to investigate and explain the relationship between root zone soil moisture anomalies and precipitation as a function of land cover variability. Vegetation-type information was obtained from the ESA's global land cover data (GlobCover 2009), which included 22 land cover classes defined by the Food and Agriculture Organization of the United Nations (UN) Land Cover Classification System (LCCS). The Climate Hazards Group Infrared Precipitation with Station (CHIRPS) (CHIRPS 2015) dataset was also adopted here to explore the spatial and temporal variability of precipitation at 0.5 degree spatial resolution. The vegetation index NDVI, obtained from Global Agricultural Monitoring System, was required to further explore the statistical relationship between ENSO and vegetation during the period from 1990 to 2013. Another vegetation dataset was the MOD13A1.006 Terra EVI Indices (Didan 2015). The harmonic analysis was applied over the EVI dataset to further model the changes of vegetation distribution over the area of SA from 2000 to 2019, as inspired by the study in the near region (Li et al. 2019).

3 Results

According to the GlobCover dataset, areas of different vegetation categories (except barren, artificial, snow, and water surfaces) in SA are presented in Table 1. The vegetation types are summarized into four main categories: cropland, mosaic vegetation, forest, and sparse vegetation. Mosaic vegetation is the mixture of cropland and other vegetation such as grassland and shrubland. The area of forest in SA is only around 0.14% of the total area, whereas the sparse vegetation occupies more than 74%, which accounts for the majority of the vegetation area. The distribution of various vegetation categories is found in Fig. 1a, where most of the vegetation habitats is located in the southwestern mountain area and some of the cropland field in the north and middle regions. Three regions marked as A, B, and C, mainly representing cropland, sparse vegetation and mosaic vegetation, respectively, are selected to explore their changes. Harmonic analysis for the whole Saudi Arabia with the EVI as the vegetation indicators is presented in Fig. 1b, with the same regions shown in Fig. 1a is also selected. The findings suggest that the different regions

Table 1 Vegetation areas for each category in SA

	Vegetation type	Area (%)
Cropland	Post-flooding or irrigated croplands	0.94
	Rainfed croplands	10.15
Mosaic vegetation	Mosaic cropland (50–70%)/vegetation (20–50%)	8.81
	Mosaic vegetation (50–70%)/cropland (20–50%)	5.00
Forest	Open (15–40%) broadleaved deciduous forest (>5 m)	0.05
	Mosaic forest-shrubland (50–70%)/grassland (20–50%)	0.09
Sparse vegetation	Close to open (>15%) shrubland (<5 m)	7.46
	Sparse (>15%) vegetation (woody, shrubs, grassland)	67.49
	Closed to open (>15%) vegetation (grassland, shrubland, woody vegetation) on regularly flooded or waterlogged soil—fresh, brackish, or saline water	0.01

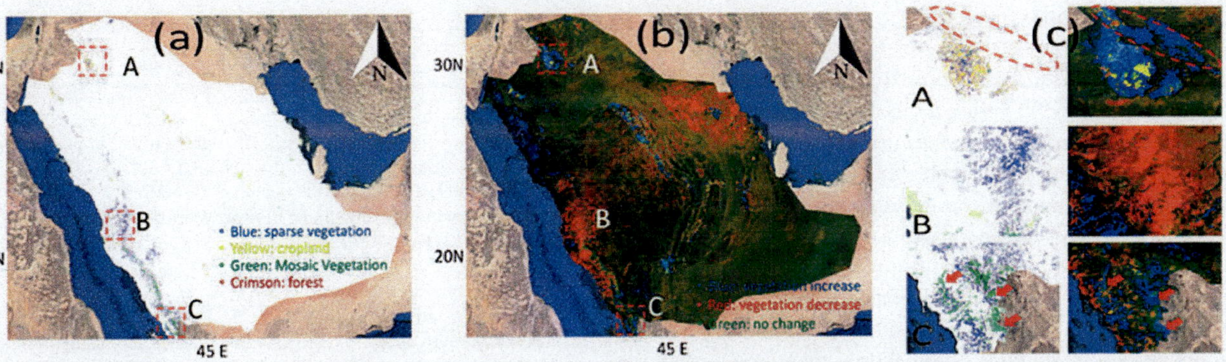

Fig. 1 **a** Vegetation distribution of Saudi Arabia in 2009, **b** EVI Trend of Saudi Arabia (in color red: decrease, blue: increase, green: no change), **c** the marked regions in (**a**) and (**b**)

within SA changes differently: regions such as coasts along northern Arabian Gulf and southern Red Sea show big area of decrease. Areas representing increase of EVI are mostly the agriculture region in the northern SA and areas near the cities. The major region of Saudi Arabia shows no change since it is a permanent dessert. Figure 1c shows vegetation changes: the increase of vegetation in the circled area near the cropland fields indicates new croplands that were reclaimed after 2009 (Fig. 1c-A). Figure 1c-B presents the overall declining trend of sparse vegetation in the mountain area. Both the increase and decrease are also shown in Fig. 1c-C, with most increasing region of mosaic vegetation, and decreasing region of sparse vegetation.

In Fig. 2a, it is found that SA's five major vegetation types, namely 'sparse vegetation,' 'rainfed croplands,' 'mosaic cropland,' 'open shrubland,' and 'mosaic vegetation,' receive most precipitation in March, April, and August. The surface soil moisture and EVI in open shrubland is most reactive to the impact of precipitation, which could be proven by the blooms in April–May, and August–September time windows. The lag correlation coefficients between SA's GIMMS NDVI standardized monthly anomalies and multiple ENSO indices are presented in Fig. 2b. For most indices (ONI, Nino 3, Nino 4 and Nino 3.4), the value is lower than 0.2, indicating weak to no relationship. Comparably, the TNI index shows a growing trend with more months till around 0.3 at lag of 11 months, which is different from the rest of ENSO indices.

4 Discussion

According to research (Deil 1998), wild vegetation of SA is divided into five broad categories: (1) vegetation of the coastal plains and Sabkhas; (2) deserts and scarcely vegetated areas; (3) dwarf shrublands; (4) woodlands and xeromorphic shrublands of high-altitude areas and (5) wadi communities, this not including the cropland. Our study shows the reclamation's impact on the vegetation changes in the SA region. Additionally, El Niño events with negative/positive TNI are classified as a "central Pacific El Niño" and "eastern Pacific El Niño," respectively. Therefore, Fig. 2 indicates the positive favor impact of eastern El Nino event to the vegetation growth in SA which needs further investigation. It is also speculated that the open shrubland is more sensitive to the possible precipitation increases altered by ENSO events for its impact on soil moisture.

Fig. 2 **a** Monthly surface soil moisture, EVI, and precipitation for the five types of vegetation in SA, **b** lag correlation from 0 to 12 months for NDVI and ENSO indices over SA

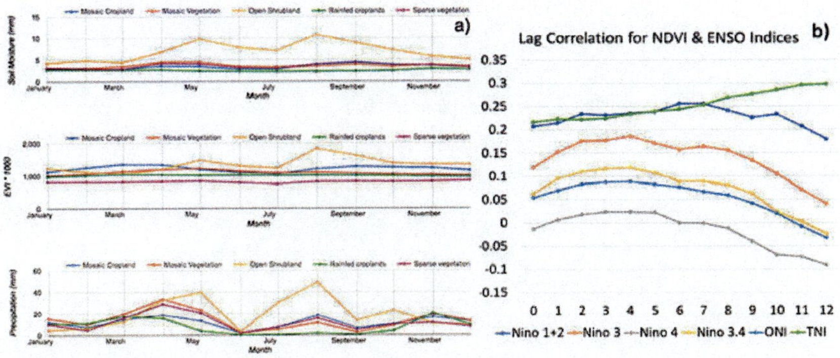

5 Conclusions

In this study, with the support from multi-sensor and harmonic modeling, the drying trend was assessed over the past 20 years for Saudi Arabia areas using EVI. The EVI images exhibited varying spatial patterns of vegetation conditions with varying magnitudes in the different areas. Overall, the vegetation in SA seemed to have a declining trend in most of their habitats, with the exception of regions with cropland reclamation. Most of the wild sparse vegetation areas (such as grassland and shrubland) suffered decreasing in the southwestern mountains. It is worth mentioning that the correlation between NDVI and TNI indicated a positive impact of eastern El Nino events to the growth of vegetation, which could be affected through the precipitation increase in strength or frequencies. The findings suggested that surface soil moisture, in relation to precipitation, varied among different vegetation types. It showed open shrubland to be the most sensitive to precipitation changes, indicating its response to the ENSO-related precipitation anomalies.

References

Deil, U.: Coastal and Sabkha vegetation. In: Ghazanfar, S.A., Fisher, M. (eds.) Vegetation of the Arabian Peninsula, pp. 209–228. Springer, Dordrecht (1998)

Didan: MOD13A1 MODIS/Terra Vegetation Indices 16-Day L3 Global 500m SIN Grid V006 (2015)

Entekhabi, D., Njoku, E.G., O'Neill, P.E., et al.: The soil moisture active passive (SMAP) mission. Proc. IEEE **98**, 704–716 (2010)

Feng, S., Fu, Q.: Expansion of global drylands under a warming climate. Atmospheric Chem. Phys. **13**, 10081–10094 (2013)

Funk, C., Peterson, P., Landsfeld, M., Pedreros, D., Verdin, J., Shukla, S., Husak, G., Rowland, J., Harrison, L., Hoell, A., Michaelsen, J.: The climate hazards infrared precipitation with stations—a new environmental record for monitoring extremes. Sci. Data **2**, 150066 (2015)

GlobCover: Global Land Cover Map. Available online: https://explorer.earthengine.google.com/#detail/ESA%2FGLOBCOVER_L4_200901_200912_V2_3. Accessed on 5 Dec 2018

Hulme, M.: Recent climatic change in the World's drylands. Geophys. Res. Lett. **23**, 61–64 (1996)

Kerr, Y.H.: Soil Moisture and Ocean Salinity SMOS. The European Space Agency, Paris, France (1998)

Kim, S.-R., Prasad, A.K., El-Askary, H., et al.: Application of the Savitzky-Golay filter to land cover classification using temporal MODIS vegetation indices. Photogramm. Eng. Remote. Sens. **80**, 675–685 (2014)

Li, W., El-Askary, H.M., Qurban, M., et al.: On the drying trends over the MENA countries using harmonic analysis of the enhanced vegetation index. In: El-Askary, H.M., Lee, S., Heggy, E., Pradhan, B. (eds.) Advances in Remote Sensing and Geo Informatics Applications, pp. 243–245. Springer International Publishing, Cham (2019)

Liu, H., Huete, A.: A feedback based modification of the NDVI to minimize canopy background and atmospheric noise. IEEE Trans. Geosci. Remote Sens. **33**, 457–465 (1995)

Narisma, G.T., Foley, J.A., Licker, R., Ramankutty, N.: Abrupt changes in rainfall during the twentieth century. Geophys. Res. Lett. **34** (2007)

Reed, S.C., Coe, K.K., Sparks, J.P., et al.: Changes to dryland rainfall result in rapid moss mortality and altered soil fertility. Nat. Clim. Change **2**, 752–755 (2012)

Sazib, N., Mladenova, I., Bolten, J.: Leveraging the Google Earth Engine for drought assessment using global soil moisture data. Remote Sens. **10**, 1265 (2018)

Whitney, K., Scudiero, E., El-Askary, H.M., et al.: Validating the use of MODIS time series for salinity assessment over agricultural soils in California, USA. Ecol. Ind. **93**, 889–898 (2018)

Urban Forest Tree Classification Using UAV-Based High-Resolution Imagery

Mojdeh Miraki and Zahra Azizi

Abstract

Unmanned aerial vehicle (UAV) remote sensing has a high potential for vegetation monitoring in complex urban landscapes. Acquiring information about tree species composition is needed for urban forest management but the field survey of these areas is time-consuming and costly. The goal of this research was to explore the ability of UAV-based RGB imagery for species classification using RGB-based vegetation indices and linear discriminant analysis. Five distinct species including two conifers and three broadleaves were selected in the study area, and the LDA algorithm was applied on raw bands, vegetation indices, and band ratios. The results show a higher accuracy for classification of conifer trees (especially *Cupressus arizonica* with user's accuracy of 0.85) rather than broadleaf species. The highest model accuracy was obtained mainly based on the red band, and the overall accuracy for LDA classification was 0.69.

Keywords

Urban forest • UAV • RGB image • Classification • Linear discriminant analysis (LDA)

1 Introduction

An urban forest inventory is often needed for planning and management purposes. However, urban forestry databases are incomplete and infrequently updated because of cost and time constrains. Traditional field measurements can provide accurate information on cost, time, and labor required for extensive field campaigns to prevent their use for biomass/carbon wall-to-wall mapping in large areas. To solve such problems, the use of remotely sensed data has emerged as a relatively cheap, fast and practical method for inventory, mapping AGC, providing information from local to global scales (Safari et al. 2017). Tree species recognition and geospatial information about tree species composition is essential for management and the estimation of the woody biomass and growing stock as well as the estimation of the monetary value of the forest (Tuominen et al. 2018). Tree species classification using multi- or hyperspectral imagery has been widely studied, and the data has mainly been captured from manned aircrafts or satellites (Nevalainen et al. 2017). The low spatial resolution of satellite images makes the detection and identification of ground features very difficult resulting in a great challenge (Gomroki et al. 2017). However, UAV imagery of very high spatial resolution is very promising for the discrimination of forest species (Lisein et al. 2015). It can produce high-resolution data at a relatively low cost (Hernandez-Santin et al. 2019). UAVs equipped with multispectral sensors provide high spectral resolution images; nevertheless, the multi- or hyperspectral sensors are more expensive than the UAV itself. For this reason, most UAVs use low-cost RGB cameras.

The main purpose of this study was to evaluate the applicability of raw bands and RGB-based vegetation indices for tree species classification in urban forests. Also, the applicability of this approach was examined for conifers and broadleaves species.

2 Materials and Methods

The studied area is Fateh Garden in Karaj, Alborz Province in Iran. The total area of the urban forest of Fateh is 24 ha at 1330 m above sea level with an east longitude of

M. Miraki
Young Researchers and Elites Club, Sanandaj Branch Islamic Azad University, Sanandaj, Iran

Z. Azizi (✉)
Department of Remote Sensing and GIS, Faculty of Natural Resources and Environment, Islamic Azad University, Science and Research Branch, 1477893855 Tehran, Iran
e-mail: zazizi@srbiau.ac.ir

H. El-Askary et al. (eds.), *Research Developments in Geotechnics, Geo-Informatics and Remote Sensing*, Advances in Science, Technology & Innovation,
https://doi.org/10.1007/978-3-030-72896-0_83

50° 58′ 19″ and north latitude of 35° 49′ 45″. Its climate is semi-arid with relatively cold winters and relatively mild summer temperatures with mean annual precipitation and temperature of 247.3 mm and 14.4 °C, respectively. In this area, the dominant species are deciduous broadleaved species and a small number of evergreen trees.

In this study, a DJI Phantom 4 pro with a 1/2.3 CMOS sensor that can capture red–green–blue (RGB) spectral information was used for imagery. A flight was conducted on May 22, 2018, at flying altitudes of 127 m above ground, ground sampling distance of 3 cm, and side and forward overlap of 80%. We collected a total of 492 raw images for the present study, Then the UAV images were processed using Agisoft PhotoScan v1.4.2 (Sadeghi and Sohrabi 2018; Kuzmin et al. 2017) to make an orthomosaic. The orthomosaic photo is an image that consists of multiple overhead images and is corrected for perspective and scale (Onishi and Ise 2018). For georeferencing and enhancing accuracy, we used 15 ground-control points (GCPs) that were collected by a GPS-RTK (Real-Time Kinematic). The field survey focuses on mature overstory trees which were clearly identifiable on the orthophoto mosaic. Based on the field survey conducted in May 2018, a total of 259 individuals of two conifer species (*Cupressus arizonica*, *Pinus eldarica*) and three broadleaved species (*Populous deltoideus*, *Juglans regia*, *Platanus orientalis*) were identified and mapped. The surveyed variables included species identification and position. The latter was measured for each tree using the azimuth–distance approach and laser distance meter as well as Suunto compass.

Individual tree crowns were manually delineated, and the spectral response was summarized at the scale of individual tree crowns by computing descriptive statistics from the orthophoto mosaics and denoted as metrics (Lisein et al. 2015). Using zonal statistics, mean, standard deviation, variance, unique, range, mode, and median statistics were calculated for raw bands (red, green, blue), vegetation indices (NRB, NGB), and band ratios (G/R, R/B) from the RGB orthophotomosaics. Finally, the LDA algorithm was applied in the mass package in R environment (3.6.0) for classification.

3 Results

Conifer species (*C. arizonica* and *P. eldarica*) were discriminated better than the broadleaved. Among the broadleaved species, the worst result was for *P. deltoideus*. *P. orientalis* was discriminated from conifers and broadleaved trees with the highest accuracy among the species. The overall accuracy of LDA-based classification of species was 69% (Fig. 1; Table 1).

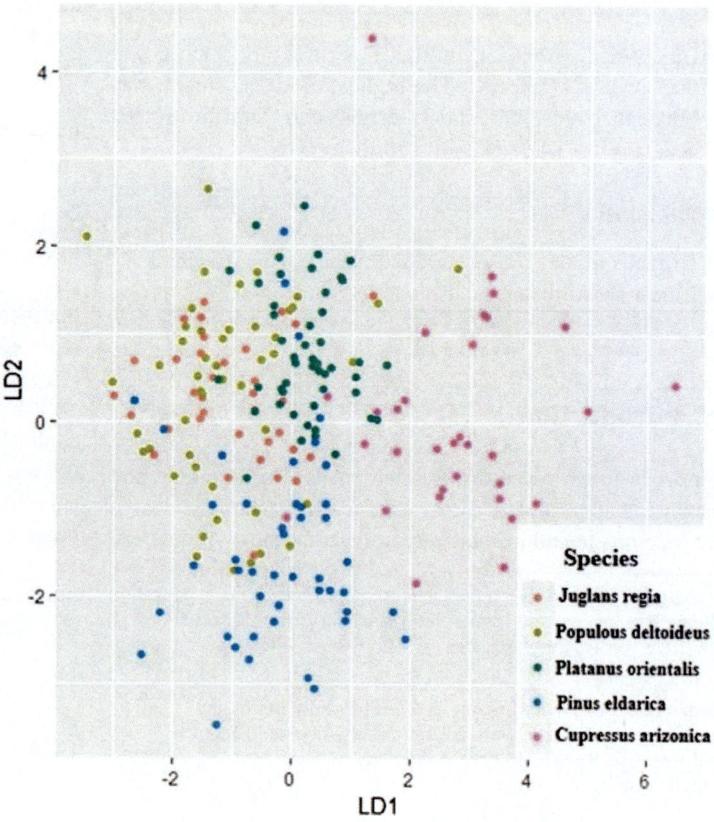

Fig. 1 Plot of tree species discrimination using LDA based on raw bands

Table 1 Confusion matrix of species classification using linear discriminant analysis

	Juglans regia	*Populous deltoideus*	*Platanus orientalis*	*Pinus eldarica*	*Cupressus arizonica*	User's accuracy
Juglans regia	28	7	4	4	0	0.65
Populous deltoideus	11	30	6	10	2	0.51
Platanus orientalis	5	3	44	4	5	0.72
Pinus edarica	2	4	8	43	0	0.75
Cupressus arizonica	0	1	3	2	33	0.85
Producer's accuracy	0.6	0.66	0.68	0.68	0.82	OA = 0.69

4 Discussion

This study investigated the ability of UAV-based photogrammetry for tree species classification. Compared to other studies, our classification results of urban forest trees is very promising. We found that conifers can be classified more accurately than broadleaves. Maschler et al. (2018) reported a similar result in which conifer trees were classified more precisely than the broadleaved ones. Also, *C. arizonica* trees had the best result of classification. This was probably due to its crown color which is quite different from the other species. Classifying different species based on RGB UAV imagery, Sadeghi and Sohrabi (2018) showed that *Populus caspica* were distinguished easily from other species due to their color. Kuzmin et al. (2017) showed that the birch segments could be separated very well based on their smaller greenness, which was expected as they did not have leaves and more elongated shape. Texture features formed a majority of the predictor variables in most of our models and contributed especially to the separation of pine and spruce, although they were still difficult to distinguish. But in the case of broadleaved trees, crown shape and spectral mixture reduced its better identification. Tree shapes are thought to be different, and leaf colors and illuminations are also different depending on season and weather. Making good use of tree shapes and seasonality of leaf colors will improve the classification accuracy. The shape and color of the studied species are similar despite the very high resolution of the aerial images; in some cases, even the visual discrimination of the species is difficult. In this case, it is better to use phenology to distinguish broadleaf species that have the same color and shape. Lisein et al. (2015), for instance, classified Birches, poplars, ash, and English oak using their phenology. In general, the classification results were acceptable in this study (OA = 0.69). This result means that our method has the potential to classify individual trees at a cost-effective way. This can be a useful tool for many forest researchers and management. Besides the data, the right choice of the classification method is also decisive for successful land use and land cover mapping. The LDA method proved to be a good classifier for the tree species recognition task, in the presence of a small sample dataset. In this regard, other studies (Kuzmin et al. 2017; Feret and Asner 2013; Bandos et al. 2009) have stated that the LDA method has acceptable results for tree classification.

5 Conclusions

The results indicated that low-cost UAV imagery is a suitable option for surveying and mapping tree species of urban forests. Also, the accuracy of species classification based on raw bands was higher than vegetation indices and band ratios.

References

Bandos, T.V.V., Bruzzone, L., Camps-Valls, G.: Classification of hyperspectral images with regularized linear discriminant analysis. IEEE Trans. Geosci. Remote Sens. **47**(3), 862–873 (2009)

Feret, J.B., Asner, G.P.: Tree species discrimination in tropical forests using airborne imaging spectroscopy. IEEE Trans. Geosci. Remote Sens. **51**(1), 73–84 (2013)

Gomroki, M., Jafari, M., Sadeghian, S., Azizi, Z.: Application of intelligent interpolation methods for DTM generation of forest areas based on LiDAR data. PFG J. Photogramm. Remote Sens. Geoinf. Sci. [Internet] [cited 2019 Aug 31] **85**(4), 227–241 (2017). Available from: https://doi.org/10.1007/s41064-017-0025-0

Hernandez-Santin, L., Rudge, M., Bartolo, R., Erskine, P., Hernandez-Santin, L., Rudge, M.L., et al.: Identifying species and monitoring understorey from UAS-derived data: a literature review and future directions. Drones [Internet] [cited 2019 Mar 1] **3**(1), 9 (2019). Available from: http://www.mdpi.com/2504-446X/3/1/9

Kuzmin, A., Korhonen, L., Manninen, T., Maltamo, M.: Automatic segment-level tree species recognition using high resolution aerial winter imagery **7254**, 238–259 (2017)

Lisein, J., Michez, A., Claessens, H., Lejeune, P.: Discrimination of deciduous tree species from time series of unmanned aerial system imagery. PLoS One [Internet] [cited 2019 Feb 23] **10**(11), e0141006 (2015) (Cristani, M. (eds)). Available from: https://doi.org/10.1371/journal.pone.0141006

Maschler, J., Atzberger, C., Immitzer, M., Maschler, J., Atzberger, C., Immitzer, M.: Individual tree crown segmentation and classification of 13 tree species using airborne hyperspectral data. Remote Sens. [Internet] [cited 2018 Dec 10] **10**(8), 1218 (2018). Available from: http://www.mdpi.com/2072-4292/10/8/1218

Nevalainen, A., Nilton, N., Antonio, M.G.: Individual tree detection and classification with UAV-based photogrammetric point clouds and hyperspectral imaging. Remote Sens. **9**(3) (2017)

Onishi, M., Ise, T.: Automatic classification of trees using a UAV onboard camera and deep learning, 2018 Apr 27 [cited 2019 Aug 31]. Available from: http://arxiv.org/abs/1804.10390

Sadeghi, S., Sohrabi, H.: Tree species discrimination using RGB vegetation indices derived from UAV images. In: UAV Small Unmanned Aerial System for Environmental Research, 6th edn, p. 5 (2018)

Safari, A., Sohrabi, H., Powell, S., Shataee, S.A.: comparative assessment of multi-temporal Landsat 8 and machine learning algorithms for estimating aboveground carbon stock in coppice oak forests. Int. J. Remote Sens. [Internet] [cited 2019 Aug 31] **38**(22), 6407–6432 (2017). Available from: https://doi.org/10.1080/01431161.2017.1356488

Tuominen, S., Näsi, R., Honkavaara, E., Balazs, A., Hakala, T., Viljanen, N., et al.: Assessment of classifiers and remote sensing features of hyperspectral imagery and stereo-photogrammetric point clouds for recognition of tree species in a forest area of high species diversity. Remote Sens. [Internet] [cited 2019 Feb 23] **10**(5), 714 (2018). Available from: http://www.mdpi.com/2072-4292/10/5/714

Land Use/Land Change Detection of the Past 30 Years (1988–2018), Using Landsat Imagery—Around the Junction Between River Nile and River Atbara (Sudan)

Ekhlas H. Ahmed, Wenbo Xu, and Basheer A. Elubid

Abstract

This study aimed to detect land use/land change around the junction area of the River Nile and River Atbara over the past 30-year period, within fifteen-year intervals (1988–2003 and 2003–2018). The remote sensing (RS) and geographic information system (GIS) are useful techniques to assess this great worldwide event. Landsat Thematic Mapper (TM)/of LT05 1988/12/07, Landsat Enhanced Thematic Mapper Plus (ETM+) of LE07 2003/12/17 and Operational Land Imager (OLI) of LC08 2018/10/31 were used for image classifications. However, the supervised classification has been used to perform powerful mapping of land use/land change classification and to categorize the study area in different land cover features. Five land use/land change classes were specified: Vegetation (cultivated land), waterbody, bare ground, built-up, and rocky areas. The final output maps have been produced using ArcGIS 10.5. The change detection analysis results proved that the vegetation of cultivated land had increased gradually during the mentioned period, i.e., in 1988, it was over 135.91 km^2 (6.25%), in 2003 169.78 km^2 (7.81%), and at 2018 has become 447.46 km^2 (20.58%). Also, the built-up areas have grown up as (8.76%), (9.88%), and (17.24%) in the years of 1988, 2003, and 2018, respectively. The land use/land change accuracy has been evaluated by using Error Matrix (ERRMAT) operation in TerrSet Geospatial Monitoring and Modeling System software, in which the results of Kappa Index of Agreement (KIA) were almost in perfect agreements.

Keywords

GIS • River Nile • Land use/Land change • River Atbara

E. H. Ahmed (✉) · W. Xu
School of Resources and Environment, University of Electronic Science and Technology of China, Chengdu, 611731, Sichuan, China

E. H. Ahmed
Ministry of Minerals, Geological Research Authority of Sudan, Khartoum, Sudan

B. A. Elubid
Department of Environmental Science and Engineering, Faculty of Geosciences and Environmental Engineering, Southwest Jiaotong University, High-Tech Zone, Chengdu, 611756, China

B. A. Elubid
Department of Hydrogeology, Faculty of Petroleum & Minerals, Al Neelain University, Khartoum, 11121, Sudan

1 Introduction

The land changes, due to several physical and artificial factors, have become a significant concern for the expansions of land resources (Geist et al. 2006; Jacobson et al. 2001). Mappings of the ground were concerned with land use/land change and continuously registered socioeconomic actions and natural episodes to explain the appearance over the land surface (Abdallah et al. 2019; Hassen and Assen 2018; Lambin and Geist 2008). The study of land use/land change patterns is essential for the planning of the land use and decision making to face the human needs expansions and welfare. However, detecting and monitoring of the land use/land change mapping and storing the output results in statistical formats and attributes data is very useful for attentive organizations of land for various issues. The RS and GIS are the useful techniques to assess and detect the land use/land change worldwide. To acquire a historical background of land use/land change, various satellite images of different acquisition times were used. To reveal land use/land change during the past 30 years from the year 1988 to 2018, we used the supervised classification (maximum likelihood). It distinguishes and locates land cover types that are known a priori through a combination of personal experience, interpretation of aerial photography, map analysis and fieldwork (Ahmad and Quegan 2012). Over the last few decades, agricultural activities and population growth

H. El-Askary et al. (eds.), *Research Developments in Geotechnics, Geo-Informatics and Remote Sensing*, Advances in Science, Technology & Innovation,
https://doi.org/10.1007/978-3-030-72896-0_84

have expanded around the two cities of Atbara and Ad Damar. The main objective of the study was to assess the land use/land change from 1988 to 2018.

2 Materials and Methods

The study area covers the extent around the junction of the River Nile and River Atbara between longitudes (34.22° E and 33.77° E) and latitudes (17.83° N and 17.41° N). It consists of two cities Atbara and Ad Damar (the capital of River Nile State—Sudan), representing the main urban cities of the River Nile State (Ahmed et al. 2019). Landsat Thematic Mapper (TM), of Landsat 5 with acquisition date (1988/12/07), Landsat Enhanced Thematic Mapper Plus (ETM) of Landsat 7 with acquisition date (2003/12/17), and Operational Land Imager (OLI) of Landsat 8 with acquisition date (2018/10/31), covering Path-173 and Row-048, were used for image classifications. However, in this case, we used the standard false-color composite (FCC) band 4, 3, and 2 as red, green, and blue (RGB) for Landsat 5 and 7, while bands 5, 4, and 3 have been used as (RGB), respectively, in Landsat 8. Vegetation looks in shades of red, urban areas are cyan blue, while soils vary from dark to light brown. Image preprocessing techniques, i.e., resampling, georeferencing, cloud removal, and spatial subset, were carried out to gain valuable information and recognize the classes and features.

The supervised classification has been used in order to perform powerful mapping of land use/land change classification and to categorize the study area in different land cover features. The supervised maximum likelihood classification (MLC) is the most widespread method used in remote sensing image classification. To observe the land use/land change due to urban growth and other human activities effects, the change detection analysis was performed using the produced annual thematic maps.

3 Results

The land use–land change of Landsat images dataset during the period of the past 30 years (1988–2018) covers an area of (2174.659 km^2) around the junction of River Nile and River Atbara, was categorized into five classes: (1) Vegetation (mainly cultivated lands), (2) waterbody, (3) bare ground, (4) built-up, and (5) rocky area, the total of 2174.659 km^2 study area (Fig. 1). The distribution of land cover classification pattern for the classified features in different years obtained from the supervised classification is shown in Table 1.

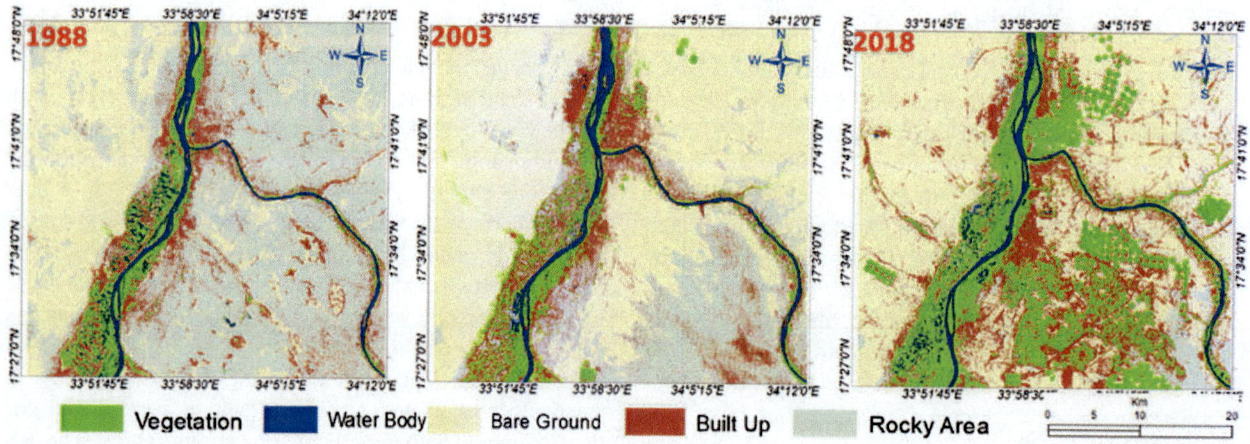

Fig. 1 Land use/land change classes through the years 1988, 2003, and 2018

Table 1 Distribution of land cover classification in the years 1988, 2003, and 2018

No	land use/Land change/Classes	1988		2003		2018	
		Area (km^2)	Area (%)	Area (km^2)	Area (%)	Area (km^2)	Area (%)
1	Vegetation	135.916	6.25	169.782	7.807	447.462	20.576
2	Waterbody	67.974	3.126	61.569	2.831	67.565	3.107
3	Bare ground	629.142	28.931	1233.013	56.699	1144.147	52.613
4	Built-up	190.456	8.758	214.886	9.881	374.883	17.239
5	Rocky area	1151.17	52.936	495.41	22.781	140.603	6.465

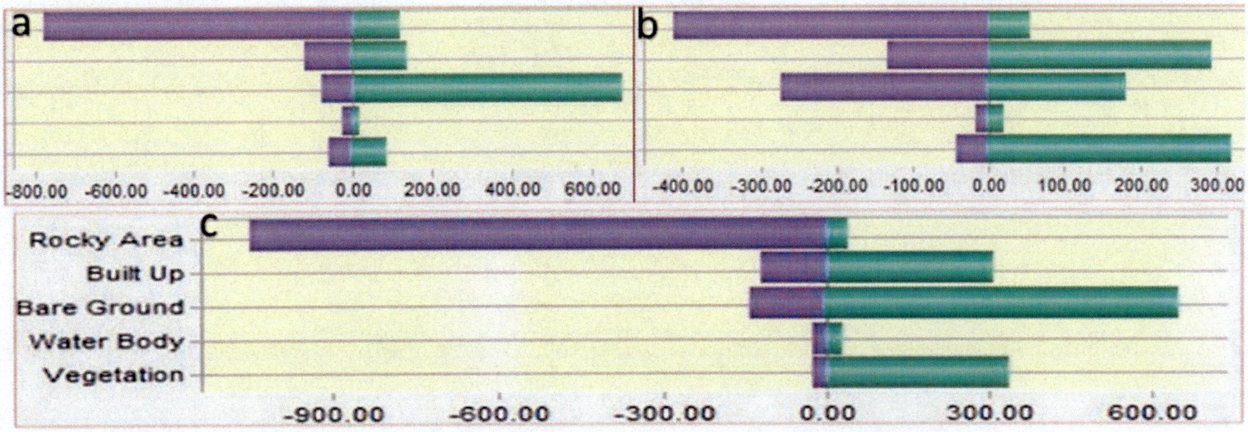

Fig. 2 **a–c** Gains and losses of classified features (km^2) between: 1988–2003, 2003–2018, and 1988–2018, respectively

The land use/land change classified images accuracy has been evaluated using Error Matrix (ERRMAT) operation in TerrSet Geospatial Monitoring and Modeling System, in which the results of Kappa Index of Agreement (KIA) were in 1988 (0.74), 2003 (0.61), and 2018 (0.71).

4 Discussion

The gains and losses between the years 1988, 2003, and 2018 have been represented in Fig. 2. The classified map of 1988 indicated that the area covered by five different classes: (viz., vegetation of agricultural land was about 6.25%, waterbody was 3.13%, bare ground cover 28.93%, built-up area covered 8.76%, and the rocky area covered most part of the study area and taken about 52.94% of the area). Furthermore, in 2003, about 7.81% of the area was covered by agricultural land compared with 6.25% area in 1988 showing an increase of about 2% in cultivation activity. Waterbody and bare ground area covered 2.83% and 56.70%, respectively, while built-up covers 9.88% followed by rocky area of 22.78%. The results interpreted from the classified images of 2018 explain that the bare ground covered more than 50% of the area; this indicates excavations of the rocky area to prepare it for agriculture activity or other issues.

5 Conclusions

This study estimated and observed the changes in land use/land change pattern around the junction area of the River Nile and River Atbara using Landsat TM-5, Landsat-7 ETM +, and Landsat-8 data from 1988 to 2018. This study confirmed that the uses of the RS and GIS approach are one of the most prominent technologies for spatiotemporal analysis. The findings of this study have significant importance for the probable land use/land change practices in the study area.

As far as we know, this study is the first to be conducted about land use/land change in the study area, so it can be recommended as a key point to the decision-makers for sustainable development of the study area.

References

Abdallah, S., Hemdan, S., Ibrahem, K.: Assessment of land use/land cover changes induced by Jizan Dam, Saudi Arabia, and their effect on soil organic carbon. Arab. J. Geosci. **12**, 350 (2019)

Ahmad, A., Quegan, S.: Analysis of maximum likelihood classification on multispectral data. Appl. Math. Sci. **6**, 6425–6436 (2012)

Ahmed, E.H., Xu, W., Elubid, B.A.: Remote sensing and geographic information system with geophysical resistivity in groundwater investigations (West Atbara Basin–River Nile State–Sudan). In: Advances in Sustainable and Environmental Hydrology, Hydrogeology, Hydrochemistry and Water Resources, pp. 229–232. Springer, Berlin (2019)

Geist, H., McConnell, W., Lambin, E.F., Moran, E., Alves, D., Rudel, T.: Causes and trajectories of land-use/cover change. In: Land-Use and Land-Cover Change, pp. 41–70. Springer, Berlin (2006)

Hassen, E.E., Assen, M.: Land use/cover dynamics and its drivers in Gelda catchment, Lake Tana watershed, Ethiopia. Environ. Syst. Res. **6**, 4 (2018)

Jacobson, R.B., Femmer, S.R., McKenney, R.A.: Land-Use Changes and the Physical Habitat of Streams: A Review with Emphasis on Studies Within the US Geological Survey Federal-State Cooperative Program, vol. 1175. US Geological Survey (2001)

Lambin, E.F., Geist, H.J.: Land-Use and Land-Cover Change: Local Processes and Global Impacts. Springer Science & Business Media (2008)

Support Vector Machine for Better Monitoring and Analyzing the City of Sharjah's Land Use/Land Cover Changes

Samy Elmahdy and Tarig A. Ali

Abstract

This study aims to monitor and analyze land use land cover (LULC) change in the Emirate of Sharjah, UAE, using the support vector machine (SVM) and image difference (ID) algorithms. The classification maps were produced by collecting 120 regions of interests (ROI) from RapidEye images with a spatial resolution of 5 m, as well as carrying out field observations, followed by post-classification algorithms such as sieving classes, majority/minority analysis, and clump classes of the produced classification maps. The accuracy assessment showed that the targeted 2000, 2005, 2010 and 2017 classification maps have 89.9%, 91.04%, 93.31% and 97.121% accuracy, respectively. The results showed that the built-up area increased by 12.22 km^2 (1.76%) between 2000 and 2005 and continues to increase even up until the present time. The assessment of changes in the periods 2000–2005 and 2010–2017 confirmed that the net vegetation area increases were less pronounced from 2000 to 2005 than from 2010 to 2017, increasing from 7.26 to 54.69 km^2, respectively. The proposed approach will offer great help to environmental engineers and decision-makers in the development of future environmental issues and investigate the impact of LULC change on groundwater level and quality.

Keywords

Sharjah • SVM • Image difference • Change detection

1 Introduction

Over the last two decades, pixel-based change detection methods, including image difference, image rationing, image regression and change vector analysis have been developed for rapid detection of changes in urban areas using remote sensing data (Huang et al. 2002). Several traditional classifiers, such as maximum likelihood, Algebra change detector, minimum distance and parallelepiped have been developed for remote sensing-based monitoring, mapping and analyzing changes (Dai and Khorram 1997; Huang et al. 2002). Other advanced classification algorithms have been applied to LULC classification and lithological mapping, including the support vector machine (SVM), artificial neural networks, machine learning decision trees, genetic algorithms and support vector machines (Foody and Mathur 2004). The algorithm showed the high performance by classifying heterogeneous landscapes using Landsat and MODIS images with limited quantity and quality of training samples (Mountrakis et al. 2011; Paneque-Gálvez et al. 2013; Senf et al. 2015). Castillo et al. (2008) modified version of the SVM classifier which is called bootstrapped SVM was designed. The integration of a genetic algorithm and SVM was employed to classify LULC and evaluate the results with a limited number of training samples (Ghoggali et al. 2009). Recently, a novel ensemble support vector machine model was applied for land cover classification (Liu and Huang 2019). However, a single method does not have the ability to precisely monitor LULC changes. This study aims to monitor and analyze using SVM and ID algorithms outputs for better monitoring and analyzing the Emirate of Sharjah LULC changes.

2 Study Area

The cities of Sharjah and Ajman are the economic capital of the Emirates of Sharjah and Ajman, and they extend between longitude 54° 59′ to 55° 36′ 42″ E and latitude 24° 42′ 30″

S. Elmahdy (✉)
Geospatial Analysis Center, American University of Sharjah, PO Box 26666 Sharjah, United Arab Emirates
e-mail: selmahdy@aus.edu

T. A. Ali
Civil Engineering Department, American University of Sharjah, PO Box 26666 Sharjah, United Arab Emirates

H. El-Askary et al. (eds.), *Research Developments in Geotechnics, Geo-Informatics and Remote Sensing*, Advances in Science, Technology & Innovation,
https://doi.org/10.1007/978-3-030-72896-0_85

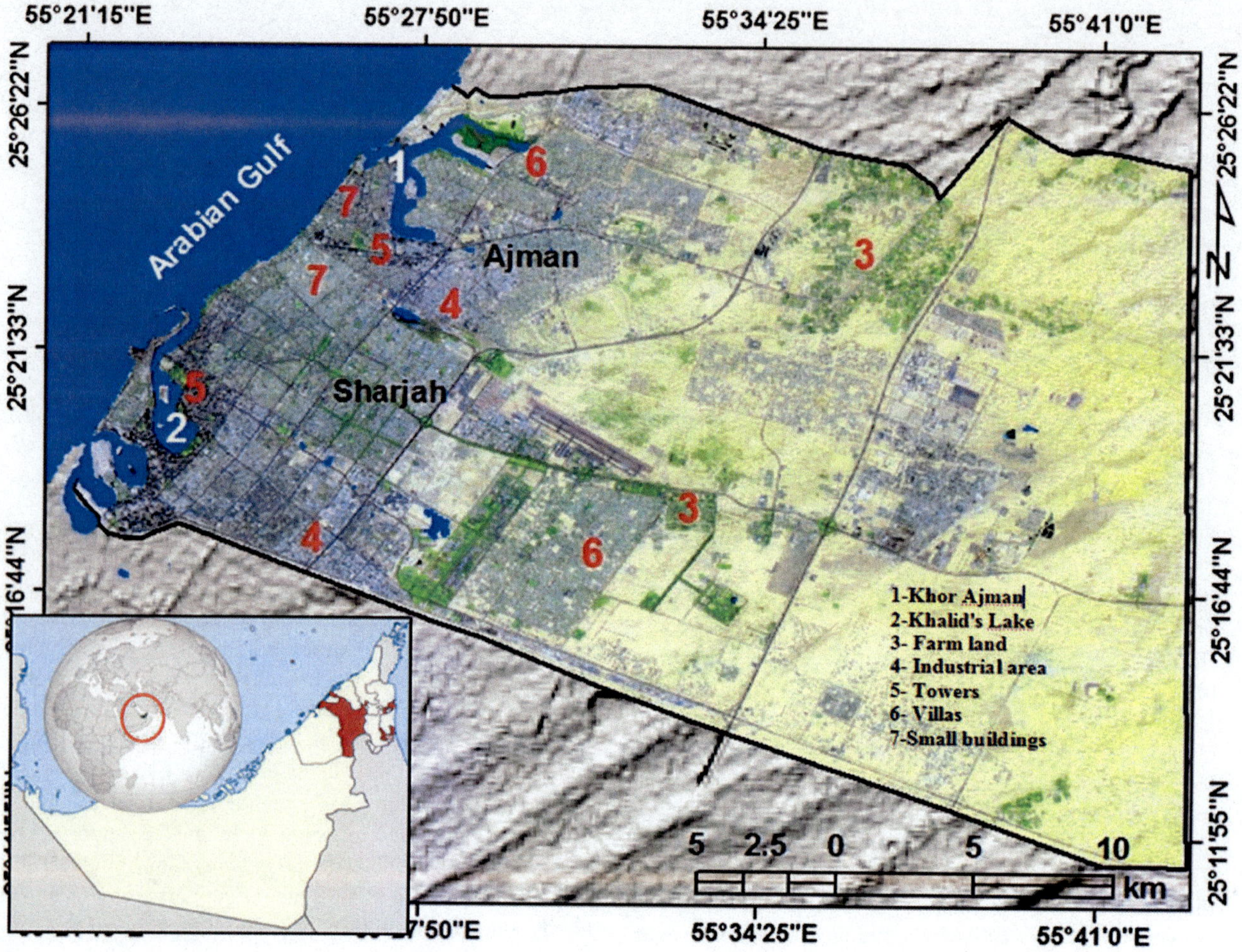

Fig. 1 RGB Landsat image of the cities of Sharjah and Ajman as of 2017

to 55° 16′ 59″ N (Fig. 1). The spatial significance of the city of Sharjah and Ajman is that they are located approximately 2 km from the Emirate of Dubai, the economic center of the UAE, and 170 km from Abu Dhabi, the national capital of the UAE. The urban growth and development of both cities is very closely associated with the Emirate of Dubai, which is one of the most rapidly developed cities in the world. The cities of Sharjah and Ajman are spread over an area of about 693.21 km^2.

3 Data and Methods

Three sets of Landsat images with a spatial resolution of 30 m and a temporal of ~5 years are downloaded from the USGS Earth Resources Observations and Science (EROS) Center through the Global Visualization Viewer (www.glovis.usgs.gov) portal. As a first step, the data were spatially and atmospherically corrected, followed by unifying the spatial and temporal resolutions. After that, an SVM classifier was applied to the 2000, 2005, 2010 and 2017 Landsat images, using 120 end members for LULC classes (30 for each LULC class), which were collected from RapidEye images and verified using field observations from the area in which the authors live. Prior to LULC process, it is important to choose the proper algorithm parameters. The suitable kernel was the RBF kernel and the best values for gamma in kernel function (γ) was 0.007, penalty parameter (C) was 120, pyramid level (P) value of 0 and classification probability threshold value of 0.05 were chosen as the finest SVM options. Then, the SVM algorithm was applied to all four Landsat images and enhanced by applying Sobel filter, Sieving and Clumping the classes using a post-classification tool implemented in Envi. v. 4.5 software. Then, the accuracy of each enhanced LULC map was assisted using a confusion matrix implemented in Envi. v. 4.5 software. Finally, a total area in km^2 of each LULC class was calculated using raster calculator implemented in Arc GIS v.

10 software (Elmahdy and Mohamed 2016). After that, change detection was performed by applying a threshold value which is equally spaced between (−1) and (+1) for simple difference to each pair of LULC maps. Monitoring LULC change was performed using change detection tool implemented in Envi. v. 4.5 software.

4 Results

The maps show that the built-up area is limited to the coastal area on both sides of the Khor of Ajman and around Khaild's Lake, while the cultivated area is distributed within the desert area, such as the sand dunes and desert plains. The maps also show that there was an increase in built-up and vegetation areas for the period from 2000 to 2017 and a sharp decrease in desert plain area (Figs. 2 and 3). The high accuracy assessment results (>87%) confirm that the enhancements applied to SVM and the obtained LULC classification maps have resulted in the improved classification mapping capability of the proposed methods. The classification image of 2017 is the best classified image, and it shows clear classification for each class with an overall accuracy of 97.17%, which is the highest accuracy percentage among the classified images (Table 1). Change detection shows that there was a significant increase (positive change) in built-up area of about 46.79 km^2 (6.74%), vegetation areas of about 61.84 km^2 (8.92%) and a negative change in desert plain area during the period from 2000 to 2017 (Figs. 2, 3 and 4; Table 2).

Unlike the traditional methods (Hamdan et al. 2013; Akinyelu and Adewumi 2017) which are time consuming

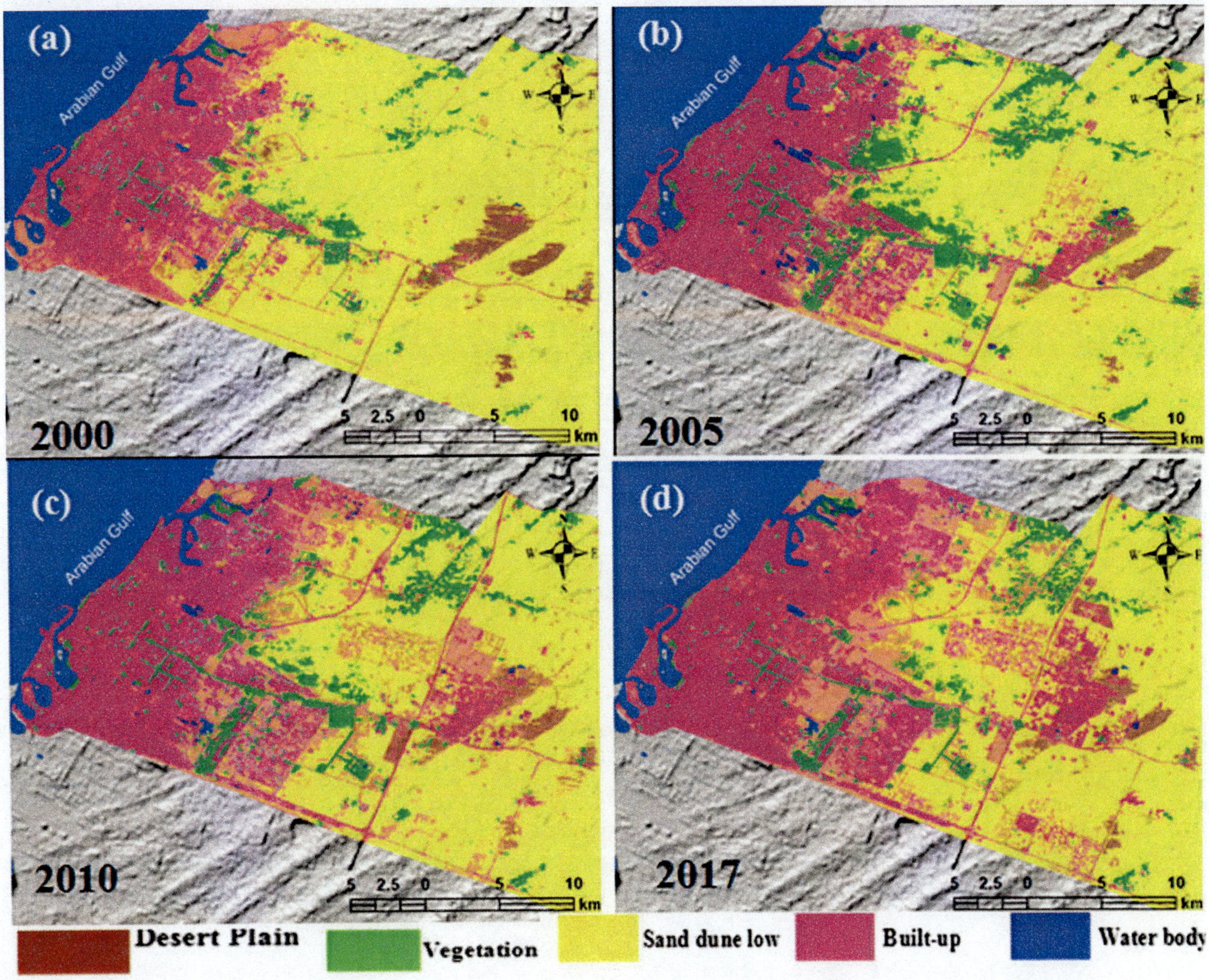

Fig. 2 Classification maps derived from Landsat ETM, ETM+ and OLI images using SVM algorithm for the years **a** 2000, **b** 2005, **c** 2010, and **d** 2017

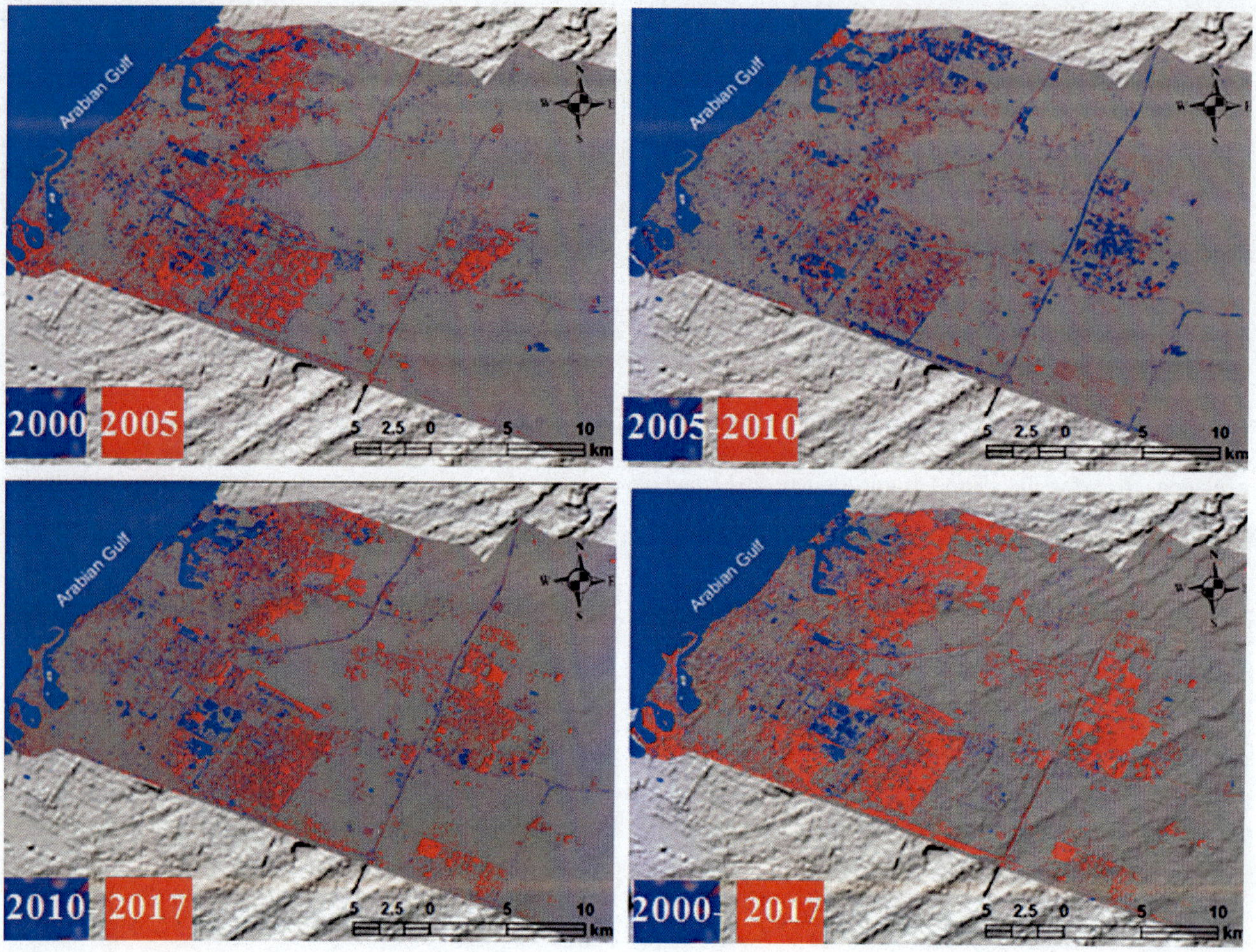

Fig. 3 Change detection maps derived from a pair of classification maps using image difference algorithm for the year intervals **a** 2000–2005, **b** 2005–2010, **c** 2010–2017 and **d** 2000–2017

Table 1 Overall accuracy (%) of the LULC maps pre- and post-enhancements

	2000	2005	2010	2017
Overall accuracy (%)	89.9	91.04	93.31	97.17
Kappa coefficient	0.82	0.84	0.88	0.93

and costly, such as the manual screen digitizing, the SVM and ID algorithms were showed to be powerful tools in mapping and monitoring LULC changes over a regional scale with low cost and in a time effective manner. The use of the multi-temporal Landsat images is useful, especially when the algorithm is trained by samples collected from high resolution images, the SVM algorithm is optimally parameterized, selecting a fixable threshold value and the interpreter has a strong back about the study area.

Total area 693.21 km^2.

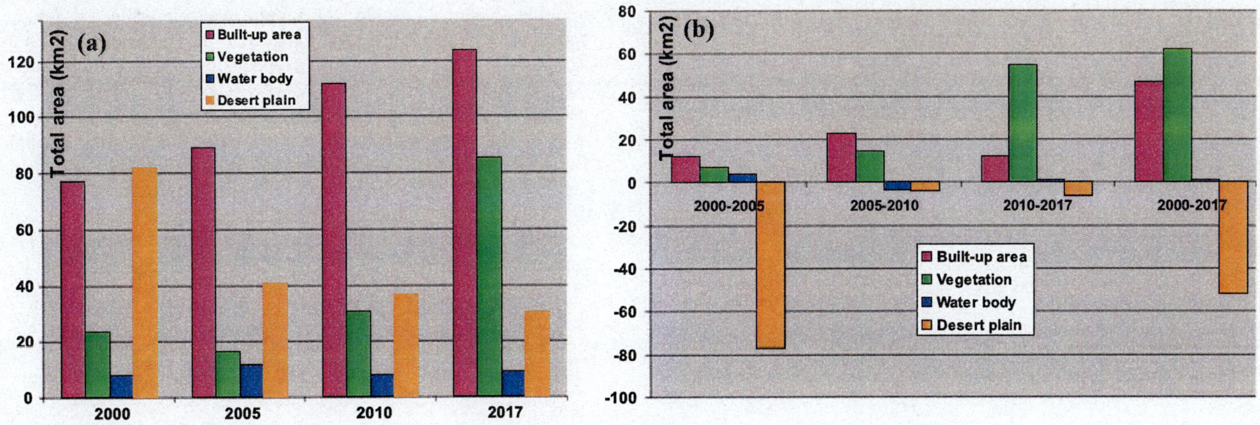

Fig. 4 Total area of each class (**a**) and LU/LC changes from 2000 to 2017 (**b**)

Table 2 Estimates of LULC classes total area and percentage and changes from 2000 to 2017

Classes	2000–2005		2005–2010		2010–2017		2000–2017	
	(km^2)	(%)	(km^2)	(%)	(km^2)	(%)	(km^2)	(%)
Built-up area	12.22	1.76	22.57	3.25	12	1.73	46.79	6.74
Vegetation	7.26	1.04	14.41	2	54.69	7.88	61.84	8.92
Water body	3.67	0.52	−3.692	0.53	0.927	0.13	0.905	0.13
Desert plain	−77.4	11.17	−4	0.57	−6.4	0.92	−51.81	7.47

5 Conclusion

In this study, a modified remote sensing approach for monitoring and analyzing LULC changes based on an integration of SVM and ID algorithms was established. The produced LULC maps were then enhanced by applying 3 × 3 Sobel filter for better discrimination of class boundaries, majority minority analysis for clean dark pixels, sieving classes and clump maps before applying change detection. Change detection was performed by applying a simple difference to each pair of LULC maps. The mapping and monitoring of LULC changes performed using the proposed approach was efficient, and it offered a reliable result, ensuring an overall accuracy of more than 97%. This value is permitting a higher level of confidence to the LULC mapping and monitoring changes obtained. The results showed that the built-up area has increased with about 46.79 km^2 (6.74%), while the vegetated area has increased with about 61.84 (8.92%) from 2000 to 2017. Conversely, the desert plain decreased about 51.81 km^2 (7.47%) during the same period. The proposed approach which uses SVM and Landsat images is timely and economically effective, especially when the SAM trained by a high number of training areas collected from satellite images with a higher spatial resolution and choice of the proper algorithm parameters.

References

Akinyelu, A.A., Adewumi, A.O.: Improved instance selection methods for support vector machine speed optimization. Secur. Commun. Netw. (2017)

Castillo, C., Chollett, I., Klein, E.: Enhanced duckweed detection using bootstrapped SVM classification on medium resolution RGB MODIS imagery. Int. J. Remote Sens. **29**(19), 5595–5604 (2008)

Dai, X., Khorram, S.: Development of a new automated land cover change detection system from remotely sensed imagery based on artificial neural networks. In: Proceedings of the International Geoscience and Remote Sensing Symposium, IGARSS/IEEE, Singapore, 3–8 Aug, pp. 1029–1031 (1997)

Elmahdy, S.I., Mohamed, M.M.: Automatic feature extraction module for change detection in Al Ain, UAE: analysis by means of multi-temporal remote sensing data. J. Indian Soc. Remote Sens. **44** (1), 1–10 (2016)

Foody, G.M., Mathur, A.: A relative evaluation of multiclass image classification by support vector machines. IEEE Trans. Geosci. Remote Sens. **42**(6), 1335–1343 (2004)

Ghoggali, N., Melgani, F., Bazi, Y.: A multiobjective genetic SVM approach for classification problems with limited training samples. IEEE Trans. Geosci. Remote Sens. **47**(6), 1707–1718 (2009)

Hamdan, N., McKnight, P., McKnight, K., Arfstrom, K.M.: The flipped learning model: a white paper based on the literature review titled a review of flipped learning. In: Flipped Learning Network. Pearson, George Mason University (2013)

Huang, C., Davis, L.S., Townshend, J.R.G.: An assessment of support vector machines for land cover classification. Int. J. Remote Sens. **23**(4), 725–749 (2002)

Liu, Y., Huang, L.: A novel ensemble support vector machine model for land cover classification.Int. J. Distrib. Sens. Netw. **15**(4), 1550147719842732 (2019)

Mountrakis, G., Im, J., Ogole, C.: Support vector machines in remote sensing: a review. ISPRS J. Photogramm. Remote Sens. **66**, 247–259 (2011). https://doi.org/10.1016/j.isprsjprs.2010.11.001

Paneque-Gálvez, J., Mas, J.-F., Moré, G., et al.: Enhanced land use/cover classification of heterogeneous tropical landscapes using support vector machines and textural homogeneity. Int. J. Appl. Earth Obs. Geoinf. **23**, 372–383 (2013). https://doi.org/10.1016/j.jag.2012.10.007

Senf, C., Leitão, P.J., Pflugmacher, D., et al.: Mapping land cover in complex Mediterranean landscapes using Landsat: improved classification accuracies from integrating multi-seasonal and synthetic imagery. Remote Sens. Environ. **156**, 527–536 (2015). https://doi.org/10.1016/j.rse.2014.10.018

Assessing the Ecological Quality of the Greater Bay Area (South China) from Landsat Data from 1998 to 2014

Ruiyang Liu, Long Li, Ting Zhang, Longqian Chen, Zhiqiang Wang, and Ziqi Yu

Abstract

The urban ecological quality is important to the living environment of urban residents and the sustainable development of cities. As a new urban ecological quality assessment method, the remote sensing ecological index (RSEI) is based on natural factors, namely greenness (NDVI), wetness (WI), dryness (NDBSI), and heat (LSTI) to quickly assess urban ecological health. We mapped the ecological quality in 1998, 2005, and 2014 of the Guangdong–Hong Kong–Macao Greater Bay Area (GBA) and characterized its spatiotemporal variability. Results show that the ecological quality of the GBA has continuously improved in the past 16 years. The RSEI value increased slightly from 0.6262 in 1998 to 0.6611 in 2005 and then to 0.6951 in 2014. This study provided insights into how ecological quality evolves in this economically important region in China.

Keywords

Ecological quality assessment • RSEI • Greater Bay Area • Landsat

1 Introduction

A good urban ecological quality is fundamental to the living of residents and guarantees the sustainability of socioeconomic development. Among the largest urban agglomerations, the Guangdong–Hong Kong–Macao Greater Bay Area (GBA) (Fig. 1) is situated in South China and consists of the special administrative regions of Hong Kong and Macao, and nine cities of Guangdong province (Guangzhou, Zhuhai, Shenzhen, Zhongshan, Foshan, Zhaoqing, Dongguan, Huizhou, and Jiangmen) (Eddie et al. 2018). It is necessary to understand the ecological quality change over time in this populated and developed region through an objective assessment. The remote sensing ecological index (RSEI) method allows urban ecological quality to be assessed without limitations such as the lack of data and human-induced bias (Xu et al. 2018). The aim of this study was to assess the urban ecological quality of the GBA through the RSEI method and characterize its spatiotemporal variability over the past years.

2 Data and Methods

Landsat images of 1998 (TM) and 2005 (TM), and 2014 (OLI/TIRS) over the GBA were freely downloaded from USGS (https://earthexplorer.usgs.gov/). Due to the large geographic area of the GBA, eight images were merged for each period. To minimize seasonal variability, all the images were acquired in or around October.

According to Xu et al. (2018), the RSEI consists of four components, namely greenness, wetness, dryness, and heat. Normalized difference vegetation index (NDVI) represents the greenness indicator, and wetness indicator (WI) is expressed by the wet component derived from tasseled cap transformation. Dryness indicator (NDBSI, normalized difference built-up soil index) is the average of bare soil index and building index. Heat indicator is represented by surface

R. Liu
School of Environmental Science and Spatial Informatics, China University of Mining and Technology, Xuzhou, 221116, China

L. Li (✉) · T. Zhang (✉) · L. Chen · Z. Wang · Z. Yu
School of Public Policy and Management, China University of Mining and Technology, Xuzhou, 221116, China
e-mail: long.li@cumt.edu.cn

T. Zhang
e-mail: tingzhang@cumt.edu.cn

L. Li
Department of Geography & Earth System Science, Vrije Universiteit Brussel, 1050 Brussels, Belgium

H. El-Askary et al. (eds.), *Research Developments in Geotechnics, Geo-Informatics and Remote Sensing*, Advances in Science, Technology & Innovation,
https://doi.org/10.1007/978-3-030-72896-0_86

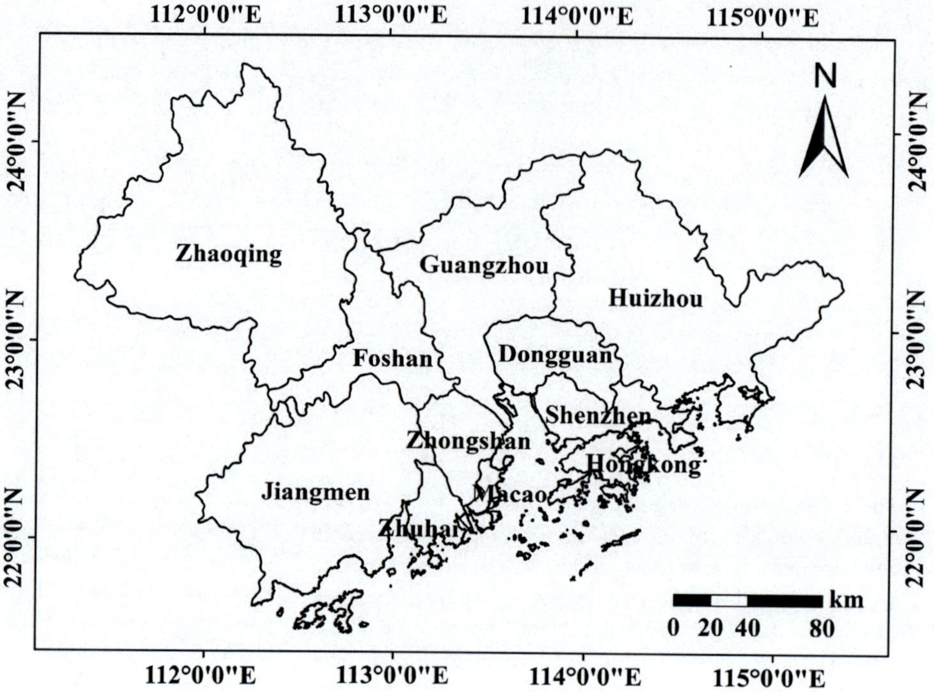

Fig. 1 Study area: the Guangdong–Hong Kong–Macao Greater Bay Area (GBA)

temperature (LSTI, land surface temperature index). The weight for each factor was determined by the principal component analysis (Eq. 1). A high RSEI value indicates a better ecological quality (Musse et al. 2018; Wang et al. 2017).

$$\text{RSEI} = 0.3185 \times \text{NDVI} + 0.2672 \times \text{WI} + 0.2205 \times \text{NDBSI} + 0.1938 \times \text{LSTI} \quad (1)$$

3 Results

Figure 2 shows the RSEI assessment results. The average RSEI of the GBA was 0.6562 in 1998, 0.6611 in 2005, and 0.6951 in 2014. The increase in RSEI from 2005 to 2014 was much larger than that from 1998 to 2005. This suggests that the ecological quality in the GBA improved over the past 16 years.

The RSEI grading maps were created (Fig. 2) through the equal interval method (Yue et al. 2019) (0–0.2 for low, 0.2–04 for medium–low, 0.4–0.6 for medium, 0.6–0.8 for medium–high, and 0.8–1.0 for high). For each period, the GBA had the largest medium–high-level area (>53%), while the low- and medium–low-level areas were very small. The high-level area saw a remarkable increase in 2014, up to 23.61%. This indicates that the ecological quality of the GBA considerably improved during the period.

4 Discussion

From Fig. 2, it is observed that central cities had a worse ecological quality than the surrounding cities. The ecological quality was better in Zhaoqing, Jiangmen, and Huizhou but poorer in Dongguan and Macao.

Through differencing the grading maps, the spatiotemporal evolution of the ecological quality was investigated (Fig. 3). The GBA was dominated by the same level (zero value), followed by a one-level increase and a one-level decrease. The ecological quality of the central cities in the GBA declined from 1998 to 2005 and the surrounding areas improved, but it improved significantly from 2005 to 2014. The continued improvement of the overall ecological quality in GBA is closely related to the "National Forest City Construction" in the region.

5 Conclusions

In this study, we performed a remote sensing-based assessment of the ecological quality in the Greater Bay Area. The ecological quality improved better over 2005–2014 period than over the 1998–2005. It is expected that the urban ecological quality of this large urban agglomeration would continue to be enhanced through national, regional, and local efforts. Further research will focus on the identification of the contributors to such increases.

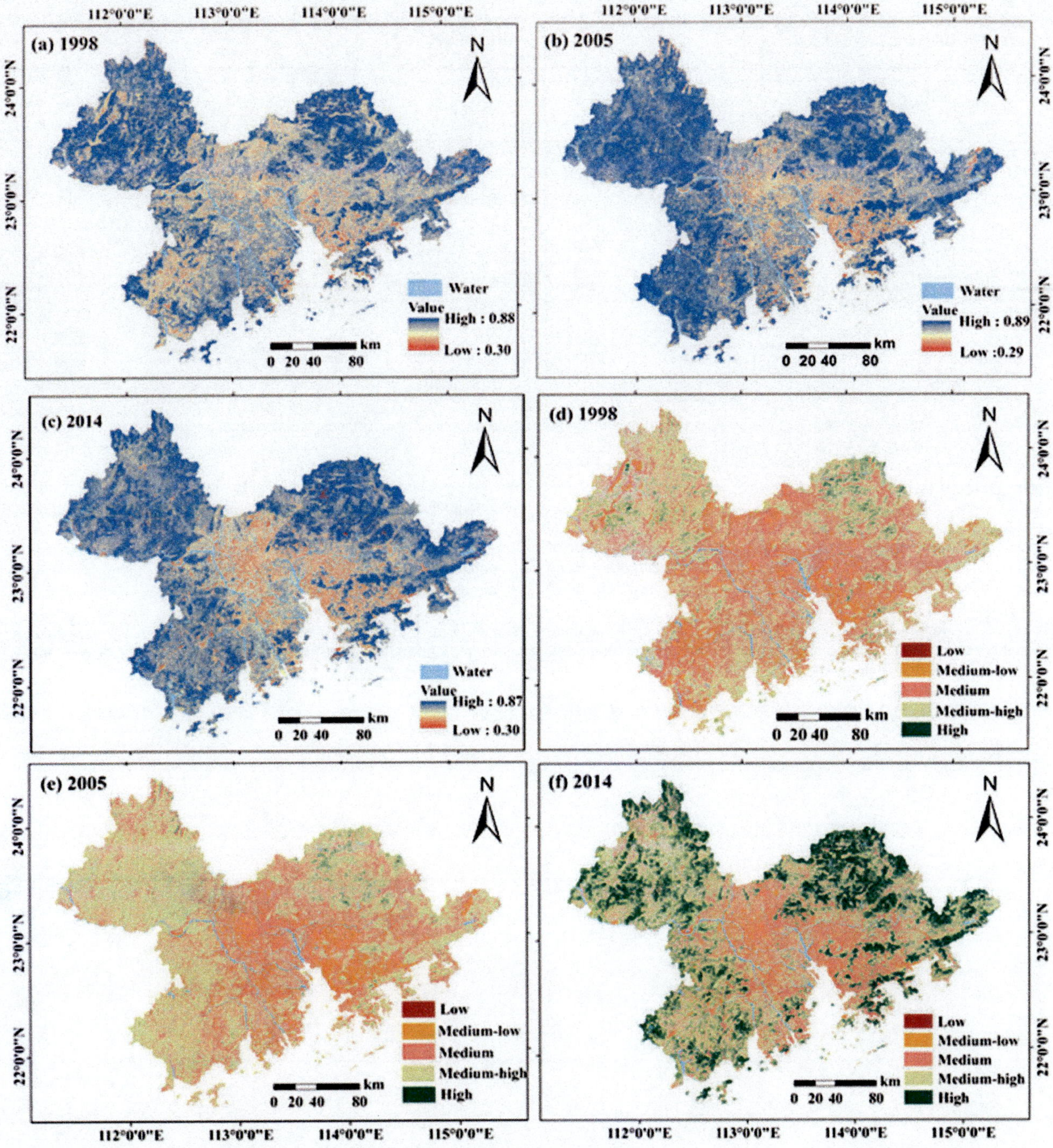

Fig. 2 RSEI maps (**a–c**) and RSEI grading maps (**d–f**) of the GBA in 1998, 2005, and 2014

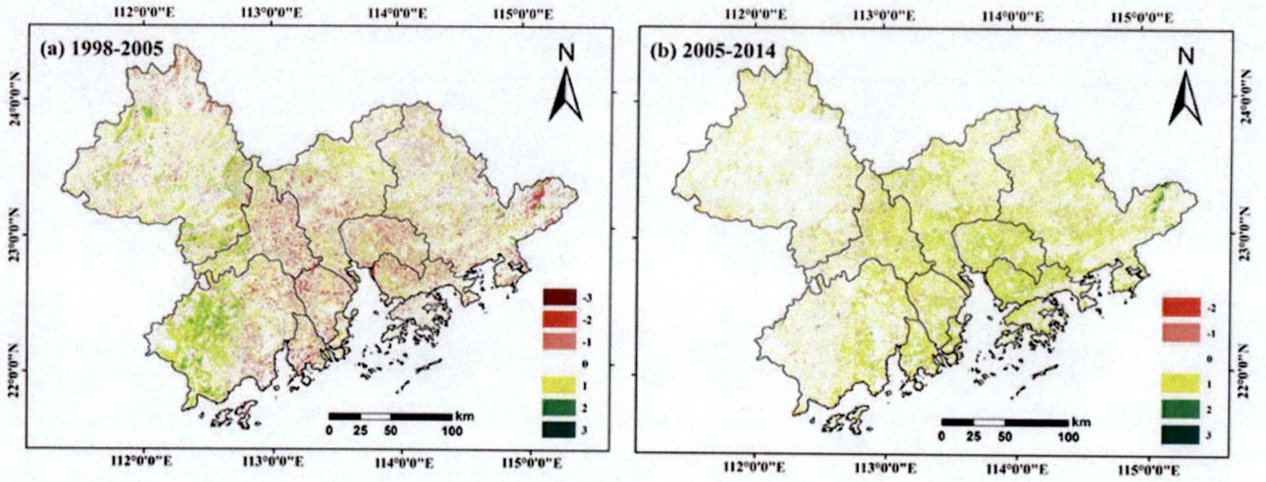

Fig. 3 Differencing analysis of the grading maps (positive values indicate an increased quality and vice versa)

References

Hui, E.C.M., Li, X., Chen, T., et al.: Deciphering the spatial structure of China's megacity region: a new bay area-The Guangdong-Hong Kong-Macao Greater Bay Area in the making. Cities **10**, 1–13 (2018)

Musse, M.A., Barona, D.A., Rodriguez, L.M.S.: Urban environmental quality assessment using remote sensing and census data. Int. J. Appl. Earth Obs. Geoinf. **71**, 95–108 (2018)

Wang, B., Chen, L., Li, L., et al.: Ecological response to land use change a case study from the Chaohu lake basin, China. Bulgarian Chem. Commun. **49**, 200–206 (2017)

Xu, H., Wang, M., Shi, T., et al.: Prediction of ecological effects of potential population and impervious surface increases using a remote sensing based ecological index (RSEI). Ecol. Ind. **93**, 730–740 (2018)

Yue, H., Liu, Y., Li, Y., et al.: Eco-environmental quality assessment in China's 35 major cities based on remote sensing ecological index. IEEE Access **7**, 295–311 (2019)

Aboveground Biomass Prediction Model Using Landsat 8 Data: A Test on Possible Approaches for Seasonally Dry Forests of Northern Ethiopia

Meley Rannestad, Tron Eid, Ole Martin Bollandsås, Terje Gobakken, and Buruh Tetemke

Abstract

Forest biomass is an important vegetation attribute for assessing forest ecosystem productivity and function. Estimation of tropical forest biomass has gained importance due to the significance of tropical forests in the global carbon balance. Using remote sensing has been suggested as a potential aboveground biomass (AGB) estimation approach for its spatial coverage and cost effectiveness. Application of such techniques for seasonally dry tropical forests in developing countries is, however, less explored. To this end, the aim of this study is to test the potential of Landsat 8 data for AGB prediction for dry forests in northern Ethiopia. A protected dry Afromontane forest (exclosure), representing a common vegetation type in the region, is selected. The forest is dominated by early successional species, mainly *Acacia etbaica*. The data necessary for ground-based AGB estimation is collected from 60 circular plots of 200 m^2. This study operates with a Landsat 8 image acquired in December 2017, corresponding to the year of vegetation inventory. After correcting the image for sensor, solar, atmospheric, and topographic effects, seven commonly used vegetation indices are derived from the spectral bands. AGB is significantly correlated to all of the Landsat 8 derived spectral variables. One important finding is a linear relationship between AGB, and all of the spectral variables, which allow the development of a linear model using a normalized difference vegetation index (NDVI) as a predictor. The root mean square error of the values predicted using the model is 4 Mg/ha. Finally, the findings allow to conclude that the remotely sensed data have the potential for predicting AGB of seasonally dry forests.

Keywords

Exclosure • NDVI • Succession

M. Rannestad (✉) · T. Eid · O. M. Bollandsås · T. Gobakken · B. Tetemke
Faculty of Environmental Sciences and Natural Resource Management, Norwegian University of Life Sciences (NMBU), Ås, Norway
e-mail: meley.araya@nmbu.no

B. Tetemke
Department of Land Resources Management and Environmental Protection, Mekelle University, P.O. Box 231 Mek'ele, Ethiopia

1 Background

Biomass, defined as the dry weight of plant material, is an important indicator of forest ecosystem productivity and function. It is a prevalent vegetation characteristic in assessing forest structure and condition, in estimating forest productivity, and in quantifying carbon sequestration and storage potential of forests (Brown 1997). The estimation, mapping, and monitoring of forest biomass have gained increasing attention during the past two decades due to the significance of forests in the global carbon balance. Biomass estimation of tropical forests in particular has become important with the recognition of the global climate change mitigation mechanism known as REDD+. REDD+ stands for reducing emissions from deforestation and degradation plus enhancing forest carbon stocks through sustainable management and conservation of forests in developing countries (UNFCCC 2010). Under this scheme, governments in developing tropical countries would receive compensation based on achieved reductions of forest-based CO_2 emissions or increased sequestration of carbon. As conditionality is a fundamental aspect of this scheme, its implementation greatly depends on reliable monitoring, reporting, and verification (MRV) protocols of carbon sequestration and storage. Many studies indicated that biomass carbon MRV protocols will be based on remote sensing techniques in the future due to their cost effectiveness. Moreover, remote sensing data allows for greater wall-to-wall areal coverage. The application of such techniques is, however,

H. El-Askary et al. (eds.), *Research Developments in Geotechnics, Geo-Informatics and Remote Sensing*, Advances in Science, Technology & Innovation,
https://doi.org/10.1007/978-3-030-72896-0_87

less explored for seasonally dry tropical forests in developing countries. Seasonally dry forests (hereafter referred to, as "dry forests") occur in tropical and subtropical areas, where rainfall is scarce and erratic, the mean annual temperature is higher than 17 °C, and the annual potential evapotranspiration is higher than precipitation for a significant part of the year (Bullock et al. 1995; Murphy 1986). These forests represent a variety of land covers including shrub lands, thickets, open woodlands, and wooded grasslands. Dry forests constitute nearly half of the world's tropical forests (Blackie et al. 2014). They play an essential role in regulating the global climate by serving as carbon reservoirs. The carbon sequestration and belowground biomass carbon storage potential of these forests are particularly high. This is explained by their widespread coverage (FAO 2015), and the fact that they are rarely at their climax (maximum growth) due to overexploitation. In addition, they have the largest root structures compared to other tropical species, which remain intact even during deforestation. Dry forests are also important sources of fuelwood, which is the primary source of energy for the majority of people in the dryland areas.

Biomass can be generally divided into aboveground living biomass (AGB) and belowground living biomass (BGB) (Lu 2016). As BGB is difficult to measure, most attempts on biomass estimation have focused on AGB (Lu 2006), as with this study. Estimation of forest biomass can be based on direct or indirect methods. Direct methods can accurately obtain forest AGB data. However, they are usually time consuming, labor-intensive, and limited in terms of spatial and temporal sampling (Chave 2014; Lu 2006; Seidel 2011; Wang 2011). Remote sensing is one of the popular techniques of the indirect method. It alternatively enables the estimation of forest biomass at multiple scales with large spatial and temporal coverage. Remote sensing generally refers to the indirect measurement of either emitted, reflected, or scattered electromagnetic energy using sensors mounted on different platforms (Ahamed 2011; Joshi 2016). The three broad categories of remote sensing platforms are ground-based (e.g., handheld devices, tripods, towers, and cranes), airborne (e.g., airplanes, and drones), and space-borne (satellite). The features of forests can be indirectly estimated using the airborne or space-borne multispectral remote sensing method (Ahamed 2011). On the one hand, airborne remote sensing data, such as aerial photographs, provide very high spatial resolution images compared to space-borne sensors (Ahamed 2011; Lu 2006). However, their area coverage is very low whereas the cost per unit area of ground coverage is very high. Space-borne sensors, on the other hand, offer unique advantages with their capability to capture large areas in a single image regularly, enabling monitoring and change detection (Ahamed 2011; Gleason 2011). Optical (spectral) and radar data are the two main types of satellite data used in remote sensing (Gleason 2011; Goetz 2009). Optical remote sensing data, with a variety of spatial and temporal resolutions, have been widely applied for forest biomass estimation due to the length of consistent datasets, spatial coverage, and the ease of availability (Cao 2018; Joshi 2016). Landsat has in particular been extensively adopted for forest mapping, monitoring, and biomass assessment (Gleason 2011). The aim of the present study is to test the potential of Landsat 8 data for AGB prediction in a dry Afromontane forest in northern Ethiopia.

2 Materials and Methods

A protected forest, an exclosure, namely Tikul, was selected from the Tigray region in northern Ethiopia. The region is located between 12° and 15° N and 36° 30′–40° 30′ E and covers about 5 million ha. The region's climate zone based on elevation is categorized as 39% semi-arid (<1500 m), 49% subtropical (1500–2300 m), and 12% temperate (>2300 m) (Birhane 2017). The study site represents the subtropical climate zone. The annual minimum and maximum temperatures of the study site are 10 and 28°C, respectively. The total annual rainfall ranges between 450 and 980 mm. Exclosures refer to areas strictly protected from human and animal interference using locally hired guards. The main purpose of establishing exclosures is to promote natural regeneration of plants on degraded communal grazing lands and thus to reclaim the degraded environment. Hence, the vegetation in most of the exclosures is dominated by early successional species such as *A. etbaica* and *Euclea schimperi*.

The data necessary for generating ground-based AGB data were collected from 60 circular plots of 200 m^2. The measurements were conducted during November in 2017. The plots were established using a systematic sampling procedure with an equal interval between the plots. AGB of each plot was calculated by applying the following allometric equations that were developed for similar vegetation in the region (Ubuy 2018).

$$\mathrm{AGB} = 0.3098 \times d_s^{1.8761}$$

AGB is total aboveground tree biomass (kg) and d_s is diameter at stump height (30 cm above the ground) (cm) (Table 1).

We found the data from images acquired in November to December, just before the dry season, to be optimal for our study. In fact, during these months, the woody species are still green while the herbaceous vegetation is already dry. These conditions make the distinction between the woody and the herbaceous vegetation easier. Accordingly, a Landsat 8 image that was acquired in December 2017,

Table 1 Summary statistics of ground-based aboveground biomass (Mg/ha)

Variable	*N*	Minimum	Mean	SE	Maximum
AGB (Mg/ha)	60	2.69	12.39	0.55	27.56

Table 2 Vegetation indices derived from the spectral bands

Vegetation index	Equation
Normalized difference vegetation index (NDVI)	NDVI = (NIR − red)/(NIR + red)
Enhanced vegetation index (EVI)	EVI = (NIR − red)/(NIR + 6 (red) − 7.5 (blue) + 1)
Soil adjusted vegetation index (SAVI)	SAVI = ((NIR − red)/(NIR + red + 0.5)) (1.5)
Modified soil adjusted vegetation index (MSAVI)	MSAVI = (2 (NIR) + 1 − sqrt ((2 (NIR) + 1)2 − 8 (NIR − red)))/2
Normalized difference moisture index (NDMI)	NDMI = (NIR − SWIR1)/(NIR + SWIR1)
Normalized burn ratio (NBR)	NBR = (NIR − SWIR2)/(NIR + SWIR2)
Normalized burn ratio-2 (NBR2)	NBR2 = (SWIR1 − SWIR2)/(SWIR1 + SWIR2)

NIR—Near infrared (band 5), *SWIR1*—short wave infrared (band 6), and *SWIR2*—short wave infrared 2 (band 7)

corresponding to the year of ground field inventory, was downloaded through the USGS Earth Explorer. The image was corrected for sensor, solar, atmospheric, and topographic effects. Moreover, we derived seven commonly used vegetation indices (Table 2) from the spectral bands to use as potential predictor variables in developing regression models. The type of the relationship between the spectral variables (seven spectral bands and seven vegetation indices) and AGB was examined using scatter plot. After the scatter plot suggested a linear relationship between the spectral variables and AGB, we estimated Pearson's correlation coefficient (ρ) to measure the degree of correlation. Based on the strength of the relationship between AGB and the spectral variables, we selected the best and most popular explanatory variable for linear model development. Root mean square error (RMSE) calculated using the leave one out cross-validation (LOOCV) method was used to evaluate the models.

3 Results

AGB was significantly correlated with all of the spectral bands (Table 3). Similarly, it was significantly correlated with NDVI, a vegetation index that is mostly used for AGB estimation (Table 4). AGB was also significantly correlated with the other vegetation indices (Table 4). The RMSE of the values predicted using the model developed was 4 Mg/ha (Table 5).

4 Discussion

The pigment in plant leaves, chlorophyll, strongly absorbs light in the red and the blue regions of the visible light spectrum. Besides, the cell structure of the leaves strongly reflects near-infrared (NIR) light. Light in the green region of the visible light spectrum is also reflected by chlorophyll, making plants look green. Hence, these lights are expected to explain variations in vegetation characteristic such as biomass density. Similarly, vegetation indices, particularly NDVI that are calculated from these lights, have long been used for predicting the extent of AGB. These indices measure greenness of the given vegetation. However, application of vegetation greenness indices for dry forests has been a challenge due to the extreme seasonality of these forests. In the study area, during the wet season, it is difficult to distinguish the woody species from grass and herbaceous plants. Correspondingly, during the dry season, it is difficult to classify the land use as a forest. Despite the abovementioned challenges, we demonstrated that these vegetation indices can be applied for such forests too. This requires choosing the season for data collection carefully when considering using remotely sensed data for AGB prediction of dry forests.

The strong linear relationship between NDVI and AGB of the studied forest supports the use of NDVI as a spectral index for indirectly measuring AGB of seasonally dry forests.

Table 3 Correlation between aboveground biomass and the spectral bands

Ultra-blue	Blue	Green	Red	NIR	SWIR1	SWIR2
−0.46***	−0.43***	−0.41**	−0.40**	−0.35**	−0.33**	−0.36**

$^{*}p < 0.1$, $^{**}p < 0.05$ and $^{***}p < 0.01$

Table 4 Correlation between aboveground biomass and the vegetation indices

NDVI	EVI	SAVI	MSAVI	NDMI	NBR	NBR2
0.36**	0.22*	0.36**	0.36**	0.22*	0.28*	0.34**

*$p < 0.1$ and **$p < 0.05$

Table 5 Parameter estimates of the linear model with one of the vegetation indices (NDVI) as a predictor

N	Predictor	α	*β*	*P*-value	RMSE (Mg/ha)	%RMSE	RMSE (LOOCV)
60	NDVI	−1.74	53.46	0.00	5.3	43	0.4439

LOOCV—Leave one out cross-validation and *RMSE*—root mean square error

5 Conclusion

Spectral variables derived from remotely sensed data are capable of predicting AGB of seasonally dry forests if the right remote sensing data collecting season is considered.

References

Ahamed, T.: A review of remote sensing methods for biomass feedstock production. Biomass Bioenerg. **35**, 2455–2469 (2011)

Birhane, E.: Exclosures as forest and landscape restoration tools: lessons from Tigray Region, Ethiopia. Int. for. Rev. **19**(4), 37–50 (2017)

Blackie, R., Baldauf, C., Gautier, D.: Tropical dry forests: the state of global knowledge and recommendations for future research. CIFOR (2014)

Brown, S.: Aboveground biomass distribution of US eastern hardwood forests and the use of large trees as an indicator of forest development. For. Ecol. Manage. **96**, 37–47 (1997)

Bullock, S.H., Mooney, H.A., Medina, E.: Seasonally Dry Tropical Forests. Cambridge University Press, Cambridge (1995)

Cao, L.: Integrating airborne LiDAR and optical data to estimate forest aboveground biomass in arid and semi-arid regions of China **10**, 532 (2018)

Chave, J.: Improved allometric models to estimate the aboveground biomass of tropical trees. Glob. Change Biol. **20**, 3177–3190 (2014)

FAO: Global forest resources assessment 2015. Desk reference. Food and Agricultural Organization of the United Nations, Rome (2015)

Gleason, C.J.: A review of remote sensing of forest biomass and biofuel: options for small-area applications. Gisci. Remote Sens. **48**, 141–170 (2011)

Goetz, S.: Mapping and monitoring carbon stocks with satellite observations: a comparison of methods. Carbon Balance Manage. **4**, 2 (2009)

Joshi, N.: A review of the application of optical and radar remote sensing data fusion to land use mapping and monitoring. Remote Sens. **8**, 70 (2016)

Lu, D.: The potential and challenge of remote sensing-based biomass estimation. Int. J. Remote Sens. **27**, 1297–1328 (2006)

Lu, D.: A survey of remote sensing-based aboveground biomass estimation methods in forest ecosystems. Int. J. Digit. Earth **9**, 63–105 (2016)

Murphy, P.G.: Ecology of tropical dry forest. Annu. Rev. Ecol. Syst. **17**, 67–88 (1986)

Seidel, D.: Review of ground-based methods to measure the distribution of biomass in forest canopies. Ann. For. Sci. **68**, 225–244 (2011)

Ubuy, M.: Aboveground biomass models for trees and shrubs of exclosures in the drylands of Tigray, northern Ethiopia. J. Arid Environ. **156**, 9–18 (2018)

UNFCCC: The Cancun agreements. Outcome of the Work of the Ad Hoc Working Group on Long-term Cooperative Action Under the Convention. Decision 1/CP.16 (2010)

Wang, G.: Uncertainties of mapping aboveground forest carbon due to plot locations using national forest inventory plot and remotely sensed data AU—Wang, Guangxing. Scand. J. for. Res. **26**, 360–373 (2011)

Vulnerability Mapping of Forest Fires Risk in the Bouhmama Region, Khenchela (East Algeria)

Bouhata Rabah, Bensekhria Aida, and Habibi Yahyaoui

Abstract

The Algerian forest, like all forests in the Mediterranean basin, annually records an average area of 34,000 ha of forests mainly destroyed by fires throughout the country, which alarmingly accelerates the degradation and threats on the ecological integrity and rural area stability. The present work aims to map the vulnerability of forest fire risk in the Bouhmama region located in the west of Khenchela (Eastern Algeria), using the Turkish model that integrates several parameters, GIS application and remote sensing. Four vulnerability classes are defined according to the main controlling factors that involve forest fire. The very high vulnerability class is mostly important and located in the northern part of the study area where a significant human presence with a dense road network and high slope is found.

Keywords

Vulnerability • Remote sensing • GIS • Forest fires • Risk • Bouhmama

1 Introduction

Forest fire is a scourge that causes forest degradation and disruption of the global ecological system, where it causes enormous damage by the extent of affected surfaces. Each year, 400,000 ha of forests and natural areas are ravaged by fire, and no less than 50,000 fire starts are declared (Tailleux 2006). Globally, the area of forests affected by fires is about 5% per year (FAO 2007a, b); in the Mediterranean basin, more than 55,000 fires cover an average of 500,000–700,000 ha of forest every year occurred, causing enormous ecological and economic damage, as well as life loss (Angelidis 1994; Alexandrian and Esnault 1998; Dimitrakopoulos and Mitsopoulos 2006) in Meddour et al. (2008). The Algerian forest, like the other forests around the Mediterranean, is ravaged by fires every year and records an average area of 34,000 ha of forest, destroyed by fires across the national territory (DPFF 2018).

In the region of Bouhmama (eastern Algeria), the climate is semi-arid with a cold winter season and a hot summer season with an average temperature of 32 marked by an irregular distribution of rainfall during the year. Forest area covers around 64% of the total area. Abundance of plant fuel, the rugged terrain and population density factors increase the risk of forest fire in this area. Thus, for this phenomenon, prevention is the only effective way to conserve forests in this region.

The objective of this study is to map the spatial variability of forest fire risk vulnerability in this region through a model that integrates and organizes five parameters for fire risk assessment, namely the type of vegetation, slope, exposure, proximity of roads and proximity to settlements.

2 Materials and Methodology

The methodology is the model that was developed by Erten et al. (2004). This model was applied in the Tiaret and Fenouane forests of Ain el Hadjer, Saida. Algeria using Eq. (1):

$$\mathrm{FRI} = 7\,\mathrm{VegI} + 5(P + E) + 3(D_\mathrm{r} + D_\mathrm{a}) \qquad (1)$$

- Fire risk index (FRI)
- Vegetation index (VegI)
- Topomorphological index composed of: P (slope) and E (exposure)

B. Rabah (✉) · B. Aida · H. Yahyaoui
Laboratory of Natural Risks and Regional Planning (LRNAT), University of Batna 2, Batna, Algeria

H. El-Askary et al. (eds.), *Research Developments in Geotechnics, Geo-Informatics and Remote Sensing*, Advances in Science, Technology & Innovation,
https://doi.org/10.1007/978-3-030-72896-0_88

- The human index composed of: D_r (distance from roads) and D_a (distance from agglomerations).

All data defining the five main layers are introduced into a geographic region under a (GIS) system (Bouhata and Kalla 2014). These layers were created in (ArcGIS 10.2) using the spatial analysis tool.

3 Results

Figure 1 represents maps of spatial variability of forest fire vulnerability. Mainly four classes were determined according to different parameters that cause this phenomenon.

- *Class of area at very high vulnerability*: is characterized by flammable vegetation, a significant human presence, a dense road network, high slope and south exposure.
- *Class of area at high vulnerability*: mainly characterized by high-risk vegetation, medium-to-high slope and exposing toward west and south.
- *Class of area at average vulnerability*: have pine-type vegetation, an average slope and exposure to the east.
- *Class of area at low vulnerability*: occupies a large part of the study area and is characterized by a slight slope, low human index with a north exposure and bare soils.

The map of risk index represents the followings (Table 1):

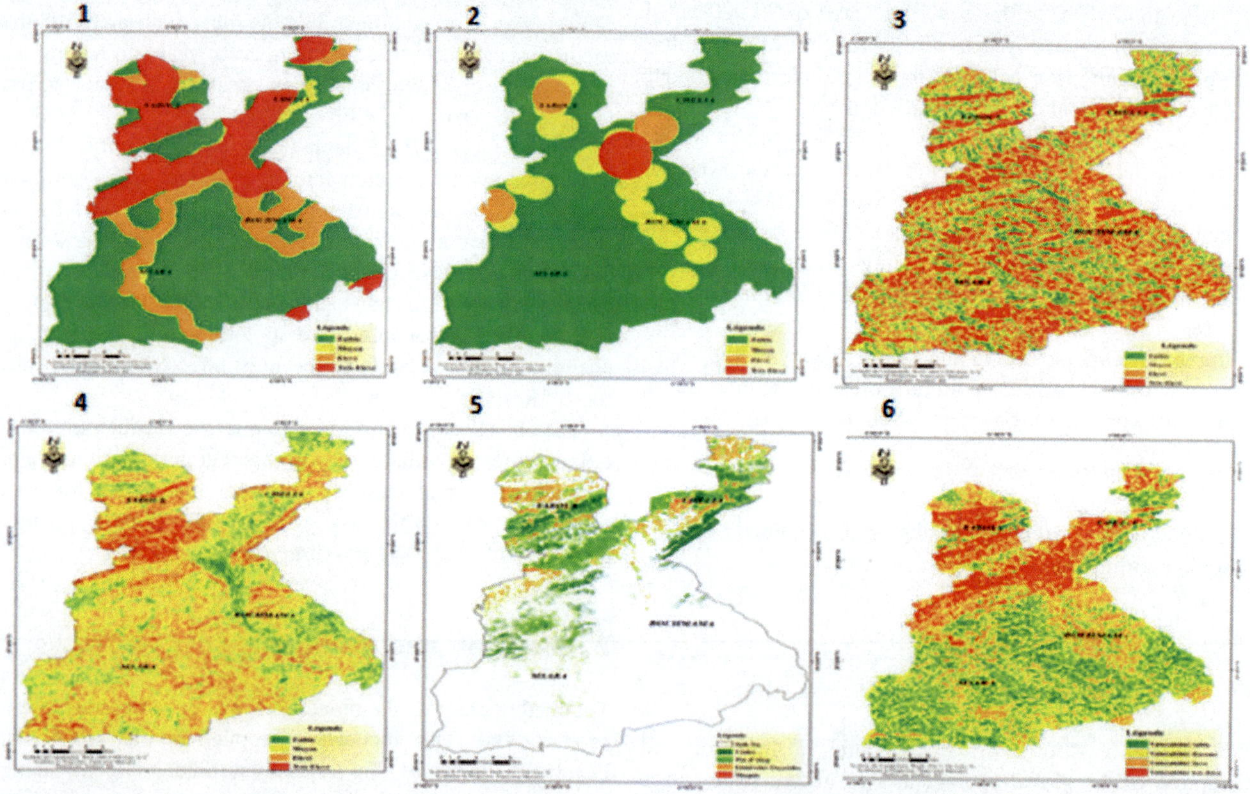

Fig. 1 Maps of respectively; (1) road proximity index, (2) dwelling proximity index, (3) exposure index, (4) slope index, (5) vegetation index and (6) vulnerability forest fire map

Table 1 Classes of risk index

Classes	Area (ha)	Percentage (%)
Low	33,580.38	26.08
Average	51,427.27	39.94
High	31,077.18	24.13
Very high	12,679.57	9.85

- 74% of the total study area is considered as an average to very-high-risk area, whereas the rest of the total area (26%) is a low-risk region.
- The area of very high to high risk percentage includes a forest with a predominance of the Aleppo pine species, while the low-risk class is essentially characterized by bare soil.
- The more the forest is close to roads and villages, the more the risk of fire can occur.

The classical methods generally used in Algeria for fire prevention took time and are not always reliable due to the complexity and diversity of the forest's ecosystems.

Works in this domain, based on remote sensing and GIS tools, have demonstrated their efficiency and speed in the development of fire risk assessment maps in forest areas through the thematic analysis of satellite images.

4 Conclusion

- Four classes of vulnerability to forest fire were defined: very high vulnerability, high vulnerability, average vulnerability and low vulnerability classes.
- The socioeconomic factor is the main controlling factor of forest fire, due to people's needs such as no controlling camping, coal for winter and business, etc.
- The obtained results are interesting, and they constitute a contribution for a better prevention against forest fires and an aid in certain decision making.

References

Alexandrian, D., Esnault, F.: Politiques nationales ayant une incidence sur les incendies de forêt dans le Bassin Méditerranéen, 15 p. Réunion FAO, 28 au 30 octobre 1998, Rome (1998)

Angelidis, A.: La politique de l'Union Européenne concernant la protection des forêts contre les incendies, 57 p. CIHEAM, IAM Zaragoza, « La protection contre les incendies de forêt », 9–20 mai 1994 (1994)

Bouhata, R., Kalla, M.: Mapping of environmental vulnerability of desertification by adaptation of the MEDALUS method in the Endoreic area of Gadaine (Eastern Algeria). Geogr. Tech. **09**(2), 1–8 (2014)

Dimitrakopoulos, A.P., Mitsopoulos, I.D.: Global Forest resources assessment 2005. Report on fires in the Mediterranean Region. Working paper FM/8/E, Forestry Department, FAO, Rome, 43 p (2006)

DPFF 2018: Directeur de la protection de la flore et de la faune (2018)

Erten, E., Kurgun, V., Musaoglu, N.: Forest fire risk zone mapping from satellite imagery and GIS a case study. In: XXth Congress of the International Society for Photogrammetry and Remote Sensing (ISPRS), 12–25 July 2004, Istanbul, Turkey (2004)

FAO: Report of the Eighteenth Session of the Committee on Forestry, 13–16 Mar 2007, Rome, Italy. COFO-2007/REP. Rome, Italy (2007a). (Disponible aussi sur ftp://ftp.fao.org/docrep/fao/meeting/012/j9643e.pdf)

FAO: Global forest resources assessment 2010—guidelines for country reporting to FRA (2007b)

Meddour, O.S., Meddour, R., Derridj, A.: Historique des feux de forêts en Algérie Analyse statistique descriptive (période 1876–2005), revue campus (2008)

Tailleux, C.: Cemagref, UR Ecosystèmes Méditerranéens et Risques. Info DFCI, Bulletin du Centre de Documentation «Forêt Méditerranéenne et Incendie» Novembre 2006 N° 57, 8 p (2006)

Grassland Classification for Livestock Management Using Multispectral Remote Sensing Imagery

Michael Li and Xiuping Jia

Abstract

Quantification of grassland monitoring and assessment using satellite multispectral remote sensing technology has the potential to significantly improve the management of livestock economy in countries, such as Australia. The aim of this study was to build an estimation model for pasture health based on multispectral imagery. Such a model can then be embedded into electronic shepherding system (ESS) so that it is able to perform livestock management automatically. Several vegetation indices (VIs) are investigated using the Sentinel-2 MSI imagery, and a class-based regression model was proposed for the 'growing', 'matured' and 'grazed' grassland, obtained from SVM classification, which led to an improved correlation between the canopy chlorophyll content index (CCCI) and the above-ground biomass (AGB). In-situ data samples were collected, and a UAV was employed for ground truthing. The developed regression model improves the coefficient of determination to 0.86, compared to 0.67 when a global data regression was adopted, enabling it to provide a good estimate of grazing intensity, guide livestock rotation cycles, and boost both scale and efficiency of livestock farming practices.

Keywords

Vegetation index • Grassland • Above-ground biomass • Regression

1 Introduction

Since the introduction of the first remote sensing satellite in 1972 with the Landsat-1, multispectral imagery together with image processing techniques have had a profound impact on land management practices. Improvements in the spatial and spectral resolutions, and wavelength coverage of the optical sensors on board remote sensing satellites have significantly widened their applications over the last four decades. Advanced image analysis techniques have been developed to distinguish between fine classes, such as various stages of vegetation growth in a paddock.

In particular, the multispectral instrument (MSI) on board the Sentinel-2 Earth Observation Mission has a good spatial resolution of 10 or 20 m. It also provides improved ability to detect fine differences in electromagnetic reflectance in the 700–800 nm wavelength range, known as the 'red-edge' range, which is sensitive to numerous vegetation markers, such as chlorophyll content and leaf canopy cover. Based on the spectral characteristics, vegetation indices (VIs) have been designed to measure vegetation health (Sibanda et al. 2016). Such indices include the normalized difference vegetation index (NDVI), the enhanced vegetation index (EVI), the leaf area index (LAI) and the canopy chlorophyll content index (CCCI). As a result, a large improvement in agricultural practices can be expected. For example, an automated e-shepherding system may be developed to direct livestock efficiently towards grazing paddocks in accordance with quantifiable vegetation measures derived from multispectral image data. In this way, land degradation can be minimized and livestock production can be increased (Wallace et al. 2017). The aim of this study was to build an estimation model for pasture health based on multispectral imagery. Such a model can then be embedded into electronic shepherding system (ESS) to improve livestock management.

M. Li · X. Jia (✉)
The University of New South Wales, Canberra, ACT 2600, Australia
e-mail: x.jia@adfa.edu.au

H. El-Askary et al. (eds.), *Research Developments in Geotechnics, Geo-Informatics and Remote Sensing*, Advances in Science, Technology & Innovation,
https://doi.org/10.1007/978-3-030-72896-0_89

2 Data Collection and Processing

2.1 Ground Sample Collection and Growth Categorization

To establish the ground truth data set to train the grassland health model, 65 grass samples were collected from various paddocks over a three-month pasture growing season from the area around Harden, NSW, Australia. The dates of sampling the collection were 03–04 September 2018, 03–04 December 2018 and 27 December 2018 when this study was conducted.

AGB was measured for each sample. A 1 m × 1 m square structure called a quadrat was used to standardize the area of grass samples to collect. All the vegetation inside of it was cut and bagged. The process was repeated three times at separate locations inside each paddock, and once weighed, the average weight of the three samples was taken as the AGB of that paddock.

In addition to measuring the AGB at each paddock, a UAV equipped with high-resolution camera and GPS was used for taking photographs of the paddock in a controlled manner, allowing qualitative analysis to be performed for each of the sampled paddocks. The samples were recognized into three broad classes according to their growth status: 'grazed', 'growing' and 'mature' as shown in Fig. 1.

2.2 Satellite Images Collection and Processing

The multispectral images from the instrument (MSI) on board the Sentinel-2 Earth Observation Mission images were used in this study due to their better spatial and spectral resolutions compared to the Landsat multispectral ones. Three sets of images captured over a three-month pasture growing season from the area around Harden, NSW, Australia, were collected. They were in line with the dates and locations as the ground samples were collected. The required pre-processing steps of image up-sampling and format conversion into GeoTIFF were performed in the Sentinel-2 Application Program and imported into MATLAB. Vegetation indices were applied to the MSI data sets. In particular, NDVI and CCCI maps were obtained using the definitions as follows:

$$\mathrm{NDVI} = (x_8 - x_4)/(x_8 + x_4), \quad (1)$$

$$\mathrm{CCCI} = [(x_8 - x_5)/(x_8 + x_5)]/[(x_8 - x_4)/(x_8 + x_4)], \quad (2)$$

where x_N is the digital number of band N. Figure 2 shows an example of an original image and its VI maps.

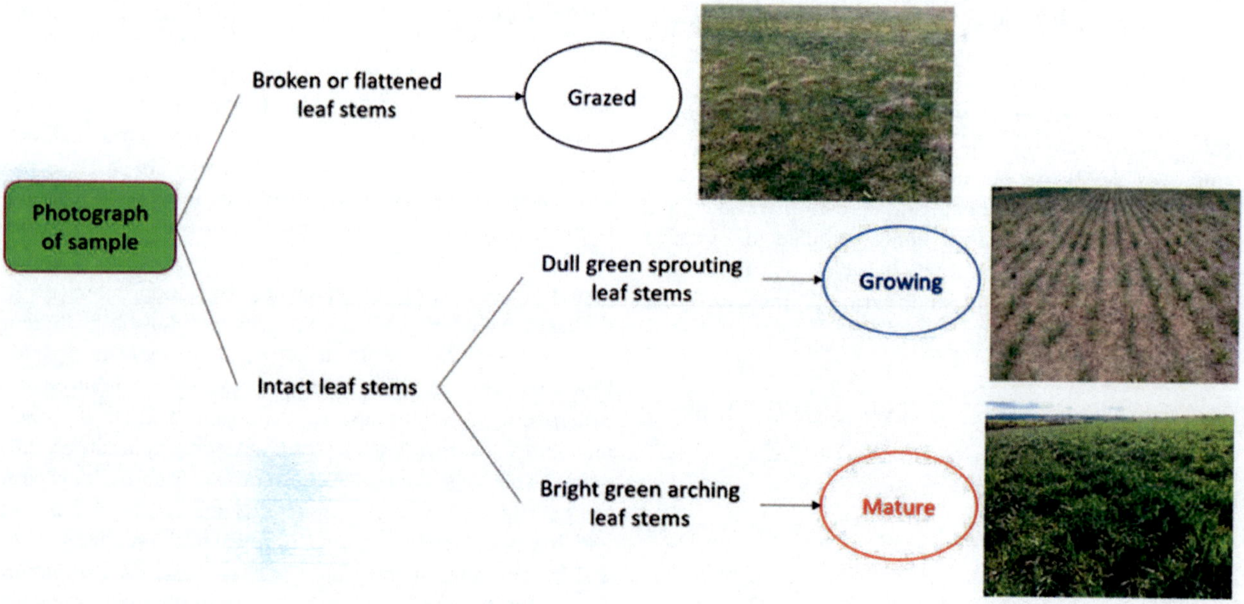

Fig. 1 An illustration of the three classes of grassland

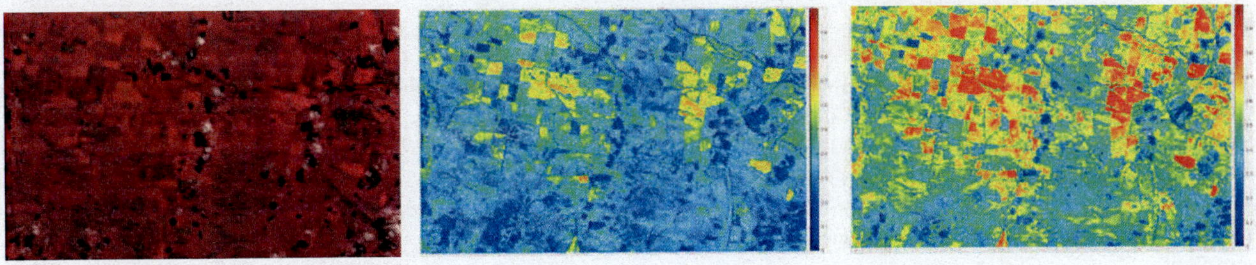

Fig. 2 Left: The false colour composite image with near-infrared, red and green bands. Centre: The NDVI map (red = high values). Right: The CCCI map (red = high values)

3 Model Development and Validation

3.1 Linear and Nonlinear Regression with Global Data

To model the relationship between a VI and AGB, various regression techniques were using different VIs. Figure 3 shows the linear and nonlinear regression with CCCI on the global data sets. Despite the use of CCCI to overcome saturation, the simple linear regression model still had some severe limitations. It produced an overall R^2 value of only 0.667, with a mean prediction error of 64.8 g/m^3. When a third-ordered polynomial was adopted, it was clear that the residuals were not randomly distributed. The model appeared to be mostly accurate for CCCI values below 0.55 and above 0.65, but produced very high residuals between CCCI values of 0.55 and 0.65. The proposed class-based regression discussed below was trigged by this observation.

3.2 Class-Based Linear Regression Model Development and Validation

A more detailed examination of the data samples was conducted and classified using the support vector machines (Richards and Jia 2006), which led to the development of class-based linear regression, as shown in Fig. 4.

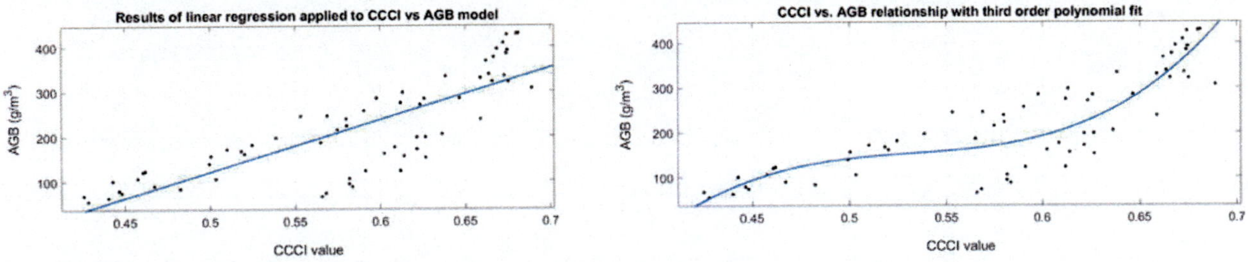

Fig. 3 Results of linear regression using a linear (left) and a cubic polynomial (right) fit

Fig. 4 Results of class-based linear regression

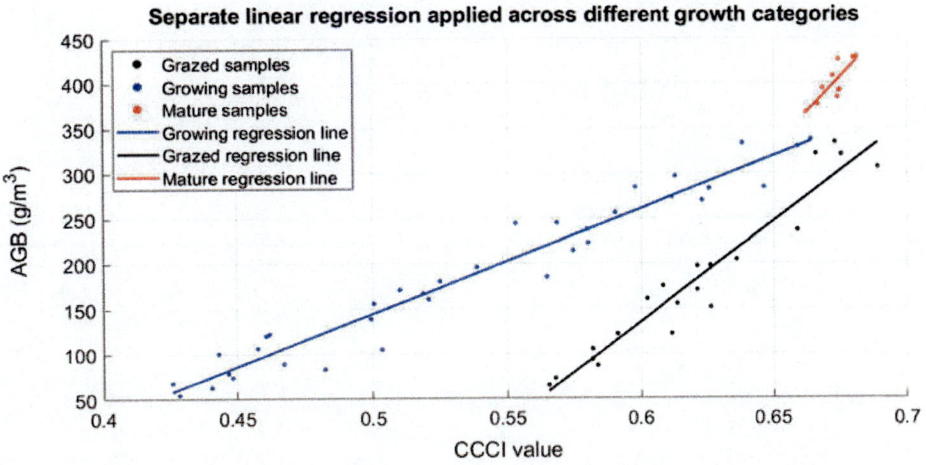

We can see once a separate regression line was used to fit the data samples of each class, the fidelity improved significantly. The overall weighted R^2 was 0.8979 with a mean prediction error of 25.53 g/m^2. 50 more grass samples were collected from pastures in a new location in the same way as the training set. A high classification accuracy of 96% was obtained. This error value of 10 g/m^3 represents a 3.125% deviation from the actual value, which can be evaluated as acceptable.

4 Discussion and Conclusion

A class-based regression was developed to model a physical quantity, AGB, which represented the pasture health of a paddock. The different slopes and magnitudes of the individual class regression lines reflect the varying spectral responses to AGB, depending on the presence of the soil component within a mixed pixel. To apply the method developed here, cloud removal is often required as a pre-processing step.

References

Richards, J.R., Jia, X.: Remote Sensing Digital Image Analysis, 4th edn. Springer, Heidelberg (2006)

Sibanda, M., Mutanga, O., Rouget, M.: Comparing the spectral settings of the new generation broad and narrow band sensors in estimating biomass of native grasses grown under different management practices. Gisci. Remote Sens. **53**(5), 614–633 (2016)

Wallace, L., Hillman, S., Reinke, K., Hally, B.: Non-destructive estimation of above-ground surface and near-surface biomass using 3d terrestrial remote sensing techniques. Methods Ecol. Evol. **8**(11), 1607–1616 (2017)

Changes in Land Use and Land Surface Temperature Over Nashik City, MS, India

Anargha Dhorde and Sweta Deka

Abstract

Rapid transformation of land use in and around major urban centers of the world has triggered a definitive change in the urban built environment. One of the consequences of this transformation is the increase in land surface temperatures (LST) in the urban areas. Nashik city also presents a peculiar case of growth in its built-up area. The present study aims at assessing the changes in land use land cover (LULC) and LST over Nashik city. This study was carried out for a period of nine years (2010–2018). Landsat ETM+ and OLI/TIRS data sets were employed to extract LULC, NDVI and LST. Changes in normalized difference vegetation index (NDVI) have also been investigated since the vegetation affects temperature. It has been observed that during the period of nine years, the built-up area increased 1.7 times and the LST increased by 2–3 °C in the city core, particularly during summer. This indicates the intensification of LST in the city core. The adjoining areas also present, more or less, a similar pattern due to the increase in the built-up area. Vegetation in the eastern part of the city has discernibly reduced to a considerable extent, which has also altered the thermal pattern of the city. A strong negative correlation is observed between the built-up areas and the LST over Nashik.

Keywords

Land use • Land surface temperature • NDVI • Nashik

A. Dhorde (✉) · S. Deka
Department of Geography, Nowrosjee Wadia College (affiliated to the Savitribai Phule Pune University), Pune, 411001, India

1 Introduction

In the present era, urban sprawl and related changes in urban built environment have led to different environmental problems in the urban sector. One of the factors that contribute in the modification of microclimate in the cities is the increase in the built-up surfaces. Concrete and tar laid surfaces are characteristic of the modern cultural landscape. These surfaces are capable of absorbing higher amounts of incident radiation and result in increasing surface temperatures, thus causing significant changes in urban thermal environments (Li et al. 2010). These changes in urban thermal environment often lead to the formation and sprawl of urban heat island (UHI), wherein higher land surface temperatures (LST) are observed at the city core as compared to that of its immediate surroundings (Landsberg 1981). Advances in remote sensing techniques and availability of thermal satellite data have enhanced the scope of urban climate studies. Considerable work has been carried out to understand the urban thermal characteristics using remote sensing data (Streutker 2002; Weng 2001). Correlation of vegetation indexes (NDVI) and built-up areas with LST are studied by a number of researchers in order to understand the impact of changing land use land cover on urban surface temperatures (Carlson et al. 1995; Barbeiri et al. 2018). Yuan and Bauer (2007), compared percentage imperviousness along with NDVI as indicators of LST, while Xiao et al. (2008) employed multiple regression and principal component analysis to establish correlation of LST with the population's density, land use and vegetation.

In India, cities are growing at faster rates due to increased demand in housing and infrastructural facilities. It has been observed in some of the Indian cities that land surface temperatures (LST) have swiftly increased with the growing urbanization. Some of these cities are Pune, Ahmedabad, Guwahati, Delhi, etc. It is therefore essential to investigate whether medium-sized cities such as Nashik are also experiencing similar phenomena. Nashik city (Fig. 1) is situated

H. El-Askary et al. (eds.), *Research Developments in Geotechnics, Geo-Informatics and Remote Sensing*, Advances in Science, Technology & Innovation,
https://doi.org/10.1007/978-3-030-72896-0_90

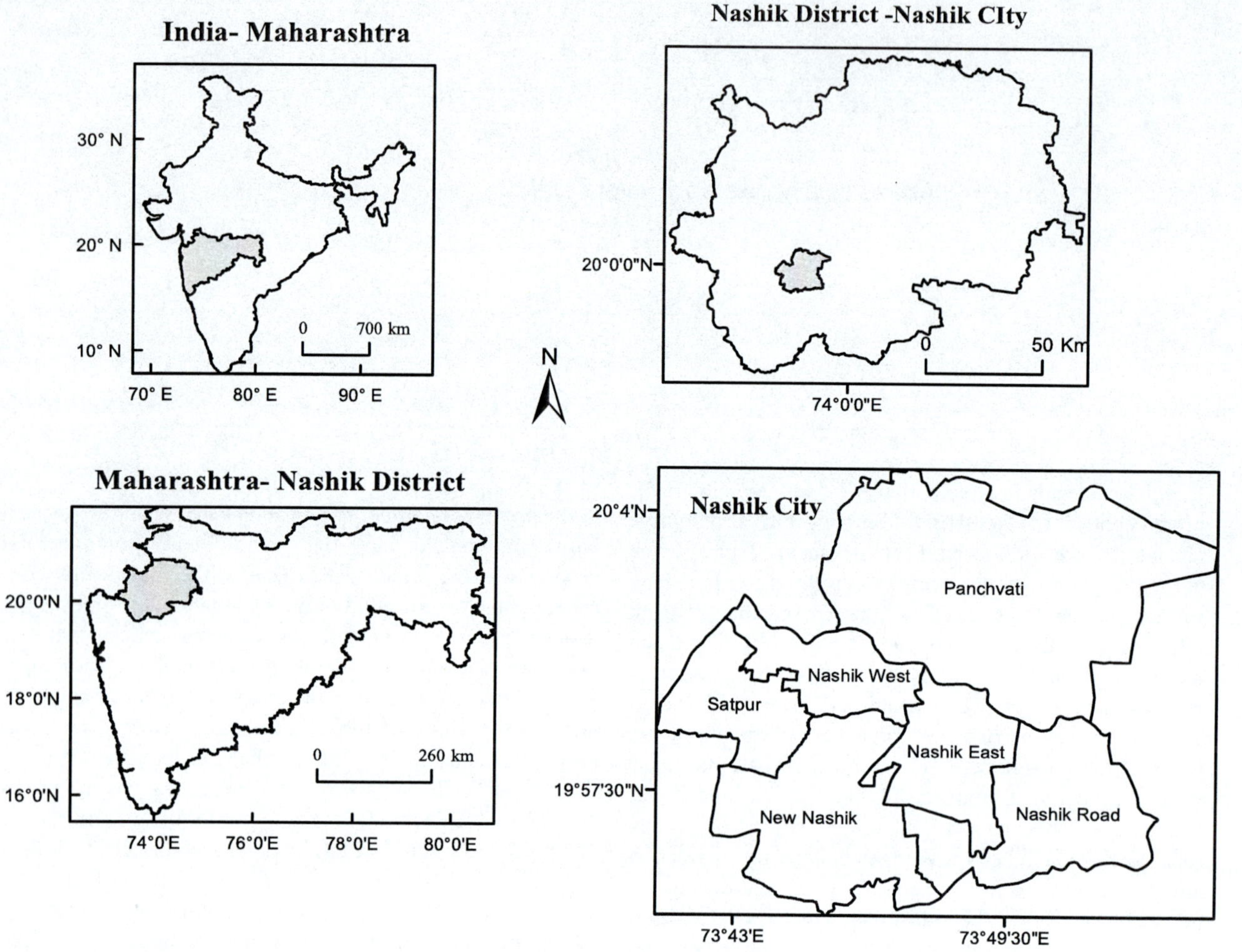

Fig. 1 Location of the study area

on the banks of River Godavari at an elevation of 584 m ASL on the western edge of the Deccan plateau. It enjoys a tropical wet and dry climate.

The main aim of the study is to bring about spatiotemporal variations in the land use and land surface temperature of the Nashik city, during the period 2010–2018.

2 Data and Methods

The details of the data set used for the present work are given in Table 1. ETM+ and OLI/TIRS data were downloaded from the global land cover facility's site. The downloaded data sets were subjected to the preprocessing procedure which included radiometric and atmospheric correction. These data were further employed for the land use land cover classification and retrieval of LST.

The satellite images obtained were primarily subjected to the basic preprocessing steps. This included the radiometric and atmospheric corrections, reprojection and enhancement of the images. In order to obtain the thematic maps for different land use types, the images were subjected to supervised classification. Eight land use classes were identified (hillsc, barren land, water bodies, roads, fallow land, built-up, vegetation and agricultural land) and processed for accuracy assessment after the classification. LST is commonly retrieved using a different set of algorithms like split window (Becker and Li 1995), mono-window (Qin et al. 2001) and single-channel technique (Jimenez-Munoz and Sobrino 2003). For the present work, LST was extracted by the single-channel algorithm. Standard procedure depicted in the Landsat 8 user handbook (2018) was adopted for the retrieval of LST. In all, LST was retrieved by the following four steps: conversion of DN (QCAL) to top of the

Table 1 Satellite data used for the study

Satellite	Sensor	**Bands** (spatial resolution in meters)	Date
Landsat 7	ETM+	**Blue**(30), **Green**(30), **Red** (30), **NIR**(30), **MIR1**(30), **TIR**(60), **MIR2**(30), **PAN**(15)	April 26, 2010
Landsat 8	OLI/TIRS	**Blue**(30), **Green**(30), **Red** (30), **NIR**(30), **SWIR1**(30), **SWIR2**(30), **PAN**(15), **Cirrus**(30), **TIR 1**(100), **TIR 2**(100)	April 5, 2013
			April 21, 2014
			April 10, 2016
			April 16, 2018

atmosphere (TOA) radiance values, conversion of spectral radiance to black body temperature (BBT), emissivity correction and LST retrieval. For the last step, LST was retrieved using Eq. 1.

$$S_t = \frac{T_B}{1 + (\lambda * T_B/\rho) \ln \varepsilon} \quad (1)$$

3 Results

3.1 Changes in LULC Over Nashik City

The output of the classified images was obtained as thematic maps, wherein the classes defined earlier were displayed using various colors (Fig. 2). The overall accuracy obtained for all the images was above 80%, and the kappa statistic was above 75%. During the last nine years, it is observed that the percentage of built-up area has increased almost by 1.7 times. In 2010, the built-up area was 26.76% which is recorded at 44.85% by 2018. This rapid increase in built-up area is mainly observed in the southern parts of the city. In the recent time, northern parts of the city also have started experiencing rise in built-up areas (Fig. 2). The vegetation (inclusive of natural, agricultural and fallow) has decreased by 7% (from 2010, (42.09%) to 2010 (35.04%)). Overall, it appears that the vegetation has not decreased much in these 9 years. A decrease by almost 5% of barren land in the last nine years' time is recorded. Most of this decrease in barren land can be attributed to the growth in settlements in these areas. Considerable decrease in hills is also noticed during the study period. 10% of the city area was covered with hills in 2010; however, by 2018, hills are attributed to a mere 4.84% of the total city area.

3.2 Changes in LST Over Nashik City (2010–2018)

Figure 3 depicts the spatiotemporal changes in LST from 2010 to 2018. Mean LSTs obtained for all these images indicated an increasing trend in LST over Nashik city (36.3 °C in 2010 to 39.41 °C in 2018). Even though LST values ranged up to 45 °C in 2010, the areas reporting this value were relatively rare. The city core region had higher LST (36–40 °C) while the immediate surroundings had lower LST (32–36 °C) in 2010. By 2013 and subsequent period, a gradual spread of higher LST values was observed moving out from the city core toward the suburbs. In all the images, lower LSTs were noted over the major water body (River Godawari) that drains through the city. The northern and southern hill spurs (reported as barren land parcels in most of the images (Fig. 2)) and the areas devoid of vegetation have recorded very high LST (>40 °C). As compared to the 2010 condition, most of the suburbs have recorded an increase in the LST values by 1–2 °C in 2018. However, most of the barren patches, hills and intermittent barren land between agricultural patches have recorded an increase in LST by almost 4 °C or even more in the last 9 years. Isolated barren patches, airport and the bare hill tops have recorded an increase by almost 6 °C or higher than that. LST for the city core appears to have increased by 2–3 °C during these nine years.

As compared to the condition of 2010, certain areas in 2016 show a decreasing trend in temperature. However, the net areas that are represented by the decreasing LST are relatively lower than the areas where the increasing trend is observed. In some localities in western parts of the city, temperatures have dropped down by 0–2 °C. There are very few patches along the river which have decreased LST values (0–2 °C).

3.3 Changes in LST Over Nashik City (2010–2018)

NDVI versus LST values were plotted for all the years; but here, they are represented for only 2010 and 2018. These plots obtained to understand the relationship between NDVI and LST (Fig. 4) indicated that there is a definite negative relationship between them, thus, indicating that a decrease in NDVI will lead to an increase in LST and vice versa. All the years recorded a negative trend and were found to be significant at the 0.05 level of significance, thus establishing the fact with 95% of confidence level.

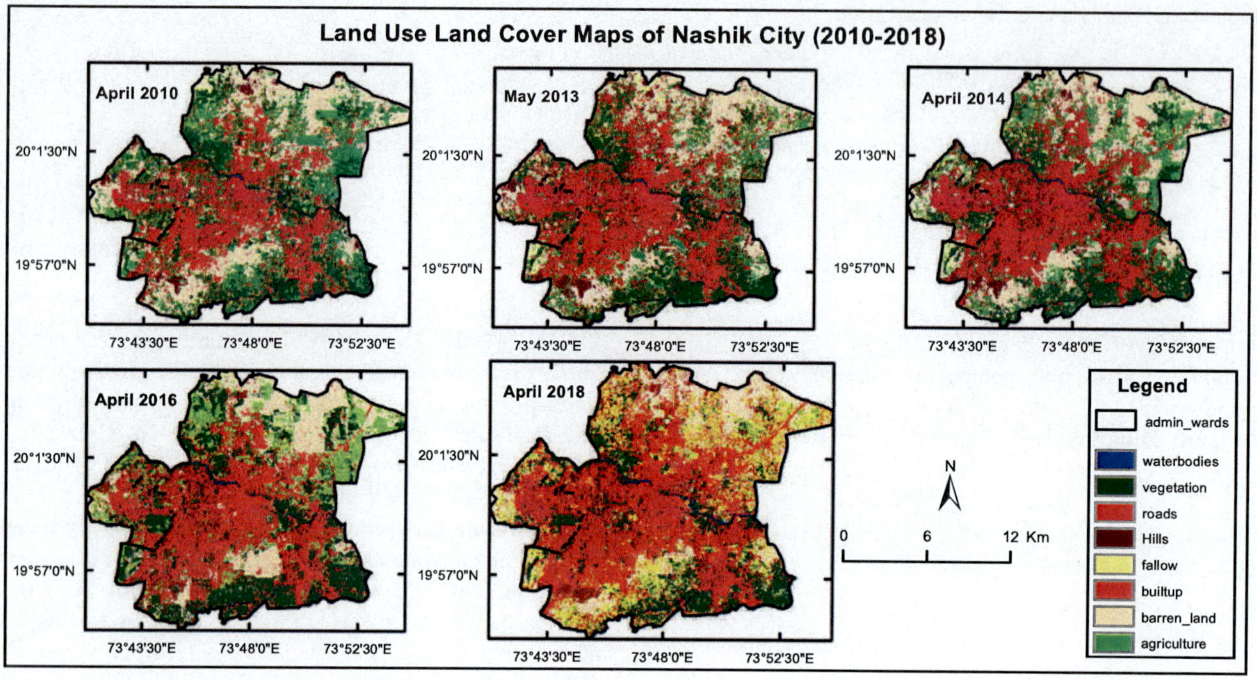

Fig. 2 Land use land cover maps of Nashik city (2010–2018)

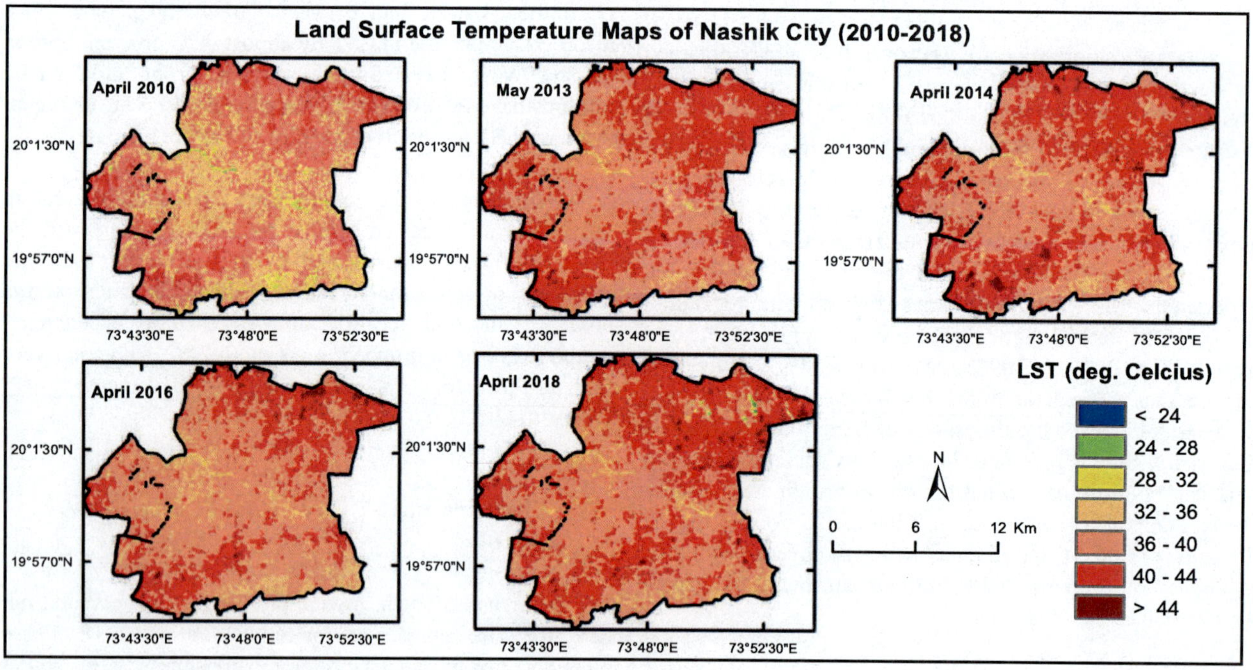

Fig. 3 Land surface temperature maps of Nashik city (2010–2018)

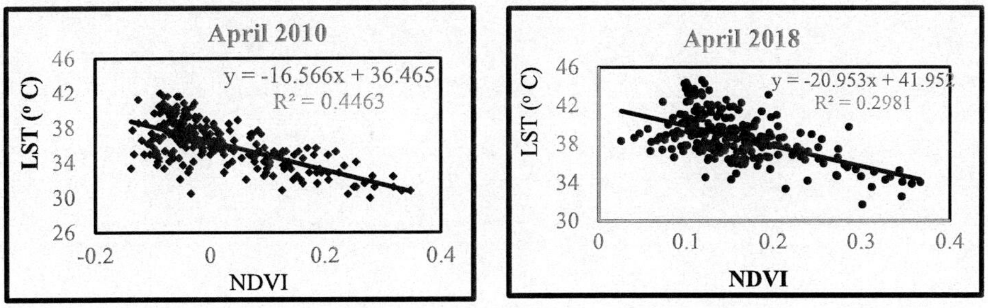

Fig. 4 NDVI–LST relationship (2010–2018)

4 Discussion

Over the period of nine years, the built-up area of Nashik city has increased from 26.7 to 44.85%. This clearly indicates that Nashik is growing at a fast rate. Overall, it is observed that for the entire city, the LST values (mean) have increased. A major increase is noted in 2013 from that of 2010 condition. In 2010, the mean LST for Nashik city was recorded as 36 °C which increased to 39.45 °C by 2013. 2014 imaging recorded a slight increase in LST (39.49 °C). While by 2016, it was noticed that the mean LST values for the city dropped down slightly by 0.78 °C and reached a value of 38.67 °C. However, again by 2018, there is an increase noticed in the mean LST of the city, where the mean LST recorded was 39.41 °C. The NDVI–LST relationship study indicated a negative relationship between NDVI and LST. While studying the LST of 2018 over 2010, it was revealed that the areas where vegetation is relatively dense recorded lower values of LST. However, areas dominated by barren land and industries recorded maximum LST. This again highlights the role of vegetation in mitigating the intensification of LST.

5 Conclusions

During a brief period of nine years, it was observed that the built-up area has gone up by almost 1.7 times in the city and the land surface temperature has increased by 2–3 °C in the city core, particularly during summer, thus, indicating an intensification of city core LST. Thermal field in the suburbs (western part of the city) has been modified, and LST has increased by 5 °C in the last 9 years due to the change in land use. Western and southwestern parts of the city show LST values as that of the city core indicating expansion of built-up area. The research also reveals higher temperatures near the city's boundary due to the barren land and hill slopes. Vegetation covers certain areas such as the eastern part of the city which has discernibly reduced to a considerable extent, which has also influenced the thermal pattern of the city. The present study has clearly brought out the spatiotemporal variations in the LST and also proved that the LST is highly (negatively) correlated to the vegetation and built-up areas. In the process of development, the city planners have to give considerable thought to maintaining the green character of the city and see to it that the brown agenda does not overwhelm the green agenda.

References

Barbieri, T., Despini, F., Teggi, S.: A multitemporal analyses of land surface temperature using landsat 8 data and open source software: the case study of Modena, Italy. Sustainability **10**, 1678 (2018)

Becker, F., Li, Z.L.: Surface temperature and emissivity at various scales: definition, measurement and related problems. Remote Sens. Rev. **12**(3–4), 225–253 (1995)

Carlson, T.N., Gillies, R.R., Schmugge, T.J.: An interpretation of methodologies for indirect measurement of soil water content and fractional vegetation cover. Agric. For. Meteorol. **77**, 191–205 (1995)

Jiménez-Muñoz, J.C., Sobrino, J.A.: A generalized single-channel method for retrieving land surface temperature from remote sensing data. J. Geophys. Res. Atmos. **108**(D22) (2003)

Landsberg, H.E.: The Urban Climate. Academic Press, New York (1981)

Li, S., Zhao, Z., Miaomiao, X., Wang, Y.: Investigating spatial non-stationary and scale-dependent relationships between urban surface temperature and environmental factors using geographically weighted regression. Environ. Model. Softw. **25**(12), 1789–1800 (2010)

Qin, Z., Karnieli, A., Berliner, P.: A mono-window algorithm for retrieving land surface temperature from Landsat TM data and its application to the Israel-Egypt border region. Int. J. Remote Sens. **22**(18), 3719–3746 (2001)

Streutker, D.R.: A remote sensing study of the urban heat island of Houston, Texas. Int. J. Remote Sens. **23**(13), 2595–2608 (2002)

Weng, Q.: A remote sensing and GIS evaluation of urban expansion and its impact on surface temperature in the Zhujiang Delta, China. Int. J. Remote Sens. **22**(10), 1999–2014 (2001)

Xiao, R., Weng, Q., Ouyang, Z., Li, W., Schienke, E.W., Zhang, Z.: Land surface temperature variation and major factors in Beijing, China. Photogram. Eng. Remote Sens. **74**(4), 451–461 (2008)

Yuan, F., Bauer, M.E.: Comparison of impervious surface area and normalized difference vegetation index as indicators of surface urban heat island effects in Landsat imagery. Remote Sens. Environ. **106**, 375–386 (2007)

Inland Water Monitoring Using Different Satellite Sensors

Elsayed Issawy and Haggag Mohamed

Abstract

The validation of satellite altimetry CryoSat-2 SARin level 2 data with the in-situ gauge data of Lake Nasser, Egypt, was performed through my contribution in evaluating the CryoSat-2 SARin level 2 data over selected inland water bodies. The validation between the two data sets approved that the waveform retracking CryoSat-2 SARin level 2 data over the lake are suitable for inland water bodies especially for Lake Nasser due to its special shape and rigid surroundings. In addition, the altimetry data of water level over the lake are downloaded, processed and analyzed for different altimetry sensors, Web sites and ground tracks to validate these output data with the in-situ gauge data of the lake. Then, the correlation between all these kinds of data was performed to provide the optimum kind of the altimetry Web sites, sensors and ground tracks which can be used for delivering altimetry water level of Lake Nasser. Moreover, the surface area of the lake was determined by using the optical satellite imagery producing a surface water extent over the lake in approached time intervals. MODIS surface reflectance 8-day is used which composites with 250 m spatial resolution (MOD09Q1) to monitor the Lake Nasser area for a certain time period. The derived images from optical satellite imagery are processed using a classification method to each image in order to distinguish between water and land. Also, the time series water surface area of the lake and the shape files with respect to time derived from the obtained surface area were performed.

Keywords

Satellite altimetry • Lake Nasser • Water level • Optical satellite imagery

1 Introduction

Nowadays, the space-borne geodetic sensors are widely used to monitor the inland water (rivers, lakes and closed sees) parameters (water level, water extent, water storage and discharge). Global data processing of satellite altimetry provides temporal and spatial time series of lakes surface height with a decimeter precision on the whole Earth. On the other hand, remote sensing techniques have proven their ability to measure different parameters within the Earth system. Satellite imagery, for instance, can provide variations in inland water areas with appropriate temporal sampling. In-situ gauge stations are limited in spatial and temporal coverage, and their number has been decreasing during the past decades. Therefore, the need was to monitor the different parameters of inland water using different satellite sensors.

Lake Nasser, Egypt, was selected as a case study to monitor the hydrological parameters of the lake. The different altimetry and remote sensing data are correlated to the in-situ data to determine the water level, surface area and volume of water in the lake (Fig. 1). The linking of these data sets provides a satisfied monitoring of the hydrological parameters (water level, surface area and volume) of Nasser Lake.

2 Materials and Methods

The procedures of different kinds of satellite data were performed as follows:

E. Issawy (✉) · H. Mohamed
National Research Institute of Astronomy and Geophysics, Helwan, Cairo, Egypt
e-mail: issawy@nriag.sci.eg

H. El-Askary et al. (eds.), *Research Developments in Geotechnics, Geo-Informatics and Remote Sensing*, Advances in Science, Technology & Innovation,
https://doi.org/10.1007/978-3-030-72896-0_91

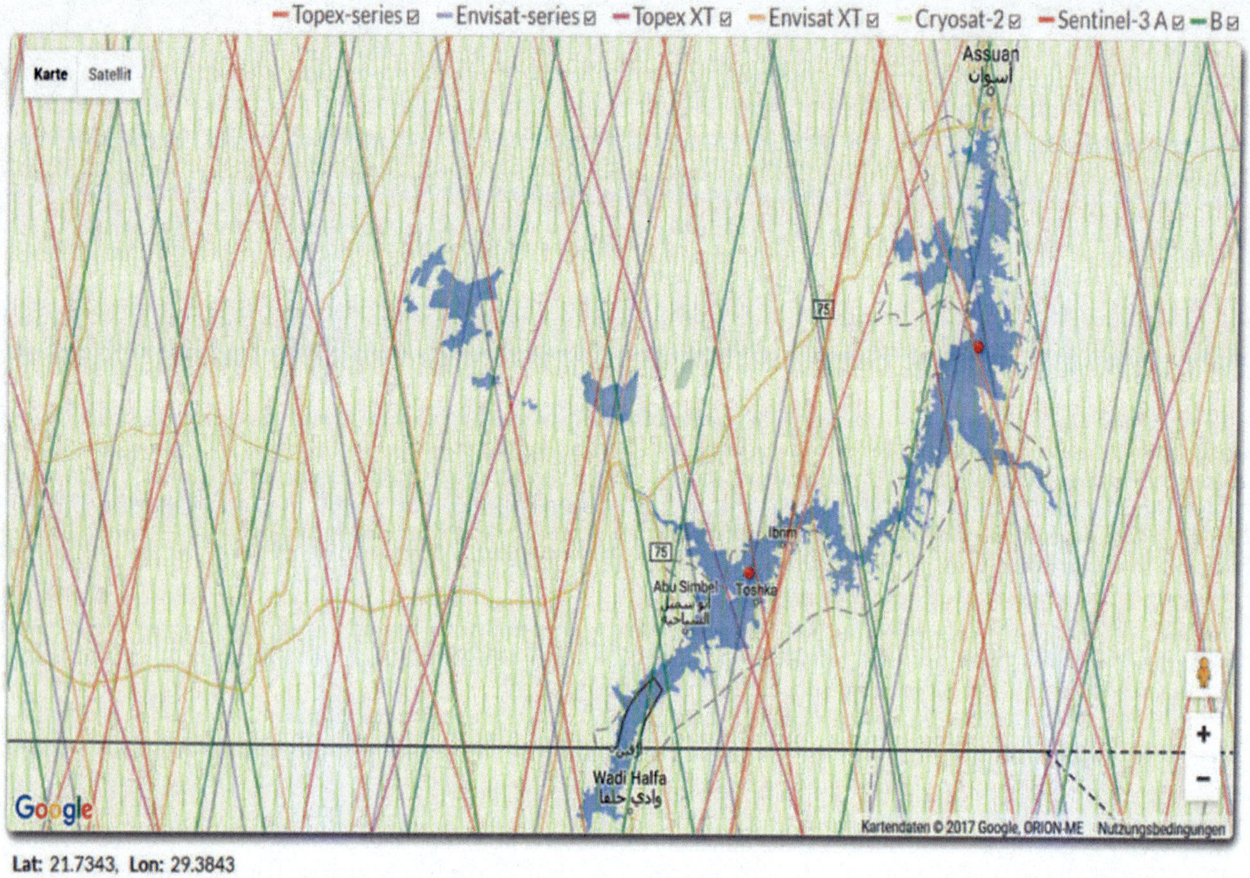

Fig. 1 Sets of altimetry sensors crossing Lake Nasser

- Downloading of altimetry data of water level over the Nasser Lake is for different altimetry sensors, Web sites and ground tracks in order to validate these output data with the in-situ gauge water level data of the lake.
- The obtained downloaded data were selected for the ground tracks covering Nasser Lake and then processed by BARAT software to read the NC files of the altimetry data produced by the producers. Then, these data were reduced for different kinds of reduction such as solid Earth tides reduction.
- The time series of all these kinds of altimetry data was produced and correlated to the in-situ gauge data of the lake to provide the optimum kind of altimetry Web sites, sensors and ground tracks which can be used for delivering the altimetry water level of Lake Nasser.
- The images from satellite imagery are also downloaded to get the different images covered the Lake and a classification method to each image are applied to distinguish between water and land.
- MODIS surface reflectance 8-day is used which composites with 250 m spatial resolution (MOD09Q1) to monitor the Lake Nasser area time period (2000–2014) as example. This level 3 product provides two bands (620–670 nm, 841–876 nm) with 250 m resolution in an 8-day gridded product using the sinusoidal map projection.
- Then, a time series of surface area of the lake was established, the shape files also created to get the shape-time maps, different kinds of data sets correlation have been performed, and special scattering diagram shows the relation between the surface area time series and the altimetry water level as well as the in-situ gauge water level data of the lake.
- The water volume calculations are derived from the water surface area's output and the in-situ data of water level data, with respect to the minimum height of water level of the lake, and the time series of water volume variation was established using a prepared MATLAB software code for water volume calculations.

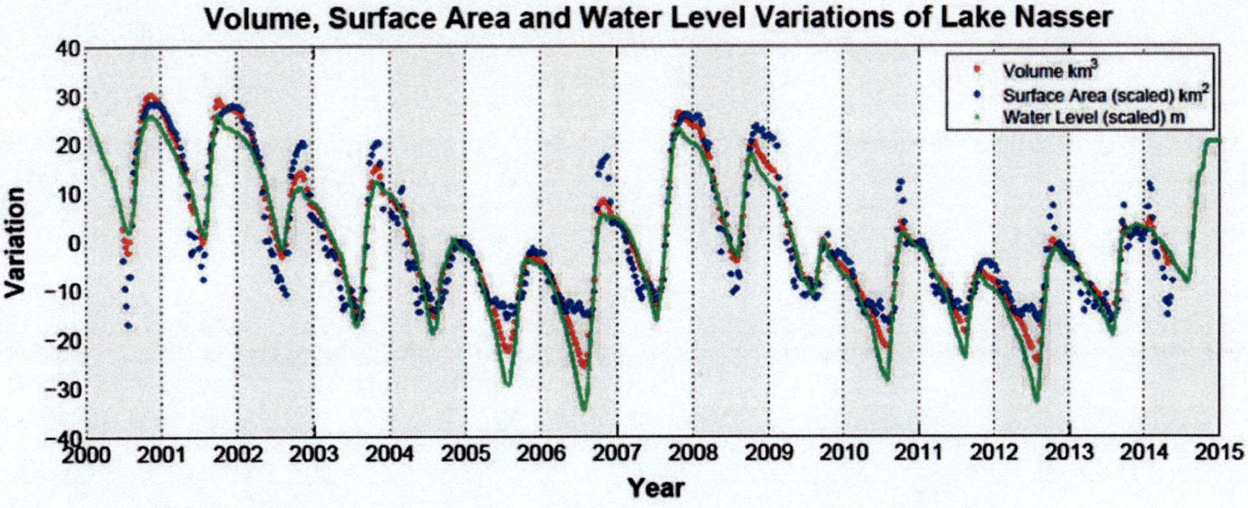

Fig. 2 Correlations between the different altimetry data sets, surface area and volume of water in Lake Nasser

3 Results

The correlation between the different other kinds of delivered altimetry sensors for water level data over the Lake Nasser (Fig. 2) in terms of different Web sites, sites, sensors, ground tracks approved that the space-borne sensors should be used for continuous monitoring of water level of the lake regardless to the in-situ water level data availability.

The calculations of the surface area from optical satellite imagery produced a surface water extent of the lake in approached time intervals. These time-shape series of the lake enable establishing time-shape maps for the surface water extent. These different kinds of time series data of surface area provide continuous and updated monitoring of area parameters (shape, water extension and time series of surface area). These continuous time-shape maps reflected the increase or decrease of the water extension of the lake at any corresponding time.

4 Conclusion

The validation of the waveform retracking CryoSat-2 SARin level 2 data for Nasser Lake approved that the waveform retracking CryoSat-2 SARin level 2 data are suitable over the lake since the correlation between the altimetry data and the in-situ gauge data has a RMSE of 28 cm although the thin and complex shape of Nasser Lake.

The optimum Web site to deal with satellite altimetry data for Lake Nasser according to correlation to in-situ data is the French Web site which is the hydroweb.

The obtained time series of water volume changes revealed that the variation of the water volume in the lake follows either the in-situ gauge data of water level or the altimetry data of water level with respect to the surface area of the lake since there is no available bathymetry data to get the true water volume of the lake.

References

Elsawwaf, M., Willems, P., Pagano, A., Berlamont, J.: Evaporation estimates from Nasser Lake, Egypt, based on three floating stations data and Bowen ratio energy budget. Theor. Appl. Climatol. **100**, 439–465 (2010)

Morton, F.I.: Operational estimates of lake evaporation. J. Hydrol. **66**, 77–100 (1993)

Myrup, L.O., Powell, T.M., Godden, D.A., Goldman, C.R.: Climatological estimate of the average monthly energy and water budgets of Lake Tahoe, California–Nevada. Water Resour. **15**(6), 1499–1508 (1979)

Omar, M.H., El-Bakry, M.M.: Estimation of evaporation from the lake of the Aswan High Dam (Lake Nasser) based on measurements over the lake. Agric. Meteorol. **23**, 293–308 (1981)

Sadek, M.F., Shahin, M.M., Stigter, C.J.: Evaporation from the reservoir of the High Aswan Dam, Egypt: a new comparison of relevant methods with limited data. Theor. Appl. Climatol. **56**, 57–66 (1997)